Aging Power Delivery Infrastructures

POWER ENGINEERING

Series Editor

H. Lee Willis

ABB Electric Systems Technology Institute
Raleigh, North Carolina

1. Power Distribution Planning Reference Book, *H. Lee Willis*
2. Transmission Network Protection: Theory and Practice, *Y. G. Paithankar*
3. Electrical Insulation in Power Systems, *N. H. Malik, A. A. Al-Arainy, and M. I. Qureshi*
4. Electrical Power Equipment Maintenance and Testing, *Paul Gill*
5. Protective Relaying: Principles and Applications, Second Edition, *J. Lewis Blackburn*
6. Understanding Electric Utilities and De-Regulation, *Lorrin Philipson and H. Lee Willis*
7. Electrical Power Cable Engineering, *William A. Thue*
8. Electric Systems, Dynamics, and Stability with Artificial Intelligence Applications, *James A. Momoh and Mohamed E. El-Hawary*
9. Insulation Coordination for Power Systems, *Andrew R. Hileman*
10. Distributed Power Generation: Planning and Evaluation, *H. Lee Willis and Walter G. Scott*
11. Electric Power System Applications of Optimization, James A. Momoh
12. Aging Power Delivery Infrastructures, *H. Lee Willis, Gregory V. Welch, and Randall R. Schrieber*

ADDITIONAL VOLUMES IN PREPARATION

Aging Power Delivery Infrastructures

H. Lee Willis
Gregory V. Welch
Randall R. Schrieber

ABB Power T&D Company Inc.
Raleigh, North Carolina

MARCEL DEKKER

MARCEL DEKKER, INC. NEW YORK • BASEL

ISBN: 0-8247-0539-4

This book is printed on acid-free paper.

Headquarters
Marcel Dekker, Inc.
270 Madison Avenue, New York, NY 10016
tel: 212-696-9000; fax: 212-685-4540

Eastern Hemisphere Distribution
Marcel Dekker AG
Hutgasse 4, Postfach 812, CH-4001 Basel, Switzerland
tel: 41-61-261-8482; fax: 41-61-261-8896

World Wide Web
http://www.dekker.com

The publisher offers discounts on this book when ordered in bulk quantities. For more information, write to Special Sales/Professional Marketing at the headquarters address above.

Current printing (last digit):
10 9 8 7 6 5 4 3 2 1

PRINTED IN THE UNITED STATES OF AMERICA

Series Introduction

Power engineering is the oldest and most traditional of the various areas within electrical engineering. However, one unfortunate consequence of that seniority with respect to other technologies is that America's electric power infrastructure is quite old. As the electric power industry de-regulates and experiences increasing levels of competition, electric utilities find themselves under growing pressure to improve their bottom line financial performance as well as customer service quality. Yet they must struggle with power systems whose basic design was drafted more than a half-century ago, and that are composed of equipment that is often up to 60 years old.

Aging power delivery infrastructures present a dilemma for electric utilities, the electric power industry, and government. In order for our society to have electric service that is both high in quality and low in price, it must have both good electric systems and healthy electric companies. However, many of these electric companies are in an unenviable position. Modernizing their aging systems would require so much expense that they risk destroying their company's financial health and market viability.

Only by innovating, by taking a new approach and changing the way they plan, engineer, operate, and in particular, manage their systems, can utilities get both the system and financial performance they need. *Aging Power Delivery Infrastructures* discusses both the problems electric utilities face and the modern solutions that they can apply to improve their systems and customer service while keeping their financial performance at competitive levels.

As both a co-author of this book and the editor of the Marcel Dekker Power

Engineering Series, I am proud to include *Aging Power Delivery Infrastructures* in this important series of books. Marcel Dekker's Power Engineering Series includes books covering the entire field of power engineering, in all of its specialties and sub-genres, all aimed at providing a comprehensive reference library of the knowledge needed to meet the electric industry's challenges in the 21st century.

Like all the books in Marcel Dekker's Power Engineering Series, this one focuses on providing modern power technology in a context of proven, practical application, and it should be useful as a reference book as well as for self-study and classroom use. Unlike most books, however, this is non-technical, aimed at the non-engineer, as well as at those power engineers who need a quick, simple overview of de-regulation and how it impacts their role in the industry.

H. Lee Willis

Preface

This book is both a reference book and a tutorial guide on aging power delivery systems and the planning, engineering, operations, and management approaches that electric utilities can take to best handle the formidable challenges they face in working with an aging electric power delivery infrastructure. Electric power is almost unique among major public industries in one important respect: the quality and usability of its product is completely determined by the characteristics of the system that delivers it, not the way in which it is manufactured. America's aging power delivery systems mean poor power quality, frequent and lengthy outages, and frustrating challenges for the customers, the owners and the employees of electric utilities. They also mean poor financial performance for the electric utility companies themselves. Aging power delivery infrastructures degrade both customer service quality and corporate profit levels.

This book exists because the authors believe strongly that in order for our society to have the economical and reliable electric service it needs to sustain growth, it must have both good electric systems, and a healthy electric utility industry. In early 1994, it first because clear to us that the aging of power delivery systems throughout the U.S. was an important and growing problem which threatened the entire power industry. Although the visible symptoms caused by aging infrastructures – frequent and lengthy interruptions of electric service and rising repair and maintenance costs – were largely incipient at that time, projections of equipment age, system performance, and utility financial performance made it clear that the problem would grow exponentially. By year 2000 it would be a major problem. Events certainly proved this projection to be accurate. During the summers of 1998 and 1999, aging utility systems throughout the United States began to experience a growing level of operating problems that led to rapidly degrading customer service and public trust in electric utilities.

In the period 1994 through 1999, we devoted considerable effort, both personally and professionally, to the problem of aging power delivery infrastructures and its possible solutions. We are fortunate enough to be with a company that is arguably the largest and the strongest in the power industry. ABB's resources, technology, and particularly its international span gave us a platform from which to study and develop solutions that work. We used that support and capability to research the problem, to determine what it was under the surface. We collected a host of innovative and proven techniques for improvement from around the world. We developed ways of combining these so they work together, and we tested them until we were confident we knew what worked, what didn't, and why. This book summarizes what we learned.

We will freely admit that our efforts, which were both expensive and frustrating at times, were motivated in large part by the desire for the competitive advantage that comes from knowing how to solve an important, complicated, and poorly understood problem. However, the health and viability of the power industry are of great concern to us, for we are part of it, too, and our futures depend on its prosperity.

What did we learn? Most importantly, aging power delivery infrastructures involve much more than just aging equipment. As important as old equipment is, the problem is also very much the result of the outdated designs, constricted sites and rights of way, outdated engineering standards and methods, and traditional operating procedures being applied in a non-traditional world. The greatest barrier to solving the problem is not financial. Many of the traditional planning, operating, and managerial prioritization methods still in use by electric utilities are completely incompatible with the ways that utility executive management must manage the financial performance of their companies.

But most important, there are viable, proven, and affordable solutions to the problem, solutions that provide results. Those, along with a thorough discussion of the problem itself, is the topic of this book.

This book is composed of three major sections based on technical focus. The first, consisting of Chapters 1 – 6, provides a series of tutorial background discussions on topics necessary to understand some of the subtleties of aging infrastructures and their solutions. The second set of Chapters, 7 – 9, looks at the various aspects of the problem, how they interrelate with one another and the many functions within a power system and an electric utility company. The third consists of seven chapters that look at various methods and technologies to solve the problem, leading up to a series of recommendations and guidelines for those who wish to put together a coherent and workable plan to deal with their aging power delivery infrastructure challenges.

The authors wish to thank their many colleagues and co-workers who have provided so much assistance and help during the time this book took to put together. In particular, we thank our colleagues Drs. Richard Brown and Andrew Hanson for their valuable assistance and good-natured skepticism and encouragement. We also thank our good long-time business associates Mike Engel of Midwest Energy, Nick Lizanich of Commonwealth Edison, and Randy Bayshore of GPU, for their friendship, good humor and valuable suggestions. We also want to thank Rita Lazazzaro and Russell Dekker at Marcel Dekker, Inc., for their involvement and efforts to make this book a quality work.

H. Lee Willis
Gregory V. Welch
Randall R. Schrieber

Contents

1 Aging Power Delivery Infrastructures

1.1 WHAT ARE AGING POWER DELIVERY INFRASTRUCTURES?

Many electric distribution utilities in the United States, Europe, and other countries around the world are experiencing problems in meeting their customer service quality targets, and in achieving the stockholder profit margin or government cost-control obligations they must meet. These problems are due to the fact that large portions of their systems consist of aging infrastructures. A power system composed of mostly old equipment near the end of its lifetime, configured in a layout that is itself quite old and not completely compatible with modern needs, produces reliability, maintenance, and budgeting challenges. If these challenges are not anticipated and brought under control, they will eventually overwhelm even the most effective organization.

As this chapter will discuss and the remainder of this book will show, aging power delivery infrastructures involve the complicated interaction of a host of managerial, operating, engineering, and design factors, some of which are subtle and obtuse in their impact. Due to this multi-faceted complexity, there is no completely satisfactory simple definition of this problem, one that will cover all situations. Some situations are unusual, a few even unique. But for the vast majority of electric utilities, an "aging power delivery infrastructure" can be described most simply as follows:

> An aging power delivery infrastructure is any area of the utility system with an average service age greater than the design lifetime of the equipment from which it is built.

It's Not Just About Old Equipment

This definition is not completely representative in one very important respect. It identifies equipment age as the active factor of aging systems and implying that it is the reason utilities have so many problems in older areas of the system. While equipment aging *is* an important factor, it is not the sole, and in many cases not even the primary factor at work in creating the service and cost problems that many utilities face. Three other major factors are involved. In all cases these other three are part of the problem and in some cases they are more important than the aging equipment. Still, overall equipment age, or "age of the system," is the simplest, most effective way to both characterize aging infrastructures and identify parts of a utility system that suffer from those problems.

Characteristics of an Aging Infrastructure Area

Situations vary, but most aging infrastructure problem areas have most of the same characteristics, which are:

- The area was originally developed, or last experienced a rebuilding boom, prior to the 1970s. Most of the existing system structure (substation sites, rights of way) dates from before that time.
- The majority of equipment in the area is more than forty years old.
- The system is well engineered and planned, in the sense that a good deal of concern and attention has been devoted to it. It fully meets the utility's minimum engineering criteria.
- The area is seeing steady but perhaps slow growth in load. It may be undergoing, or about to undergo a rebuilding boom.
- The area is plagued by above-average equipment failure rates. Overtime is high due to large amounts of unscheduled repair and restoration.
- SAIFI degradation leads SAIDI degradation by a few years. Although invariably both the frequency and the duration of customer service interruptions increase. Typically, frequency of interruption begins rising from five to eight years prior to big increases in SAIDI.
- When a major "event" such as a widespread interruption of service to customers occurs, it is due to a series of events that are quite unusual; or something truly bizarre. These can result in the failure of a relay/breaker. This normally would have caused only a minor

inconvenience, except that it failed during the brief time when the system was configured into a non-standard configuration in order to support the outage of some other unit of equipment – a contingency-switched mode where there was no backup.

- Things go bad a lot.

1.2 POWER DELIVERY, NOT POWER T&D

The authors have deliberately used the term "Power Delivery" rather than "T&D" throughout this book in order to make a point that is both crucial to success in dealing with aging infrastructures, but very often overlooked. The infrastructure involved in the issues being discussed here includes a good deal of the T&D system, but not all of it. In order to understand the problem, and appreciate how to solve it effectively, but at the lowest possible cost, a person must appreciate that both problems and solutions must be examined from the standpoint of function – what does the equipment involved do?

Modern power T&D systems can be divided into two parts, or levels, based on the function of that part of the power system under a de-regulated structure. In terms of the new de-regulated industry paradigm these are most easily distinguished as the wholesale level and the retail level. *The point of this section, and the focus of this entire book, is that the problems of aging infrastructure, as they impact reliability of customer service, are almost exclusively limited to the retail level.*

This is not to say that the wholesale power transmission level is unimportant. It is absolutely critical, and problems at that level can cause customer service interruptions. What is most important at that level, however, is system security, the continued ability of the system to remain "up and running." That is slightly different than reliability.

But frankly, the consumer does not care or even recognize the difference. Whether his lights are out because the wholesale grid cannot supply all the power needed regionally, or because a part of the local power delivery chain has failed and power cannot be moved to his site, the result is the same – he is without electric power. But this book is about understanding and fixing the problems caused by aging power delivery infrastructures, and eradicating the customer service quality problems they cause. In that context, the distinctions between wholesale and retail matters a great deal. The record shows that the problems caused by aging T&D infrastructures are limited almost exclusively to the retail level. The vast majority of customer service quality issues seen throughout the industry are created at this level, and the solution to those problems lies there as well. Problems at the wholesale level are generation shortages and transmission congestion – very serious indeed, but that is an entirely different set of concerns, and not involved with aging infrastructures.

The Wholesale Grid

At the top – closest to the generation – is the high voltage grid. This is often called the wholesale grid or the bulk power system. These names are particularly apt here, because they distinguish that part of the utility system from the power delivery functions associated with the customers – the retail level. The high voltage grid exists to:

1. Interconnect all generation to create a single regional system.
2. Provide economic dispatch capability for the system (traditional view), or open market access for every generator (de-regulated viewpoint).
3. Provide bulk power transmission capability throughout and across the region (from boundary to boundary).
4. Provide system security – the ability to tolerate the transient loss of any generator or unit of equipment while still maintaining a fully synchronized interconnection of the system and full capability to meet the needs for bulk transmission between regions.

Throughout the remainder of this book, the authors' will use the term "wholesale grid" for this part of the system. It is a critical element of the power industry, and to some people and from some perspectives, more important than the remainder. However, the wholesale grid is not connected to any energy consumers (except the few, large, wholesale consumers of electric power, many of whom are also merchant generators). Failures or other problems on the wholesale grid level do not cause poor electrical service at the retail level, whether that be defined as unacceptable voltage, high harmonic content, or lack of availability.[1]

The distinction between the power delivery system and the wholesale grid is one of function and exclusivity of purpose. While the wholesale grid is involved in getting power to the consumer, it exists mainly or at least in large part for other reasons (see list above). Furthermore, any one part of it may or may not be involved in delivering power to a particular consumer at any one moment. This is due *only* to decisions with regard to economy of supply that the electric Distribution Company or its customer energy service companies have made.

---

[1] A good deal of attention is being given to the wholesale transmission level, and with good reason. It has its own set of serious problems. Among them are how to maintain interconnected security under open access when demand levels are high, and how to manage and alleviate congestion problems on the grid caused by highly localized demand-supply mismatches. Aging equipment is an issue in some of these cases, but not in most, nor is it in any situation the key factor of concern, in the authors' opinion.

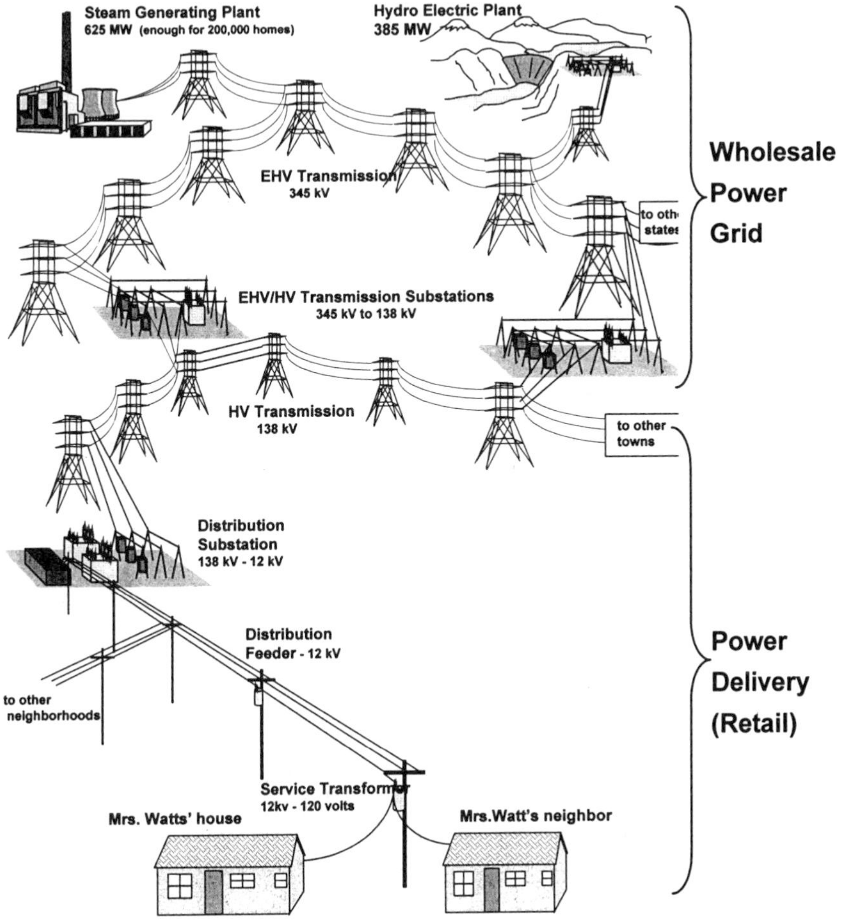

Figure 1.1 The electric power industry, including generation (now de-regulated), the high-voltage transmission network (the open-access wholesale grid), and the power delivery level, which consists of sub-transmission, substations, feeders and laterals, and the utilization voltage system leading to all customers.

The power Mrs. Rose uses to heat her home this week might come from parts of the wholesale grid north of her town, because the utility/service provider found a bargain with respect to generation located there. Next week it might flow in from the west, for similar reasons. This flexibility of flow at the wholesale level is not just considered "business as usual," it is in fact *a key design goal* of the wholesale grid (see number 2 in list above) and one important aspect that distinguishes it from the retail level. Factors, inefficiencies, and outright failures at the wholesale level most often affect market demand and supply economics and transmission congestion, not customer reliability. These are price, not reliability, issues.

Power Delivery: Getting Power to the People

By contrast to the wholesale level, retail, or power delivery, system exists *solely* to route power to consumers. It provides none of the functions, even in small part, listed above for the wholesale grid, but exists only because power must be routed to consumers if they are to buy it. Furthermore, and most importantly to this discussion, any particular consumer's power *always* flows through the same part of the delivery system, except on those rare occasions when there is a contingency. The power that Mrs. Rose uses each day is intended to *always* come through the same set of service drops, through the same service transformer, lateral branch, feeder trunk, substation switchgear, substation transformer, high-side switchgear, and substation transmission-level feed line. Only during an equipment failure will an alternate route be used for some part of this supply chain. During that time the utility may be willing to accept considerably higher than normal loads and lower-than-ideal power quality in order to maintain service.

> "Power delivery infrastructures," aged or not, are therefore those portions of the power system whose purpose is to deliver power to energy consumers. They are justified solely on the basis of the need to perform that function.

Power delivery is accomplished by an "infrastructure" that includes the power distribution system and a portion of what traditionally was called the transmission system. That portion of the transmission-level system is what was traditionally called the "sub-transmission lines – those lines that delivery power to distribution substations. Figure 1.1 indicates the portion of the power system identified as the power delivery system. This is normally the portion that falls under FERC's distinction of distribution, as opposed to transmission (Table 1.2).

From the perspective used throughout this book, the wholesale grid is part of the supply (generation system). Operating problems at that level can be serious and can potentially lead to widespread customer interruptions. In a few cases aging equipment and infrastructures may be a contributing factor.

Table 1.1 Major System Outages Identified By US DOE in 1999

Area	When	Interruption Cause	Power Delivery?	Aging Infra?
New England	Jun 7-8	Wholesale generation shortage		
Chicago	Aug 12	Multiple power delivery failures	x	x
	Jul 30	Multiple power delivery failures	x	x
New York	Jul 6-7	Multiple power delivery failures	x	x
Long Island	Jul 3-8	Power delivery and grid problems	x	x
Mid Atlantic	July 6-19	Wholesale generation & grid problems	x	
N. New Jersey	Jul 5 - 8	Multiple power delivery failures	x	x
So. Central US	Jul 23	Wholesale generation shortages		x
Delmarva	Jul 6	Wholesale grid and generation shortages		

Table 1.2 Seven Characteristics of Distribution (Power Delivery) Systems as Opposed to Wholesale Transmission, according to the FERC

- Local distribution is normally in close proximity to retail customers.
- Local distribution system is primarily radial in character.
- Power flows into local distribution. It rarely, if ever, flows out.
- Power on the local distribution system is not re-consigned or transported to another market.
- Power on the local distribution system is consumed in a comparatively restricted geographic area.
- Meters at the interface from transmission to distribution measure the flow onto the distribution system.
- Local distribution is of reduced voltage compared to the transmission grid.

However, the record shows that the customer service interruptions and system events which brought aging infrastructures to the attention of the industry were all power delivery related, not wholesale grid, related. Table 1.1 lists several of the major customer interruption events in 1999. Of eight major events, five were involved and stayed limited to problems with the power system level below the high voltage grid. All five were power delivery system and aging infrastructure issues. Aging (of generation) was mentioned by DOE as a contributing factor in one wholesale grid event, and was cited as a possible concern in one other. However, factors at the wholesale level most often include market demand, supply economics and transmission constraints due to congestion of poor reservation of firm capacity by users.

Beyond those major events shown in Table 1.1, there were several dozen other events of less significant impact (in terms of total customers out of service, public recognition, and/or time involved) throughout the summer of 1999. A number of utilities experienced continuing problems with large numbers of small outages, which cumulatively caused high levels of cumulative customer-hours out of service. All of these problems involved only the power delivery level, and a majority was related to aging infrastructures.

1.3 THE BUSINESS ENVIRONMENT HAS CHANGED

If this were not enough, the technical, social and political environment in which utilities must function has changed dramatically in the last fifty years. Actually, the environment thirty years ago was not radically different from fifty years ago. Certainly the world is a far different place than it was when the paradigm for system planning and design was developed.

In 1970 most practicing engineers still used slide rules or mechanical calculators for their "desk top" calculations. The electronic calculator was still in its infancy; desktop computers were not yet widely deployed. Large computationally intensive tasks such as power system simulations were run on mainframe computers from input provided by trays of punched cards. Similarly, at home most consumers did not have any appliances or other products that relied heavily on electronics other than perhaps their radio, television or stereo system.

The transistor had been invented in 1947. Discrete component technology found its way into consumer products in the 1960s with the introduction of the "Transistor Radio". However, the proliferation of electronics in consumer mass-market products did not begin to occur on a large scale until the development of the integrated circuit. Intel introduced the 8008™ microprocessor in 1972. The capability to support computational complexity and the sensitivity of products to disturbances on the electric power supply system have been increasing at accelerating rates ever since. Today one can find integrated circuit technology throughout the home. Televisions, clock radios, home video game systems, desk top computers, programmable microwaves, programmable ovens, programmable dish washing machines, programmable VCRs and an increasing number of other appliances incorporating programmable controllers are becoming common place.

Office technology has seen a similar evolution to greater complexity and sensitivity. Mechanical adding machines and calculators were replaced with first electronic calculators and more recently desktop computers. Mechanical typewriters were replaced with electric models, then the stand-alone word processor. Now word processing programs run on the same desktop computers used for accounting, spreadsheet analysis and computationally intensive engineering calculations or graphic design. Many businesses own or lease their own telephone switching systems, another application of microprocessor technology.

Likewise, production and manufacturing facilities have evolved. Manual control of machinery gave way to analog controls. Analog controls have given way to numerical control, first to punched tape, then to process control computers. Electronics have made possible variable-speed AC motor drives. Efforts to increase productivity and efficiency in the factory have introduced many of the same systems used in the modern office to the factory, such as desktop computers and customer owned telephone systems. Additionally, the internal power distribution systems of the plants have become more demanding as they provided faster protection of valuable equipment. Today's microprocessor controlled relays, motor control systems, and recording and monitoring systems are much more sensitive than the electromechanical relays, analog instrumentation and motor control centers of 50 years ago.

While the microprocessor revolution was taking place in homes, offices and factories, little was changing in the construction of the power delivery system. Granted, solid-state relays were developed and deployed, improving the security of the bulk power system. The computer revolution also improved the ability of the utility engineer to model and analyze the operation of the power delivery system further optimizing technical aspects of system design in accord with the planning and design parameters developed in the 50s and 60s. Advances in measurement technology based on the microprocessor also allow technicians to monitor the operation of the system in near real time. Digital event recorders capture the transient operation of the system for later analysis. With the right instruments and communication channels, one can monitor various attributes of electric service for many points on the power delivery system.

However, the system itself has not changed in any significant way. Although there have been improvements in the quality of the material and equipment, the system itself remains essentially the same. The power delivery system is composed of poles, wires, insulated cables, switches, fuses and transformers. It is subject to damage inflicted by man and nature in addition to the random and rather infrequent failure of the components. Finally, the way in which the system reacts when subjected to a tree limb falling on an overhead wire or a back hoe digging into a cable is the same today as it was 50 years ago. A fuse or a breaker or other protective device opens to isolate the damaged segment of the system. In the time required for the protective equipment to operate, high fault currents flow depressing the voltage for all upstream customers and customers on other circuits served from the same source until the protective device operates. In many systems, fast acting reclosers or circuit breaker relay schemes open and reclose before the protective device can operate in an attempt to clear temporary faults subjecting all customers on the affected circuit to a series of momentary interruptions.

The end result of the mismatch between the sensitivity of the equipment used by customers and the operation of the delivery system has been a steady decline of the customer's perception of system performance. Residential customers may arrive home and find all of the electronic clocks blinking or with the words "power failure" showing on the display. The initiating event may have been a momentary interruption due to a tree branch brushing against an overhead conductor, or even a voltage sag due to a short circuit on another circuit from the same substation. Fifty years ago with electric clocks driven by small motors, the event would have gone unnoticed by anyone not at home. For those at home, the most they would have noted was a blink or a momentary dimming of the lights.

In the office and the factory the situation is similar. In the 1980s it was not unusual to encounter problems associated with the application of new technology that was internal to the facility and blamed on the utility. In one instance a manufacturer had installed microprocessor-controlled robots on an arc-welding line. The controls were supplied from the same circuit in the plant that supplied the welders. It was relatively easy to demonstrate to the manager that the controls worked fine when the welders did not strike an arc. In another instance a plastics injection molding plant replaced all of the electromechanical voltage and current protection devices with new solid state devices. Because he was now able to do so, the plant electrician had set them at maximum sensitivity and zero time delay. The end result was that they would trip the supply to the compressor motors any time there was a switching operation on the supply line, despite being only several hundred yards from the substation. The problem was resolved by reducing the sensitivity of the protective devices to the same level of those they replaced.

Today's problems are not as easy to solve. Sensitive process control equipment may be built into the system that it is controlling. Manufacturers of some products can not tolerate even a momentary interruption due to the high degree of process integration. Sensitivity of equipment throughout a plant may be such that even routine voltage sags are intolerable. A single momentary or voltage sag may cause the loss of hundreds of thousands of dollars. Solutions to these problems have been available at utilization voltages for some time. There are un-interruptable power supplies (UPS), motor generator sets and constant voltage transformers to name just a few. Yet, it may be that there is not enough space available for the manufacturer to install the necessary mitigation equipment. Installation of low voltage mitigation equipment at multiple locations throughout the plant may not be cost effective.

It is in this context that the aging power delivery infrastructure must be viewed. Not only does the performance suffer, but also the customers have become much less tolerant of any disturbance on the power system. Taken together, the poorer performance during a period of rising customer expectations has put the power delivery companies in the spotlight.

1.4 FOUR FACTORS CONTRIBUTE TO AGING INFRASTRUCTURE PROBLEMS

Throughout the United States, and in many other places around the world, portions of electric utility systems have become aged: in the sense as described above. The average age of equipment in the region or area of the system exceeds the design lifetime for electrical equipment. Figure 1.2 shows the distribution of average ages of substation power transformers throughout the power industry, estimated by the authors. The mean is about 31 years, meaning that the average transformer is roughly ¾ of the way through its normal lifetime. The figure indicates that more than a few transformers have been in service over 60 years.

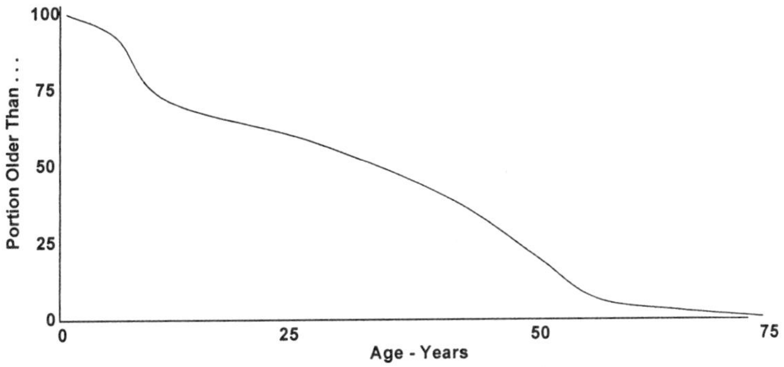

Figure 1.2 Estimated distribution of age for distribution substation power transformers in the United States. Virtually none are over 75 years in service. About 25% are over 47 years in service and 50% over 35 years of age.

The fact that a transformer, circuit breaker, or other piece of electrical equipment is old does not mean that can no longer do its job. Old equipment will often still function, and although not as efficiently as modern units, nonetheless get the job done. But what age does mean is that the unit is much more likely to fail in the next year, as compared to a new unit of similar size and design. As a result, a system composed of old equipment has a much higher incidence of equipment failures. Equipment failures often lead to interruptions of service to customers, and that means the system provides relatively poor service – at least poor service as compared to a system that is composed of newer equipment.

This perspective provides the easiest way to define and identify an aging power delivery infrastructure – simply look at the age of its equipment. However, it provides only a partially view of the problem. It is correct in that that equipment failures do cause customer service interruptions, and that a disproportionately high number of these interruptions are due to the failure of older equipment. But what this perspective misses, is that three other factors common to older parts of the utility infrastructure heavily interact with equipment age to exacerbate its effects and contribute to the overall problems of poor customer reliability and high operating costs. Essentially, the customer service quality problems created by aging power infrastructures are due to, and the solutions they must address; the combined interaction of *four factors*, only one of which is associated with equipment. Table 1.3 summarizes these

Table 1.3 Four Contributing Factors to Aging Infrastructure Problems

Factor	% of Problem	Description
Aging power equipment	20% - 40%	Older equipment has higher failure rates, leading to higher customer interruption rates. It creates higher inspection maintenance costs and much more unexpected O&M costs due to repair and restoration work it creates.
Obsolete system layouts	25% - 50%	Older areas of the system often are in bad need of additional room for substation sites and rights of way that cannot be obtained, and must make do" with only that obtained decades ago
Outdated engineering	12% - 25%	Traditional tools for power delivery planning and engineering are ineffective in recognizing and correcting some of the problems that result for use of aged equipment, obsolete system layouts, and modern de-regulated loading levels.
Old cultural value	5% - 50%	Planning, engineering, and operation of systems using concepts and procedures that worked fine in the vertically integrated industry that exacerbates problems when applied under de-regulation.

four interrelated, contributing factors, which are all at work in an aging power delivery infrastructure. Together, they work against efficient, reliable operation of the older power system. These are discussed in more detail below, and in much greater detail elsewhere in this book.

Contributing Factor 1: Aging Equipment

As cited above, old equipment is a serious contributing factor to poor reliability and high costs in many utility systems. In fact, there is usually some area in any electric distribution utility's system where old equipment is a worrisome factor in providing good customer service quality.

All power system equipment will eventually fail. At some point, after a period of service, something will wear out or deteriorate to the point that it no longer adequately performs its role. The equipment will cease to provide electrical power delivery to those customers downstream of it. The nature of electrical equipment is that very often failures lead to a fault, requiring protective equipment action that isolate a larger portion of the system than just the customers served by that failed unit.

Regardless, electrical equipment does not have a uniform probability of failure with age. Usually electrical equipment fails only after many years of completely satisfactory service. Thus, old equipment is much more likely to fail than new equipment (roughly three to ten times as likely). As a result, a system or area of a system composed of old equipment will have a much higher failure rate than one composed of new equipment. This has three consequences, all of which are undesirable:

1a. *Equipment failure* causes lower customer service quality. Customers are more likely to have interruptions of their electrical service due to the unexpected failures of this older equipment. The frequency of interruption rises – SAIFI increases.[2]

1b. *Higher repair and restoration efforts.* More failures mean more and more unpredicted events requiring emergency or contingency mitigation efforts. The impact of this on all but the most efficiently managed and staffed organizations is to increase duration of equipment outages (and hence customer interruptions). There are more numerous outages to repair, taxing the restoration and repair resources to the limit and often requiring repair of one to wait on another. SAIDI increases slightly as a result.

Also quite important for the utility's bottom line, these higher repair and restoration costs increase costs.

---

[2] SAIFI, System Average Interruption Frequency Index, the average number of interruptions of service experienced annually per customer. For a discussion of this and other reliability-related indices, see Chapter 4.

1c. *More inspection and testing* efforts required. Equipment more prone to fail often benefits from more frequent and comprehensive inspection and testing, which in many cases can reveal developing problems before they cause failures. Catching trouble early allows what would have become unscheduled emergency repair work to be done in a more efficient, less costly scheduled manner. This reduces the utility's costs, but the inspection and testing increases required increased costs. Compared to newer systems, costs are higher.

Chapters 6 and 7 will focus on various aspects of equipment aging, its management, and the mitigation of its effects in more detail. Usually, a utility cannot afford to replace all of its aging equipment. As a result, the solution to this particular problem source is a combination of partial measures that seek only to reduce failure rate slightly and control its effects:

a) Careful assessment and tracking of condition,

b) Good utilization of capacity and expected reliability of equipment,

c) Artful management of restoration and replacement policies,

d) Design of the system for failure – planning that assumes that things will go wrong.

Chapters 10 and 11 will discuss these in more detail.

Contributing Factor 2: Obsolete System Layouts

This, not aging equipment *per se,* is perhaps the strongest contributor to poor reliability in a majority of aging infrastructure cases. This is often not recognized or identified as an aging issue, but the facts speak for themselves.

Usually, the design of the power system in an area of a power system, its layout of substations and rights of way and the configuration of how they are interconnected and supported by one another during contingencies, dates from the time the equipment now in service was installed, or before. For example: the central part of the power system in many urban areas of the U.S. The "downtown power system" in most large cities is based on substation sites and rights of way that were laid out in the 1940s through the 1960s, when the post WWII boom created a period of sustained economic growth. These designs were executed with great skill to meet both the needs of the moment and reasonable amounts of load growth two or more decades into the future.

But between four and six (not just two) decades have passed since these systems were first laid out and built. In many urban areas throughout the United States, the electric demand has doubled or tripled since the power system last had a major revision in *layout.* By default, these systems had their layout "locked in:" As the city continued to grow, the utility found it increasingly difficult to obtain new substation sites or rights of way in the crowded, growing downtown area. Both economic and political pressures stood in the way of expanding its existing sites or adding new ones. Adding capacity at existing sites instead accommodated load growth. The utility often had to compromise preferred configuration (switchgear type and interconnection) and/or maintainability (quick accessibility) in order to squeeze every last MVA of capacity possible into crowded spaces.

At some point, equipment additions and renewals notwithstanding, the power system reached the limits of the existing sites and rights of way. Every bit of room had been used, every innovative trick within design standards had been exercised, and there was simply no more room. From thereon out, the utility either had to find some way to get the "unattainable" addition sites, or face the only other choice left: load the existing equipment in its sites and rights of way to higher levels than it would prefer.

In many cases, the rather extreme situations lead the utilities to overcome significant economic and political pressures and obtain some additional sites or space. But in all cases, the major manner of accommodation of load growth was by simply allowing loading on existing equipment to rise above traditional limits. A certain amount of this increase is by design, and does not necessarily have a negative impact. Since the mid 1980s, equipment utilization ratios (the ratio of peak loading to equipment capability) has risen through the power industry as electric distribution utilities sought to increase efficiency of equipment usage and cut back on expenses. It was not uncommon to see an electric distribution utility raise its target utilization ratio for transformers from 66%, about typical for the mid-1970s, to 70 – 83% during the last 15 years of the 20th century. This increase was by design.

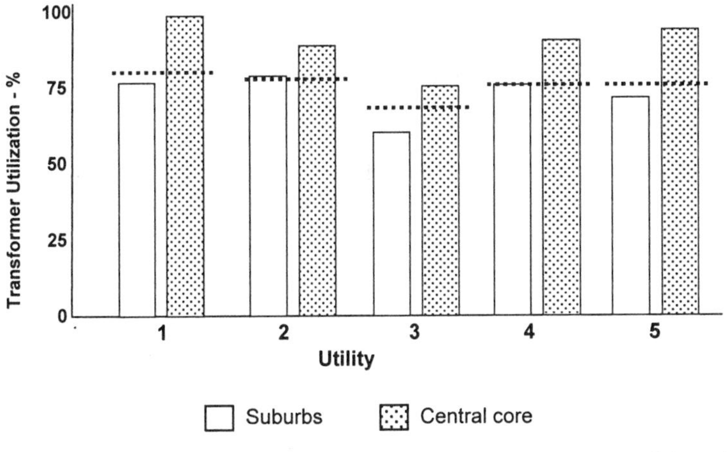

Figure 1.3 Substation transformer utilization ratios in the central core and the suburbs of five major metropolitan areas. Dotted line shows the utility's own design guidelines.

But the increases in loading forced on aging systems due to a combination of obsolete system layouts and other factors (to be discussed below) went far beyond these limits. In the central core of many cities in the U.S., power delivery equipment such as sub-transmission cables, substation power transformers, and primary distribution circuits are all loaded to their maximum rating at time of peak demand, leaving next to nothing for contingencies or emergencies. In a few cases, loadings at peak average close 100% throughout the downtown area and exceed that at a few places. Several of the utilities that experience serious problems during the summer of 1999 had some major delivery equipment in their central systems that was loaded to over 110% of nameplate rating during peak period, not by accident, but by intention.

There is plenty of evidence that this trend is widespread. Figure 1.3 compares 1999 utilization ratios in the central core and the suburbs of five major metropolitan systems, all among the largest power delivery systems in the U.S. Not all urban utilities follow this pattern, but a majority seems to match this profile. This is not to say that the utility's engineers preferred this higher loading level or that management and operators were entirely comfortable with the high loadings. But it was accepted, not as an emergency situation, but as the "normal" plan for serving peak demand.

High loadings: a complicated interaction with design

These higher loading levels were not accepted without a good deal of preparation and often only with changes to the system to keep it, at least ostensibly, within engineering criteria. In all cases the authors are aware of the utility planning and engineering departments' applied intense engineering methods to make arrangements so that the system could, at least based on traditional analysis, tolerate equipment outages in spite of the potential reliability problems such high loadings could cause.

However, the full implications of higher loadings are difficult to evaluate fully with traditional power system engineering tools. As a result, some of the potential problems of accommodating load growth in aging system areas by using higher loading levels were not fully appreciated. (This is part of contributing cause number 3 and will be discussed later in this chapter). While the systems in question met traditional reliability criteria, the systems failed to perform to the level the criteria were supposed to assure.

Configuration limitations

In addition, it is important to realize that the obsolete system presented the utility with challenges beyond just those caused by the higher loadings that had to result from the limited amounts of substation and rights of way space. The utility had to distribute a growing amount of power out of the same limited number of substation sites. This had detrimental effects on the performance and reliability of the primary feeder system. Beyond a certain point, a primary feeder system simply cannot be reinforced to deliver more power out of a limited number of sites, without accepting a certain amount of degradation in reliability and/or cost as a consequence of that limitation. This is due to host of secondary, but important effects related to primary feeder configuration and performance that will be discussed in Chapters 7, 8, 11, and 12.

Symptoms of the obsolete system layout

The net result of obsolete infrastructure is that higher loading levels have three undesirable consequences:

2a. Accelerated aging. Time to failure for most electrical equipment is often to a large degree a function of its loading. The higher utilization rates forced on utilities in these areas of the system cause electrical equipment there, particularly cables and transformers, to "age" or deteriorate faster and fail sooner. Thus, in those areas where age and failure rate is already a serious concern, the obsolete system layout forces the utility to exacerbate those problems just to get by. Over time, this tends to increase the frequency of service interruptions (i.e., SAIFI goes up).

2b. Limited contingency capability. High loading levels make it very difficult to provide service to all customers when a unit has failed. To begin with, a system with high utilization has "a lot of eggs in each basket." When a failure does occur, if service to consumers is to be maintained, a lot of load has to be re-distributed to other, still-functioning equipment nearby. But those remaining nearby units are already at high loading levels, meaning they have little available capacity to pick up additional load. The result is that interruptions, when they do occur, are difficult to restore, involving more load transfers, more emergency measures, and taking more time. While this does not increase the number of consumer interruptions to any large degree, it tends to increase the average duration of interruptions that do occur. At some point, SAIDI increases – the number of interruptions has gone up (due to age and the accelerated aging cited above) *and* now the length of interruptions goes up due to this reason. CAIDI (essentially their product) increases considerably.

2c. Higher delivery operating costs. The distribution systems in question operate at high loading levels. One consequence of this is the need in places for the utility to replace existing lines with a larger conductor simply to handle the MW-miles of delivery burden. Regardless, such systems typically have relatively high electrical losses. In addition, voltage drops are extreme. Tap changers and voltage regulators operate frequently, both fail more often and both require more preventive maintenance. These reasons and a host of similar secondary effects increase operating cost at the distribution level by 10%–20%.

In Table 1.2 and the discussion above, the authors have indicated that this particular cause among the four presented here typically contributes as much or more to a utility's reliability and cost problems than any of the other three. This is because the obsolete layout interacts so much with the other aging issues. It exacerbates equipment aging. Solutions require analysis very sensitive to the weaknesses of traditional engineering methods. And, on its own, the obsolete system layout creates operating and reliability challenges.

Both the problem and its solutions are complicated because they interact so strongly with both the other three causes of the aging infrastructure problem, and with many other aspects of the power delivery utility's operation. Generally, the solution is some combination of:

a) Somehow obtaining a few new sites and ROW to improve the system layout,

b) Use of innovative means of squeezing more capacity into existing sites, and most important,

c) Optimized configuration of substation switching and buswork, and primary feeder-level switching, based on sound reliability-based engineering (to see discussed below).

Chapter 8 focuses on obsolete system layout and its effects in more detail.

Contributing Cause Number 3: Old Engineering Methods

Many T&D utilities are using engineering methods that worked well in the 1970s, but cannot fully guarantee reliability operation in a world where substation and line loadings are pushing beyond traditional levels. As a result, engineering planning functions at many utilities cannot identify precisely what the problems are or prescribe the most effective and economical remedies. This particular subsection will provide a particularly comprehensive summary of this subject. Engineering tools are in an area where utilities can make immediate improvements and one where improvement must be made if the aging infrastructure issue is to be properly addressed. Chapters 9, 12, and 13 provide more detail on this subject.

N-1 criterion

Traditionally, reliability was designed into a power system by making certain that there was a "backup" of sufficient capability for every major element of the power system. If that element failed, negative consequences of that failure were covered. This is called "N–1" planning, or use of the N–1 criterion. Any one element of the power system, composed of N elements, can be removed, and the system will still perform its job, even under the most stressful (peak) conditions. Simply stated, one bought reliability with capacity

N–1 engineering is done by running N separate studies of a power system, one in which each of the N elements is removed (outaged) and the system switched as appropriate to have other equipment pick up the load it was serving. Computer programs that use a load flow algorithm (the basic power system simulator which estimates voltage, current, and loading throughout a power system for a specified set of conditions) have been written to do this automatically. The power system engineer inputs a description of the system and its loads, and the program both solves the "normal" (everything working) load flow case and all N "single contingency" (one element out) cases, and reports all with any problems such as overloads, low voltages, poor power factor, etc.

More generally, this approach is called the "N–X" criterion – a power system of N elements can be designed to tolerate the loss of any X elements. This requires N^X cases to be run by the automatic analysis program.

The use of this N–1 approach is dogma throughout the power industry. Developed in the 1960s, it proved very successful: systems designed with it worked well. Every major utility developed the design of its high voltage transmission grid, sub-transmission and substation systems, and other critical elements of its system, using this method. Nearly every utility modified the method in numerous small respects to fit its needs or preferences, but overall, this approach was universally applied. The reasons were:

1) The basic concept makes tremendous sense – design the system so that if anything goes wrong, nothing bad happens as a result.
2) Experience showed that this method worked: traditional power systems designed with this approach provided good reliability.
3) There were no alternatives, at least originally.

N–1's limitations

The N–1 approach is a sound engineering method, in fact one that the authors recommend as a part of any power system planning and engineering procedure. But like all engineering methods it makes assumptions, uses approximations, and take short cuts to gain efficiency – in other words, it has its limits. There is no doubt (based on publications from the 1960s and 1970s) that the engineers who developed and first applied N–1 recognized these limitations. However, because the traditional applications of N–1 seldom encountered these limitations, and because institutional memories are short, many utility planners and engineers working today never considered that as conditions change, their systems might run into those limitations.

By any practical standpoint, N–1 capability is a *necessary requirement* for power system reliability. To be reliable, a power system must have the ability to tolerate the loss of any one element of the system and still do its job. However, that does not make it a *necessary and sufficient* criterion. One that, if met, assures adequate reliability of service, which is how the industry often has applied the criterion and methods based upon it: design a system to meet N–1 and it was by definition "reliable enough." The fact that a power system meets N–1 criterion is not sufficient to assure that it will provide satisfactory reliability, or even fairly good reliability. As proof, it is only necessary to look at the system outages that occurred in summer of 1999. All of the systems that experienced severe reliability problems fully met N–1 and many cases met N-2, and even N–3 design criteria in critical places.

N–1's "ragged edge" as a dependable planning tool revolves around its lack of probabilistic expectation in its computations. N–1 methods and the N–1 criteria assure power system planners and engineers that there is *some feasible way* to back up every unit in the system, should it fail. However, they make no assessment of how likely the system is to get into situations that go beyond that. N–1 methods do not provide any analysis of:

- How likely is it that backup will be needed?
- How reasonable is the feasible plan for each contingency situation? Is the planner is actually building a "house of cards" by expecting "too many things to go right" once one thing has gone wrong?
- How much stress might the system be under during such contingency situations, and what are the long-term implications for both equipment life and operating feasibility of the system?
- How often do conditions occur which cannot be backed up (e.g., multiple failures) and how bad could the situation become when that is the case?

As a result, systems that meet the N–1 criteria can be less reliable than needed engineers might expect, even though the N–1 criteria "guarantees" there is a way to back up every unit in the system. This is much more likely to happen in modern power systems than it was in traditional, regulated power systems, due to changes in utilization and design made in the period 1990–2000, and the more volatile operating environment of de-regulation.

The reason is that modern power systems are much "leaner" than those of thirty or forty years ago, when N–1 was first developed and applied. In the 1960s through the early 1980s, electric utilities typically loaded key equipment, such as substation power transformers and downtown sub-transmission cables, to only about 2/3 or a little more (typically about 66%) of their capacity, even during peak periods. The remaining capacity was kept as "operating reserve" or "contingency margin." Engineers and planners at distribution utilities designed their power systems using the N–1 and other criteria, while counting on this margin. In this environment, N–1 was both necessary and sufficient to assure good reliability.

Beginning in the 1980s and increasingly throughout the 1990s, utilities pushed equipment utilization upwards to where, in some systems, the average substation transformer was loaded to 83% of its rating during peak periods. As will be discussed in Chapter 8, in aging areas of these systems, which suffered from "obsolete system layout," utilization rates often averaged close to 100% under "normal" peak conditions.

Whenever a power delivery utility increased its planned equipment utilization rates, changes were carefully engineered into the system, so that it could fully met the N–1 criteria everywhere, and even attain N–2 capability in critical places. This despite the greatly reduced contingency margin levels now available: *there were feasible ways to back up any and all failures, despite the higher utilization rates.*

However, as will be discussed in Chapter 9, this fell far short of providing traditional levels of reliability. N–1 is not sufficient to assure traditional levels of reliability, when the system is operated at high utilization rates. Basically the reasons are that such systems are much more likely to experience "high stress events" to the point that situations that go beyond the N–1 backup capability and are far more likely to occur. Such systems may have from three to five times the potential for customer service quality problems as traditional systems. Coupled with the higher failure rates caused by aging equipment, this creates a serious service quality handicap for the utility.

Other Engineering and Planning Related Problems

Power systems that operate with either an obsolete system layout and/or at high utilization rates are sensitive to more than just probability-related increases in interruption rates. The sensitivity of the system service quality to other problems increases exponentially. In particular, problems are encountered with partial failures, configuration complexity and load forecast sensitivity, which are discussed in further detail in the following paragraphs.

Partial failures

Contingency-based methods such as N–1 are basically "zero-one" models of failure: a unit of equipment is modeled as completely in, or completely out of service. But modern power systems often encounter partial failures or operating constraints of some components:

- A transformer may be in service but its tap changer has been diagnosed as problematic and is locked in one position.

- An oil-filled UG cable's pumps are disabled and the cable has been de-rated.
- A bus ground that failed tests has dictated opening a tiebreaker to balance fault duties.

These and other similar problems impact an area of the system with obsolete system layout or an area of the system operating at high utilization ratios, far more than they impact a traditional (66% – 70% utilization) power system. The obsolete layout is stressed with respect to voltages, flows, etc., and limitations in its full equipment usage due to such partial failures adds stress and often "uses up" contingency capability. A power transformer loaded to 85% at peak, whose tap changer is locked into one position, is subject to voltage regulation problems that can easily reduce its ability to handle load by more than 15%. The contingency margin (85%-100%) that the typical N–1 method assumes is there, is in fact gone.

Configuration complexity

At high utilization levels a system needs additional switching flexibility to accommodate its need to backup equipment operating at high loading levels with other equipment that is also already operating at high loading levels. Traditional N-1engineering methods are capable of analyzing the additional complexity this additional switching adds to the system design, at least from the standpoints of both electrical flows and feasibility evaluation of contingency operation. However, configuration decisions at substations (i.e., ring-bus vs. breaker-and-a-half vs. double bus design) and feeder layout and branching/switching design (see Chapter 10) affect reliability – how likely and how often – failures or outages will lead to interruptions. When configuration becomes more complex, as it does in most cases where adequate levels of reliability are being sought when utilization rates are high, the interaction of configuration and reliability is non-trivial and not intuitively obvious. Candidate plans that look like they will provide an improvement can, in fact be less reliable overall.

N-1 methods cannot assess, nor provide good feedback to engineers on the interaction of configuration with actual customer-level reliability. For this reason they leave a gap in the engineer-planner's understanding of the system being designed. Generally, systems laid out only with N-1 methods have configurations, which will, providing feasible contingency capability, offer far less operating flexibility than really necessary for reliable operation.

Load Forecast Sensitivity

A system that is under a good deal of stress, either because it has a system layout that is less than ideal for its load levels, or because equipment is being operated at high utilization rates, is particularly sensitive to errors in the load forecast used in its planning. Poor load forecasts used in the planning or operation of a power delivery system effectively "use up" its contingency capability [Willis and Powell, 1985]. This will be discussed in more detail in section 9.2. Poor normalization of weather data for forecasting or poor spatial forecasting (poor correlation of loads with areas and equipment) result in deterioration of a systems contingency to withstand capability. This greatly exacerbates the reliability-of-service problems discussed up to this point in this chapter.

Contributing Cause 4: Non-Optimum Use of the Primary Distribution Level

Averaged over the entire power industry, the primary distribution system is without a doubt both the most important and least well-used part of the power delivery chain, for several reasons. First, it accomplishes more of the power delivery function than any other level of the system, in the sense that power is routed into a few locations in fairly large chunks. Then the distribution system routes and sub-divides that power into very small allotments delivered to many different locations. No other layer (See Figure 2.3) of the power system accomplishes so much dispersion over area or so much sub-division of power.

Secondly, the primary distribution level has more impact on customer power quality than any other layer. In the vast majority of power systems, it is both immediately downstream of the lowest voltage regulation equipment (substation voltage regulators or load-tap changers) in the system, and by the nature of its topography, the source of the majority of voltage drop from that regulation point to the customer. That topography also guarantees that the primary distribution system sees the majority of outages that lead to customer service interruptions. Thus, in most systems, the distribution system is the source of the majority of both the voltage-related and availability-related power quality problems seen by energy consumers.

Not well used

At many power distribution utilities, the primary distribution system is planned and engineered using principles and layout concepts that were developed in the 1920s and 1930s and that have been handed down through generations of planners and engineers without substantial revision or re-examination. Foremost among these long-time "habits" are the overall layout style of the primary feeder system: whether the large trunk, multi-branch of the hierarchical subdivision, often-called "feathered" template for radial feeder layout, guides that.

Additional factors include switching zone and contingency backup policy, voltage regulation and 'reach" design, and conductor set selection and tapering policy.

Any reasonable layout style, conductor size set, switching zone, and conductor sizing rule base, will result in a distribution system that does its basic job as interpreted from the traditional perspective: route power from substations to customers without violation of voltage, loading, or power factor standards. However, this interpretation of the distribution system's job is obsolete, and incomplete. It looks at the distribution function only from the perspective of moving power and neglects its role in and capability for improvement of customer-service reliability.

Very few primary distribution systems are planned and engineered to make maximum use of the distribution system as a resource for customer reliability. As mentioned above, the guiding rules for primary distribution planning and design at most utilities were developed in the 1930s, and last seriously re-examined and revised in the late 1960s. Additionally, they are applied in a "cookie cutter" manner, using rules-of-thumb, informal guidelines for the larger issues of layout, switch planning (along with copious formal guidelines about the details), and rote rules for overall design. In a majority of cases, these rules are often too conservative, not allowing the full use of distribution as a resource.

An untapped resource

As a result, distribution systems in most utilities are underutilized in two ways:

1. Economics – when viewed just from their standpoint of their traditional function moving and dispersing power – they cost more than is necessary.

2. Reliability – the potential of the primary system to maximally contribute to improved reliability is partially untapped in the vast majority of cases.

Table 1.xx gives statistics on the results of "distribution optimization" studies from six utilities, which the authors' have recently worked. These utilities are representative of all the utilities the authors have worked with over the past ten years. They are an average set, not a set picked to show particularly good results. They demonstrate both the margin of improvement that is possible by updating and modernizing primary distribution planning and design paradigms, and the variation that is seen among different utilities.

Table 1.4 Improvements in Distribution System Cost and Performance Derived from Revision of Traditional Design Guidelines and Engineering Methods

Case	Utility System Type	Electrical $	Existing $	Reliability $	System Reli.
1	Urban UG, Atlantic coast	8%	2%	79%	3.4:1
2	Rural, small towns, Central US	12%	0%	25%	.8:1
3	Metropolitan, Midwest US	6%	3%	40%	1.8:1
4	Entire system, IOU, Central US	2%	3%	36%	1.5:1
5	Rural, small towns, Southern US	12%	11%	25%	1.6:1
6	Entire system, N.E. US	22%	16%	33%	.92:1
Averages		10%	6%	40%	1.7:1

Table 1.4 lists four statistics for each system. The first, "Electrical - $," is the reduction in overall (lifetime) cost of distribution that could be wrought with respect to the traditional primary distribution paradigm of moving power. This is the savings in *new* capital additions that result from improving how the distribution system is designed to do its traditional job – moving power.

The second statistic, "Existing - $" shows how much the useful utilization of an *existing* distribution system can be improved with respect to that traditional paradigm. Existing systems encompass both the good and bad points of their past engineering and design, and those systems cannot be thrown away and re-designed based on new and improved rules. However, as shown, some improvement can be wrought in the MW/$ capability of these systems. However, improvement is only useful where needed (i.e., if one improves a 5 MVA feeder so it can carry 5.4 MVA, that is useless unless it is in an area where one *needs* .4 MVA more capability). These values reflect both what is possible, and what was found useful, on the respective systems.

The third statistic, "Reliability - $," represents the reduction in the cost of improving reliability that was attained by revising design guidelines and applying the most effective reliability-based planning and engineering methods to these systems. These improvements while rather dramatic, are typical. As stated earlier, distribution systems are simply not designed from a reliability standpoint. When this is done and the engineering is optimized, the result is a very considerable improvement.

The final statistic, "System Rel.," represents the margin in "bang for the buck" that the revised distribution reliability improvement made over the existing cost of improving reliability that was in effect before the revision. This statistic goes to the heart of the use of distribution as a resource in aging infrastructure areas, and of the authors' overall theme of using all of the power system chain optimally to obtain the greatest "bang for the buck." In four of the six cases listed, reliability improvement was between 20% and 40% less expensive to buy at the primary level than at other levels of the system. In these cases, spending on the distribution system could deliver reliability improvement for less cost than spending on other levels of the system.

In two other cases, this was not the case, and the value shown is negative. Of note here is that the utilities were unaware of this entire issue, because the traditional tools they were using for the design of all three levels were not capable of directly engineering reliability or of even measuring such things as reliability improvement per dollar. Thus, in all cases, improvements of between 9% (1/ 92 – see final column) and 70% were possible by adopting reliability-based engineering methods and using them at the distribution system and for coordination of design among the distribution and sub-transmission-substation levels.

Table 1.4 makes it clear than revision of distribution system guidelines and adoption of reliability-based engineering methods should be viewed as a valuable improvement in terms of the traditional (electrical) perspective on distribution. The improvements in cost effectiveness of 10% and 6% are very noticeable and quite valuable.

However, from what might be termed a "modern perspective" – one that views the system's role as one of both performing the electrical duties and providing reliability of power availability, the improvement represents a quantum leap in cost effectiveness. By looking at the distribution system as a reliability resource, and optimizing its use for that purpose, an average 40% increase in cost effectiveness of reliability improvement is obtained at the distribution level. Coordination of that with the planning and design of other levels of the system achieve a 70% increase in overall effectiveness of money spent on reliability improvement.

The authors are aware that many readers, particularly some with long experience as planners at power delivery utilities, will dispute that these improvements shown are theoretical and were not implement-able (they were), or that that their system is different (not likely), or that "it won't work here" (it will). Chapter 10 will go into more detail on these improvements.

It <u>is</u> rocket science

The rather significant improvement in effectiveness that is possible by revising guidelines and adopting newer engineering methods at the primary distribution level brings up a key point about why the industry clung to what are essentially outdated engineering practices for so long. Distribution planning and engineering at most utilities is kept at a rather low level.

Engineering of a distribution system so that it performs the basic electrical functions that were traditionally required of it is rather straightforward. It can be done and was done for decades with nothing more than rough calculations (slide rules at best), nomographs and tables, rote-memorized rules of thumb, and institutionalized memory informally handed down from one generation of engineers to another. When digital computers are used, that level of engineering becomes quite simple. For this reason, one often hears terms such as "it isn't rocket science" or "distribution planning and design are quite simple." This cultural paradigm is exacerbated by the following facts: 1) Distribution is of relatively low voltage (compared to other parts of the power system; 2) the equipment involved is rather small and inexpensive (compared to other levels) and; 3) the system is normally laid out in radial form (which makes power flow easy to calculate).

In fact, the power distribution system is by far the most difficult of all layers in the power system to plan and engineer to a truly high standard. It is rocket science, because distribution systems are like piranha. Yes, all the equipment is low voltage, all of it is small, and each unit is some type of very common, "commodity" design.

But a distribution system consists of tens or hundreds or thousands of thousands of these components. In aggregate it is large, expensive, and most importantly, *very* complicated. A distribution system represents a tremendous challenge in *combinatorial decision-making* – in deciding how to select and arrange equipment, line routes and capacity tapering, interconnection, normal and contingency flow patterns, and a myriad of other factors to best achieve the required goals.

It is possible to select and arrange this equipment based on standardized layout rules, traditional (and rather straightforward) guidelines on equipment and engineering methods, so that it works well enough from an electrical standpoint.

Beyond this, distribution systems present a number of other challenges which ultimately create a large gap between the quality of a plan that works, and that of a really good plan. First, the distribution system has to handle the vast majority (roughly 95%) of the load coincidence (diversity of peak) dynamics on the system. Secondly, distribution is closer to the consumer than any other part of the power system: it has more than its share of constraints on routing, equipment placement, and design. Third, distribution must deal with a good deal of 'fuzziness' in terms of the demand levels it serves. Unlike at the transmission level, accurate load measurements are far and few between. Usually loads are estimated from metered, kWh data, and assumed values of diversity.

For all of these reasons, getting the most value possible out of an existing distribution system, or from new investment about to be made in it, is more difficult than in any other level of a power systems. Distribution planning, done to the most effective manner possible, is rocket science.

Contributing Cause 5: Old Cultural Identities and Ideas

Old habits die hard. At some utilities, the managers, planners, engineers and operators still tend to think in terms of the culture and goals that were appropriate for a purely regulated, vertically integrated industry. New goals may have been articulated, but for various reasons a change in thinking, in priorities, and in method has not followed, one which makes the planning, engineering, and operating compatible with those new goals.

Certainly, to some extent, the use of the out-dated engineering methods cited earlier could be characterized as part of this problem. Particularly, the considerable resistance to new ideas that prevents many utilities from fully adopting new reliability approaches (Contributed Cause 3, above) and distribution planning practices and guidelines (Contributing Cause 4, above) is mostly due to cultural resistance to a change in institutionalized values and methods.

However, the authors have identified those as separate issues – methodology, rather than culture -- in order to focus attention on broader issues. In a number of cases, perhaps half of the aging infrastructure problems the authors have analyzed – retention of outmoded priorities, corporate values, and concepts - has been a contributing factor in the problems the utilities experience with aging infrastructures. This is often not a result of resisting change as much as one of recognizing when, how and where to change.

Executive management's perspective: a more business-driven focus

Since the power industry was de-regulated in the mid 1990s, the 21st century power companies and distribution utilities that fell out of the dis-aggregation of traditional vertically integrated utilities, began taking a more "business driven" approach to their decision-making. Investments and operating policies were justified on the basis of their business case: if and how much they helped the bottom line of the business. This perspective is a huge departure from the values traditionally used at many utilities. Executives at many modern utilities drive all priorities into their organization starting from their financial model. "What will it take for us to achieve our financial goals?" Budgets and priorities are set by that requirement.

While electric utilities always recognized they were businesses, many of the values they used in decision-making were based on a "public stewardship" concept that does the right thing, regardless of cost. This attitude made sense within the traditionally regulated utility business model. Any expenditure justified within that framework could be put into the rate base. Ultimately the utility would recover its expenditures and earn a reasonable return on invested capital.

Deregulation, and shifts in attitudes within the industry and regulatory agencies, has changed that perspective. No longer are capital expenditures and operating expenses acceptable to a utility's management just because projected revenues will cover all projected future costs. Executives insist that spending be contained for a variety of business-related reasons. Regulators are concerned about price as much as quality. The result is a system that is much more focused on the business implications of each project, not just the assurance that a project is the minimum revenue requirements solution to a given "problem," as defined by the planning guides.

In this new paradigm, traditional values such as good customer service, high utilization of physical assets, and efficient operation are all-important priorities, because they contribute to that end. It is interesting that there seems to be more focus on these "virtues" of good utility operation under deregulation than there was under a traditionally regulated operation. Certainly utilization rates are higher, attention paid to reliability is greater than ever, and more effort is being directed at improving efficiency than ever before. The reason is that when one has less of something, more attention is given and more effort is put into getting the most out of it. Modern utilities understand that all of the virtues they desire boil down to one resource: money, and that they will never have enough to meet all possible needs based on traditional planning standards.

Additionally, in aging infrastructure areas, a power distribution utility will have nowhere near enough money to take the "obvious" approach to the aging infrastructure: replace all the old equipment and upgrade the obsolete layout. There simply is not enough money. This makes the whole planning situation budget-constrained. The overall, executive management perspective is thus,

> "The distribution system is a physical asset that we purchase and maintain to provide good customer service and stockholder profits. It is doing its job when it supports our business model."

Changes in cultural "standards" may be needed

Many of today's power delivery planning, engineering and operating concepts were developed by or for vertically integrated utilities in a regulated, non-competitive industry. Some are quite inappropriate for a business-driven utility environment. They use procedures that are rule- rather than results-based procedures, and thus not easily adaptable to an environment, in which decisions are ultimately driven by financial, rather than standards or policy, considerations. Required changes are scattered throughout this book, but in general, the utility needs to re-examine:

- *Standards and guidelines* – one utility that encountered severe problems in 1999 had not re-examined and revised its distribution planning and engineering guidelines and standards since 1978. Forgetting entirely the issues of de-regulation and competition, load characteristics, capital vs. operating cost considerations, and a host of other changes in system requirements have occurred in the intervening 28 years.

- *Budget base.* Many utilities operate from year-to-year by revising the current year budget incrementally for the following year. Budget allocations and functions among departments are effectively institutionalized along traditional lines.

- *Protection and restoration practices* at many utilities follow practices developed in the late 60s and early 70s. These are often incompatible with modern consumer needs of fewer interruptions and do not make use of the most modern equipment and techniques for elimination of interruption propagation through the network whenever an outage occurs.

- *Prioritization of projects* is still based on models and concepts developed while utilities were regulated, vertically integrated companies. Most of these prioritization methods are entirely incapable of allocating a limited (constrained) budget among needed projects in the best possible manner [Willis, 1997, Chapter 18].

- *Reliability engineering* at many utilities is mostly reactive. Reliability analysts review operating records and experience in order to identify poor performing areas or equipment. Once identified, these areas are turned over to Engineering and Planning for rehabilitation, using what are basically traditional contingency-based tools.

- *System design* at many utilities throughout the United States still follows specific styles and guidelines developed decades ago. Nearly every utility in the industry has slightly different preferences in how it lays out and specifies equipment for substations, how it routes, sizes, and switches feeders, how it designs its protection schemes and c0ontingency backup plans, and how it operates its system at times of high stress. Every utility insists that its methods are best. But they can't all be correct – in fact few have it right, and all can learn by reviewing what other utilities do best.

In these cases, and a host of others, utilities can make improvements by re-examining what are often long cherished practices and unique characteristics or policies, which they firmly believed in. This fresh approach should both look at the changes in customer loads and needs since the last time practices were examined, and should make certain that the new standards, guidelines, methods and practices are maximally supportive of business-driven focus of the utility.

Inappropriate paradigms reduce the effectiveness of efforts the utility makes to improve or maintain performance, particularly in areas where reliability is relatively costly to maintain (aging areas). The impact is raised costs, which translates to increases in SAIFI or SAIDI, whichever the utility needs, because money is used inefficiently, which would otherwise go to make improvements as needed.

1.5 CONCLUSION AND SUMMARY

An aging power delivery infrastructure is an area of an electric utility system that is composed of mostly old equipment near the end of its lifetime, configured in a layout that is itself quite old and not completely compatible with modern needs. Situations vary from one location to another, but most aging infrastructure areas share similar characteristics, exhibiting a type of "signature" as listed in Table 1.5 and discussed earlier in section 1.1. Table 1.5 summarizes eight of the most typical characteristics of aging infrastructure areas.

For a variety of reasons, the aged equipment and layout create reliability, maintenance, and budgeting challenges that will eventually overwhelm even the most effective and efficient utility company if not anticipated and brought under control. The most easily identifiable characteristic of aging delivery

Table 1.5 Characteristics of the Typical Aging Infrastructure Area

1. The system layout and design was first put in place more than forty years ago.
2. The majority of equipment in the area is more than forty years old.
3. The system is well engineered and fully meets minimum engineering criteria.
4. The area is seeing steady, if perhaps low, load growth in load.
5. The area is plagued by above average equipment failure rates. Overtime is high due to large amounts of unscheduled repair and restoration.
6. SAIFI began rising some years ago; SAIDI is beginning to rise now.
7. Major interruption events always occur due to a bizarre series of outages.
8. Things go bad a lot.

Table 1.6 Summary of Suggested "Fixes" for Aging Infrastructure Problems

Priority	Action	Comments
1	Cross-functional results-centered management	Optimizes and coordinates overall utility focus. Accommodates other planning, engineering, operation and management methods aimed at improving reliability.
2	Zero base all budgets	Make everything re-justify itself, at least once.
3	Augment N –1 with reliability-based methods	Optimizing reliability (obtaining the most for the less cost) requires using techniques that explicitly measure the expected reliability and engineer it.
4	Use budget-constrained prioritization methods	Proven effective in allocating budgets most effectively when there is not nearly enough money for all "worthy causes."
5	Apply Reliability-Centered Maintenance (RCM)	It works, reducing maintenance costs and improving results.
6	Optimize planning & operation against PBR	Let regulatory rules define the specific targets for customer service quality.
7	Revise standards & guidelines to modern needs	Make equipment usage and system design rules compatible with themselves and with modern needs.
8	Adopt results- rather than rule-based standards	Pragmatic and flexible standards are necessary if maximum results are to be obtained.
9	Use IT systems focused on customer service	Focus "if-y" information systems on improving customer service, not reducing cost.
10	Implement TLM type equipment tracking	Periodic equipment evaluation and produce of "jeopardy lists" can improve effectiveness and cut failure impact.
11	Optimize equipment loading for service value	Load equipment differently depending on the situation
12	Use innovative compact substation design	Innovative substation layouts and equipment can pack up to three times the capacity and ten times the reliability into the same site size.
13	Optimize substation configuration for reliability	Reliability engineering intensely applied to high and low side design of substations can greatly enhance overall customer service quality
14	Minimize operate voltage at the primary level	Reduction in primary distribution voltage to the lowest possible level consistent with good service can reduce equipment aging rates by up to 10%

infrastructures is that the equipment in the area is old. It will have an *average* service age greater than its design lifetime. However, this chapter presented five interrelated problem areas that all contribute to the poor customer service levels seen in aging delivery infrastructure areas. Table 1.6 lists these four problem areas and specific factors within each, and describes whether they impact mostly the frequency of customer service interruptions (SAIFI) the duration (SAIDI) or the costs a utility sees in trying to provide good service.

What Can Be Done

The conclusion of this book, Chapter 16, summarizes the recommendations and guidelines developed in Chapters 2 through 15. Central to the entire approach, however, are the six "guiding principles" of result-centered approach.

1. *Use a results-focused approach.* The two results desired are customer service quality and low cost. Thus, the guiding principle above all others is to maximize their ratio over the entire system. This is discussed nearly everywhere throughout this book.

2. *Zero-based* all decisions and budgets. *Nothing* is scared, and no level of expense is considered acceptable just because they represent a traditional level of spending in that category.

3. *Marginal benefit/cost ratio* should be used as the basis for all evaluation of "bang for the buck." This provides far superior results to prioritization done on the basis of benefit/cost ratio or other means. This concept is discussed and developed in detail in Chapters 5 (section 5.3), 11, and 14.

4. *Macro-comparison of intra-project alternatives* using that marginal benefit/cost ratio should be used across departmental and functional boundaries to prioritize spending decisions among candidate projects and budgets. This rather lengthy, "technical title" actually describes a rather simple but critical procedure, in which the keys to making results-centered management work, is covered in Chapters 11 and 14.

5. *Reliability*-based planning, engineering, operating and maintenance methods should be employed throughout the organization and a customer reliability-focused approach should guide all decisions on IT.

6. *Planning for failure.* The best results come from a management

perspective that recognizes that equipment failures are not the result of an "our failure" and that they cannot be avoided. They can, however, be managed. Failures are a natural consequence of equipment use, and they will happen. Their likelihood and consequences can be engineering and managed.

Table 1.6 summarizes the major specific operational recommendations developed from this guiding principle, as given elsewhere in this book and summarized in Chapter 16. The central theme is "results-centered management," the use of a combined business-engineering prioritization focus on reduced customer service problems at a minimum cost. This umbrella approach encompasses the other recommendations and coordinates them for maximum effectiveness and channels all the utility's recourses in the most effective manner possible. Application to aging infrastructure areas (the authors have seen it applied successfully to improve the "bang for the buck" obtained in very new systems, too). It is described here in the context of and combined with other solutions aimed at aging areas with poor service reliability and cost performance, within traditional electric power systems.

REFERENCES

R.E.Brown, et al, "Spatial Load Forecasting Using Non-Uniform Areas" in *Proceedings of the IEEE T&D Conference,* April, 1999,

R.E. Brown, et al, "Reliability and Capacity: A Spatial Load Forecasting Method for a Performance-Based Regulatory Environment," in *Proceedings of the IEEE PICA Conference,* May, 1999, San Jose

M. V. Engel et al., editors, *Tutorial on Distribution Planning*, IEEE Course Text EHO 361-6-PWR, Institute of Electrical and Electronics Engineers, Hoes Lane, NJ, 1992.

H. N. Tram et al., "Load Forecasting Data and Database Development for Distribution Planning,"*IEEE Trans. on Power Apparatus and Systems,* November 1983, p. 3660.

H. L. Willis, "Load Forecasting for Distribution Planning, Error and Impact on Design," *IEEE Transactions on Power Apparatus and Systems,* March 1983, p. 675.

H. L. Willis and R. W. Powell, "Load Forecasting for Transmission Planning," *IEEE Transactions on Power Apparatus and Systems,* August 1985, p. 2550.

H. L. Willis, *Spatial Electric Load Forecasting,* Marcel Dekker, New York, 1996.

H. L. Willis, *Power Distribution Planning Reference Book,* Marcel Dekker, New York, 1997.

H. L. Willis and J. E. D. Northcote-Green, "Spatial Load Forecasting -- A Tutorial Review," *Proceedings of the IEEE,* Feb. 1983, p. 232.

H. L. Willis and H. N. Tram, "Load Forecasting for Transmission Planning," *IEEE Trans. on Power Apparatus and Systems,* paper 83 SM 379-5, March 1984.

P. Carrier (Chairman), *Interim Report of the U.S. Department of Energy's Power Outage Study Team: Findings from the summer of 1999*, United States Department of Energy, Office of Policy, Washington, January 2000

2
Power Delivery Systems

2.1 INTRODUCTION

Retail sale of electric energy involves the delivery of power in ready to use form to the final consumers. Whether marketed by a local utility, load aggregator, or direct power retailer, this electric power must flow through a power delivery system on its way from power production to consumer. This transmission and distribution (T&D) system consists of thousands of transmission and distribution lines, substations, transformers, and other equipment scattered over a wide geographical area and interconnected so that all function in concert to deliver power as needed to the utility's consumers.

This chapter is a quick tutorial review on T&D systems, their mission, characteristics, and design constraints. It examines the natural phenomena that shape T&D systems and explains the key physical relationships and their impact on design and performance. For this reason experienced planners are advised to scan this chapter, or at least its conclusions, so they understand the perspective upon which the rest of the book builds.

In a traditional electric system, power production is concentrated at only a few large, usually isolated, power stations. The T&D system moves the power from those often-distant generating plants to the many consumers who consume the power. In some cases, cost can be lowered and reliability enhanced through the use of distributed generation (DG): numerous smaller generators placed at strategically selected points throughout the power system in proximity to the consumers.[1] This and other distributed resources - so named because they are distributed throughout the system in close proximity to consumers - including storage systems and demand-side management, often provide great benefit.

But regardless of the use of distributed generation or demand-side management, the T&D system is the ultimate distributed resource, consisting of thousands, perhaps millions, of units of equipment scattered throughout the service territory, interconnected and operating in concert to achieve uninterrupted delivery of power to the electric consumers. These systems represent an investment of billions of dollars, require care and precision in their operation, and provide widely available, economical, and reliable energy.

This chapter begins with an examination of the role and mission of a T&D system: why it exists and what it is expected to do, in section 2.2. Section 2.3 looks at several fundamental physical "laws" that constrain T&D systems design. The typical hierarchical system structure that results and the costs of its equipment are summarized in sections 2.4, 2.5 and 2.6. In section 2.7, a number of different ways to lay out a distribution system are covered, along with their advantages and disadvantages. Section 2.8 covers the "systems approach," already mentioned in Chapter 1, and perhaps the single most important concept in the design of retail electric delivery systems, which aim to be both inexpensive and reliable.

2.2 T&D SYSTEM'S MISSION

A T&D system's primary mission is to *deliver* power to electrical consumers at their *place* of consumption and in *ready-to-use* form. This means it must be dispersed throughout the utility service territory in rough proportion to consumer locations and demand (Figure 2.1). This is the primary requirement for a T&D system -- *the system must cover ground* -- reaching every consumer with an electrical path of sufficient strength to satisfy that consumer's demand.

That electrical path must be *reliable*, too, so that it provides an uninterrupted flow of stable power to the utility's consumers. Reliable power delivery means delivering all of the power demanded not just some of the power needed, and doing so all of the time. Anything less than near perfection in meeting this goal is considered unacceptable. While 99.9% reliability of service may sound impressive, it means nearly nine hours of electric service interruption each year, an amount that would be unacceptable in nearly any first-world country.

---

[1] See *Distributed Power Generation – Planning and Evaluation,* by H. L. Willis and W. G. Scott, Marcel Dekker, 2000.

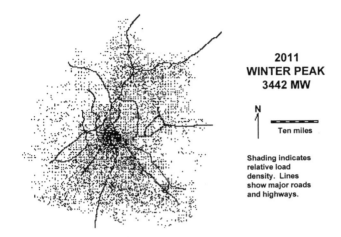

Figure 2.1 Map of electrical demand for a major US city shows where the total demand of more than 2,000 MW peak is located. Degree of shading indicates electric load distribution. The T&D system must cover the region with sufficient capacity at every location to meet the consumer needs there.

Beyond the need to deliver power to the consumer, the utility's T&D system must also deliver it in ready-to-use form - at the utilization voltage required for electrical appliances and equipment, and free of large voltage fluctuations, high levels of harmonics or transient electrical disturbances (Engel et al., 1992).

Most electrical equipment in the United States is designed to operate properly when supplied with 60 cycle alternating current at between 114 and 126 volts, a plus or minus five percent range centered on the nominal *utilization voltage* of 120 volts (RMS average of the alternating voltage). In many other countries, utilization standards vary from 230 to slightly over 250 volts, at either 50 or 60 cycles AC. [2] But regardless of the utilization voltage, a utility must maintain the voltage provided to each consumer within a narrow range centered within the voltages that electric equipment is designed to tolerate.

A ten- percent range of delivery voltage throughout a utility's service area may be acceptable, but a ten- percent range of fluctuation in the voltage supplied to any one consumer is not. An instantaneous shift of even three

---

[2] Power is provided to consumers in the United States by reversed alternating current legs (+120 volts and -120 volts wire to ground). This scheme provides 240 volts of power to any appliance that needs it, but for purposes of distribution engineering and performance, acts like only 120 volt power.

percent in voltage causes a perceptible (and to some people, disturbing) flicker in electric lighting. More importantly, voltage fluctuations can cause erratic and undesirable behavior of some electrical equipment.

Thus, whether high or low within the allowed range, the delivery voltage of any one consumer must be maintained at about the same level all the time - normally within a range of three to six percent - and any fluctuation must occur slowly. Such *stable voltage* can be difficult to obtain, because the voltage at the consumer end of a T&D system varies inversely with electric demand, falling as the demand increases, rising as it decreases. If this range of load fluctuation is too great, or if it happens too often, the consumers may consider it poor service.

Thus, a T&D system's mission is to:

1. Cover the service territory, reaching all consumers
2. Have sufficient capacity to meet the peak demands of its consumers
3. Provide highly reliable delivery to its consumers
4. Provide stable voltage quality to its consumers

And of course, above everything else, achieve these four goals at the *lowest cost* possible.

2.3 THE "LAWS OF T&D"

The complex interaction of a T&D system is governed by a number of physical laws relating to the natural phenomena that have been harnessed to produce and move electric power. These interactions have created a number of "truths" that dominate the design of T&D systems:

1. *It is more economical to move power at high voltage.* The higher the voltage, the lower the costs per kilowatt to move power any distance.

2. *The higher the voltage, the greater the capacity and the greater the cost of otherwise similar equipment.* Thus, high voltage lines, while potentially economical, cost a great deal more than low voltage lines, but have a much greater capacity. They are only economical in practice if they can be used to move a lot of power in one block - they are the giant economy size, but while always giant, they are only economical if one truly needs the giant size.

3. *Utilization voltage is useless for the transmission of power.* The 120/240 volt single-phase utilization voltage used in the United States, or even the 250 volt/416 volt three-phase used in "European systems" is not equal to the task of economically moving power more than a few hundred yards. The application of these lower

voltages for anything more than very local distribution at the neighborhood level results in unacceptably high electrical losses and high costs.

4. *It is costly to change voltage level* - not prohibitively so, for it is done throughout a power system (that's what transformers do) - but voltage transformation is a major expense, which does nothing to move the power any distance in and of itself.

5. *Power is more economical to produce in very large amounts.* There *is* a significant economy of scale in generation, notwithstanding the claims by advocates of modern distributed generators. Large generators produce power more economically than small ones. Thus, it is most efficient to produce power at a few locations utilizing large generators.[3]

6. *Power must be delivered in relatively small quantities at low (120 to 250 volt) voltage level.* The average consumer has a total demand equal to only 1/10,000th or 1/100,000th of the output of a large generator.

An economical T&D system builds upon these concepts. It must "pick up" power at a few, large sites (generating plants), and deliver it to many, many more small sites (consumers). It must somehow achieve economy by using high voltage, but only when power flow can be arranged so that large quantities are moved simultaneously along a common path (line). Ultimately, power must be subdivided into "house-sized" amounts, reduced to utilization voltage, and routed to each business and home via equipment whose compatibility with individual consumer needs means it will be relatively inefficient compared to the system as a whole.

---

[3] The issue is more complicated than just a comparison of the cost of big versus small generation, as will be addressed later in this book. In some cases, distributed generation provides the lowest cost overall, regardless of the economy of scale, due to constraints imposed by the T&D system. Being close to the consumers, distributed generation does not carry with it the costs of adding T&D facilities to move the power from generation site to consumer. Often this is the margin of difference, as will be discussed later in this book.

Hierarchical Voltage Levels

The overall concept of a power delivery system layout that has evolved to best handle these needs and "truths" is one of hierarchical voltage levels as shown in Figure 2.2.

As power is moved from generation (large bulk sources) to consumer (small demand amounts) it is first moved in bulk quantity at high voltage - this makes particular sense since there is usually a large bulk amount of power to be moved out of a large generating plant. As power is dispersed throughout the service territory, it is gradually moved down to lower voltage levels, where it is moved in ever smaller amounts (along more separate paths) on lower capacity equipment until it reaches the consumers. The key element is a "lower voltage and split" concept.

Thus, the 5 kW used by a particular consumer -- Mrs. Rose at 412 Oak Street in Metropolis City - might be produced at a 750 MW power plant more than three hundred miles to the north. Her power is moved as part of a 750 MW block from plant to city on a 345 kV transmission line, to a switching substation. Here, the voltage is *lowered* to 138 kV through a 345 to 138 kV transformer, and immediately after that, the 750 MW block is *split* into five separate flows in the switching substation buswork, each of these five parts

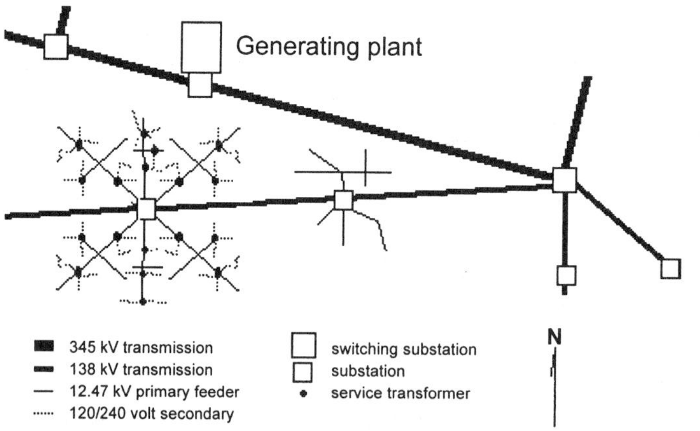

Figure 2.2 A power system is structured in a hierarchical manner with various voltage levels. A key concept is "lower voltage and split" which is done from three to five times during the course of power flow from generation to consumer.

being roughly 150 MW. Now part of a smaller block of power, Mrs. Rose's electricity is routed to her side of Metropolis on a 138 kV transmission line that snakes 20 miles through the northern part of the city, ultimately connecting to another switching substation. This 138 kV transmission line feeds power to several distribution substations along its route,[4] among which it feeds 40 MW into the substation that serves a number of neighborhoods, including Mrs. Rose's. Here, her power is run through a 138-kV/12.47kV-distribution transformer.

As it emerges from the low side of the substation distribution transformer at 12.47 kV (the primary distribution voltage) the 40 MW is split into six parts, each about 7 MW, with each 7 MVA part routed onto a different distribution feeder. Mrs. Rose's power flows along one particular feeder for two miles, until it gets to within a few hundred feet of her home. Here, a much smaller amount of power, 50 kVA (sufficient for perhaps ten homes), is routed to a service transformer, one of several hundred scattered up and down the length of the feeder. As Mrs. Rose's power flows through the service transformer, it is reduced to 120/240 volts. As it emerges, it is routed onto the secondary system, operating at 120/240 volts (250/416 volts in Europe and many other countries). The secondary wiring splits the 50 kVA into small blocks of power, each about 5 kVA, and routes one of these to Mrs. Rose's home along a secondary conductor to her service drops - the wires leading directly to her house.

Over the past one hundred years, this hierarchical system structure has proven a most effective way to move and distribute power from a few large generating plants to a widely dispersed consumer base. The key element in this structure is the "reduce voltage and split" function - a splitting of the power flow being done simultaneously with a reduction in voltage. Usually, this happens between three and five times as power makes its way from generator to consumers.

2.4 LEVELS OF THE T&D SYSTEM

As a consequence of this hierarchical structure, a power delivery system can be thought of very conveniently to be composed of several distinct levels of equipment, as illustrated in Figure 2.3. Each level consists of many units of fundamentally similar equipment, doing roughly the same job, but located in different parts of the system. For example, all of the distribution substations

---

[4] Transmission lines whose sole or major function is to feed power to distribution substations are often referred to as "sub-transmission" lines.

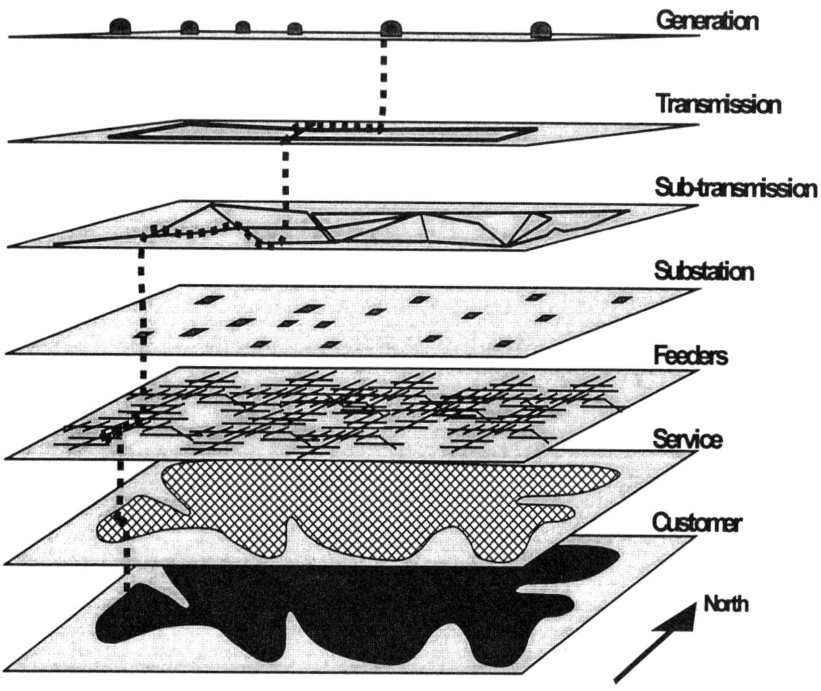

Figure 2.3 A T&D system consists of several levels of power delivery equipment, each feeding the one below it.

are planned and laid out in approximately the same manner and do roughly the same job. Likewise, all feeders are similar in equipment type, layout, and mission, and all service transformers have the same basic mission and are designed with similar planning goals and to similar engineering standards.

Power can be thought of as flowing "down" through these levels, on its way from power production to consumer. As it moves from the generation plants (system level) to the consumer, the power travels through the transmission level, to the sub-transmission level, to the substation level, through the primary feeder level, and onto the secondary service level, where it finally reaches the consumer. Each level takes power from the next higher level in the system and delivers it to the next lower level in the system. While each level varies in the types of equipment it has, its characteristics, mission, and manner of design and planning, all share several common characteristics:

- Each level is fed power by the one above it, in the sense that the next higher level is electrically closer to the generation.

- Both the nominal voltage level and the average capacity of equipment drops from level to level, as one moves from generation to consumer. Transmission lines operate at voltages of between 69 kV and 1,100 kV and have capacities between 50 and 2,000 MW. By contrast, distribution feeders operate between 2.2 kV and 34.5 kV and have capacities somewhere between 2 and 35 MW.

- Each level has many more pieces of equipment in it than the one above. A system with several hundred thousand consumers might have fifty transmission lines, one hundred substations, six hundred feeders, and forty thousand service transformers.

- As a result, the net capacity of each level (number of units times average size) increases as one moves toward the consumer. A power system might have 4,500 MVA of substation capacity but 6,200 MVA of feeder capacity and 9,000 MVA of service transformer capacity installed.[5]

- Reliability drops as one moves closer to the consumer. A majority of service interruptions are a result of failure (either due to aging or to damage from severe weather) of transformers, connectors, or conductors very close to the consumer, as shown in Figure 2.4.

---

[5] This greater-capacity-at-every-level is deliberate and required both for reliability reasons and to accommodate coincidence of load, which will be discussed in Chapter 3.

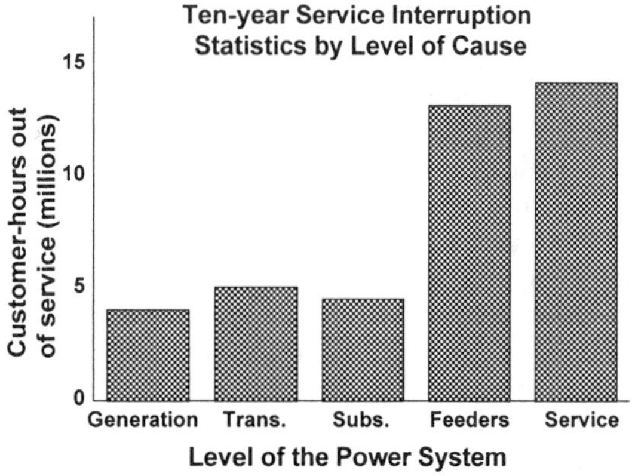

Figure 2.4 Ten years of consumer interruptions for a large electric system, grouped by level of cause. Interruptions due to generation and transmission often receive the most attention because they usually involve a large number of consumers simultaneously. However, such events are rare whereas failures and interruptions at the distribution level create a constant background level of interruptions.

Table 2.1 gives statistics for a typical system. The net effect of the changes in average size and number of units is that each level contains a greater total capacity than the level above it. The service transformer level in any utility system has considerably more installed capacity (number of units times average capacity) than the feeder system or the substation system. Total capacity increases as one heads toward the consumer because of non-coincidence of peak load (which will be discussed in Chapter 3) and for reliability purposes.

Table 2.1 Equipment Level Statistics for a Medium-Sized Electric System

Level of System	Voltage kV	Number of Units	Avg. Cap. MVA	Total Cap MVA
Transmission	345, 138	12	150	1,400
Sub-transmission.	138, 69	25	65	1,525
Substations	139/23.9, 69/13.8	45	44	1,980
Feeders	23.9, 13.8	227	11	2,497
Service Trans.	.12, .24	60,000	.05	3,000
Secondary/Service	.12, .24	250,000	.014	3,500
Consumer	.12	250,000	.005	1,250

The Transmission Level

The transmission system is a network of three-phase lines operating at voltages generally between 115 kV and 765 kV. Capacity of each line is between 50 MVA and 2,000 MVA. The term "network" means that there is more than one electrical path between any two points in the system (Figure 2.5). Networks are laid out in this manner for reasons of reliability and operating flow--if any one element (line) fails, there is an alternate route and power flow is (hopefully) not interrupted.

In addition to their function in moving power, portions of the transmission system, the largest elements, namely its major power delivery lines, are designed, at least in part, for stability needs. The transmission grid provides a strong electrical tie between generators, so that each can stay synchronized with the system and with the other generators. This arrangement allows the system to operate and to function evenly as the load fluctuates and to pick up load smoothly if any generator fails - what is called stability of operation. A good deal of the equipment put into transmission system design, and much of its cost, is for these stability reasons, not solely or even mainly, for moving power.

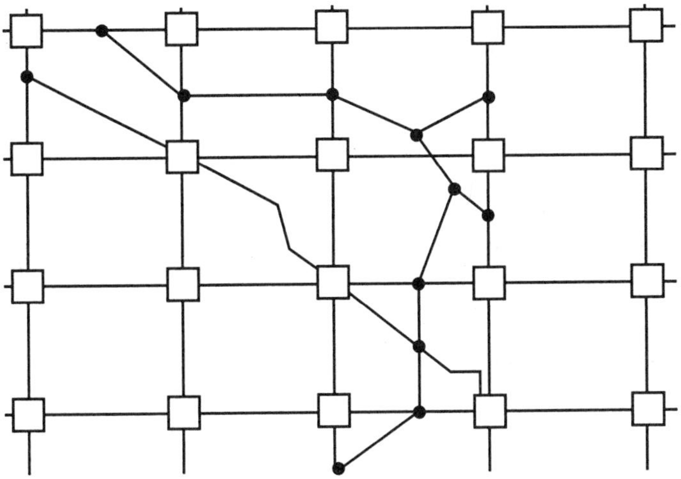

Figure 2.5 A network is an electrical system with more than one path between any two points, meaning that (if properly designed) it can provide electrical service even if any one element fails.

The Sub-transmission Level

The sub-transmission lines in a system take power from the transmission switching stations or generation plants and deliver it to substations along their routes. A typical sub-transmission line may feed power to three or more substations. Often, portions of the transmission system, bulk power delivery lines, lines designed at least in part for stability as well as power delivery needs to do this too. The distinction between transmission and sub-transmission lines becomes rather blurred.

Normally, sub-transmission lines are in the range of capacity of 30 MVA up to perhaps 250 MVA, operating at voltages from 34.5 kV to as high as 230 kV. With occasional exceptions, sub-transmission lines are part of a network grid - they are part of a system in which there is more than one route between any two points. Usually, at least two sub-transmission routes flow into any one substation, so that feed can be maintained if one fails.[6]

The Substation Level

Substations, the meeting point between the transmission grid and the distribution feeder system, are where a fundamental change takes place within most T&D systems. The transmission and sub-transmission systems above the substation level usually form a network, as discussed above, with more than one power flow path between any two parts. But from the substation on to the consumer, arranging a network configuration would simply be prohibitively expensive. Thus, most distribution systems are radial - there is only one path through the other levels of the system.

Typically, a substation occupies an acre or more of land on which the necessary substation equipment is located. Substation equipment consists of high and low voltage racks and busses for the power flow, circuit breakers for both the transmission and distribution level, metering equipment, and the "control house," where the relaying, measurement, and control equipment is located. But the most important equipment, what gives this substation its capacity rating, are the substation *transformers*. These transformers convert the incoming power from transmission voltage levels to the lower primary voltage for distribution.

Individual substation transformers vary in capacity, from less than 10 MVA to as much as 150 MVA. They are often equipped with tap-changing

---

[6] Radial feed - only one line - is used in isolated, expensive, or difficult transmission situations, but for reliability reasons is not recommended.

mechanisms and control equipment to vary their winding ratio so that they maintain the distribution voltage within a very narrow range, regardless of larger fluctuations on the transmission side. The transmission voltage can swing by as much as 5%, but the distribution voltage provided on the low side of the transformer stays within a narrow band, perhaps only ± .5%.

Very often, a substation will have more than one transformer. Two is a common number, four is not uncommon, and occasionally six or more are located at one site. Having more than one transformer increases reliability. In an emergency, a transformer can handle a load much over its rated load for a brief period (e.g., perhaps up to 140% of rating for up to four hours). Thus, the T&D system can pick up the load of the outaged portions during brief repairs and in emergencies.

Equipped with one to six transformers, substations range in "size" or capacity from as little as five MVA for a small, single-transformer substation, serving a sparsely populated rural area, to more than 400 MVA for a truly large six-transformer station, serving a very dense area within a large city.

Often T&D planners will speak of a *transformer unit,* which includes the transformer and all the equipment necessary to support its use - "one fourth of the equipment in a four-transformer substation." This is a much better way of thinking about and estimating cost for equipment in T&D plans. While a transformer itself is expensive (between $50,000 and $1,000,000); the buswork, control, breakers, and other equipment required to support its use can double or triple that cost. Since that equipment is needed in direct proportion to the transformer's capacity and voltage, and since it is needed *only* because a transformer is being added, it is normal to associate it with the transformer as a single planning unit. Add the transformer and add the other equipment along with it.

Substations consist of more equipment and involve more costs than just the electrical equipment. The site has to be purchased and prepared. Preparation is not trivial. The site must be excavated, a grounding mat - wires running under the substation to protect against an inadvertent flow during emergencies - laid down, and foundations and control ducting for equipment must be installed. Transmission towers to terminate incoming transmission must be built. Feeder getaways - ducts or lines to bring power out to the distribution system - must be added.

The Feeder Level

Feeders, typically either overhead distribution lines mounted on wooden poles or underground buried or ducted cable sets, route the power from the substation throughout its service area. Feeders operate at the primary distribution voltage.

The most common primary distribution voltage in use throughout North America is 12.47 kV, although anywhere from 4.2 kV to 34.5 kV is widely used. Worldwide, there are primary distribution voltages as low as 1.1 kV and as high as 66 kV. Some distribution systems use several primary voltages - for example 23.9 kV, 13.8 kV and 4.16 kV.

A feeder is a small transmission system in its own right, distributing between 2 MVA to more than 30 MVA, depending on the conductor size and the distribution voltage level. Normally between two and 12 feeders emanate from any one substation, in what has been called a *dendrillic* configuration - repeated branching into smaller branches as the feeder moves out from the substation toward the consumers. In combination, all the feeders in a power system constitute the *feeder system* (Figure 2.6). An average substation has between two and eight feeders, and can vary between one and forty feeders.

The main, three-phase trunk of a feeder is called the *primary trunk* and may branch into several main routes, as shown in the diagram on the next page. These main branches end at open points where the feeder meets the ends of other feeders - points at which a *normally open switch* serves as an emergency tie between two feeders.

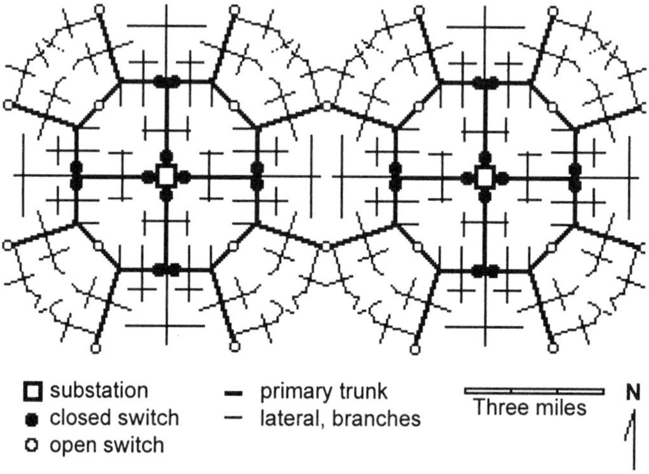

Figure 2.6 Distribution feeders route power away from the substation, as shown (in idealized form - configuration is never so evenly symmetric in the real world) for two substations. Positions of switches make the system electrically radial, while parts of it are physically a network. Shown here are two substations, each with four feeders.

Additionally, normally closed switches into several switchable elements will divide each feeder. During emergencies, segments can be re-switched to isolate damaged sections and route power around outaged equipment to consumers who would otherwise have to remain out of service until repairs were made.

By definition, the feeder consists of all primary voltage level segments between the substations and an open point (switch). Any part of the distribution level voltage lines - three-phase, two-phase, or single-phase - that is switchable is considered part of the primary feeder. The primary trunks and switchable segments are usually built using three phases. The largest size of distribution conductor (typically this is about 500-600 MCM conductor, but conductor over 1,000 MCM is not uncommon, and the author has feeders with 2,000 MCM conductor) being used for reasons other than just maximum capacity (e.g., contingency switching needs). Often a feeder has excess capacity because it needs to provide back up for other feeders during emergencies.

The vast majority of distribution feeders used worldwide and within the United States are overhead construction, wooden pole with wooden crossarm or post insulator. Only in dense urban areas, or in situations where esthetics are particularly important, can the higher cost of underground construction be justified. In this case, the primary feeder is built from insulated cable, which is pulled through concrete ducts that are first buried in the ground. Underground feeder costs are three to ten times that of overhead costs.

Many times, however, the first several hundred yards of an overhead primary feeder are built underground even if the system is overhead. This underground portion is used as the *feeder get-away.* Particularly at large substations, the underground get-away is dictated by practical necessity, as well as by reliability and esthetics. At a large substation, ten or twelve 3-phase, overhead feeders leaving the substation mean from 40 to 48 wires hanging in mid-air around the substation site, with each feeder needing the proper spacing for electrical insulation, safety, and maintenance.

One significant aging infrastructure issue is that at many large-capacity substations in a tight location, there is simply not enough overhead space for enough feeders to be routed out of the substation. Even if there is, the resulting tangle of wires looks unsightly, and perhaps most importantly, is potentially unreliable. One broken wire falling in the wrong place can disable a lot of power delivery capability. The solution to this dilemma is the underground feeder getaway, usually consisting of several hundred yards of buried, ducted cable that takes the feeder out to a riser pole, where it is routed above ground and connected to overhead wires. Very often, this initial underground link sets the capacity limit for the entire feeder. The underground cable ampacity is the limiting factor for the feeder's power transmission.

The Lateral Level

Laterals are short stubs or line-segments that branch off the primary feeder and represent the final primary voltage part of the power's journey from the substation to the consumer. A lateral is directly connected to the primary trunk and operates at the same nominal voltage. A series of laterals tap off the primary feeder as it passes through a community, each lateral routing power to a few dozen homes.

Normally, laterals do not have branches, and many laterals are only one or two-phase. All three phases are used only if a relatively substantial amount of power is required, or if three-phase service must be provided to some of the consumers. Normally, single and two-phase laterals are arranged to tap alternately different phases on the primary feeder. An attempt by the Distribution Planning Engineer to balance the loads as closely as possible is shown below.

Typically, laterals deliver from as little as 10 kVA for a small single-phase lateral, to as much as 2 MVA. In general, even the largest laterals use small conductors (relative to the primary size). When a lateral needs to deliver a great deal of power, the planner will normally use all three phases, with a relatively small conductor for each, rather than employ a single-phase and use a large conductor. This approach avoids creating a significant imbalance in loading at the point where the lateral taps into the primary feeder. Power flow, loadings and voltage are maintained in a more balanced state if the power demands of a "large lateral" are distributed over all three phases.

Laterals (wooden poles) are built overhead or underground. Unlike primary feeders and transmission lines, single-phase laterals are sometimes buried directly. In this case, the cable is placed inside a plastic sheath (that looks and feels much like a vacuum cleaner hose). A trench is dug, and the sheathed cable is unrolled into the trench and buried. Directly buried laterals are no more expensive than underground construction in many cases.

The Service Transformers

Service transformers lower voltage from the primary voltage to the utilization or consumer voltage, normally 120/240-volt two-leg service in most power systems throughout North America. In overhead construction, service transformers are pole mounted and single-phase, between 5-kVA and 166 -kVA capacity. There may be several hundred scattered along the trunk and laterals of any given feeder; since power can travel efficiently only up to about 200 feet at utilization voltage, there must be at least one service transformer located reasonably close to every consumer.

Passing through these transformers, power is lowered in voltage once again, to the final utilization voltage (120/240 volts in the United States) and routed to the secondary system or directly to the consumers. In cases where the system is supplying power to large commercial or industrial consumers, or the consumer requires three-phase power, between two and three transformers may be located together in a transformer bank, and be interconnected in such a way as to provide multi-phase power. Several different connection schemes are possible for varying situations.

Padmount, or vault-type, service transformers provide underground service, as opposed to overhead pole-mounted service. The concept is identical to overhead construction, with the transformer and its associated equipment changed to accommodate incoming and outgoing lines that are underground.

The Secondary and Service Level

Secondary circuits fed by the service transformers, route power at utilization voltage within very close proximity to the consumer. Usually this is an arrangement in which each transformer serves a small radial network of utilization voltage secondary and service lines. These lead directly to the meters of consumers in the immediate vicinity.

At most utilities, the layout and design of the secondary level is handled through a set of standardized guidelines and tables. Engineering technicians use these and clerks to produce work orders for the utilization voltage level equipment. In the United States, the vast majority of this system is single-phase. In European systems, much of the secondary is 3-phase, particularly in urban and suburban areas.

What is Transmission and what is Distribution?

Definitions and nomenclature defining "transmission" and "distribution" vary greatly among different countries, companies, and power systems. Generally, three types of distinction between the two are made:

1. By voltage class: transmission is anything above 34.5 kV; distribution is anything below that.
2. By function: distribution includes all utilization voltage equipment, plus all lines that feed power to service transformers.

3. By configuration: transmission includes a network; distribution is all the radial equipment in the system.

Generally, all three definitions apply simultaneously since in most utility systems any transmission above 34.5 kV is configured as a network and does not feed service transformers directly. On the other hand, all distribution is radial, built of only 34.5 kV or below, and does feed service transformers. Substations – the meeting places of transmission lines (incoming) and distribution lines (outgoing) – are often included in one or the other category, and sometimes are considered as separate entities.

2.5 UTILITY DISTRIBUTION EQUIPMENT

The preceding section made it clear that a power delivery system is a very complex entity, composed of thousands, perhaps even millions, of components, which function together as a *T&D system.* Each unit of equipment has only a small part to play in the system, and is only a small part of the cost, yet each is critical for satisfactory service to at least one or more consumers or it would not be included in the system.

T&D system planning is complex because each unit of equipment influences the electrical behavior of its neighbors. It must be designed to function well in conjunction with the rest of the system, under a variety of different conditions, regardless of shifts in the normal pattern of loads or the status of equipment nearby. While the modeling and analysis of a T&D system can present a significant challenge, individually its components are relatively simple to understand, engineer, and plan. In essence, there are only two major types of equipment that perform the power delivery function:

1. transmission and distribution lines which move power from one location to another

2. transformers which change the voltage level of the power

Added to these three basic equipment types are two categories of equipment used for a very good reason:

1. Protective equipment which provides safety and "fail safe" operation

2. Voltage regulation equipment, which is used to maintain voltage within an acceptable range as the load, changes. Monitoring and

control equipment used to measure equipment and system performance and feed this information to control systems so that the utility knows what the system is doing and can control it, for both safety and efficiency reasons.

Transmission and Distribution Lines

By far the most omnipresent part of the power distribution system is the portion devoted to actually moving the power flow from one point to another. Transmission lines, sub-transmission lines, feeders, laterals, secondary and service drops, all consist of electrical conductors, suitably protected by isolation (transmission towers, insulator strings, and insulated wrappings) from voltage leakage and ground contact. It is this conductor that carries the power from one location to another.

Electrical conductors are available in various capacity ranges, with capacity generally corresponding to the metal cross section (other things being equal, thicker wire carries more power). Conductors can be all steel (rare, but used in some locations where winter ice and wind loadings are quite severe), all aluminum, or copper, or a mixture of aluminum and steel. Underground transmission can use various types of high-voltage cable. Line capacity depends on the current-carrying capacity of the conductor or the cable, the voltage, the number of phases, and constraints imposed by the line's location in the system.

The most economical method of handling a conductor is to place it overhead, supported by insulators on wooden poles or metal towers suitably clear of interference or contact with persons or property. However, underground construction, while generally more costly, avoids esthetic intrusion of the line and provides some measure of protection from weather. It also tends to reduce the capacity of a line slightly due to the differences between underground cable and overhead conductor. Suitably wrapped with insulating material in the form of underground cable, the cable is placed inside concrete or metal ducts or surrounded in a plastic sheath.

Transmission/sub-transmission lines are always 3-phase. There are three separate conductors for the alternating current, sometimes with a fourth neutral (un-energized) wire. Voltage is measured between phases. A 12.47 kV distribution feeder has an alternating current voltage (RMS) of 12,470 volts as measured between any two phases. Voltage between any phase and ground is 7,200 volts (12.47 divided by the square root of three). Major portions of a distribution system - trunk feeders - are as a rule built as three-phase lines, but lower-capacity portions may be built as either two-phase, or single-phase.[7]

Regardless of type or capacity, every electrical conductor has impedance (a resistance to electrical flow through it) that causes voltage drop and electrical losses whenever it is carrying electric power. Voltage drop is a reduction in the voltage between the sending and receiving ends of the power flow. Losses are a reduction in the net power, and are proportional to the square of the power. Double the load and the losses increase by four. Thus, 100 kilowatts at 120 volts might go in one end of a conductor, only to emerge at the other as 90 kilowatts at 114 volts at the other end. Both voltage drop and losses vary in direct relation to load - within very fine limits if there is no load, there are no losses or voltage drop. Voltage drop is proportional to load. Double the load and voltage drop doubles. Losses are *quadratic*, however. Double the load and losses quadruple.

Transformers

At the heart of any alternating power system are transformers. They change the voltage and current levels of the power flow, maintaining (except for a very small portion of electrical losses), the same overall power flow. If voltage is reduced by a factor of ten from the high to low side, then Current is multiplied by ten, so that their overall product (Voltage times Current equals Power) is constant in and out.

Transformers are available in a diverse range of types, sizes, and capacities. They are used within power systems in four major areas:

1. At power plants, where power which is minimally generated at about 20,000 volts is raised to transmission voltage (100,000 volts or higher)
2. At switching stations, where transmission voltage is changed (e.g., from 345,000 volts to 138,000 volts before splitting onto lower voltage transmission lines)
3. At distribution substations, where incoming transmission-level voltage is reduced to distribution voltage for distribution (e.g., 138 kV to 12.47 kV)
4. And at service transformers, where power is reduced in voltage from the primary feeder voltage to utilization level (12.47 kV to 120/240 volts) for routing into consumers' homes and businesses.

---

[7] In most cases, a single-phase feeder or lateral has two conductors: the phase conductor and the neutral.

Larger transformers are generally built as three-phase units, in which they simultaneously transform all three phases. Often these larger units are built to custom or special specifications, and can be quite expensive - over $3,000,000 per unit in some cases. Smaller transformers, particularly most service transformers, are single-phase - it takes three installed sides by side to handle a full three-phase line's power flow. They are generally built to standard specifications and bought in quantity.

Transformers experience two types of electrical losses - no-load losses (often called core, or iron, losses) and load-related losses. No-load losses are electrical losses inherent in operating the transformer - due to its creation of a magnetic field inside its core. They occur simply because the transformer is connected to an electrical power source. They are constant, regardless of whether the power flowing through the transformer is small or large. No-load losses are typically less than one percent of the nameplate rating. Only when the transformer is seriously overloaded, to a point well past its design range, will the core losses change (due to magnetic saturation of the core).

Load-related losses are due to the current flow through the transformer's impedance and correspond very directly with the level of power flow. Like those of conductors and cables they are proportional to current squared or quadrupling whenever power flow doubles. The result of both types of losses is that a transformer's losses vary as the power transmitted through it varies, but always at or above a minimum level set by the no-load losses.

Switches

Occasionally, it is desirable to be able to vary the connection of line segments within a power delivery system, particularly in the distribution feeders. Switches are placed at strategic locations so that the connection between two segments can be opened or closed. Switches are planned to be normally closed (NC) or normally open (NO), as was shown in Figure 2.6.

Switches vary in their rating (how much current they can vary) and their load break capacity (how much current they can interrupt or switch off), with larger switches being capable of opening a larger current. They can be manually, automatically, or remotely controlled in their operation.

Protection

When electrical equipment fails, for example if a line is knocked down during a storm, the normal function of the electrical equipment is interrupted. Protective equipment is designed to detect these conditions and isolate the damaged equipment, even if this means interrupting the flow of power to some consumers. Circuit breakers, sectionalizers, and fused disconnects, along with control relays and sensing equipment, are used to detect unusual conditions and interrupt the power flow whenever a failure, fault, or other unwanted condition occurs on the system.

These devices and the protection engineering required to apply them properly to the power system are not the domain of the utility planners and will not be discussed here. Protection *is* vitally important, but the planner is sufficiently involved with protection if he or she produces a system design that *can* be protected within standards, and if the cost of that protection has been taken into account in the budgeting and planning process. Both of these considerations are very important.

Protection puts certain constraints on equipment size and layout - for example, in some cases a very large conductor is too large (because it would permit too high a short circuit current) to be protected safely by available equipment and cannot be used. In other cases, long feeders are too long to be protected (because they have too low a short circuit current at the far end). A good deal of protective equipment is quite complex, containing sensitive electro-mechanical parts, (many of which move at high speeds and in a split-second manner), and depending on precise calibration and assembly for proper function. As a result, the cost of protective equipment and control and the cost of its maintenance are often significant. Differences in protection cost *can* make the deciding difference between two plans.

Voltage Regulation

Voltage regulation equipment includes line regulators and line drop compensators, as well as tap changing transformers. These devices vary their turns-ratio (ratio of voltage in to voltage out) to react to variations in voltage drop. If voltage drops, they raise the voltage, if voltage rises, they reduce it to compensate. Properly used, they can help maintain voltage fluctuation on the system within acceptable limits, but they can only reduce the range of fluctuation, not eliminate it altogether.

Capacitors

Capacitors are a type of voltage regulation equipment. By correcting power factor they can improve voltage under many heavy loads (hence large voltage drop) cases. Power factor is a measure of how well voltage and current in an alternating system are in step with one another. In a perfect system, voltage and current would alternately cycle in conjunction with one another - reaching a peak, then reaching a minimum, at precisely the same times. But on distribution systems, particularly under heavy load conditions, current and voltage fall out of phase. Both continue to alternate 60 times a second, but during each cycle voltage may reach its peak slightly ahead of current - there is a slight lag of current versus voltage, as shown in Figure 2.7.

It is the precise, simultaneous peaking of both voltage and current that delivers maximum power. If out of phase, even by a slight amount, effective power drops, as does power factor - the ratio of real (effective) power to the maximum possible power, if voltage and current were locked in step.

Power engineers refer to a quantity called VARs (Volt-Amp Reactive) that is caused by this condition. Basically, as power factors worsen (as voltage and current fall farther apart in terms of phase angle) a larger percent of the electrical flow is VARs, and a smaller part is real power. The frustrating thing is that the voltage is still there, and the current is still there, but because of the shift in their timing, they produce only VARS, not power. The worse the power factor, the higher the VAR content. Poor power factor creates considerable cost and performance consequences for the power system: large conductor is still required to carry the full level of current even though power delivery has dropped. And because current is high, the voltage drop is high, too, further degrading quality of service.

Unless one has worked for some time with the complex variable mathematics associated with AC power flow analysis, VARs are difficult to picture. A useful analogy is to think of VARs as "electrical foam." If one tried to pump a highly carbonated soft drink through a system of pipes, turbulence in the pipes, particularly in times of high demand (high flow) would create foam. The foam would take up room in the pipes, but contribute little value to the net flow - the equivalent of VARs in an electrical system.

Poor power factor has several causes. Certain types of loads create VARs - loads that cause a delay in the current with respect to voltage as it flows through them. Among the worst offenders are induction motors - particularly small ones as almost universally used for blowers, air conditioning compressors, and the powering of conveyor belts and similar machinery. Under heavy load conditions, voltage and current can get out of phase to the point that power factor can drop below 70%. Additionally, transmission equipment itself can often create this lag and "generate" a poor power factor.

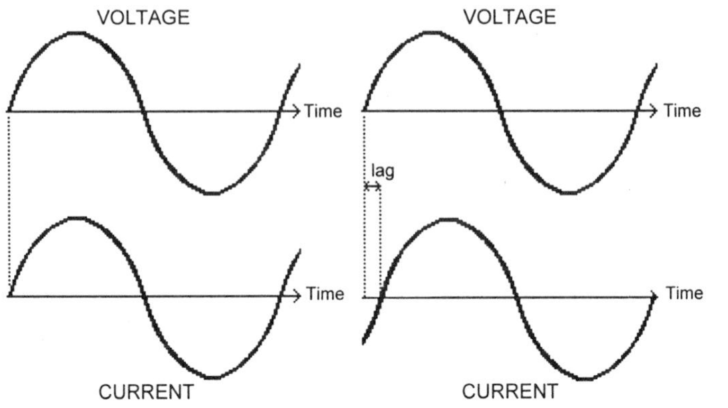

Figure 2.7 Current and voltage in phase deliver maximum power (left). If current and voltage fall out of phase (right), actual power delivered drops by very noticeable amounts -- the *power factor* falls.

Capacitors correct the poor power factor. They "inject" VARs into a T&D line to bring power factor close to 1.0, transforming VAR flow back into real power flow, regaining the portion of capacity lost to poor power factor. Capacitors can involve considerable cost depending on location and type. They tend to do the most good if put on the distribution system, near the consumers, but they cost a great deal more in those locations than if installed at substations.

2.6 T&D COSTS

A T&D system can be expensive to design, build, and operate. Equipment at every level incurs two types of costs. Capital costs include the equipment and land, labor for site preparation, construction, assembly and installation, and any other costs associated with building and putting the equipment into operation. Operating costs include labor and equipment for operation, maintenance and service, taxes and fees, as well as the value of the power lost to electrical losses. Usually, capital costs are a one-time cost (once it's built, the money's been spent). Operating costs are continuous or periodic.

Electrical losses vary depending on load and conditions. While these losses are small by comparison to the overall power being distributed (seldom more than 8%), they constitute a very real cost to the utility. The present worth of the lifetime losses through a major system component such as a feeder or transformer can be a significant factor impacting its design and specification, often more than the original capital cost of the unit. Frequently, a more costly type of transformer will be selected for a certain application because its design leads to overall savings due to lower losses. Or a larger capacity line (larger conductor) will be used when really needed due to capacity requirements, purely because the larger conductor will incur lower cost losses.

Cumulatively, the T&D system represents a considerable expense. While a few transmission lines and switching stations are composed of large, expensive, and purpose-designed equipment, the great portion of the sub-transmission-substation-distribution system is built from "small stuff" - commodity equipment bought mostly "off the shelf" to standard designs. Individually inexpensive, they amount to a significant cost when added together.

Transmission Costs

Transmission line costs are based on a per-mile cost and a termination cost at either end of the line associated with the substation at which it is terminated. Costs can run from as low as $50,000/mile for a 46 kV wooden pole sub-transmission line with perhaps 50 MVA capacity ($1 per kVA-mile) to over $1,000,000 per mile for a 500 kV double circuit construction with 2,000 MVA capacity ($.5/kVA-mile).

Substation Costs

Substation costs include all the equipment and labor required to build a substation, including the cost of land and easements/ROW. For planning purposes, substations can be thought of as having four costs:

1. *Site cost* - the cost of buying the site and preparing it for a substation.

2. *Transmission cost* - the cost of terminating the incoming sub-transmission lines at the site.

3. *Transformer cost* - the transformer and all metering, control, oil spill containment, fire prevention, cooling, noise abatement, and other transformer related equipment, along with typical buswork, switches, metering, relaying, and breakers associated with this type of transformer, and their installation.

4. *Feeder buswork/Getaway costs* - the cost of beginning distribution at the substation, includes getting feeders out of the substation.

To expedite planning, estimated costs of feeder buswork and getaways are often folded into the transformer costs. The feeders to route power out of the substation are needed in conjunction with each transformer and in direct proportion to the transformer capacity installed, so that their cost is sometimes considered together with the transformer as a single unit. Regardless, the transmission, transformer, and feeder costs can be estimated fairly accurately for planning purposes.

Cost of land is another matter entirely. Site and easements or ROW into a site have a cost that is a function of local land prices, which vary greatly depending on location and real estate markets. Site preparation includes the cost of preparing the site which includes grading, grounding mat, foundations, buried ductwork, control building, lighting, fence, landscaping, and an access road.

Substation costs vary greatly depending on type, capacity, local land prices and other variable circumstances. In rural settings where load density is quite low and minimal capacity is required, a substation may involve a site of only several thousand square feet of fenced area. This single area includes a single incoming transmission line (69 kV), one 5 MVA transformer; fusing for all fault protection; and all "buswork" built with wood poles and conductor, for a total cost of perhaps no more than $90,000. The substation would be applied to serve a load of perhaps 4 MW, for a cost of $23/kW. This substation in conjunction with the system around it would probably provide service with about ten hours of service interruptions per year under average conditions.

However, a typical substation built in most suburban and urban settings would be fed by two incoming 138 kV lines feeding two 40 MVA, 138 kV to 12.47 kV transformers. These transformers would each feed a separate low side (12.47 kV) bus, each bus with four outgoing distribution feeders of 9 MVA peak capacity each, and a total cost of perhaps $2,000,000. Such a substation's cost could vary from between about $1.5 million and $6 million, depending on land costs, labor costs, the utility equipment and installation standards, and other special circumstances. In most traditional vertically integrated, publicly regulated electric utilities, this substation would have been used to serve a peak load of about 60 MVA (75% utilization of capacity), meaning that at its nominal $2,000,000 cost works out to $33/kW. In a competitive industry, with tighter design margins and proper engineering measures taken beforehand, this could be pushed to a peak loading of 80 MVA (100% utilization, $25/kW). This substation and the system around it would probably provide service with about two to three hours of service interruptions per year under "normal," average conditions.

Feeder System Costs

The feeder system consists of all the primary distribution lines, including three-phase trunks and their lateral extensions. These lines operate at the primary distribution voltage - 23.9 kV, 13.8 kV, 12.47 kV, 4.16kV or whatever – and may be three, two, or single-phase construction as required. Typically, the feeder system is also considered to include voltage regulators, capacitors, voltage boosters, sectionalizers, switches, cutouts, fuses, and any intertie transformers (required to connect feeders of different voltage at tie points - for example: 23.9 and 12.47 kV) that are installed on the feeders (not at the substations or at consumer facilities).

As a rule of thumb, construction of three-phase overhead, wooden pole crossarm type feeders using a normal, large conductor (about 600 MCM per phase) at a medium distribution primary voltage (e.g., 12.47 kV) costs about $150,000/mile. However, cost can vary greatly due to variations in labor, filing and permit costs among utilities, as well as differences in design standards, and terrain. Where a thick base of topsoil is present, a pole can be installed by simply auguring a hole for the pole. In areas where there is rock close under the surface, holes have to be jack hammered or blasted, and costs go up accordingly. It is generally less expensive to build feeders in rural areas than in suburban or urban areas. Thus, while $150,000 is a good average cost, a mile of new feeder construction could cost as little as $55,000 in some situations and as much as $500,000 in others.

A typical distribution feeder (three-phase, 12.47 kV, 600 MCM/phase) would be rated at a thermal (maximum) capacity of about 15 MVA and a recommended economic (design) peak loading of about 8.5 MVA peak, depending on losses and other costs. At $150,000/mile, this capacity rating gives somewhere between $10 to $15 per kW-mile as the cost for basic distribution line. Underground construction of a three-phase primary feeder is more expensive, requiring buried ductwork and cable, and usually works out to a range of $30 to $50 per kW-mile.

Lateral lines, short primary-voltage lines working off the main three-phase circuit, are often single or two-phase and consequently have lower costs but lower capacities. Generally, they are about $5 to $15 per kW-mile overhead, with underground costs of between $5 to $15 per kW-mile (direct buried) to $30 to $100 per kW-mile (ducted).

Cost of other distribution equipment, including regulators, capacitor banks and their switches, sectionalizers, line switches, etc., varies greatly depending on the specifics of each application. In general, the cost of the distribution system will vary from between $10 and $30 per kW-mile.

Service Level Costs

The service or secondary system consists of the service transformers that convert primary voltage to utilization voltage, the secondary circuits that operate at utilization voltage and the service drops that feed power directly to each consumer. Without exception these are very local facilities, meant to move power no more than a few hundred feet at the very most and deliver it to the consumer "ready to use."

Many electric utilities develop cost estimates for this equipment on a per-consumer basis. A typical service configuration might involve a 50 MVA pole-mounted service transformer feeding ten homes, as shown in Figure 2.8. Costs for this equipment might include:

Heavier pole & hardware for transformer application	\$250
50 kW transformer, mounting equipment, and installation	\$750
500 feet secondary (120/240 volt) single-phase @ \$2/ft.	\$1,000
10 service drops including installation at \$100	\$1,000
	\$3,000

A cost of about \$300 per consumer, or about \$60/kW of coincident load.

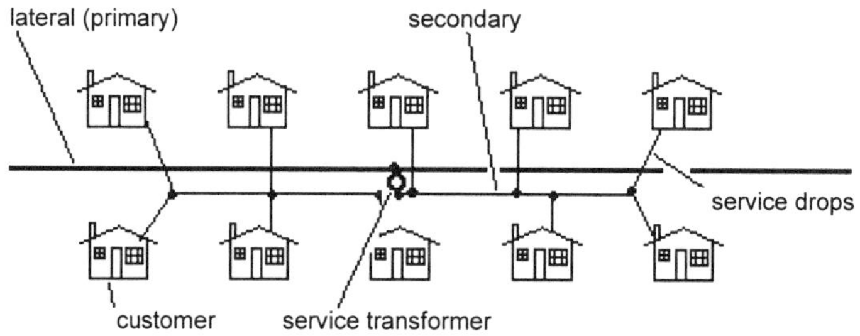

Figure 2.8 Here, a service transformer, fed from a distribution primary-voltage lateral, feeds in turn ten homes through secondary circuit operating at utilization voltage.

Maintenance and Operating Costs

Once put into service, T&D equipment must be maintained in sound, operating condition, hopefully in the manner intended and recommended by the manufacturer. This will require periodic inspection and service, and may require repair due to damage from storms or other contingencies. In addition, many utilities must pay taxes or fees for equipment (T&D facilities are like any other business property). Operating, maintenance, and taxes are a continuing annual expense.

It is very difficult to give any generalization of O&M&T costs, partly because they vary so greatly from one utility to another, but mostly because utilities account for and report them in very different ways. Frankly, the authors have never been able to gather a large number of comparable data sets from which to produce even a qualitative estimate of average O&M&T costs.[8] With that caveat, a general rule of thumb: O&M&T costs for a power delivery system probably run between 1/8 and 1/30 of the capital cost, annually.

The Cost to Upgrade Exceeds the Cost to Build

One of the fundamental factors affecting design of T&D systems is that it costs more to upgrade equipment to a higher capacity than to build to that capacity in the original construction. For example, a 12.47 kV overhead, three-phase feeder with a 9 MW capacity (336 MCM phase conductor) might cost $120,000/mile to build ($13.33 per kW-mile). Building it with 600 MCM conductor instead, for a capacity of 15 MVA, would cost in the neighborhood of $150,000 ($10/kW-mile).

However, upgrading an existing 336 MCM, 9 MW capacity line to 600 MCM, 15 MVA capacity could cost $200,000/mile - over $30 per kW-mile for the 6 MW of additional capacity. This is more expensive because it entails removing the old conductor and installing new conductor along with brackets, cross-arms, and other hardware required supporting the heavier new conductor. Typically, this work is done hot (i.e., with the feeder energized), which means

---

[8] For example, some utilities include part of O&M expenses in overhead costs, others do not. A few report all repairs (including storm damage) as part of O&M, others accumulate major repair work separately. Still others report certain parts of routine service (periodic rebuilding of breakers) as a type of capital cost because it extends equipment life or augments capacity. Others report all such work as O&M, even when the rebuilding upgrades capacity or voltage class.

the work must be undertaken with extreme care and following a number of safety-related restrictions on equipment and labor.

Thus, T&D planners have an incentive to look at their long-term needs carefully, and to "overbuild" against initial requirements if growth trends show future demand will be higher. The cost of doing so must be weighed against long-term savings, but often T&D facilities are built with considerable margin (50%) above existing load to allow for future load growth.

The very high cost per kW for upgrading a T&D system in place creates one of the best-perceived opportunities for DSM and DG reduction. Note that the capital cost/kW for the upgrade capacity in the example above ($33/kW) is nearly three times the cost of similar new capacity. Thus, planners often look at areas of the system where slow; continuing load growth has increased load to the point that local delivery facilities are considerably taxed, as areas where DSM and DG can deliver significant savings.

In some cases, distributed resources *can* reduce or defer significantly the need for T&D upgrades of the type described above. However, this does not assure significant savings for the situation is more complicated than an analysis of capital costs to upgrade may indicate. If the existing system (e.g., the 9 MW feeder) needs to be upgraded, then it is without a doubt highly loaded, which means its losses may be high, even off-peak. The upgrade to a 600 MCM conductor will cut losses 8,760 hours per year. Financial losses may drop by a significant amount, *enough in many cases to justify the cost of the upgrade alone.* The higher the annual load factor in an area, the more likely this is to occur, but it is often the case even when load factor is only 40%. However, DSM and in some cases DG also lowers losses, making the comparison quite involved, as will be discussed later in this book.

Electrical Losses Costs

Movement of power through any electrical device, be it a conductor, transformer, regulator or whatever, incurs a certain amount of electrical loss due to the impedance (resistance to the flow of electricity) of the device. These losses are a result of inviolable laws of nature. They can be measured, assessed, and minimized through proper engineering, but never eliminated completely.

Losses are an operating cost

Although losses do create a cost (sometimes a considerable one) it is not always desirable to reduce them as much as possible. Perhaps the best way to put them in proper perspective is to think of T&D equipment as *powered by electricity* - the system that moves power from one location to another runs on electric energy itself. Seen in this light, losses are revealed as what they are - a necessary operating expense to be controlled and balanced against other costs.

Consider a municipal water department, which uses electric energy to power the pumps that drive the water through the pipes to its consumers. Electricity is an acknowledged operating cost, one accounted for in planning, and weighed carefully in designing the system and estimating its costs. The water department could choose to buy highly efficient pump motors. Motors that command a premium price over standard designs but provide a savings in reduced electric power costs, and use piping that is coated with a friction-reducing lining to promote rapid flow of water (thus carrying more water with less pump power), all toward reducing its electric energy cost. Alternatively, after weighing the cost of this premium equipment against the energy cost savings it provides, the water department may decide to use inexpensive motors and piping and simply pay more over the long run. The point is that the electric power required to move the water is viewed merely as one more cost that had to be included in determining the lowest "overall" cost. *It takes power to move power.*

Since its own delivery product powers electric delivery equipment, this point often is lost. However, in order to do its job of delivering electricity, a T&D system must be provided with power itself, just like the water distribution system. Energy must be expended to move the product. Thus, a transformer consumes a small portion of the power fed into it. In order to move power 50 miles, a 138 kV transmission line similarly consumes a small part of the power given to it.

Initial cost of equipment can always be traded against long-term financial losses. Highly efficient transformers can be purchased to use considerably less power to perform their function than standard designs. Larger conductors can be used in any transmission or distribution line, which will lower impedance, and thus losses for any level of power delivery. But both examples here cost more money initially - the efficient transformer may cost three times what a standard design does. The larger conductor might entail a need for not only large wire, but also heavier hardware to hold it in place, and stronger towers and poles to keep it in the air. In addition, these changes may produce other costs - for example, use of a larger conductor not only lowers losses, but a higher fault duty (short circuit current), which increases the required rating and cost for circuit breakers. Regardless, initial equipment costs can be balanced against long-term financial losses through careful study of needs, performance, and costs, to establish a minimum overall (present worth) worth.

Load-related losses

Flow of electric power through any device is accompanied by what are called load-related losses, which increase as the power flow (load) increases. These are due to the impedance of the conductor or device. Losses increase as the *square* of the load - doubling the power flowing through a device quadruples the losses. Tripling power flow increases the losses by nine.

With very few exceptions, larger electrical equipment always has lower impedance, and thus lower load-related losses, for any given level of power delivery. Hence, if the losses inherent in delivering 5 MW using a 600 MCM conductor are unacceptably large, the use of a 900 MCM conductor will reduce them considerably. The cost of the larger conductor can be weighed against the savings in reduced losses to decide if it is a sound economic decision.

No-load losses

"Wound" T&D equipment - transformers and regulators - have load-related losses as do transmission lines and feeders but in addition, they have a type of electrical loss that is constant, regardless of loading. No-load losses constitute the electric power required to establish a magnetic field inside these units, without which they would not function. Regardless of whether a transformer has any load - any power passing through it at all - it will consume a small amount of power, generally less than 1% of its rated full power, simply because it is energized and "ready to work." No-load losses are constant, and occur 8,760 hours per year.

Given similar designs, a transformer will have no load losses proportional to its capacity. A 10 MVA substation transformer will have twice the no-load losses of a 5 MVA transformer of similar voltage class and design type. Therefore, unlike the situation with a conductor, selection of a larger transformer does not always reduce net transformer losses, because while the larger transformer will always have lower load-related losses, it will have higher no-load losses, and this increase might outweigh the reduction in load-related losses. Again, low-loss transformers are available, but cost more than standard types. Lower-cost-than-standard but higher-loss transformers are also available (often a good investment for backup and non-continuous use applications).

The costs of losses

The electric power required to operate the T&D system - the electrical losses - is typically viewed as having two costs, demand and energy. Demand cost is the cost of providing the peak capacity to generate and deliver power to the T&D equipment. A T&D system that delivers 1,250 MW at peak might have losses during this peak of 100 MW. This means the utility must have generation, or buy power at peak to satisfy this demand, whose cost is calculated using the utility's power production cost at time of peak load This is usually considerably above its average power production cost.

Demand cost also ought to include a considerable T&D portion of expense. Every service transformer in the system (and there are many) is consuming a small amount of power in doing its job at peak. Cumulatively, this might equal 25 MW of power - up to 1/4 of all losses in the system. That power must not only be generated by the utility but also transmitted over its transmission system, through its substations, and along its feeders to *reach* the service transformers. Similarly, the power for electrical losses in the secondary and service drops (while small, are numerous and low voltage, so that their cumulative contribution to losses is noticeable) has to be moved even farther, through the service transformers and down to the secondary level.

Demand cost of losses is the total cost of the capacity to provide the losses and move them to their points of consumption.

Energy losses occur whenever the power system is in operation, which generally means 8,760 hours per year. While losses vary as the square of load, so they drop by a considerable margin off-peak. Their steady requirement every hour of the year imposes a considerable energy demand over the course of a year. This cost is the cost of the energy to power the losses.

Example: Consider a typical 12.47 kV, three-phase, OH feeder, with 15 MW capacity (600 MCM phase conductor), serving a load of 10 MW at peak with 4.5% primary-level losses at peak (450 kW losses at peak), and having a load factor of 64% annually. Given a levelized capacity cost of power delivered to the low side bus of a substation of $10/kW, the demand cost of these losses is $4,500/year. Annual energy cost, at 3.5¢ /kWhr, can be estimated as:

$$450 \text{ kW losses at peak} \times 8{,}760 \text{ hours} \times (64\% \text{ load factor})^2 \times 3.5¢ = \$56{,}500$$

Thus, the financial losses (demand plus energy costs) for this feeder are nearly $60,000 annually. At a present worth discount factor of around 11%, this means losses have an estimated present worth of about $500,000. This computation used a simplification - squaring the load factor to estimate load factor impact on losses - which tends to underestimate losses slightly. Actual losses probably would be more in the neighborhood of $565,000 PW. If the peak load on this feeder were run up to its maximum rating (about 15 MW instead of 10 MW) with a similar load factor of 64%, annual losses and their cost would increase to $(15/10)^2$ or $1,250,000 dollars.

This feeder, in its entirety, might include four miles of primary trunk (at $150,000/mile) and thirty miles of laterals (at $50,000/mile), for a total capital cost of about $2,100,000. Thus, *total losses are on the order and magnitude of original cost of the feeder itself,* and in cases where loading is high, can approach that cost. Similar loss-capital relations exist for all other levels of the T&D system, with the ratio of losses capital cost increasing as one nears the consumer level (lower voltage equipment has higher losses/kW).

Total T&D Costs

What is the total cost to deliver electric power? Of course this varies from system to system as well as from one part of a system to another - some consumers are easier to reach with electric service than others (Figure 2.9). Table 2.2 shows the cost of providing service to a "typical" residential consumer.

2.7 TYPES OF DELIVERY SYSTEM DESIGN

There are three fundamentally different ways to lay out a power distribution system used by electric utilities, each of which has variations in its own design. As shown in Figure 2.10, radial, loop, and network systems differ in how the distribution feeders are arranged and interconnected about a substation.

Most power distribution systems are designed as *radial distribution systems.* The radial system is characterized by having only one path between each consumer and a substation. The electrical power flows exclusively away from the substation and out to the consumer along a *single path,* which, if interrupted, results in complete loss of power to the consumer. Radial design is by far the most widely used form of distribution design, accounting for over ninety-nine

Table 2.2 Cost of Providing Service to Typical Residential Consumers

Level	Cost Components	Cost
Transmission	4 kW x 100 miles x $.75/kW mile	$300
Substation	4 kW x $60/kW	$240
Feeder	4 kW x 1.5 miles x $10/kW-mile	$60
Service	1/10th of 50 kVA local service system	$300
	Total Initial cost (Capital)	$900
	Operations, Maintenance, and Taxes (PW next 30 years) =	$500
	Cost of electrical losses (PW next 30 years) =	$700
	Estimated cost of power delivery, 30 years, PW	$2,100

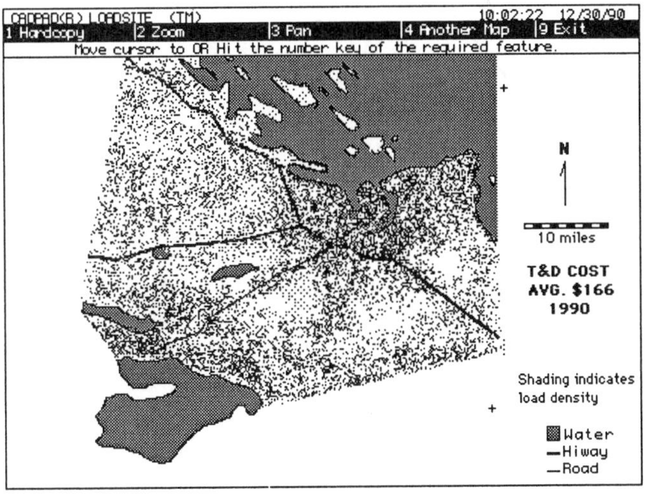

Figure 2.9 Cost of power delivery varies depending on location. Shown here are the annual capacity costs of delivery evaluated on a ten-acre basis throughout a coastal city of population 250,000. Cost varies from a low of $85/kW to a high of $270/kW.

percent of all distribution construction in North America. Its predominance is due to two overwhelming advantages: it is much less costly than the other two alternatives and it is much simpler in planning, design, and operation.

In most radial plans, both the feeder and the secondary systems are designed and operated radially. Each radial feeder serves a definite service area (all consumers in that area is provided power by only that feeder).

Most radial feeder systems are built as networks, but operate radially by opening switches at certain points throughout the physical network (shown earlier in Figure 2.6), so that the resulting configuration is electrically radial. The planner determines the layout of the network and the size of each feeder segment in that network and decides where the open points should be for proper operation as a set of radial feeders.

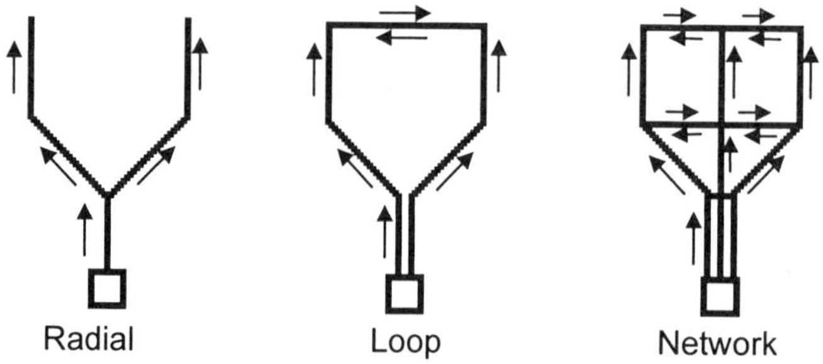

Figure 2.10 Simplified illustration of the concepts behind three types of power distribution configuration. Radial systems have only one electrical path from the substation to the consumer, loop systems have two, and networks have several. Arrows show most likely direction of electric flows.

A further attribute of most radial feeder system designs, but not an absolutely critical element, is the use of single-phase laterals. Throughout North America, most utilities use single- and two-phase laterals to deliver power over short distances, tapping off only one or two phases of the primary feeder. This minimizes the amount of wire that need be strung for the short segment required to get the power in the general vicinity of a dozen or so consumers. These laterals are also radial, but seldom, if ever, end in a switch (they just end). There are many utilities, particularly urban systems in Europe, Africa, and Asia that build every part of the radial distribution system, including laterals, with all three phases.

Each service transformer in these systems feeds power into a small radial system around it, basically a single electrical path from each service transformer to the consumers nearby.

Regardless of whether it uses single-phase laterals or not, the biggest advantages of the radial system configuration, in addition to its lower cost, is the simplicity of analysis and predictability of performance. Because there is only one path between each consumer and the substation, the direction of power flow is absolutely certain. Equally important, the load on any element of the system can be determined in the most straightforward manner - by simply adding up the entire consumer loads "downstream" from that piece of equipment.

Before the advent of economical and widely available computer analysis, this alone was an overwhelming advantage, for it allowed simple, straightforward, "back of the envelope" design procedures to be applied to the distribution system with confidence that the resulting system would work well.

The simplicity of analysis, and confidence that operating behavior is strictly predictable are still great advantages.

Because load and power flow direction are easy to establish, voltage profiles can be determined with a good degree of accuracy without resorting to exotic calculation methods. Equipment capacity requirements can be ascertained exactly: fault levels can be predicted with a reasonable degree of accuracy; and protective devices - breaker-relays and fuses - can be coordinated in an absolutely assured manner, without resorting to network methods of analysis. Regulators and capacitors can be sized, located, and set using relatively simple procedures (simple compared to those required for similar applications to non-radial designs in which the power flow direction is not a given).

On the debit side, radial feeder systems are less reliable than loop or network systems because there is only one path between the substation and the consumer. Thus, if any element along this path fails, a loss of power delivery results. Generally, when such a failure occurs, a repair crew is dispatched to re-switch temporarily the radial pattern network, transferring the interrupted consumers onto another feeder, until the damaged element can be repaired. This minimizes the period of outage, but an outage still occurred because of the failure.

Despite this apparent flaw, radial distribution systems, if well designed and constructed, generally provide very high levels of reliability. For all but the most densely populated areas, or absolutely critical loads (hospitals, important municipal facilities, the utility's own control center) the additional cost of an inherently more reliable configuration (loop or network) cannot possibly be justified for the slight improvement that is gained over a well-designed radial system.

An alternative to purely radial feeder design is a *loop system* consisting of a distribution design with two paths between the power sources (substations, service transformers) and every consumer. Such systems are often called "European," because this configuration is the preferred design of many European utilities, as well as European electrical contractors when called upon to lay out a power distribution system anywhere in the world. Equipment is sized and each loop is designed so that service can be maintained regardless of where an open point might be on the loop. Because of this requirement, whether operated radially (with one open point in each loop), or with closed loops, the basic equipment capacity requirements of the loop feeder design do not change.

Some urban areas in Europe and Asia are fed by multiple hierarchical loop systems. A 100+kV sub-transmission loop routes power to several substations, out of which several loop feeders distribute power to service transformers, which each route powers through a long loop secondary.

In terms of complexity, a loop feeder system is only slightly more complicated than a radial system. Power usually flows out from both sides toward the middle, and in all cases can take only one of two routes. Voltage drop, sizing, and protection engineering are only slightly more complicated than for radial systems.

But if designed thus, and if the protection (relay-breakers and sectionalizers) is also built to proper design standards, the loop system is more reliable than radial systems. Service will not be interrupted to the majority of consumers whenever a segment is outaged, because there is no “downstream” portion of any loop.

The major disadvantage of loop systems is capacity and cost. A loop must be able to meet all power and voltage drop requirements when fed from only one end, not both. It needs extra capacity on each end, and the conductor must be large enough to handle the power and voltage drop needs of the entire feeder if fed from either end. This makes the loop system inherently more reliable than a radial system, but the larger conductor and extra capacity increase costs.

Distribution networks are the most complicated, the most reliable, and in very rare cases also the most economical method of distributing electric power. A network involves multiple paths between all points in the network. Power flow between any two points is usually split among several paths, and if a failure occurs it instantly and automatically re-routes itself.

Rarely does a distribution network involve primary voltage-level network design, in which all or most of the switches between feeders are closed so that the feeder system is connected between substations. This is seldom done because it proves very expensive and often will not work well.[9] Instead, a "distribution network" almost always involves "interlaced" radial feeders and a network secondary system - a grid of electrically strong (i.e., larger than needed to feed consumers in the immediate area) conductor connecting all the consumers together at utilization voltage. Most distribution networks are underground simply because they are employed only in high-density areas, where overhead space is not available.

In this type of design, the secondary grid is fed from radial feeders through service transformers, basically the same way the secondary is fed in radial or loop systems. The feeders are radial, but laid out in an interlaced manner - none has a sole service area, but instead they overlap. The interlaced configuration

---

[9] The major reason is this puts feeder network paths in parallel with transmission between substations, which results in unacceptable loop and circular flows as well as large dynamic shifts in load on the distribution system.

means that alternate service transformers along any street or circuit path are fed from alternate feeders, as shown in Figure 2.11.

While segments from two feeders always run parallel in any part of the system, the same two feeders never overlap for all of their routing. The essence of the interlaced system (and a design difficulty in any practical plan) is to mix up feeders so that each feeder partially parallels a few other feeders. Thus, if the feeder fails, it spreads its load out over a few other feeders, overloading none severely (Figure 2.12).

At a minimum, distribution networks use an interlacing factor of two, meaning that two feeders overlap in any one region, each feeding every other service transformer. But such a system will fail when any two feeders are out of service. Interlacing factors as high as five (four overlapping feeders, each feeding every fourth consecutive service transformer) have been built. Such systems can tolerate the loss of any three feeders (the other two in any area picking up the remaining load, although often very overloaded) without any interruption of consumer service. If an element fails, the power flow in the elements around it merely re-distributes itself slightly.

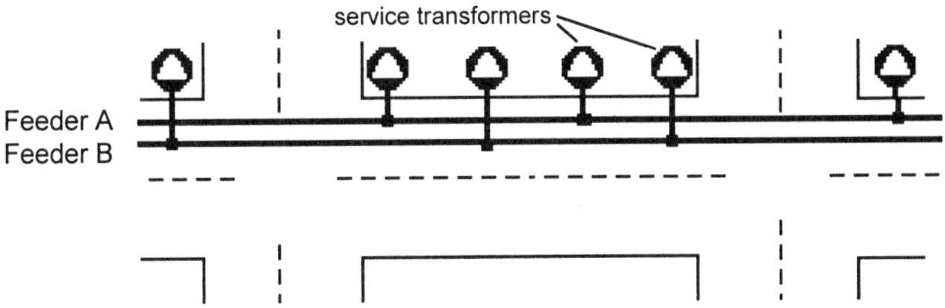

Figure 2.11 To obtain an interlacing factor of 2, two feeders are routed down each street, with alternating network transformers fed from each.

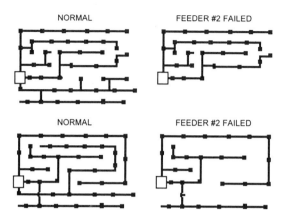

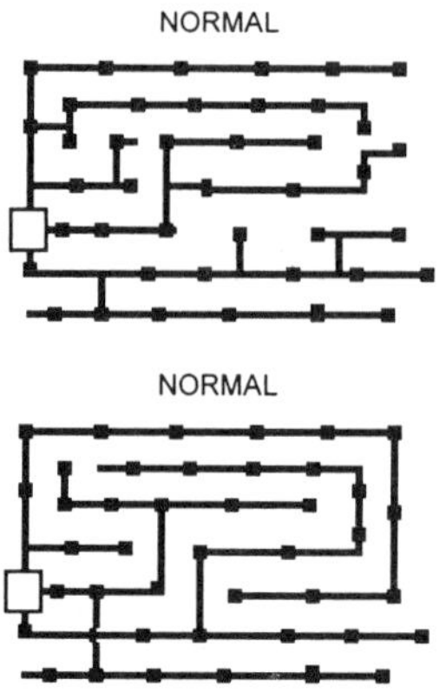

Figure 2.12 Top: a non-interlaced feeder system experiences the loss of one feeder, and all transformers in the lower right part of the system are lost - service is certain to be interrupted. Bottom: the same system interlaced. Loss of the feeder is a serious contingency, but can be withstood because the feeder losses are distributed in such a way that transformers still in service surround each transformer out of service.

Networks are more expensive than radial distribution systems, but not greatly so in some instances. In dense urban applications, where the load density is very high, where the distribution must be placed underground, and where repairs and maintenance are difficult because of traffic and congestion, networks may cost little more than loop systems. Networks require only a little more conductor capacity than a loop system. The loop configuration required "double capacity" everywhere to provide increased reliability. A distribution network is generally no worse and often needs considerably less maintenance than if built to a clever design.

Networks have one major disadvantage. They are *much* more complicated, than other forms of distribution, and thus much more difficult to analyze. There is no "downstream" side to each unit of equipment in a radial or loop system, so that the load seen by any unit of equipment cannot be obtained by merely adding up consumers on one side of it, nor can the direction of power flow through it be assumed. Loadings and power flow, fault currents and protection must be determined by network techniques such as those used by transmission planners. However, even more sophisticated calculation methods than those applied to transmission may be required, because a large distribution network can consist of 50,000 nodes or more - the size of the very largest transmission-level power pool. Distribution network load flows are often more difficult to solve than transmission systems because the range of impedance in the modeled circuits is an order of magnitude wider.

In densely populated regions, such as the center of a large metropolitan area, networks are not inherently more expensive than radial systems designed to serve the same loads. Such concentrated load densities require a very large number of circuits anyway, so that their arrangement in a network does not inherently increase the number of feeder and secondary circuits, or their capacity requirements. It increases only the complexity of the design.

But in other areas, such as in most cities and towns, and in all rural areas, a network configuration will call for some increase (in kVA-feet of installed conductor) over that required for a radial or loop design. The excess capacity cost has to be justifiable on the basis of reliability.

Large-Trunk vs. Multi-Branch Feeder Layout

Figure 2.13 illustrates the basic concept behind two different ways to approach the layout of a radial distribution system. Each has advantages and disadvantages with respect to the other in certain situations, and neither is superior to the other in terms of reliability, cost, ease of protection, and service quality. Either can be engineered to work in nearly any situation. Most planning utilities have an institutionalized preference for one or the other. Beyond showing that there are significantly different ways to layout a distribution system, this brings to light an important point about distribution design: major differences in standards exist among electric utilities, as a result of which comparison of statistics or practice from one to the other is often not valid. Feeder layout types and practices are discussed in much greater detail in Chapters 8 and 9 of the *Power Distribution Planning Reference Book.*

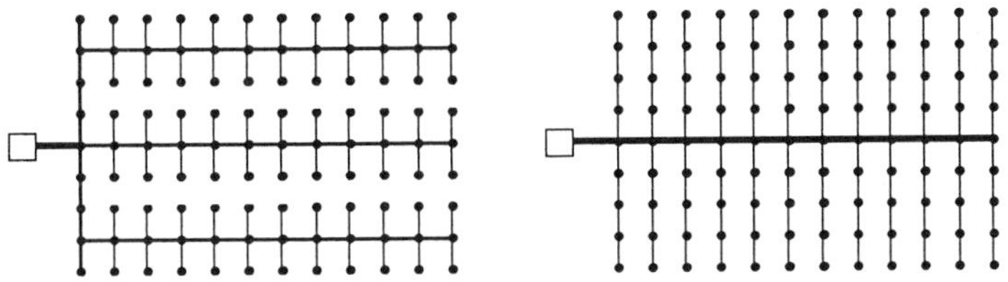

Figure 2.13 Two ways to route a radial feeder to 108 service transformers. Left, a "multi-branch" configuration. Right, a "large trunk" design.

Service Areas

As mentioned earlier, in most power systems, each substation is usually the sole provider of electrical service to the region around it - its service area. Similarly, feeders and distribution networks also have distinct service areas. Usually, the service area for a substation, feeder, or other unit of equipment is the immediate area surrounding it, and usually these service areas are contiguous (i.e. not broken into several parts) and exclusive - no other similar distribution unit serves any of the load in an area. As an example, Figure 2.14 shows a map of substation service areas for a rectangular portion of a power system. Each distribution substation exclusively serves all consumers in the area containing it.

Cumulatively, the consumers in a substation or feeder's service territory determine its load, and their simultaneous peak demand defines the maximum power the substation must serve. Within a power system, each individual part, such as a substation or service transformer, will see its peak load at whatever time and in whatever season the consumers in its service area generate their cumulative peak demand. One result of this is that the peak loads for different substations often occur at different seasons of the year or hours of the day. But whenever the peak occurs, it defines the maximum power the unit is required to deliver. Peak demand is one of the most important criteria in designing and planning distribution systems. Usually it defines the required equipment capacity.

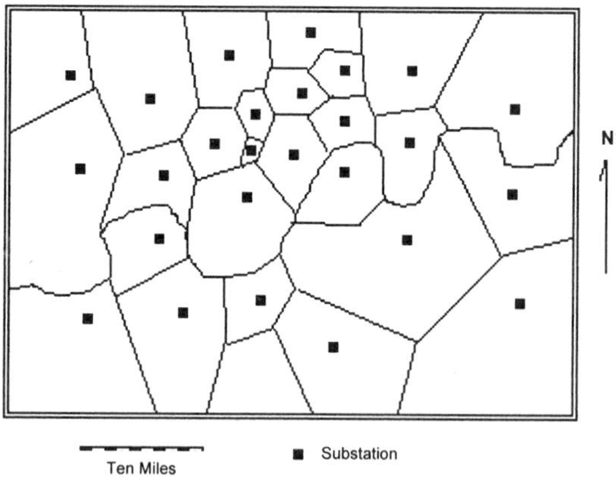

Figure 2.14 A power system is divided by substation service boundaries into a set of substation service areas, as shown.

Dynamic Service Area Planning

By making switching changes in the distribution system, it is possible to expand or shrink a substation or feeder's service area significantly, increasing or decreasing its net load. This is an important element of T&D planning, as shown in Figure 2.15, which illustrates a very typical T&D expansion situation. Two neighboring substations, A and B, each have a peak load near the upper limit of their reliable load-handling range. Load is growing slowly throughout the system, so that in each substation annual peak load is increasing at about 1 MW per year. Under present conditions, both will need to be upgraded soon. Approved transformer types, required to add capacity to each, are available only in 25 MVA or larger increments, costing $500,000 or more.

Both substations do not need to be reinforced. A new 25 MVA transformer and associated equipment are added to substation A, increasing its ability to handle a peak load by about 20 MVA. Ten MW of substation B's service area is then transferred to A. The result is that each substation has 10 MW of margin for continued load growth - further additions are not needed for 10 years.

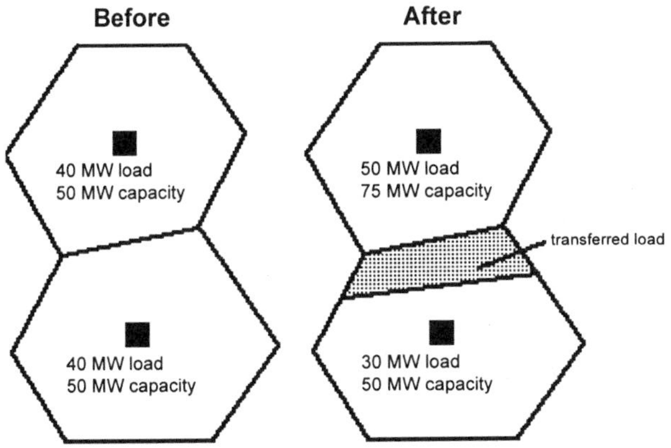

Figure 2.15 Load in both substations is growing at about 1 MW per year. Each substation has sufficient capacity to handle present load within contingency criteria (a 25% margin above peak) but nothing more. By transferring load as shown above, only one substation has to be reinforced with an additional (25 MVA) capacity, yet both end up with sufficient margin for another ten years' growth. Service area shifts keep expansion costs down in spite of the fact that equipment like transformers are available only in large, discrete sizes.

This type of planned variation in service areas is a major tool used to keep T&D expansion costs low, a key element in building a reliable, economical schedule of expansion. Optimization of this particular aspect of planning can be a challenge, not because of any inherent difficulty in analysis, but simply because there are so many parts of the system and constraints to track at one time. This is one reason for the high degree of computerization in distribution planning at many utilities. Balancing the myriad requirements of many substations and their design and operating constraints is an ideal problem for numerical optimization techniques.

The Systems Approach

One complication in determining the most economical equipment for a power system is that its various levels - transmission, substation, and distribution - are interconnected, with the distribution in turn, connected to the consumers and hence a function of the local load density. The best size and equipment type at each level of the system is a function of the types of equipment selected for the other levels of the system and the load density. In general, the equipment is so interconnected that it is impossible to evaluate any one aspect of design without taking all others into account.

Consider the question of substation spacing - determining how far apart substations should be, on average for best utilization and economy. Within any service territory, if substations are located farther apart, there will be fewer of them, saving the utility the cost of buying and preparing some substation sites, as well as reducing the cost of building a large number of substations. Thus, overall substation cost is a function of spacing, but how it varies depends on land and equipment costs, which must be balanced carefully. With fewer substations however, each substation must serve a larger area of the system. Hence, each substation will have a larger load and thus require a larger capacity, meaning it must have more or larger transformers.

The larger substations will also require a larger amount of power to be brought to each one, which generally calls for a higher sub-transmission voltage. Yet there will be fewer sub-transmission lines required because there are fewer substations to which power must be delivered. All three aspects of layout are related - greater substation spacing calls for larger substations with bigger transformers, and a higher transmission voltage, but fewer lines are needed - and all three create better economies of scale as spacing is increased. Thus, transmission costs generally drop as substation spacing is increased.

Nevertheless, there is a limit to this economy. The distribution feeder system is required to distribute each substation's power through its service area, moving power out to the boundary between each substation's service area and that of its neighbors. Moving substations farther apart means that the distribution system must move power, on average, a greater distance. Distributing power over these longer distances requires longer and more heavily loaded feeders. This situation increases voltage drop and can produce higher losses, all of which can increase cost considerably. Employing a higher distribution voltage (such as 23.9 kV instead of 13.8 kV) improves performance and economy, but it costs more to distribute power from a few larger substations farther apart, than to distribute it from many smaller substations close together. Feeder costs go up rapidly as substation spacing is increased.

The major point of this section is that all four of the above aspects of system design are interconnected - 1) substation spacing in the system, 2) size and number of substations, 3) transmission voltage and design and, 4) distribution feeder voltage and design. One of these factors cannot be optimized without close evaluation of its interrelationship with the other three. Therefore, determining the most cost-effective design guideline involves evaluating the transmission-substation-feeder system design as a whole against the load pattern, and selecting the best combination of transmission voltage, substation transformer sizes, substation spacing, and feeder system voltage and layout.

This economic equipment sizing and layout determination is based on achieving a balance between two conflicting cost relationships:

1. Higher voltage equipment is nearly always more economical on a per-MW basis.

2. Higher voltage equipment is available only in large sizes (lots of MW).

In cases where the local area demands are modest, higher voltage equipment may be more expensive simply because the minimum size is far above what is required - the utility has to buy more than it needs. How these two cost relationships play against one another depends on the load and the distances over which power must be delivered. Other factors to be considered are those unique to each power system, such as the voltages at which power is delivered from the regional power pool, and whether the system is underground or overhead.

Figure 2.16 illustrates the difference that careful coordination of system design between levels of the power system can have in lowering overall cost. Shown are the overall costs from various combinations of a T&D system layout for a large metropolitan utility in the eastern United States. Each line connects a set of cost computations for a system built with the same transmission and distribution voltages (e.g., 161 kV transmission and 13.8 kV distribution) but varying in substation sizes (and hence, implicitly, their spacing).

In all cases, the utility had determined it would build each substation with two equally sized transformers (for reliability), with none over 75 MVA (larger transformers are too difficult to move along normal roads and streets, even on special trailers). Either 161 kV or 69 kV could be used as sub-transmission, either 23.9 kV or 13.8 kV could be used as distribution voltage. Any size transformer, from 15 MVA to 75 MVA could be used, meaning the substation could vary from 30 MVA to 150 MVA in size. (Peak load of such substations can normally be up to 75% of capacity, for a peak load of from 23 to 100 MW). Substation spacing itself is implicit and not shown. Given the requirement to cover the system, determining transmission voltage, distribution, and substation size defines the system design guidelines entirely.

Overall, the ultimate and lowest cost T&D system guidelines are to build 120 MVA substations (two 60 MVA transformers) fed by 161 kV sub-transmission and distributing power at 23.9 kV. This has a levelized cost (as computed for this utility) of about \$179/kW. In this particular case, a high distribution voltage is perhaps the most important key to good economy - if 13.8 kV is used instead of 23.9 kV as the primary voltage, minimum achievable cost rises to \$193/kW.

The very worst design choices plotted in Figure 2.16, from an economic standpoint, would be to build 25 MVA substations fed by 161 kV sub-transmission and feeding power to 23.9 kV feeders (\$292/kW). This would require many small substations, each below the effective size of both the transmission and distribution voltages, and lead to high costs. Overall, 161 kV and 23.9 kV are the correct choices for economy, but only if used in conjunction with a few, large substations. If substations are to be 25 MVA, then 69 kV and 13.8 kV do a much more economical job (\$228/kW), but still don't achieve anything like the optimum value. The point: Achieving economy in power delivery involves coordinating the interactions, performance, and economies of the multiple system levels. Chapters 11 and 12 discuss the issues of and techniques for such coordinated multi-level planning.

2.8 CONCLUSION

A transmission and distribution system moves power from a utility's power production and purchase points to its consumers. The T&D system's mission of

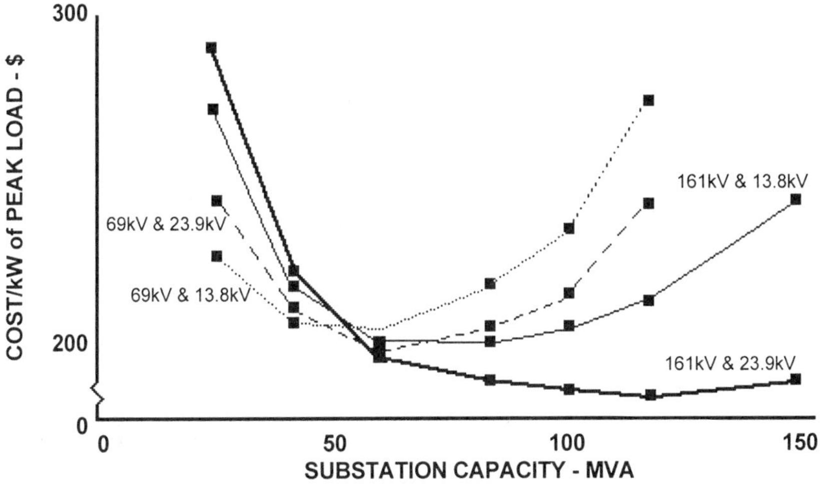

Figure 2.16 Overall cost of T&D system depends on the balanced design of the sub-transmission, substation, and feeder level, as described in the text. Cost can vary by significant margins depending on how well performance and economy at various levels of the system are coordinated.

delivering power to the consumers means that it must be composed of equipment spread throughout the service territory and arranged locally so that capacity is always in proportion to local electrical demand. The facilities need to be in each neighborhood, sized and configured both to fit well into the whole and to serve the local needs. In the authors' opinion, engineering a system to meet this challenge is seldom easy, but engineering a system to do so at the minimum possible cost is always extremely challenging.

A T&D system is composed of several interconnected, hierarchical levels, each of which is required for completion of the power delivery task. These levels are:

1. Transmission
2. Sub-transmission
3. Substation
4. Feeder
5. Secondary
6. Consumer

To a certain extent, the burden of power delivery, and costs, can be shifted from one level to another through changes in the specifications of equipment, layout standards, and design of the various levels. For example, costs can be pared at the substation level by using fewer, larger substations. This means, feeders in each substation area will have to carry power farther, and perhaps carry more power per feeder as well, increasing feeder costs. Low overall design cost is achieved by balancing these factors.

The performance and economy and thus the design, of a power system are dominated by a number of constraints due to physical laws, and further shaped by a number of practical considerations with regard to equipment, layout, and operating requirements. The more important of these are:

1. *The T&D System Must Cover Ground.* This is *the* rule about T&D - ultimately the electric system must "run a wire" to every consumer. A significant portion of the cost of a distribution system is due to this requirement alone, independent of the amount of peak load or energy supplied.

2. *Interconnectivity of Levels.* The sub-transmission/substation-feeder triad comprises a highly interconnected system with the electrical and economic performance at one level *heavily* dependent on design, siting, and other decisions at another level. To a certain extent the T&D system must be planned and designed as a whole - its parts cannot be viewed in isolation.

3. *Discrete Equipment Sizes.* In many cases equipment is available only in certain discrete sizes. For example, 69 kV/12.47kV transformers may be available only in 5, 10, 20, and 22 MVA sizes. Usually, there is a large economy of scale - the installed cost of a 20 MVA unit being considerably less than two times that for a 10 MVA unit.

4. *Dynamic Service Area Assignment is a Powerful Planning Tool.* Maximization of equipment utilization and minimization of cost per unit of capacity can be obtained despite the fact that capacity upgrades can be built only in one of several discrete sizes. To accomplish this the planner changes the boundaries of the service area during the planning process. When reinforced with new capacity additions, a substation might pick up load from surrounding substations, effectively spreading the capacity addition among several substations.

5. *Losses Cost Can Be Significant.* In some cases, the present worth of future losses on a line or heavily loaded transformer can exceed its total capital cost. In most cases, this does not happen. But in most cases, the present worth of losses is greater than the difference in cost between most of the choices available to the planner. This means loss is a major factor to be considered in achieving overall least-cost design.

6. *Cost to Upgrade is Greater than the Cost to Build.* For example, one mile of a 12.47 kV, three-phase feeder using 600 MCM conductor (15 MW thermal capacity) might cost $150,000 to build, and one mile of 9 MW feeder (336 MCM conductor) might cost only $110,000. But the cost to upgrade the 336 MCM feeder to 600 MCM wire size at a later date would be about $130,000, for a cumulative total of $240,000. Therefore, one aspect of minimizing cost is to determine size for equipment not on the basis of immediate need, but by assessing the eventual need and determining whether the present worth of the eventual savings warrants the investment in larger size now.

7. *Coincidence of Peaks.* Not all consumer loads occur at the same time. This has a number of effects. First, peak load in different parts of the system may occur at different times. Second, the system peak load will always be less than the sum of the peak loads at any one level of the system. For example, in more power systems the sum of all substation peak loads usually exceeds system total peak by 3% to 8%. Diversity of peak loads means that considerable attention must be paid to the pattern and timing of electric load if equipment needs (and consequently costs) are to be minimized.

8. *Reliability is obtained through Contingency Margin.* Traditionally, T&D reliability in a static system (one without automation, and hence incapable of recognizing and reacting to equipment outages and other contingencies by re-configuring and/or re-calibrating equipment) is assured by adding an emergency, or contingency margin throughout the system. At the T&D level this means that an average of 20% to 50% additional capacity is included in most equipment, so that it can pick up additional load if neighboring equipment fails or is overloaded.

A Complicated System

In conclusion, a power T&D system provides the delivery of an electric utility's product. It must deliver that product to every consumer's site, with high reliability and in ready-to-use form. While composed of equipment that is individually straightforward, most T&D systems are quite complex due to the interactions of thousands, even millions, of interconnected elements. Achieving economy and reliability means carefully balancing a myriad of mutual interactions with conflicting cost tradeoffs.

REFERENCES

J. J. Burke, *Power Distribution Engineering – Fundamentals and Applications,* Marcel Dekker, New York, 1994.

M. V. Engel et al., editors, *Tutorial on Distribution Planning*, IEEE Course Text EHO 361-6-PWR, Institute of Electrical and Electronics Engineers, Hoes Lane, NJ, 1992.

A. C. Monteith and C. F. Wagner, *Electrical Transmission and Distribution Reference Book*, Westinghouse Electric Company, Pittsburgh, PA, 1964.

H. L. Willis, *Power Distribution Planning Reference Book,* Marcel Dekker, New York, 1997

H. L. Willis and W. G. Scott, *Distributed Power Generation – Planning and Evaluation,* Marcel Dekker, New York, 2000

3
Customer Demand for Power and Reliability of Service

3.1 THE TWO Qs: QUANTITY AND QUALITY OF POWER

Electric consumers require power, whether delivered from the utility grid or generated locally by distributed sources, in order to help accomplish the uses for which they need energy. Their need for electric power, and the value they place upon its delivery to them, has two interrelated but fundamentally separate dimensions. These are the two Qs: Quantity, the amount of power needed, and Quality, the most important aspect of which is usually dependability of supply (reliability of power supply, or availability as it is often called). The relative importance of these two features varies from one consumer to another depending on their individual needs, but each consumer finds value in both the amount of power he obtains, and its availability as a constant, steady source that it will be there whenever needed.

This chapter discusses demand and use of electric power as seen from the customer's standpoint: the utility's job is to satisfy consumer needs as fully as possible within reasonable cost constraints. Cost is very much an important aspect to consumers too, so both the utility and the consumer must temper their plans and desires with respect to power and reliability based on real world economics. Consumers do not get everything they want; only what they are willing to pay for. Utilities should not aim to provide flawless service, which would be prohibitively expensive, but instead aim to provide the highest level possible within economic constraints of the customers' willingness to pay.

This chapter begins, in section 3.2, with a discussion of consumer use of electricity and includes the quantity of electric demand as seen from an "end-use" perspective. It continues with how demand varies as a function of consumer type and end-use, and how power demand is represented in electric system studies using load curves and load duration curves. Section 3.3 then discusses reliability and availability as seen by the customers and ways this can be characterized and studied. Section 3.4 briefly reviews Two-Q analysis and planning concepts and their application to customer load analysis. Finally, section 3.5 provides a summary of key points.

3.2 ELECTRIC CONSUMER NEED FOR QUANTITY OF POWER

No consumer wants electric energy itself. Consumers want the *products* it can provide – a cool home in summer, hot water on demand, compressed air for manufacturing, electronic robotic factory control, cold beer in the 'fridge and football on color TV. Electricity is only an intermediate means to some end-use.

These different goals and needs are called *end-uses,* and they span a wide range of applications. Some end-uses are unique to electric power (the authors are not aware of any manufacturer of natural gas powered TVs, stereos, or computers). For many other end-uses, electricity is only one of several possible energy sources (water heating, home heating, cooking, or clothes drying). In many other end-uses, electricity is so convenient that it enjoys a virtual monopoly, even though there are alternatives, e.g., gasoline-powered refrigerators, and natural gas for interior lighting and for air conditioning.

Each end-use is satisfied through the application of appliances or devices that convert electricity into the desired end product. For example, with lighting a wide range of illumination devices are used, including incandescent bulbs, fluorescent tubes, sodium vapor, high-pressure monochromatic gas-discharge tubes, and in special cases, lasers. Each type of lighting device has differences from the others that give it an appeal to some consumers or for certain types of applications. Regardless, each requires electric power to function, creating an electric load when it is activated. Similarly, for other end-uses, such as space heating, there are various types of appliances, each with advantages or disadvantages in initial cost, operating efficiency, reliability and maintenance, noise and vibration, or other aspects. Each produces an electric load when used to produce heat for a home or business.

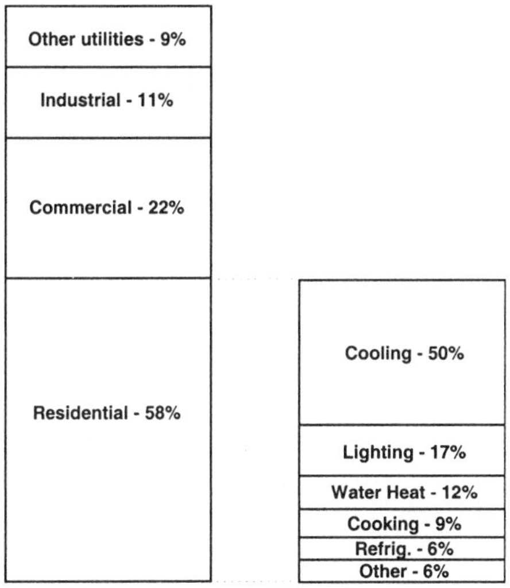

Figure 3.1 Electric peak demand of a utility in the southeastern United States broken down by consumer class and within the residential class, by contribution to peak for the major uses for which electricity is purchased at time of peak by the residential class.

Consumer Classes

Different types of consumers purchase electricity. About half of all electric power is used in residences, which vary in the brands and types of appliances they own, and their daily activity patterns. Another fourth is consumed by Commercial businesses both large and small, that buy electricity, having some similar end-uses to residential consumers (heating and cooling, and illumination), but that have many needs unique to commercial functions (cash register/inventory systems, escalators, office machinery, neon store display lighting, parking lot lighting). Finally, industrial facilities and plants buy electricity to power processes such as pulp heating, compressor and conveyor motor power, and a variety of manufacturing applications. As a result, the load on an electric system is a composite of many consumer types and appliance applications. Figure 3.1 shows this breakdown of the peak electrical load for a typical power system by consumer class and end-use category within one class.

Appliances Convert End Uses into Electric Load

The term *load* refers to the electrical demand of a device connected to and drawing power from the electric system in order to accomplish some task, e.g., opening a garage door, or converting that power to some other form of energy, such as a light. Such devices are called *appliances*, whether they are a commonly regarded household item, e.g., a refrigerator, lamp, garage door opener, paper shredder, electric fence to keep cattle confined, etc. To the consumer, these appliances convert electricity into the end product. But the electric service planner can turn this relation around and view an appliance (e.g., a heat pump) as a device for transforming a demand for a particular end-use - warm or cool air - into electric load.

The level of power they need usually rates electrical loads, measured in units of real volt-amperes, called *watts*. Large loads are measured in kilowatts (thousands of watts) or megawatts (millions of watts). Power ratings of loads and T&D equipment refer to the device at a specific *nominal voltage*. For example, an incandescent light bulb might be rated at 75 watts and 1,100 lumens at 120 volts, at which voltage it consumes 75 watts and produces 1,100 lumens of light. If provided with less voltage, its load (and probably its light output) will fall.

Load Curves and Load Curve Analysis

The electric load created by any one end-use usually varies as a function of time. For example, in most households, demand for lighting is highest in the early evening, after sunset but before most of the household members have gone to bed. Lighting needs may be greater on weekends, when activity often lasts later into the evening, and at times of the year when the sun sets earlier in the day. Some end-uses are quite seasonal. Air-conditioning demand generally occurs only in summer, being greatest during particularly hot periods and when family activity is at its peak, usually late afternoon or very early evening. Figure 3.2 shows how the demand for two products of electric power varies as a function of time.

The result of this varying demand for the products of electricity application, is a variation in the demand for power as a function of time. This is plotted as a load *curve,* illustrated in Figure 3.3. Typically, the values of most interest to the planners are the peak load (maximum amount of power that must be delivered). This defines directly or indirectly, the capacity requirements for equipment; the minimum load and the time it occurs; the total energy, area under the curve that must be delivered during the period being analyzed; and the load value at the time of system peak load.

Consumer Class Load Curves

While all consumers differ in their electrical usage patterns, consumers within a particular class, such as residential, tend to have broadly similar load curve patterns. Those of different classes tend to be dissimilar in their demand for

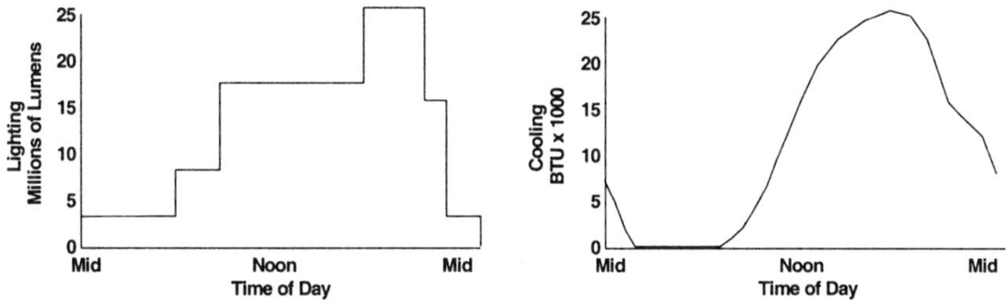

Figure 3.2 End-use demand. (Left) Average demand for BTU of cooling among houses in one of the authors' neighborhoods on a typical weekday in June. Right, lighting lumens used by a retail store on a typical day in June.

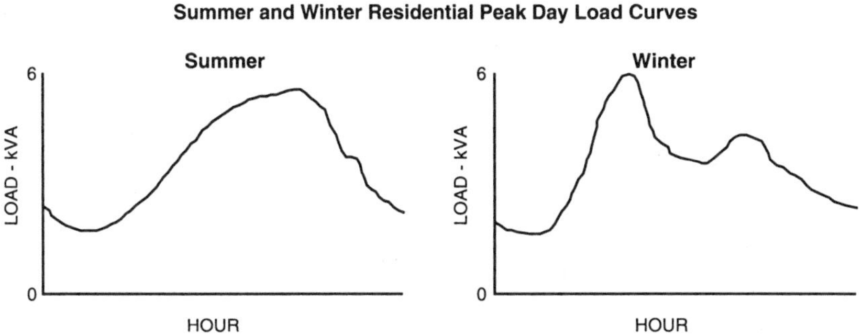

Figure 3.3 Electric demand for each class varies hourly and seasonally, as shown here, with a plot of average coincident load for residential users in central Florida.

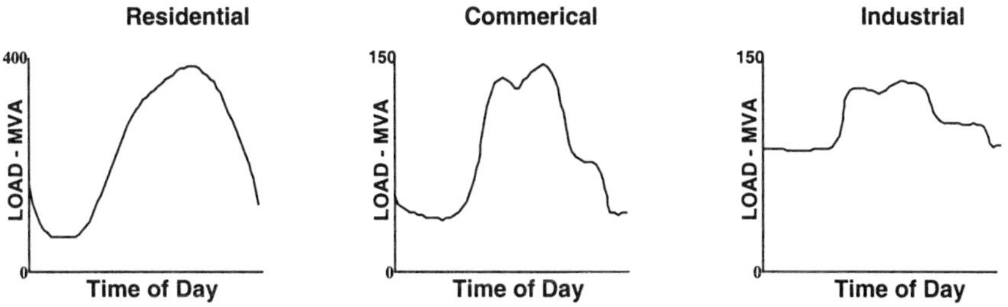

Figure 3.4 Different consumer classes have different electrical demand characteristics, particularly with regard to how demand varies with time. Here are summer peak day load curves for the three major classes of consumer from the same utility system.

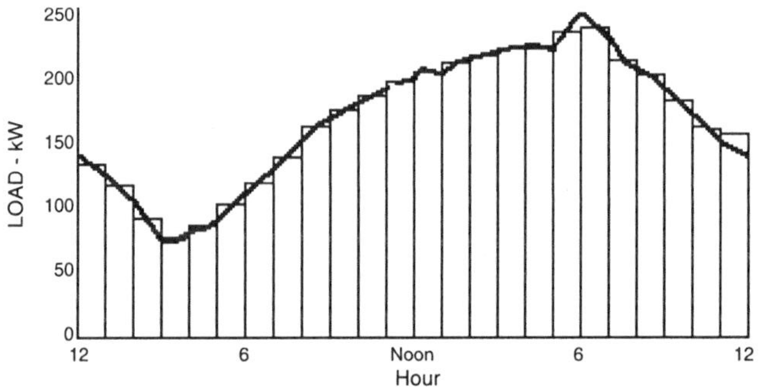

Figure 3.5 Demand on an hourly basis (blocks) over 24 hours, and the actual load curve (solid black line) for a feeder segment serving 53 homes. Demand measurement averages load over each demand interval (in this case each hour) missing some of the detail of the actual load behavior. In this case the actual peak load (263 kW at 6 PM) was not seen by the demand measuring, which "split the peak," averaging load on an hourly basis and seeing a peak demand of only 246 kW, 7% lower. As will be discussed later in this chapter, an hourly demand period is too lengthy for this application.

both quality and quantity and the time of day and year when their demand is highest. Therefore, most electric utilities distinguish load behavior on a class-by-class basis, characterizing each class with a "typical daily load curve," showing the average or expected pattern of load usage for a consumer in that class on the peak day, as shown in Figure 3.4. These "consumer class load curves" describe how the demand varies as a function of time. While often an entire 8,760-hour record for the year is available, usually only key days – perhaps one representative day per season – are used for studies.

The most important points concerning the consumers' loads from the distribution planner's standpoint are:

1. *Peak demand* and its time and duration
2. *Demand at time of system peak*
3. *Energy* usage (total area under the annual load curve)
4. *Minimum* demand, its time and duration

Details of Load Curve Measurement

Demand and demand periods

"Demand," as normally used in electric load analysis and engineering, is the average value of electric load over a period of time known as the *demand interval.* Very often, demand is measured on an hourly basis as shown in Figure 3.5, but it can be measured on any interval basis – seven seconds, one minute, 30 minutes, daily, and monthly. The average value of power during the demand interval is given by dividing the kilowatt-hours accumulated during the demand interval by the length of the interval. Demand intervals vary among power companies, but those commonly used in collecting data and billing consumers for "peak demand" are 15, 30, and 60 minutes.

Load curves may be recorded, measured, or applied over some specific time. For example, a load curve might cover one day. If recorded on an hourly demand basis, the curve consists of 24 values, each the average demand during one of the 24 hours in the day, and the peak demand is the maximum hourly demand seen in that day. Load data is gathered and used on a monthly basis and on an annual basis.

Load factor

Load factor is the ratio of the average to the peak demand. The average load is the total energy used during the entire period (e.g., a day, a year) divided by the number of demand intervals in that period (e.g., 24 hours, 8,760 hours). The average is then divided by the maximum demand to obtain the load factor:

$$\text{Load Factor} = \frac{\text{kWh/Hrs}}{\text{Peak Demand kW}} = \frac{\text{Average Demand kW}}{\text{Peak Demand kW}} \tag{3.1}$$

$$= \frac{\text{KWh}}{(\text{kW Demand}) \times (\text{Hr})} \tag{3.2}$$

Load factor gives the extent to which the peak load is maintained during the period under study. A high load factor means the load is at or near peak a good portion of the time.

Load Duration Curves

A convenient way to study load behavior is to order the demand samples from greatest to smallest, rather than as a function of time, as in Figure 3.6. The two diagrams consist of the same 24 numbers, in a different order. Peak load, minimum load, and energy (area under the curve) are the same for both.

Load duration curve behavior will vary as a function of the level of the system. Load duration curves for small groups of consumers will have a greater ratio of peak to minimum than similar curves for larger groups. Those for very small groups (e.g., one or two consumers) will have a pronounced "blockiness," consisting of plateaus – many hours of similar demand level (at least if the load data were sampled at a fast enough rate). The plateaus correspond to combinations of major appliance loads. The ultimate "plateau" would be a load duration curve of a single appliance, for example a water heater that operated a total of 1,180 hours during the year. This appliance's load duration curve would show 1,180 hours at its full load, and 7,580 hours at no load, without any values in between.

Annual load duration curves

Most often, load duration curves are produced on an annual basis, reordering all 8,760 hourly demands (or all 35,040 quarter hour demands if using 15-minute demand intervals) in the year from highest to lowest, to form a diagram like that in Figure 3.7. The load shown was above 26 kW (demand minimum) 8,760 hours in the year, never above 92 kW, but above 40 kW for 2,800 hours.

Spatial Patterns of Electric Demand

An electric utility must not only produce or obtain the power required by its consumers, but also must deliver it to their locations. Electric consumers are scattered throughout the utility service territory, and thus the electric load can be thought of as distributed on a *spatial* basis as depicted in Figure 3.9. Just as load curves show how electric load varies as a function of time, (and can help identify when certain amounts of power must be provided), so has Spatial Load Analysis been used since its development in the 1970s (Scott, 1972) to identify where load is located and how much capacity is needed in each locality.

The electric demand in an electric utility service territory varies as a function of location depending on the number and types of consumers in each locality, as shown by the load map in Figure 3.9. Load densities in the heart of a large city can exceed 1 MW/acre, but usually average about 5 MW per square mile over the entire metropolitan area. In sparsely populated rural areas, farmsteads can be as far as 30 miles apart, and load density as low as 75 watts per square mile. Regardless of whether an area is urban, suburban, or rural, electric load is a function of the types of consumers, their number, their uses for electricity, and the appliances they employ. Other aspects of power system performance, including capability, cost, and reliability, can also be analyzed on a location-

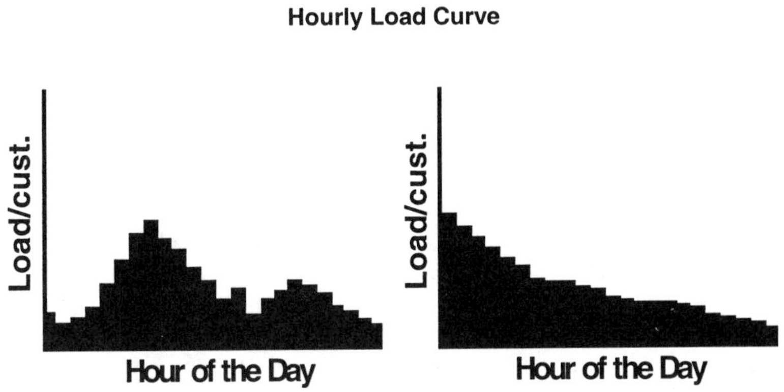

Figure 3.6 The hourly demand samples in a 24-hour load curve are reordered from greatest magnitude to least to form a daily load duration curve.

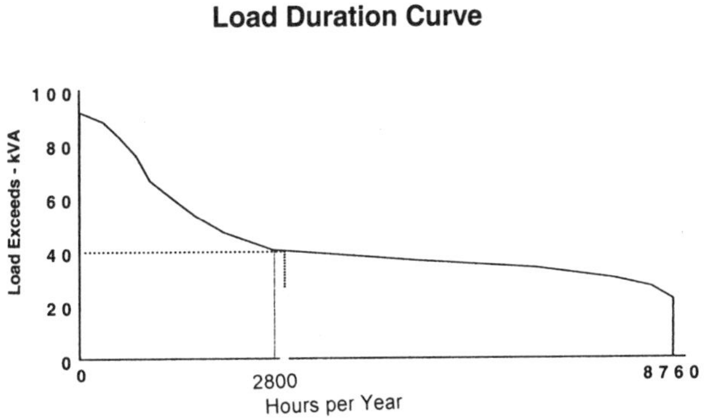

Figure 3.7 Annual load duration curve for a commercial site with a 90 kW peak demand, from an example DG reliability study case in Chapter 16. As shown here, in that analysis the demand exceeds 40 kW, the capacity of the smaller of two DG units installed at the site, 2,800 hours a year. During those 2,800 hours, adequate service can be obtained only if both units are operating.

basis, an often important aspect of siting for facilities, including DG (Willis, 1996).

Most often, load duration curves are produced on an annual basis, reordering all 8,760 hourly demands (or all 35,040 quarter hour demands if using 15-minute demand intervals) in the year from highest to lowest, to form a diagram like that in Figure 3.7. The load shown was above 26 kW (demand minimum) 8,760 hours in the year, never above 92 kW, but above 40 kW for 2,800 hours.

Spatial Patterns of Electric Demand

An electric utility must not only produce or obtain the power required by its consumers, but also must deliver it to their locations. Electric consumers are scattered throughout the utility service territory, and thus the electric load can be thought of as distributed on a *spatial* basis as depicted in Figure 3.9. Just as load curves show how electric load varies as a function of time, (and can help identify when certain amounts of power must be provided), so has Spatial Load Analysis been used since its development in the 1970s (Scott, 1972) to identify where load is located and how much capacity is needed in each locality.

The electric demand in an electric utility service territory varies as a function of location depending on the number and types of consumers in each locality, as shown by the load map in Figure 3.9. Load densities in the heart of a large city can exceed 1 MW/acre, but usually average about 5 MW per square mile over the entire metropolitan area. In sparsely populated rural areas, farmsteads can be as far as 30 miles apart, and load density as low as 75 watts per square mile. Regardless of whether an area is urban, suburban, or rural, electric load is a function of the types of consumers, their number, their uses for electricity, and the appliances they employ. Other aspects of power system performance, including capability, cost, and reliability, can also be analyzed on a location-basis, an often important aspect of siting for facilities, including DG (Willis, 1996).

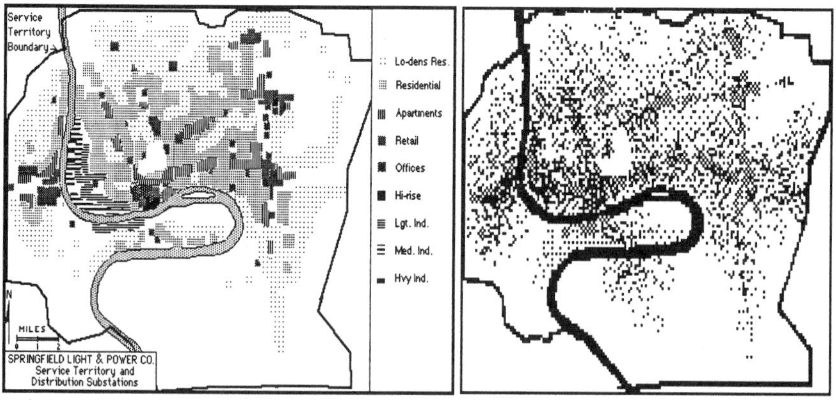

Figure 3.8 (Left) Map showing types of consumer by location for a small city. (Right) map of electric demand for this same city.

3.3 ELECTRIC CONSUMER NEED FOR QUALITY OF POWER

As mentioned in this chapter's introduction, a central issue in customer value of service analysis is matching availability and power quality against cost. T&D systems with near perfect availability and power quality can be built, but their high cost will mean electric prices the utility customers may not want to pay, given the savings an even slightly less reliable system would bring. All types of utilities have an interest in achieving the correct balance of quality and price. The traditional, franchised monopoly utility, in its role as the "electric resource manager" for the customers it serves, has a responsibility to build a system whose quality and cost balances its customers' needs. A competitive retail distributor of power wants to find the best quality-price combination - only in that way will it gain a large market share.

While it is possible to characterize various power quality problems in an engineering sense, characterizing them as interruptions, voltage sags, dips, surges, or harmonics the customer perspective is somewhat different. Customers are concerned with only two aspects of service quality:

1. They want power when they need it.
2. They want the power to do the job.

If power is not available, neither aspect is provided. However, if power is available, but quality is low, only the second is not provided.

Assessing Value of Quality by Studying the Cost of a Lack of It

In general, customer value of reliability and service quality are studied by assessing the "cost" that something less than perfect reliability and service quality creates for customers. Electricity provides a value, and interruptions or poor power quality decrease that value. This value reduction - cost - occurs for a variety of reasons. Some costs are difficult, if not impossible to estimate: re-scheduling of household activities or lack of desired entertainment when power fails;[1] or flickering lights that make reading more difficult.

But often, very exact dollar figures can be put on interruptions and poor power quality. Food spoiled due to lack of refrigeration; wages and other operating costs at an industrial plant during time without power; damage to product caused by the sudden cessation of power; lost data and "boot up" time for computers; equipment destroyed by harmonics; and so forth. Figure 3.9 shows two examples of such cost data.

[1] No doubt, the cost of an hour-long interruption that began fifteen minutes from the end of a crucial televised sporting event, or the end of a "cliffhanger" movie, would be claimed to be great.

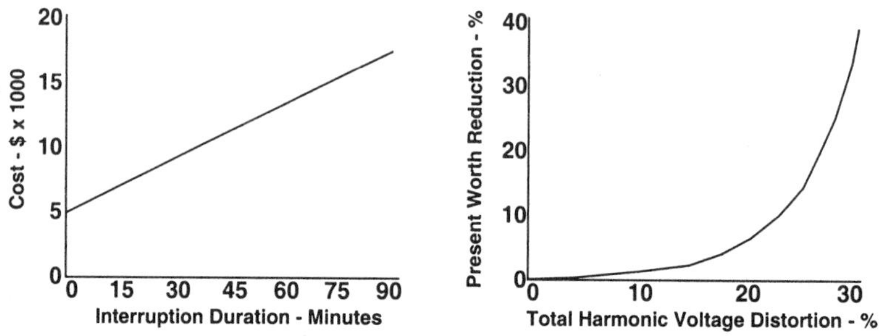

Figure 3.9 Left, cost of a week-day interruption of service to a pipe rolling plant in the southeastern United States, as a function of interruption duration. An interruption of any length costs about $5,000 - lost wages and operating costs to unload material in process, bring machinery back to "starting" position and restart - and a nearly linear cost thereafter. At right, present worth of the loss of life caused by harmonics in a 500 horsepower three-phase electric motor installed at that same industrial site, as a function of harmonic voltage distortion.

Value-Based Planning

To be of any real value in utility planning, information of the value customers put on quality must be usable in some analytical method that can determine the best way to balance quality against cost. Value-based planning (VBP) is such a method: it combines customer-value data of the type shown in Figure 3.9 with data on the cost to design the T&D system to various levels of reliability and power quality, in order to identify the optimum balance. Figure 3.10 illustrates the central tenet of value-based planning. The cost incurred by the customer due to various levels of reliability or quality, and the cost to build the system to various levels of reliability, are added to get the total cost of power delivered to the customer as a function of quality.[2] The minimum value is the optimum balance between customer desire for reliability and aversion to cost. This approach can be applied for only reliability aspects, i.e., value-based reliability planning, or harmonics, or power quality overall. Generally, what makes sense

---

[2] Figure 3.10 illustrates the concept of VBP. In practice, the supply-side reliability curves often have discontinuities and significant non-linearities that make application difficult. These and other details will be discussed in Chapter 5.

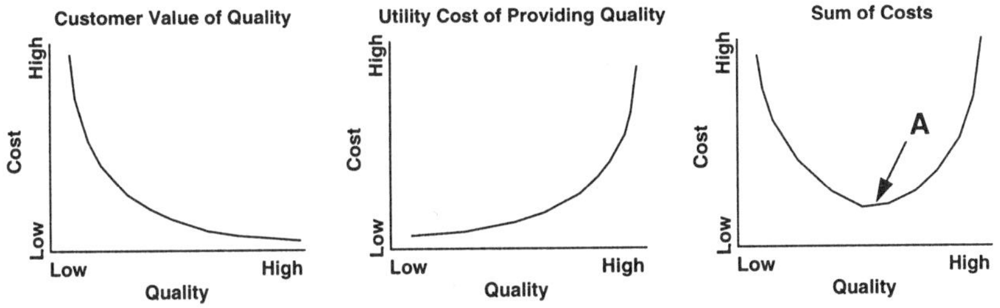

Figure 3.10 Concept of value-based planning. The customer's cost due to poorer quality (left) and the cost of various power delivery designs with varying levels of quality (center) are computed over a wide range. When added together (right) they form the total cost of quality curve, which identifies the minimum cost reliability level (point A).

is to apply it on the basis of whatever qualities (or lack of them) impact the customer - interruptions, voltage surges, harmonics, etc., in which case it is comprehensive value-based quality of service planning.

Cost of Interruptions

The power quality issue that affects the most customers, and which receives the most attention, is cessation of service often termed "service reliability." Over a period of several years, almost all customers served by any utility will experience at least one interruption of service. By contrast, a majority will never experience serious harmonics, voltage surge, or electrical noise problems. Therefore, among all types of power quality issues, interruption of service receives the most attention from both the customers and the utility. A great deal more information is available about cost of interruptions than about cost of harmonics or voltage surges.

Voltage Sags Cause Momentary Interruptions

The continuity of power flow does not have to be completely interrupted to disrupt service: If voltage drops below the minimum necessary for satisfactory operation of an appliance, power has effectively been "interrupted" as illustrated in Figure 3.11. For this reason many customers regard voltage dips and sags as momentary interruptions - from their perspective these are interruptions of the end-use *service they desire,* if not of voltage.

Much of the electronic equipment manufactured in the United States, as well as in many other countries, have been designed to meet or exceed the CBEMA

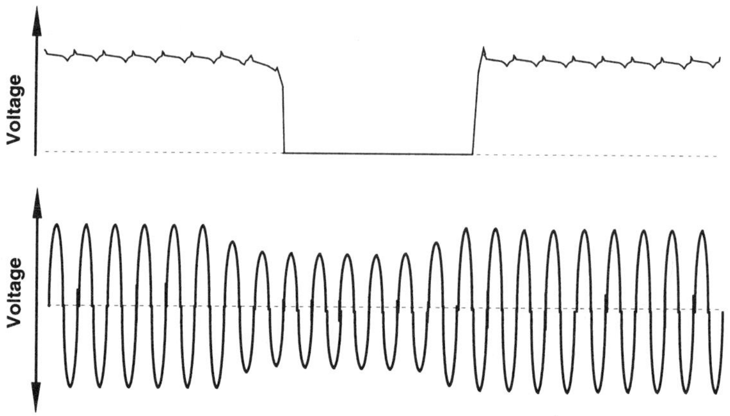

Figure 3.11 Output of a 5.2 volt DC power supply used in a desktop computer (top) and the incoming AC line voltage (nominal 113 volts). A voltage sag to 66% of nominal causes power supply output to cease within three cycles.

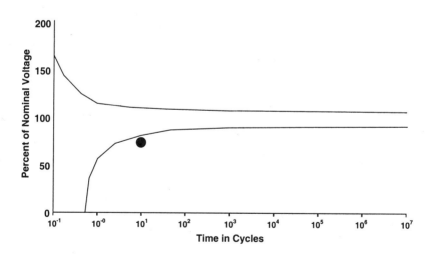

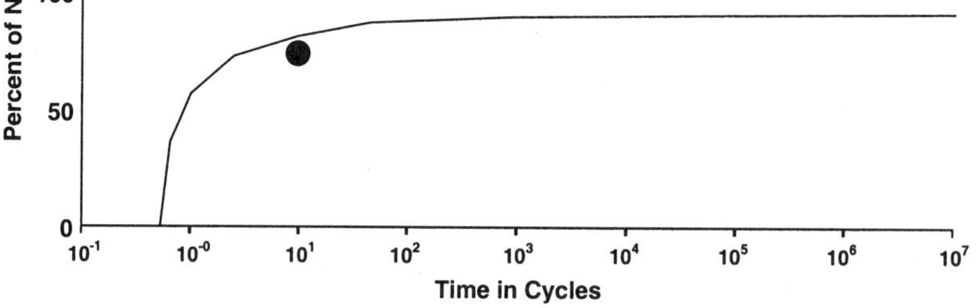

Figure 3.12 CBEMA curve of voltage deviation versus period of deviation, with the sag shown in Figure 3.11 plotted (black dot).

(Computer and Business Equipment Manufacturer's Association) recommended curves for power continuity, shown in Figure 3.12. If a disturbance's voltage deviation and duration characteristics are within the CBEMA envelope, then normal appliances should operate normally and satisfactorily.

However, many appliances and devices in use will not meet this criterion at all. Others will fail to meet it under the prevailing ambient electrical conditions (i.e., line voltage, phase unbalance power factor and harmonics may be less than perfect).

The manner of usage of an appliance also affects its voltage sag sensitivity. The voltage sag illustrated in Figure 3.11 falls just within the CBEMA curve, as shown in Figure 3.12. The manufacturer probably intended for the power supply to be able to withstand nearly twice as long a drop to 66% of nominal voltage before ceasing output. However, the computer in question had been upgraded with three times the standard factory memory, a second and larger hard drive, and optional graphics and sound cards, doubling its power usage and the load on the power supply. Such situations are common and means that power systems that deliver voltage control within recommended CBEMA standards may still provide the occasional momentary interruption.

For all these reasons, there are often many more "momentary interruptions" at a customer site than purely technical evaluation based on equipment specifications and T&D engineering data would suggest. Momentary interruptions usually cause the majority of industrial and commercial interruption problems. In addition, they can lead to one of the most serious customer dissatisfaction issues. Often utility monitoring and disturbance recording equipment does not "see" voltage disturbances unless they are complete cessation of voltage, or close to it. Many events that lie well outside the CBEMA curves and definitely lead to unsatisfactory equipment operation are not recorded or acknowledged. As a result, a customer can complain that his power has been interrupted five or six times in the last month, and the utility will insist that its records show power flow was flawless. *The utility's refusal to acknowledge the problem irks some customers more than the power quality problem itself.*

Frequency and Duration of Interruptions Both Impact Cost

Traditional Power System Reliability Analysis recognizes that service interruptions have both frequency and duration (See Chapter 4). Frequency is the number of times during some period (usually a year) that power is interrupted. Duration is the amount of time power is out of service. Typical values for urban/suburban power system performance in North America are 2.2 interruptions per year with 100 minutes total duration.

Both frequency and duration of interruption impact the value of electric service to the customer and must be appraised in any worthwhile study of customer value of service. A number of reliability studies and value-based planning methods have tried to combine frequency and duration in one manner or another into "one dimension." A popular approach is to assume all interruptions are of some average length (e.g., 2.2 interruptions and 100 minutes is assumed to be 2.2 interruptions per year of 46 minutes each). Others have assumed a certain portion of interruptions are momentary and the rest of the duration is lumped into one "long" interruption (i.e., 1.4 interruptions of less than a minute, and one 99-minute interruption per year). Many other approaches have been tried (see References and Bibliography). But all such methods are at best an approximation, because frequency and duration impact different customers in different way. No single combination of the two aspects of reliability can fit the value structure of all customers.

Figure 3.13 shows four examples of the author's preferred method of assessing interruption cost, which is to view it as composed of two components, a fixed cost (Y intercept) caused when the interruption occurred, and a variable cost that increases as the interruption continues. As can be seen in Figure 3.13, customer sensitivity to these two factors varies greatly. The four examples are:

1. A pipe-rolling factory (upper left). After an interruption of any length, material in the process of manufacturing must be cleared from the welding and polishing machinery, all of which must be reset and the raw material feed set up to begin the process again. This takes about 1/2 hour and sets a minimum cost for an interruption. Duration longer than that is simply a linear function of the plant operating time (wages and rent, etc., allocated to that time). Prior to changes made by a reliability study, the "re-setting" of the machinery could not be done until power was restored (i.e., time during the interruption could not be put to use preparing to re-start once it was over). The dotted line shows the new cost function after modifications to machinery and procedure were made so that preparations could begin *during* the interruption.

2. An insurance claims office (upper right) suffers loss of data equivalent to roughly one hour's processing when power fails. According to the site supervisor, an unexpected power interruption causes loss of about one hour's work as well as another estimated half hour lost due to the impact of any interruption on the staff. Thus, the fixed cost of each interruption is equivalent to about ninety minutes of work. After one-half hour of interruption, the supervisor's policy is to put the staff to work "on other stuff for a while," making cost impact lower (some productivity); thus, variable interruption cost goes down. The dotted line shows the cost impact

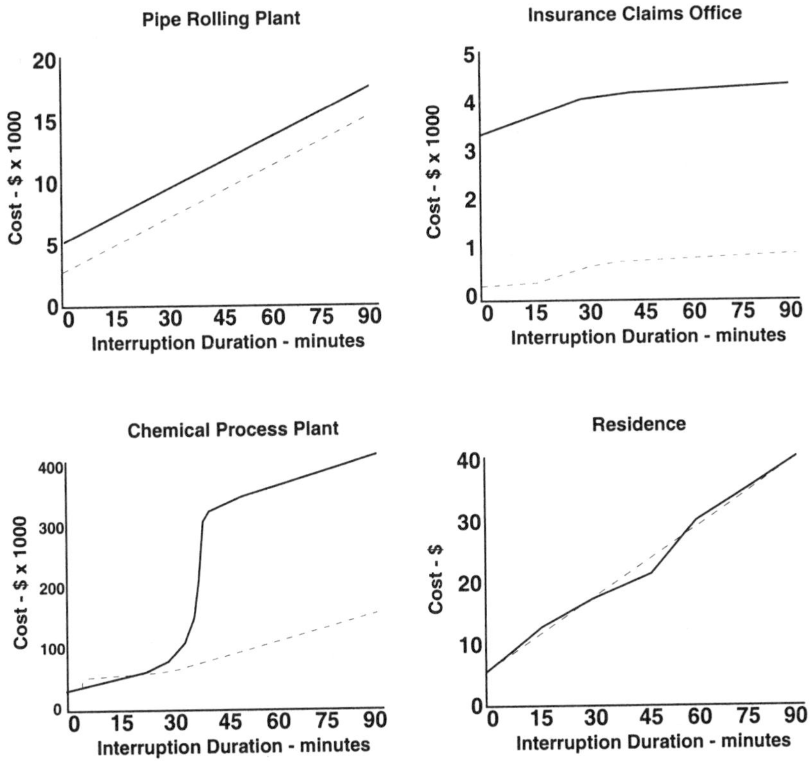

Figure 3.13 The author's recommended manner of assessing cost of interruptions includes evaluation of service interruptions on an event basis. Each interruption has a fixed cost (Y-intercept) and a variable cost, which increases as the interruption continues. Examples given here show the wide range of customer cost characteristics that exist. The text gives details on the meaning of solid versus dotted lines and the reasons behind the curve shape for each customer.

of interruptions after installation of UPS on the computer system, which permits orderly shutdown in the event of an interruption.

3. An acetate manufacturing and processing plant (lower left) has a very non-linear cost curve. Any interruption of service causes $38,000 in lost productivity and after-restoration set-up time. Cost rises slowly for about half an hour. At that point, molten feedstock and interim ingredients inside pipes and pumps begins to cool, requiring a day-long process of cleaning sludge and hardened stock out of the system. The dotted line shows the plant's interruption cost function after installation of a diesel generator, started whenever interruption time exceeds five minutes.

4. Residential interruption cost function (lower right), estimated by the authors from a number of sources including a survey of customers made for a utility in the northeastern United States in 1992, shows roughly linear cost as a function of interruption duration, except for two interesting features. The first is the fixed cost equal to about eight minutes of interruption at the initial variable cost slope which reflects "the cost to go around and re-set our digital clocks," along with similar inconvenience costs. Secondly, a jump in cost between 45 and 60 minutes, which reflect inconsistencies in human reaction to outage, time on questionnaires. The dotted line shows the relation the authors' uses in their analysis, which makes adjustments thought reasonable to account for these inconsistencies.

This recommended analytical approach, in which cost is represented as a function of duration on a per event basis, requires more information and more analytical effort than simpler "one-dimensional" methods, but the results are more credible.

Interruption Cost is Lower if Prior Notification is Given

Given sufficient time to prepare for an interruption of service, most of the momentary interruption cost (fixed) and a great deal of the variable cost can be eliminated by many customers. Figure 3.14 shows the interruption cost figures from Figure 3.13 adjusted for "24 hour notification given."

Cost of Interruption Varies by Customer Class

Cost of power interruption varies among all customers, but there are marked distinctions among classes, even when cost is adjusted for "size" of load by computing all cost functions on a per kW basis. Generally, the residential class has the lowest interruption cost per kW and commercial the highest. Table 3.14

Table 3.1 Typical Interruption Costs by Class for Three Utilities – Daytime, Weekday (dollars per kilowatt hour)

Class	1	2	3
Agricultural	3.80	4.30	7.50
Residential	4.50	5.30	9.10
Retail Commercial	27.20	32.90	44.80
Other Commercial	34.00	27.40	52.10
Industrial	7.50	11.20	13.90
Municipal	16.60	22.00	44.00

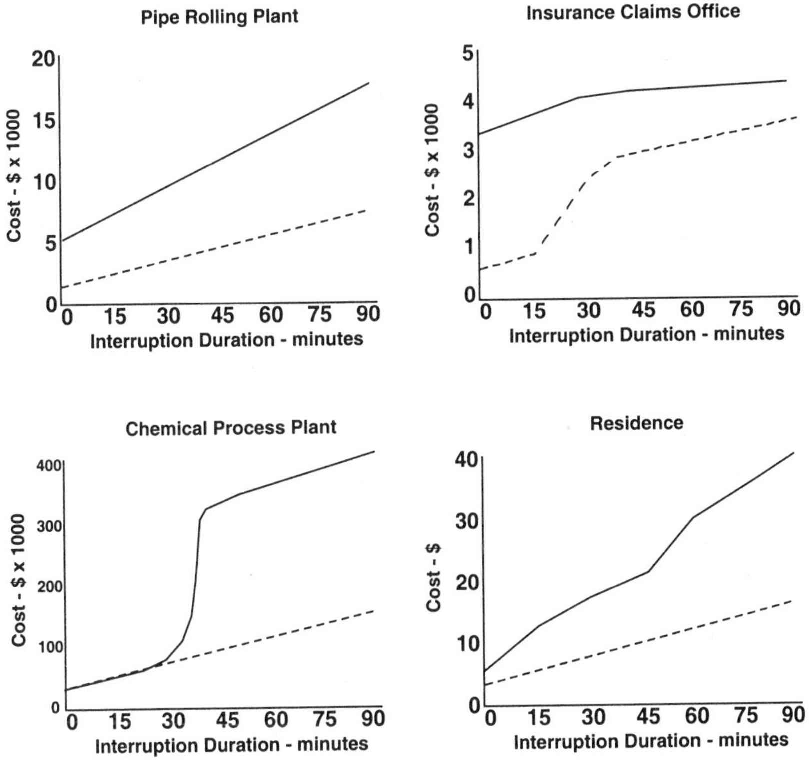

Figure 3.14 If an interruption of service is expected customers can take measures to reduce its impact and cost. Solid lines are the interruption costs (the solid lines from Figure 3.13). Dotted lines show how 24-hour notice reduces the cost impact in each case.

gives the cost/kW of a one-hour interruption of service by customer class, obtained using similar survey techniques for three utilities in the United States:

1. A small municipal system in the central plains,
2. an urban/suburban/rural system on the Pacific Coast and,
3. an urban system on the Atlantic coast.

Cost Varies from One Region to Another

Interruption costs for apparently similar customer classes can vary greatly depending on the particular region of the country or state in which they are located. There are many reasons for such differences. The substantial difference (47%) between industrial costs in utilities 1 and 3 shown in Table 3.14 is due to differences in the type of industries that predominate in each region. The differences between residential costs of the regions shown reflect different demographics and varying degrees of economic health in their respective regions.

Cost Varies among Customers within a Class

The figures given for each customer class in Table 3.14 represent an average of values within those classes as surveyed and studied in each utility service territory. Value of availability can vary a great deal among customers within any class, both within a utility service territory and even among neighboring sites. Large variations are most common in the industrial class, where different needs can lead to wide variations in the cost of interruption, as shown in Table 3.2. Although documentation is sketchy, indications are *the major differing factor is the cost of a momentary interruption* - some customers are very sensitive to any cessation of power flow, while others are impacted mainly by something longer than a few cycles or seconds.

Cost of Interruption Varies as a Function of Time of Use

Cost of interruption will have a different impact depending on the time of use, usually being much higher during times of peak usage, as shown in Figure 3.17. However, when adjusted to a per-kilowatt basis, the cost of interruption can sometimes be higher during off-peak than during peak demand periods, as shown. There are two reasons. First, the data may not reflect actual value. A survey of 300 residential customers for a utility in New England revealed that customers put the highest value on an interruption during early evening (Figure 3.17). There could be inconsistencies in the values people put on interruptions (data plotted were obtained by survey).

Table 3.2 Interruption Costs by Industrial Sub-Class for One hour, Daytime, Weekday (dollars per kilowatt)

Class	$/kW
Bulk plastics refining	38
Cyanide plant	87
Weaving (robotic loom)	72
Weaving (mechanical loom	17
Automobile recycling	3
Packaging	44
Catalog distribution center	12
Cement factory	8

However, there is a second, and valid, reason for the higher cost per kilowatt off-peak: only essential equipment, such as burglar alarms, security lighting and refrigeration is operating - end-uses that have a high cost of interruption. While it is generally safe to assume that the cost of interruption is highest during peak, the same cannot be assumed about the cost per kilowatt.

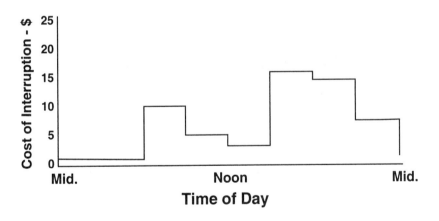

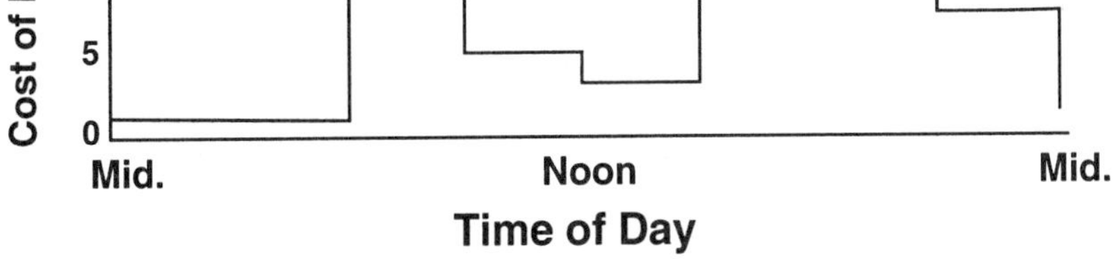

Figure 3.15 Cost of a one-hour interruption as a function of when it occurs, as determined by surveys and interviews with 300 customers of a utility in the northeastern United States, was determined on a three-hour period basis. A high proportion of households in this survey have school age children at home and thus perhaps weighed interruption costs outside of school hours more than during school hours. However in general, most households rate interruption cost is higher in early morning and early evening.

Recommended Method of Application of Customer Interruption Cost Data

As mentioned earlier, the recommended analytical approach to value-based planning of power delivery includes assessment of customer costs using functions that acknowledge both a fixed cost for any interruption, no matter how brief, and a variable cost as a function of duration. It is also important to acknowledge the differences in value of service among the different customer classes, and the time differences within those customer classes. In the case of large industrial customers, interruption cost can be significant enough that specific site studies of the customer's location, T&D system, and delivery history can be justified.

Ideally, reliability and service quality issues should be dealt with using a value-based reliability or power-quality planning method with customer interruption cost data obtained through statistically valid and unbiased sampling of a utility's own customers (see Sullivan et al.). Not with data taken from a reference book, report, or technical paper describing data on another utility system. However, in many cases for initial planning purposes the cost and time of data collection are not affordable.

Figure 3.18 provides a set of costs of typical interruption curves that the author has found often match overall customer values in a system. It is worth stressing that major differences can exist in seemingly similar utility systems, due to cultural and economic differences in the local customer base. These are not represented as average, or "best" for use in value-based planning studies, but they are illustrative of the type of cost functions usually found. They provide a guide for the preparation of approximate data from screening studies and initial survey data.

Cost in Figure 3.18 is given in terms of "one hundred times nominal price." *It is worth noting that in many surveys and studies of interruption cost, the cost per kW of interruption is on the order of magnitude one hundred times the normal price (rate) for a kWh.* Generally, if a utility has low rates its customers report a lower cost of interruption than if it has relatively higher rates. No reliable data about why this correlation exists has been forthcoming.[3]

[3] It could be that value of continuity is worth more in those areas where rates are high (generally, more crowded urban areas). However, the authors believe that a good part of this correlation is simply because in putting a value on interruptions, respondents to surveys and in focus groups base their thinking on the price they pay for electricity. Given that a typical residential customer uses roughly 1,000 kWh/month, they may simply be valuing an interruption as about "one tenth of my monthly bill."

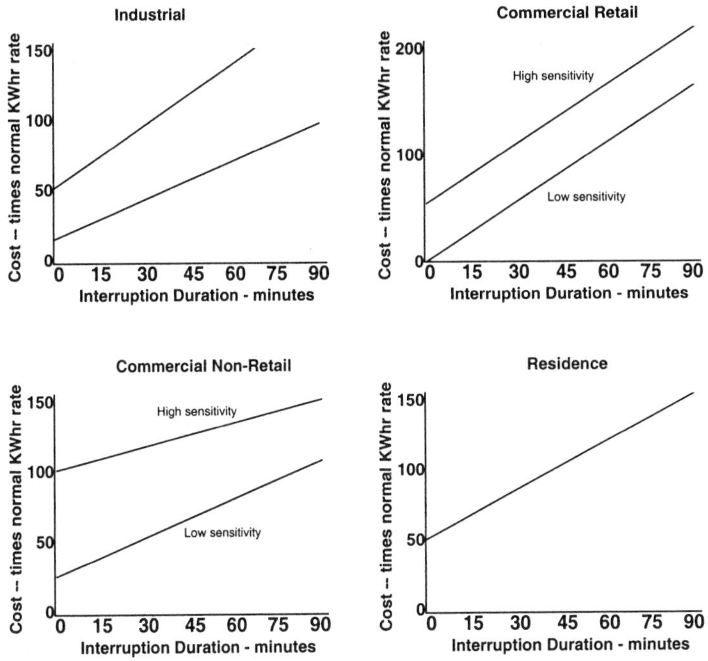

Figure 3.16 Typical interruption cost characteristics for customer classes.

Cost of Surges and Harmonics

Far less information is available on customer costs of harmonics and voltage surges as compared to that available on the cost of interruptions. What data is made available in publications, and most of the reported results and interpretations in the technical literature, were obtained in very customer-specific case studies, most done on a single customer basis. As a result there is very little information available on average of typical costs of voltage surge sensitivity and harmonics sensitivity.

Few customers suffer from voltage dip (voltage problems lasting minutes or hours), surge, harmonics, electrical noise, and similar problems. For this reason most studies of cause, effect, and curve are done on a single-customer or specific area basis. Here, results reported in technical literature are the best guide.

End-Use Modeling of Consumer Availability Needs

The customer-class, end-use basis for analysis of electric usage, discussed in Chapters 2 and 15, provides a reasonably good mechanism for study of the service reliability and power quality requirements of customers, just as it provides a firm foundation for analysis of requirements for the amount of power. Reliability and power quality requirements vary among customers for a number of reasons, but two predominate:

1. *End-usage patterns differ.* The timing and dependence of customers' need for lighting, cooling, compressor usage, hot water usage, machinery operation, etc., varies from one to another.

2. *Appliance usage differs.* The appliances used to provide end-uses will vary in their sensitivity to power quality. For example, many fabric and hosiery manufacturing plants have very high interruption costs purely because the machinery used (robotic looms) is quite sensitive to interruption of power. Others (with older mechanical looms) put a much lower cost on interruptions.

End-use analysis can provide a very good basis for detailed study of power quality needs. For example, consider two of the more ubiquitous appliances in use in most customer classes: the electric water heater and the personal computer. They represent opposite ends of the spectrum from the standpoint of both amount of power required and cost of interruption. A typical 50-gallon storage electric water heater has a connected load of between 3,000 and 6,000 watts, a standard PC a demand of between 50 and 150 watts. Although it is among the largest loads in most households, an electric water heater's ability to provide hot water is not impacted in the least by a one-minute interruption of power. In most cases a one-hour interruption does not reduce its ability to satisfy the end-use demands put on it.[4] On the other hand, interruption of power to a computer, for even half a second, results in serious damage to the "product." Often there is little difference between the cost of a one-minute outage and a one-hour outage.

It is possible to characterize the sensitivity of most end-uses in most customer classes by using an end-use basis. This is in fact how detailed studies

---

[4] Utility load control programs offer customers a rebate in order to allow the utility to interrupt power flow to water heaters at its discretion. This rebate is clearly an acceptable value for the interruption, as the customers voluntarily take it in exchange for the interruptions. In this and many other cases, economic data obtained from market research for DSM programs can be used as a starting point for value analysis of customer reliability needs on a value-based planning basis.

of industrial plants are done in order to establish the cost-of-interruption statistics, which they use in VBP of plant facilities and in negotiations with the utility to provide upgrades in reliability to the plant. Following the recommended approach, this requires distinguishing between the fixed cost (cost of momentary interruption) and variable cost (usually linear-ized as discussed above) on an end-use basis.

A standard end-use model used to study and forecast electric demand can be modified to provide interruption cost sensitivity analysis, which can result in "two-dimensional" appliance end-use models as illustrated in Figure 3.17. Generally, this approach works best if interruption costs are assigned to appliances rather than end-use categories. In commercial and industrial classes different types of appliances within one end-use can have wildly varying power reliability and service needs. This requires an "appliance sub-category" type of an end-use model. Modifications to an end-use simulation program to accommodate this approach are straightforward (see Willis, 1996, Chapter 11), and not only provide accurate representation of interruption cost sensitivity, but produce analysis of costs by time and location, as shown in Figures 3.20 and 3.21.

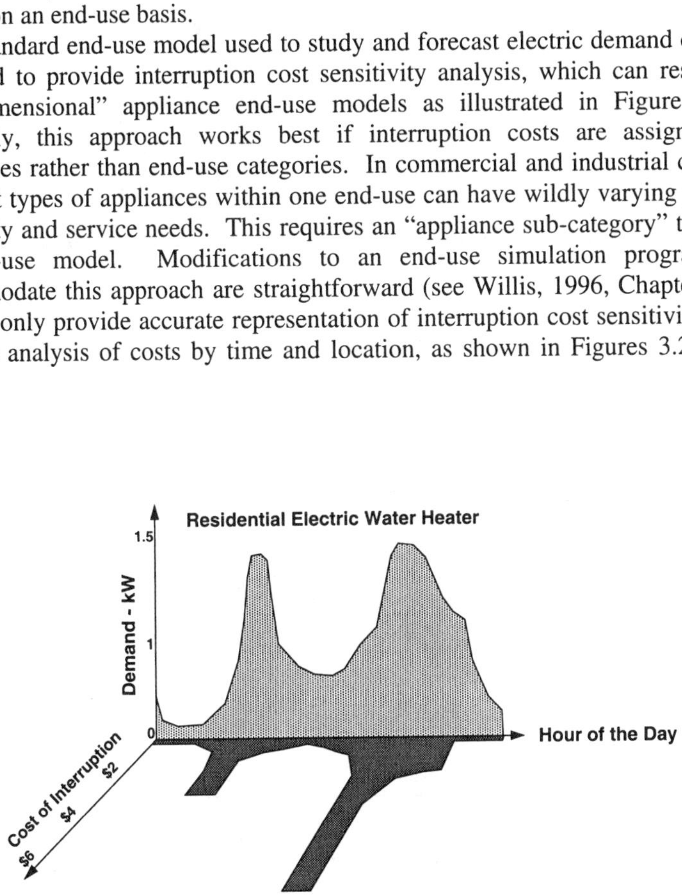

Figure 3.17 The simulation's end-use model is modified to handle "two-dimensional" appliance curves, as shown here for a residential electric water heater. The electric demand curve is the same data used in a standard end-use model of electric demand. Interruption cost varies during the day, generally low prior to and during periods of low usage and highest prior to high periods of use (a sustained outage prior to the evening peak usage period would result in an inability to satisfy end-use demand).

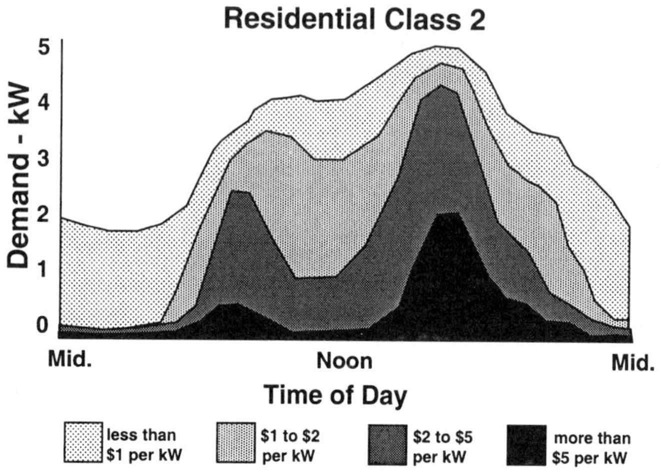

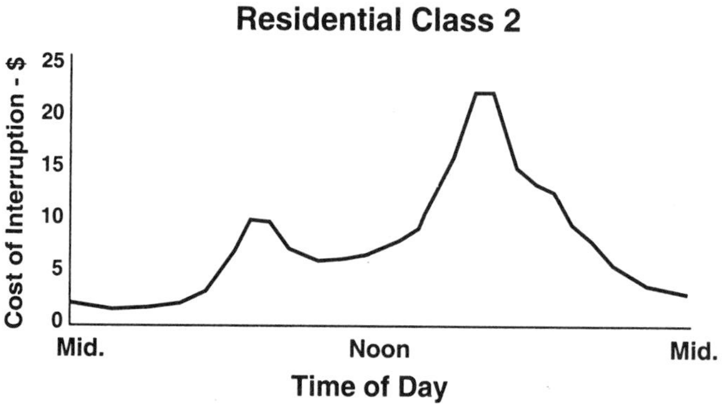

Figure 3.18 Result of a power quality evaluation, using an end-use model. Top: the daily load curve for single family homes segmented into four interruption-cost categories. High-cost end-uses in the home are predominantly digital appliances (alarm clocks, computers) and home entertainment and cooking. The bottom figure shows total interruption cost by hour of the day for a one-hour outage - compare to Figure 3.19.

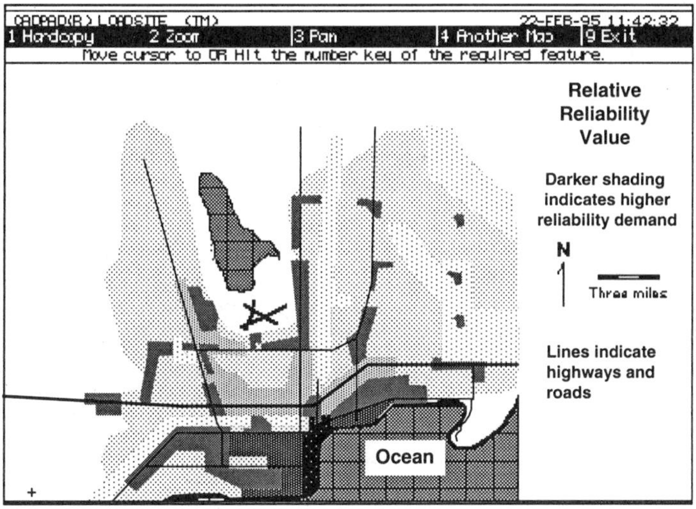

Figure 3.19 Map of average reliability needs computed on a 10-acre small area grid basis for a port city of population 130,000, using a combination of an end-use model and a spatial customer simulation forecast method, of the type discussed in Chapter 15. Shading indicates general level of reliability need (based on a willingness-to-pay model of customer value).

3.4 Two-Q Analyses: Quantity and Quality of Power Are Both Important Aspects of Consumer Value

The amount of power used, and the dependability of its availability for use is both key attributes in determining the *value* of the electric power to each consumer. The values attached to both quantity and quality by most consumers are linked, but somewhat independent of the value attached to the other. In order to demonstrate this, it is worth considering two nearly ubiquitous appliances, which happen to represent opposite extremes of valuation in each of these dimensions: the electric water heater and the personal computer.

A typical residential storage water heater stores about 50 gallons of hot water and has a relatively large connected load compared to most household appliances: about 5,000 watts of heating element controlled by a thermostat. Its contribution to coincident peak hourly demand in most electric systems is about 1,500 watts. The average heater's thermostat has it operating about 30% of the time (1,500/5,000) during the peak hour.

Despite its relatively high demand for quantity, this appliance has a very low requirement for supply quality. A momentary interruption of electric flow to this device – on the order of a minute or less – is literally undetectable to anyone using it at that moment. Interruptions of up to an hour usually create little if any disruption in the end-use the device provides, which is why direct load control of water heaters is, an "economy discount" option offered by many utilities. Similarly, the water heater is not critically dependent on other aspects of power quality. It will tolerate, and in fact turn into productive heat, harmonics, spikes, and other "unwanted" contamination of its supply.

By contrast, a typical home computer requires much less quantity, about 180 watts. But unlike the water heater, it is rendered useless during any interruption of its power supply. Just one second without power will make it cease operation. Battery backup power supplies, called *un-interruptible power supplies,* can be used to avoid this problem, but at an increase in the computer's initial and operating cost. In addition to its sensitivity to supply availability, a computer is also more sensitive to harmonics, voltage dips, and other power quality issues than a water heater.

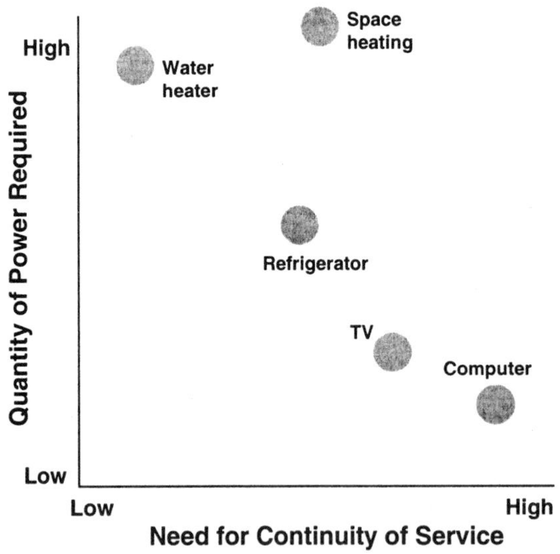

Figure 3.20 Electrical appliances vary in the amount of electricity they demand, and the level of continuity of service they require to perform their function adequately. Scales shown here are qualitative. Typically quantitative scales based on the most applicable customer values are used in both dimensions.

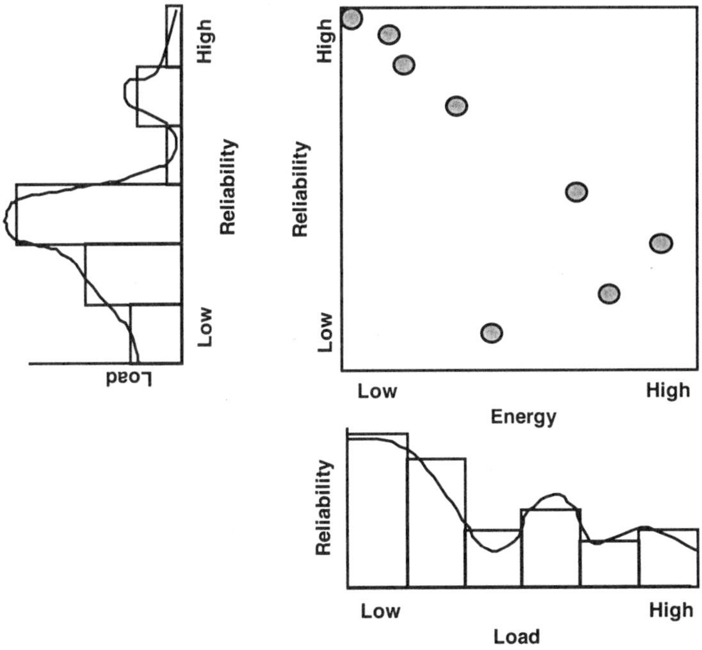

Figure 3.21 A Two-Q diagram of a consumer's needs can be interpreted in either of two dimensions to identify the total (area under the curve) and range of needs in either Q direction, producing Q profiles of the importance of demand, or reliability to his particular needs.

Figure 3.20 shows these two appliances, and several other household energy uses plotted on a two-dimensional basis according to the relative importance of both quantity and quality. This Two-Q plot is a type of two-dimensional demand and capability analysis that treats both Qs as equally important. Traditionally, "demand" has been viewed only as a requirement for quantity of power, but as can be seen the true value of electric service depends on both dimensions of delivery. Recognition of this fact has been the great driver in moving the electric industry toward "tiered rates," "premium rates," and other methods of pricing power not only by amount and time of use, but by reliability of delivery. Either or both dimensions in the plot can be interpreted as single-dimensional profiles for quantity or quality, as the case may be, providing useful information in the analysis of consumer needs and how to best meet them. (Figure 3.21).

Two-Q analysis using these same two dimensions also applies to the design of utility power systems, as will be discussed in Chapter 4 (section 4.4). Similarly dimensioned plots as in Figures 3.22 and 3.23 provide details on the capability of power systems to provide both quantity and quality of power, identification of shortfalls in capability against need, and information on how the system design can be altered to increase capability in either or both dimensions.

3.5 CONCLUSION AND SUMMARY

The whole purpose of electric utility systems is to serve the electric demands of individual households, businesses, and industrial facilities. Two useful mechanisms for evaluation of consumer requirements and planning for their provision are end-use analysis and Two-Q planning, both of which provide a useful basis for the study and analysis of both quantity and quality.

In addition to having a demand for electric energy, electric consumers value availability and quality of power too, seeking satisfactory levels of both the amount of power they can obtain, and the availability of its supply. Each customer determines what is satisfactory based on their specific needs. An interruption that occurs when a customer does not need power does not concern them. Harmonics or surges that make no impact on appliances are not considered of consequence. Among the most important points in customer value analyses are:

1. *Interruption of service* is the most widely affecting power quality problem. A majority of electric customers in most utility systems experience unexpected interruptions at least once a year.

2. *Voltage dips and sags create momentary interruptions* for many customers that are indistinguishable in impact on their needs with short-term interruptions of power flow.

3. *Analysis of customer needs for power quality is generally done by looking at the cost which less than perfect quality creates, not the value that near-perfect quality provides.* Thus, availability of power is usually studied by analyzing the cost incurred due to interruptions, and determining the costs created by those interruptions assesses harmonics.

4. *Power quality costs and values vary greatly* depending on customer class, time of day and week, region of the country, and individual characteristics.

5. *Price concerns often outweigh quality concerns.* Throughout the late 1980s and early 1990s, several independent studies indicated that

roughly "30%-40% of commercial and industrial customers are willing to pay more for higher power quality," or something similar, a statistic that seems quite believable and realistic. This led to considerable focus on improving power system reliability and quality. However, it is worth bearing in mind that these same studies indicate that 20-30% is quite content with the status quo, while the remaining 30-40% would be interested in trading some existing quality for lower cost.

REFERENCES

P. F. Albrecht and H. E Campbell, "Reliability Analysis of Distribution Equipment Failure Data," EEI T&D Committee Meeting, January 20, 1972.

R. N. Allan, et al., "A Reliability Test System for Educational Purposes – Basic Distribution System Data and Results," *IEEE Transactions on Power Systems,* Vol. 6, No. 2, May 1991, pp. 813-821.

R. E. Brown, and J. R. Ochoa, "Distribution System Reliability: Default Data and Model Validation," paper presented at the 1997 IEEE Power Engineering Society Summer Meeting, Berlin.

"Guide for Reliability Measurement and Data Collection," EEI Transmission and Distribution Committee, October 1971, Edison Electric Institute, New York.

W. F. Horton, et al., "A Cost-Benefit Analysis in Feeder Reliability Studies," *IEEE Transactions on Power Delivery,* Vol. 4, No. 1, January 1989, pp. 446 - 451.

Institute of Electrical and Electronics Engineers, *Recommended Practice for Design of Reliable Industrial and Commercial Power Systems,* The Institute of Electrical and Electronics Engineers, Inc., New York, 1990.

A. D. Patton, "Determination and Analysis of Data for Reliability Studies," IEEE *Transactions on Power Apparatus and Systems,* PAS-87, January 1968.

N. S. Rau, "Probabilistic Methods Applied to Value-Based Planning," *IEEE Transactions on Power Systems,* November 1994, pp. 4082 - 4088.

4
Power System Reliability and Reliability of Service

4.1 INTRODUCTION

"Reliability" as normally applied by power distribution-utilities means continuity of service to the electric consumers. A 100% reliable power system provides power without interruption, power that is available all of the time regardless of how much and if the consumer is using it at any given moment. Almost anything less than perfection in this regard will garner complaints that the power supply is "unreliable." An availability of 99.9% might sound impressive, but it means eight and three quarter hours without electric service each year, nearly all electric consumers in North America would consider that a level of service unsatisfactory. That is why availability averages approximately 99.98%, or roughly 1.75 hours per year. Although even this level of performance is unacceptable to some customers.

This chapter reviews power delivery reliability concepts in general and introduces the key components used in reliability analysis. Sections 4.2 and 4.3 cover equipment outages and service interruptions, the reliability indices used to evaluate reliability performance, and their application to electric utility systems. Section 4.4 covers various ways of setting reliability standards and design criteria. Section 4.5 discusses the cost of reliability, and the various types of tradeoffs made in design. Two-Q analysis of power systems is covered in section 4.6. Section 4.7 concludes the chapter with a summary of the most important concepts in the chapter.

Table 4.1 Reliability Terms and Definitions

Availability:	The fraction of the time that service is available. The (steady-state) probability that power will be available.
Duration:	The total elapsed time of an interruption or outage.
Failure rate:	The average number of failures of a component or unit of the system in a given time (usually a year).
Failure probability:	The probability that a unit or system will be in a "failure mode" at a time selected at random.
Frequency:	How often interruptions or outages, as the case may be, occur.
Interruption:	A cessation of service to one or more customers, whether power was being used at that moment or not.
Mean time between failures (MTBF):	The average or expected time a unit of equipment is in service between failures.
Mean time to repair (MTTR):	The average or expected time to repair an outaged component or unit of the system once failed.
Outage:	The unavailability of part of the power supply system, whether due to unexpected or planned circumstances.
Forced outage:	The outage due to unexpected and unscheduled event (usually a failure or impending failure).
Scheduled outage:	An outage that was planned and scheduled in advance (usually Maintenance).
Reliability:	The fraction of time that a unit of equipment or portion of the system is capable of performing the required function. The (steady state) probability that it will be in service in a condition where it can function as required.
Reporting period:	The period of time over which reliability, interruption, and outage statistics have been gathered or computed.
Restoration:	Return of electric service after an interruption, because of repair of the outage causing the interruption, or because of re-switching of the supply source, or the starting of an alternate DG source.
Restoration time:	The time required restoring service to an interrupted customer or portion of the system.
Service:	Supply of electric power in *sufficient amount and quality* to satisfy customer demand.

4.2 OUTAGES CAUSE INTERRUPTIONS

Table 4.1 provides a list of key reliability definitions. Among the most important are "outage" and "interruption." Although often used interchangeably, they have far different meanings. "Outage" means a failure of part of the power supply system. For example, a line down, transformer out of service, or a breaker that opens when it shouldn't. "Interruption" means a cessation of service to one or more customers. Interruptions are almost always caused by outages, and this leads many people to confuse the two words and their meanings. However, to understand power supply reliability properly, and more importantly, to set criteria and design a system to achieve high reliability, the distinction between outage (the cause of service problems) and interruption (the result) is critical. Electric service to a customer is interrupted whenever equipment in the line of supply to that customer fails or is otherwise out of service. Interruptions are caused by outages, but not all outages cause interruptions.

Frequency and Duration

Two different aspects of reliability receive attention in any type of power supply reliability analysis, whether of interruptions or outages. These are the *frequency* (how often something occurs) and *duration* (how long it lasts). With respect to interruption, frequency refers to the number of times service is interrupted during a period of analysis – once a decade, two times a year, five times a month, or every afternoon. Duration refers to the length of interruptions - some last only a few cycles, others for hours, even days.

Extent

Frequency and duration are important factors in reliability analysis and design, but there is a third factor equally important to the electric service planner because it is the one over which the planner has considerable control, particularly in the design phase. The *extent* of an outage's impact on service – how many customers or what portion of the plant's loads are interrupted by the outage of a particular unit of equipment – is the key to the outage-interruption relationship. The design of the electric power supply system greatly influences just how much interruption occurs when an equipment outage occurs.

The Distribution Planner has control over extent to which outages lead to customer interruptions. Among other things it is a function of how the distribution system is laid out. In general, a system laid out to minimize extent of outages from any failure will be more expensive than one that is not, but for the planner this means that consideration of configuration and its influence on reliability is just one more alternative to be considered in planning. In many cases a higher-reliability layout may be less expensive than any other way of achieving the particular reliability goal.

Most distribution systems use some form of radial feeder and secondary layout, which means that a single failure anywhere in the electrical path between substation and customer will cause service interruptions to all customers downstream. One way to improve reliability in such systems is to lay out the feeder trunk and branches, and arrange the protection (fusing) so that outages have a low extent (i.e. so that the average feeder segment failure does not interrupt service to very large numbers of customers).

Types of Interruptions

Interruptions are classified by their duration. Table 4.2 lists typical interruption definitions (see IEEE Gold Book for more information), with the times given being averages obtained by the authors in a survey of definitions and usage throughout the power industry. Electric service planners should keep in mind that practice and interpretation of reliability definitions vary greatly from industry to industry, from utility to utility, and from engineer to engineer.

In the authors' opinion, the definitions shown are all somewhat unsatisfactory. The names imply that they relate to interruption duration, yet they are defined by equipment and operating factors, such as whether an interruption is restored automatically or manually. This, along with differences in exactly how the definitions are interpreted, means that the application of a particular definition, for example "momentary interruption," varies greatly both in definition and therefore in *how it is used.*

Table 4.2 Traditional Classification for Power Supply Interruptions

Type	Definition
Instantaneous:	An interruption restored immediately by completely automatic equipment, or a transient fault that causes no reaction by protective equipment. Typically less than 15 sec.
Momentary:	An interruption restored by automatic, supervisory, or manual switching at a site where an operator is immediately available. Often defined as less than three minutes, but a proposal by an IEEE Task Force suggests standardizing on five minutes.
Temporary:	An interruption restored by manual switching by an operator who is not immediately available. Typically, 30 minutes.
Sustained:	Any interruption that is not instantaneous, momentary, or temporary. Normally more than an hour.

Table 4.3 Duration Limits Used to Define Instantaneous, Momentary, and Temporary Outages at Eight Utilities

Utility	Instantaneous	Momentary	Temporary
1	not used	< 5 minute	5 min. to 45 min.
2	< 15 sec.	15 sec. to 1 min.	1 min. to 45 min.
3	not used	up to 30 sec.	30 sec. to 30 min.
4	< 15 sec.	15 sec. to 2 min.	2 min. to 1 hour
5	< 60 sec.	60 sec. to 5 min.	5 min. to 2 hours
6	not used	< 1 min.	1 min. to 52 min.
7	< 1 min.	not used	1 min. to 2 hours
8	"cleared by inst. relay"	up to 2 min.	1 min. to 1 hour

Perhaps most importantly, since these distinctions mean nothing to electric consumers, they do little to help an electric service planner improve quality from the consumer's perspective. A four-minute interruption is a four-minute interruption, whether classified as momentary or temporary or permanent. Beyond the distinctions shown in Table 4.2, interruptions are classified into scheduled interruptions (the utility scheduled the interruption for maintenance purposes) or forced interruptions (the interruption was not expected and not scheduled). Implicit in these definitions, and the most important factor in the authors' minds, is whether customers are informed reasonably far in advance of the details of any "scheduled interruption."

Every planner should pay close attention to the definitions used by his company and by utilities and other players in any planning situation of collaboration. What appear to be identical indices can vary because of differences in how interruptions are defined. Many utilities do not include scheduled outages, storms, and other "non-failure-related" interruptions in their reporting procedures.[1] Others leave momentary interruptions out of reported averages, or report them as a special category.

Every power system and DG Planner should bear in mind when working with reliability statistics: that utilities that report a certain type of interruption statistic, such as number of momentary operations, vary widely in how they interpret that and other definitions. This is shown by Table 4.3, which lists interruption definitions from eight utilities in North America. They should also

[1] Most utilities remove "storms" from their statistics, arguing that natural disasters should be removed from statistics that are meant to measure their performance. However this practice has been carried to extremes and abused by some utilities. As Samuel Clemmens once said, "There are three kinds of lies: Lies, damned lies and statistics."

interruption definitions from eight utilities in North America. They should also remember that utility customers and electric consumers do not really care about definitions and nomenclature. For them, an "outage" or "interruption" is anything that causes their equipment and facilities to stop working.

Such differences in definitions, interpretations, and reporting practices can frustrate planners who try to find some real meaning in comparisons between the reliability figures reported by different electric utility companies or by different industrial facilities. Worse, these differences often cause confusion and frustration among customers, regulators, and standards organizations.

Voltage Sags Often Look Like Interruptions

Many electric consumers report a momentary or "very short' interruption when all that really occurred was a brief "voltage sag". For example, during a fault (short circuit) on a distribution line, voltage on nearby circuits might drop to less than 70% of normal. Typically the fault would be cleared within less than a second, but this "blink" may be sufficient to halt the operation of computers, robotic equipment, and other digital systems. In that case, this short "voltage sag" is indistinguishable from a short, total cessation of voltage (an interruption). Either way, some critical equipment stops functioning.

The Computer and Business Equipment Manufactures Association (CBEMA) has a recommended guideline on what voltage sags and surges should be tolerable to office and business equipment, as shown in Figure 4.1. The "CBEMA envelope" establishes a relationship between degree of voltage deviation and time.

The curve's application is simple: if an event's voltage deviation and duration characteristics are within the CBEMA envelope, then normal appliances should operate normally and satisfactorily despite it. But many appliances and devices in use are more sensitive than specified by the CBEMA curves, and will not meet this criterion at all. Others may fail to meet it under the prevailing electrical conditions (i.e., line voltage, phase imbalance, power factor, and harmonics may be less than perfect).

The manner of usage of an appliance also affects its voltage sag sensitivity. The small circle in Figure 4.1 indicates an event that "crashed" a local net server (computer) at a hosiery plant. The event falls just within the CBEMA curve, and the computer's manufacturer probably intended for it to withstand that particular deviation-time combination. However, the computer in question had been upgraded with three times the standard memory, a larger hard drive, and optional network cards, doubling its power usage and the load on its power supply. Such situations are common and mean that power systems that deliver voltage control within recommended CBEMA standards may still create problems for computers, robotic assembly tools, and other sensitive equipment.

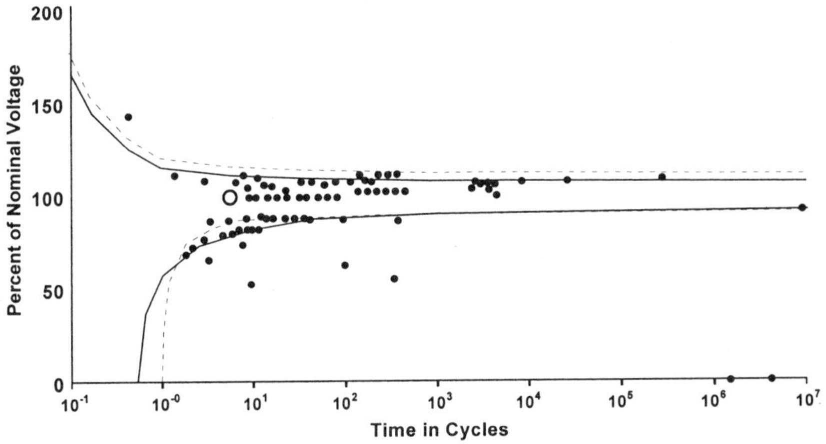

Figure 4.1 Data from 67 disturbances which occurred over a two-year period at a clothing factory, plotted against the CBEMA curve (solid line) and the actual requirement envelope of a digitally-controlled hosiery loom (dotted line). In many systems, about 40% of all voltage sags and 10% of all voltage surges lie outside of the CBEMA envelope. Small circle refers to event mentioned in text.

4.3 RELIABILITY INDICES

In order to deal meaningfully with reliability as a design criterion for distribution, it is necessary to be able to measure it and set goals. A bewildering range of reliability indices is in use within the power industry. Some measure only frequency of interruption, others only duration. A few try combining both frequency and duration into a single value, which proves to be nearly an impossible task. Some measures are system-oriented, looking at reliability averaged over the entire customer base. Others are customer or equipment-oriented, meaning that they measure reliability only with regard to specific sets of customers or equipment. To complicate matters further, most utilities use more than one reliability index to evaluate their reliability. The large number of indices in use is compelling evidence that no one index is really superior to any other. In fact, the authors are convinced that no one index alone is particularly useful.

The basic problem in trying to measure reliability is how to relate the two quantities, frequency and duration. Are two one-hour interruptions equivalent to one two-hour interruption, or are two one-hour interruptions twice as bad as the one two-hour interruption? Most people conclude that the correct answer lies somewhere in between, but no two electric service planners, and no two electric consumers for that matter will agree on exactly how frequency and duration "add up" in importance. For one thing, the importance of frequency and duration vary tremendously from one electric consumer to another. There are some industries where a one-minute interruption of power will cause over ninety minutes of lost productivity. Computer and robotic systems must be reset and restarted, or a long production process must be cleared and bulk material reloaded, before the plant can restart at the beginning of a production cycle. For these types of consumers, five one-minute outages are much more serious than a single outage, even if it is five hours duration.

But there are other customers for whom short outages cause no significant problems, but who experience inconvenience during a sustained outage. This category includes factories or processes such as pulp paper mills, where production is only damaged if equipment has time to "cool down." Interruption of power to a heater or tank pressurized for a minute or two is not serious (although it might be inconvenient), but sustained interruption allows a cool-down or pressure drop that results in lost production.

Over the last two decades of the 20th century, frequency of interruption became increasingly important to a larger portion of the electric power market. Before the widespread use of digital clocks and computerized equipment, few residential or commercial customers cared if power was interrupted briefly (for a few minutes at most) in the early morning (e.g., 3:00 AM) while the electric utility performed switching operations. It was common practice for utilities to perform minor maintenance and switching operations during that period to minimize any inconvenience to their customers. Today, this practice leads to homeowners who wake up to a house full of blinking digital displays, and wake up later than they expected, because their alarm clocks did not operate.

The amount of power interrupted is also an important factor. Some reliability calculations and indices are weighted proportionally to customer load at the time of interruption, or to the estimated energy (kWh) *not* delivered during the interruption. However, the four most popular indices used in reliability analysis make no distinction of demand size. They treat all customers in a utility system alike regardless of peak demand, energy sales, or class. These indices are SAIFI and CAIFI, which measure only frequency, and SAIDI and CTAIDI, which measure only duration. Usually these four are used in conjunction – four numbers that give a rough idea of what is really happening to reliability system-wide.

These four indices, and most others, are based on analysis of customer interruptions during some *reporting period,* usually the previous month or year. All count a customer interruption as the interruption of service to a single customer. If the same customer is interrupted three times in a year, that constitutes three customer interruptions. If one equipment outage caused simultaneous interruption of service of three customers, that too is three customer interruptions.

System Average Interruption Frequency Index (SAIFI) is the average number of interruptions per utility customer during the period of analysis.

$$\text{SAIFI} = \frac{\text{number of customer interruptions}}{\text{total customers in system}} \tag{4.1}$$

Customer Average Interruption Frequency Index (CAIFI) is the average number of interruptions experienced by customers who had at least one interruption during the period.

$$\text{CAIFI} = \frac{\text{number of customer interruptions}}{\text{number of customers who had at least one interruption}} \tag{4.2}$$

The "S" in SAIFI means it averages the interruption statistic over the entire customer base (the system), the "C" in CAIFI means it refers to only customers who experienced interruptions. Customers who had uninterrupted service during the period are precluded from CAIFI.
System Average Interruption Duration Index (SAIDI) is the average duration of all interruptions, obtained by averaging over all of the utility customers – those who had outages and those who did not.

$$\text{SAIDI} = \frac{\text{sum of the durations of all customer interruptions}}{\text{total customers in system}} \tag{4.3}$$

Customer Total Average Interruption Duration Index (CTAIDI) is the average total duration of all interruptions, but averaged only over the number of utility customers who had at least one outage.

$$\text{CTAIDI} = \frac{\text{sum of the durations of all customer interruptions}}{\text{number of customers who had at least one interruption}} \tag{4.4}$$

Momentary Interruption Index

Momentary Average Interruption Frequency Index (MAIFI) is the average number of momentary (and sometimes instantaneous) interruptions per utility customer during the period of analysis,

$$\text{MAIFI} = \frac{\text{number of customer momentary interruptions}}{\text{total customers in system}} \tag{4.5}$$

Often, momentary interruptions are not included in the SAIFI value reported by an electric utility. In such cases, total interruptions (from the consumer standpoint) are best estimated by the sum: SAIFI + MAIFI.

Load Curtailment Indices

In addition, a number of indices try to relate reliability to the size of customer loads, in order to weight the interruption of a large load more than a small load,

$$\text{Customer load curtailment} = \text{duration of outage} \times \text{kVA unserved} \qquad (4.6)$$

Customer Average Load Curtailment Index (CALCI) is the average kVA x duration interrupted per affected customer, per year

$$\text{CALCI} = \frac{\text{sum of all customer load curtailments}}{\text{number of customers who had at least one interruption}} \qquad (4.7)$$

Normal usage is often sloppy and informal

Even though the "S" in the terms SAIDI and SAIFI means, "averaged over the entire system," some utilities internally report "SAIDI" and "SAIFI" values by feeder. While strictly speaking, SAIDI and SAIFI are the same for all feeders (they are computed by being averaged over the entire system), it is understood that what they mean is the average annual duration and frequency for customers on that feeder. Similarly, engineers at these and other utilities often use the two terms when they mean "duration" and "frequency" of interruptions for an area of the system or a small group of customers.

Analysis Using Reliability Indices

Reliability indices are used to evaluate historical and recent data to reveal trends and patterns, expose problems, and reveal how and where reliability can be improved. Any of the reliability indices discussed above can be tracked over time to identify trends that indicate developing problems. Figure 4.5 shows frequency of customer interruption (SAIFI) for one suburban operating district of a metropolitan utility on an annual basis over a 35-year period. The operating district was open farmland (no tree-caused trouble) up to 1962. At that time suburban growth of a nearby metropolitan region first spilled into the area, with much construction and with it the typically high interruption rates characteristic of construction areas (trenchers digging into underground lines, early equipment failures of new distribution facilities, etc.).

Over the following ten years, as the area plotted in Figure 4.5 gradually filled in with housing and shopping centers and construction and system additions stabilized, frequency of interruption dropped. Twenty years after the area first began developing; interruption rate again began to rise because of two factors:

1. Gradually increasing cable failure rates due to age.
2. The effects of trees, planted when the area first developed, reaching a height, where they brushed against bare conductors on overhead lines.

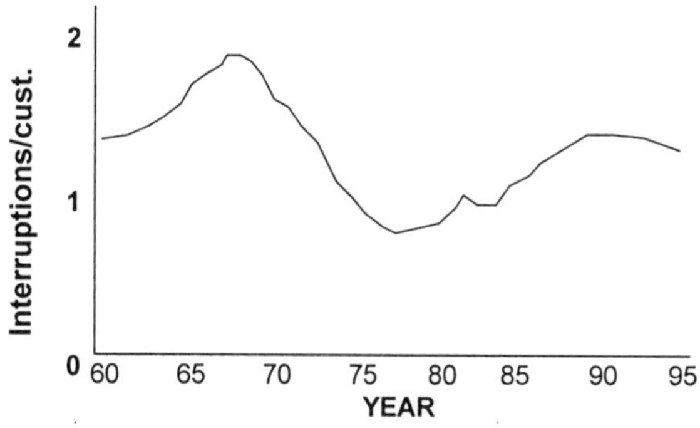

Figure 4.2 Three-year moving average of the annual number of interruptions for a region of a utility in the central United States. This area grew from open farmland into a developed metropolitan suburb in the period from 1962 to 1970, during which time interruption rate rose due to the problems common to areas where new construction is present (digging into lines, initial above-average failure rates for newly installed equipment). Twenty years later, interruption rate began to rise again, a combination of the effects of aging cable (direct buried laterals in some residential areas) and trees, planted when the area was new, finally reaching a height where they brushed against overhead conductor. The anomaly centered on 1982 was due to a single, bad storm.

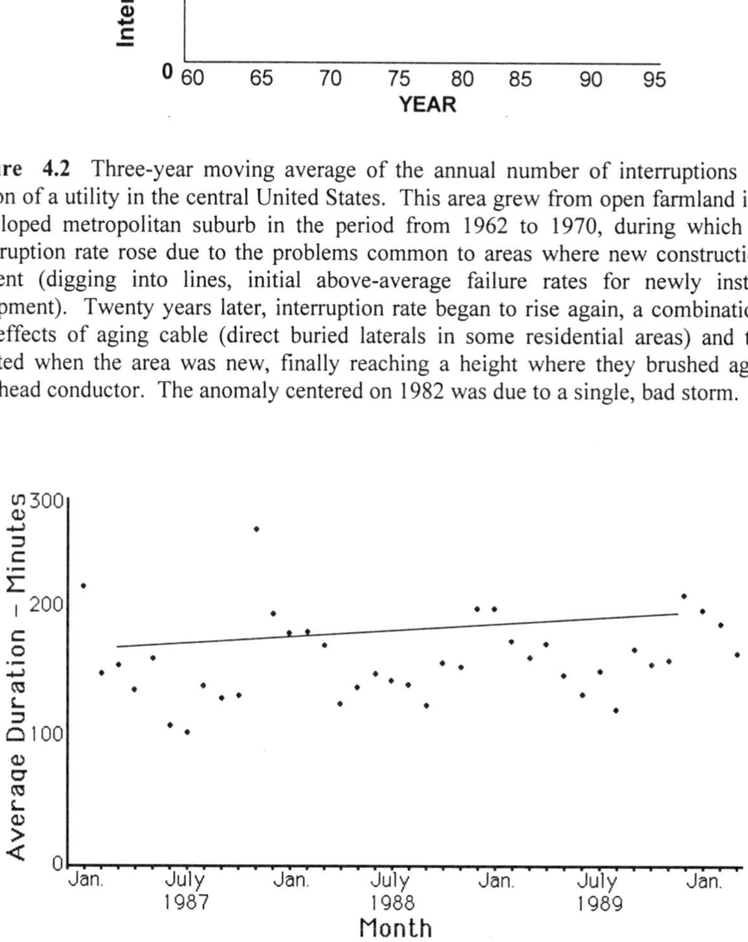

Figure 4.3 Average hours out of service per interruption on a monthly basis for a rural utility in the northern United States varies due to random fluctuations and changes in weather. It appears that duration of interruptions (dotted line) is slowly increasing over time, but further study is necessary before this conclusion can be confirmed.

Another obvious comparison to look at is the average duration of an outage, which is simply the total duration of all interruptions divided by the total number of interruptions that occurred during a period. Average duration of interruption provides an indication of how the utility is managing the entire process from trouble call to service restoration, as shown in Figure 4.3. The increase in outage duration during winter for this utility in the northern United States is clearly visible (winter storms stress repair resources and make travel to repair sites lengthier). A trend upward is discernible. Without additional study, it is not possible to determine whether this is due to an increase in the number of outages, a change in type (perhaps "serious" outages are occurring more often as equipment ages), or a less effective (slower) repair process.

4.4 RELIABILITY AND CONTINGENCY CRITERIA FOR PLANNING

Indirect Reliability Engineering: Contingency and Margin Criteria

The vast majority of electric utilities engineer reliability of service into their system indirectly, through the use of *contingency design criteria.* The utility maintains reliability goals and still measures and reports SAIDI, SAIFI and similar indices, but these values are not used directly as goals in the design process. Rather, adequate performance is achieved by applying contingency margin criteria which have been interpreted from those SAIDI and SAIFI goals and which are hoped to be appropriate. For example, distribution system engineers will apply equipment loading standards, design and layout guidelines, and switching and fusing rules, which limit the extent of outages and afford extra capacity to pick up outaged load by alternative means.

Contingency margin standards are designed to allow sufficient leeway for any unit of equipment to pick up load interrupted due to the outage of a nearby unit, and to permit reasonably quick (e.g., 15 or 30 minutes to an hour) switching. For example, a utility may establish a policy saying that its substations will always have two transformers of equal size each loaded to no more than 80% of its rating. During the outage of either transformer, the other can pick up both loads with an overload of 60%, an amount it should be able to sustain for two to four hours, time enough, hopefully, for the utility to take steps to handle the problem over a longer period if necessary. Similarly, a utility may ensure that every feeder will have three switch-able portions, or zones, each of which can be divided and transferred to other feeders if its primary feed is out of service.

Such a policy establishes a certain type of design (two transformers per substation), loading standard (a transformer cannot exceed 80% of its rating) and feeder layout (three switch-able zones) in order to leave sufficient margin for equipment failure and switching. Similar and compatible criteria are established for conductor loading, switch placement, and other design practices, all rules and standards that *imply* certain reliability results.

Contingency-based Planning Methods have Weaknesses for Aging Infrastructure Planning

This type of engineering, called *contingency-based engineering,* is heavily involved with the entire aging infrastructure issue. Contingency based engineering has sensitivities to high utilization rates for system equipment, and other factors that make it not as dependable as it was traditionally with respect to use as a necessary and sufficient design criteria. Chapter 8 will discuss this is considerably more detail, with examples.

Direct Reliability Engineering

By contrast, reliability engineering achieves the targeted level for availability in the system by actually computing the expected reliability of the power system based on detailed analysis of its equipment and configuration. Areas of equipment that do not meet the criteria are re-engineered until they do. In this approach, frequency and duration of service interruption are computed for each potential alternative design and compared to planning reliability targets to assure compliance with all new designs, additions, and operating changes. Use of reliability values directly as design criteria is not a widely used practice at the time of this writing. Instead, most utilities still use indirect criteria based on contingency margin, switching zone counts, etc., as described above. However, it can be expected that in the future many will choose reliability directly as a criterion for three reasons:

1) It addresses reliability needs directly from the consumer standpoint, rather than indirectly from the utility (equipment criteria standpoint), leading to improved quality assurance.

2) It imposes no restrictions on design or equipment selection other than the final result meet criteria. This can lead to lower costs.

3) Dependable, proven methods to accommodate reliability-based design are becoming widely available.

Duration Is the Most Popular Criterion

Regardless of whether an electric utility tries to ensure adequate customer service availability by using indirect or direct reliability engineering methods, almost all maintain specific reliability targets as their performance goal. In order of importance or popularity as a design rule, the most widely used are:

SAIDI = the expected total time in a year that the average customer in the area being studied will be without available power

MaxD = the maximum expected total time in a year that any customer in the area being studied will be without available power

SAIFI = the expected number of times in a year an average customer in the system or area being studied will have power availability interrupted

MaxF = the expected maximum number of times in a year that any customer in the area being studied will have power availability interrupted

These reliability targets are applied within any study or design area as general targets that must be satisfied, much like standards of voltage drop, power factor, etc. In the design of any feeder, substation layout, or other addition to the distribution system, the planners use these criteria as design constraints. Any acceptable plan must have an expected average frequency and duration of interruption equal to or better than the corporate goals. Otherwise, the plan would dilute the system's overall reliability. In addition, the expected maximum frequency and duration for any customer covered in the planning case must be less than MaxF and MaxD.

In general, Distribution Planning studies focus on small portions of the system, at most a few substations at a time. If every such area of the system is planned so that its frequency and duration of interruption fall close to the corporate SAIDI and SAIFI targets, reliability will be as evenly distributed as is practical throughout the system. The utility might still compute CTAIDI and CAIFI in its assessment of actual operating experience, to make certain results match its goals, but planners do not need to apply more than these four criteria.

Duration Is Equivalent to "Probability of Unavailable Power"

Some utilities use a simple "probability of being without power" factor as a design target. This is equivalent, in practice, to planning with duration of interruption as the target. For example, a criterion that calls for the outage probability to be no more than .025%, means that the customer will be without power no more than

$$8760 \text{ hours x } .025\% = 2.16 \text{ hours/year} \quad (4.8)$$

Reliability Targets for DG Planning

Generally, distributed generation is not "distributed" over a large system like a distribution system, but is focused on one or a small number of customers. Thus, SAIDI and SAIFI are not really meaningful statistics to use in DG planning. Their equivalents in DG studies are the expected duration and frequency of unavailability. Again, often planners simply compute the expectation of total hours of outage, using this as the figure of merit in planning.

Tiered Reliability Areas

In a few cases, a utility will designate different frequency and duration targets for urban and rural areas. For example, its reliability criteria may specify part of the service territory as "urban" with SAIFI and SAIDI targets of 1.0 interruptions and 50 minutes/year for the system within that region. The rest of the system would be characterized as "rural" and designated with SAIFI and SAIDI rates of 2.2 and 150 minutes respectively. Such a tiered reliability design structure merely recognizes a common and long-standing characteristic of most utility systems – service reliability is lower in sparsely populated areas than in densely populated ones.

This disparity between urban and rural standards does not imply that rural customers are less important, or is it due to inattention on the part of the utility or to poorer design practices applied to rural areas. The cost of reliability enhancement in rural areas is higher, because of the great distances involved, and the fact that the higher cost must be spread over fewer customers per mile of line. Most rate-design and regulatory processes do not consider it equitable to spread the cost of upgrading service in these rural areas to the same standards achievable in urban areas over the entire rate base (i.e., to those who live in the cities, too). Thus, a higher frequency of interruption, and longer interruptions in electric service, are among the expected pleasures of rural life.

Predictive Reliability Analysis Methods

For planning purposes, reliability analysis normally focuses on comparing the expected reliability performance of different candidate designs. Plans are evaluated and compared to determine which is less likely to fall short of required capacity. Reliability is only one aspect of planning evaluation: cost and other factors enter into the final decision too, often with reliability vs. cost being the major design compromise the planner must consider.

Table 4.4 Reliability Evaluation Methods

Method	Reference
Simple State Enumeration	Used here
Explicit State Enumeration	Brown et al
Monte Carlo Simulation	Patton et al
Cut-Set Analysis	Billinton
Adequacy Equivalent	Zhang and Billinton
Markov Chain Analysis	Patton
Multi-Criteria Analysis	Atanackovic et al

Failure: when capacity is less than load

Generally failure states for a power system include all those conditions in which the demand for power (load) at a site exceeds the available supply of power to that site. At the distribution level, failure conditions are generally easy to identify. Since distribution circuits are usually radial, a simple "loss of connectivity" analysis is sufficient to identify outages that lead to failure. For distributed generation, the situation is far more difficult to analyze. Given several generators at a site, service might be maintained even if one or two are out of service. Various combinations of units in service, and their probability of occurrence, have to be evaluated against load curve shape. Methods to do so will be demonstrated in Chapters 13 and 16. Reliability evaluation of an entire power system is more challenging still, due to the number of combinations of sources and paths. Even a small power system with less than 70 generators and 100 lines, can become a considerable challenge (Zhang and Billinton).

Reliability Evaluation Methods

Literally hundreds of technical papers, and quite a few books, have been devoted to power system reliability analysis (see For Further Reading at the end of this chapter). Table 4.7 provides a summary of the methods along with references on each. But despite this diversity in the technical, numerical means of approach, reliability analysis for planning falls into two functional categories:

1. *Historical Assessment Methods* analyze historical operating results to determine the cause of past reliability performance problems and to infer the base equipment failure rates and reliability factors. E.g., "Based on historical analysis overhead lines are likely to fail once

every five years and require an average of five hours to repair".

2. *Predictive Reliability Methods* estimate the reliability expected from a candidate design or system configuration. These are used by Planners to determine if a change in design, for example a "fix" to a reliability problem, will give sufficient availability of power.

Application of Probabilities

Almost all reliability prediction methods apply some form of analysis based on probabilities. A few rare methods based on adaptive pattern recognition do not. But the basis for what most methods apply as reliability, the way they apply it, and the detail and representations used in that application, vary greatly among methods. This makes it crucial to pick a method tailored to a situation's particular needs. For instance, some methods analyze the probability or the expected amount of time that power is unavailable at a particular point in the power network. They estimate the total amount of time that power will be unavailable (e.g., .03%, equal to 2.6 hours per year). Others predict the mean time between failures and the expected time to repair or restore power in each case.

The best of these methods can estimate both the expected average frequency and duration of interruption for any particular candidate system /DG design (e.g., power is expected to fail on average twice a year and take an average of 1.3 hours to restore each time). Still more complicated methods predict the expected probability for both frequency and duration of interruptions. E.g. power is expected to fail twice a year, but there is a 5% likelihood it could fail four times in a year - the average duration is expected to be 1.3 hours, but there is a 25% likelihood an interruption could last more than ten hours. But almost without exception, the more comprehensive methods require more data, involve more involved mathematical methods, and are more sensitive to details or implicit assumptions.

These various reliability prediction methods differ in the degree of approximation; the type of system they can analyze; and the type of reliability evaluation they provide; their complexity of analysis; and data requirements. In addition many methods are most appropriate for application to one type of system design (e.g., network, radial) or under certain conditions (only one source of supply but many available power transmission paths). The choice of reliability assessment method depends on all these factors, as well as the degree of detail sought by the planner.

In most places throughout this book, the authors will use a simple reliability assessment method that computes only the expected time power is unavailable, not the frequency and duration. This method is general enough for our examples, and much simpler. In many cases, merely adding together the examples, and much simpler. In many cases, merely adding together the probabilities of rare occurrences, rather than considering that they could overlap further approximates reliability analysis.

4.5 COST IS KING

Quantity and quality *both* have value for the electric consumer. But so does the consumer's money and as a result, no factor in the decision about energy source and energy usage is more important to most consumers than cost. Chapters 4 and 5 will deal with planning, cost evaluation, engineering economics and decision-making. The authors have included this short section to emphasize the point that quantity, quality, and above all, *cost,* form the basis for the market's evaluation of suppliers and technologies, and that a competitive, de-regulated electric industry will no doubt respond by offering options with respect to reliability and power quality.

The three dimensions of power – quantity, quality, and cost – form a "value cube" that determines the overall benefit consumers see from the use of electric power (Figure 4.4). The importance of each in determining what is "best" will vary from one consumer to another.

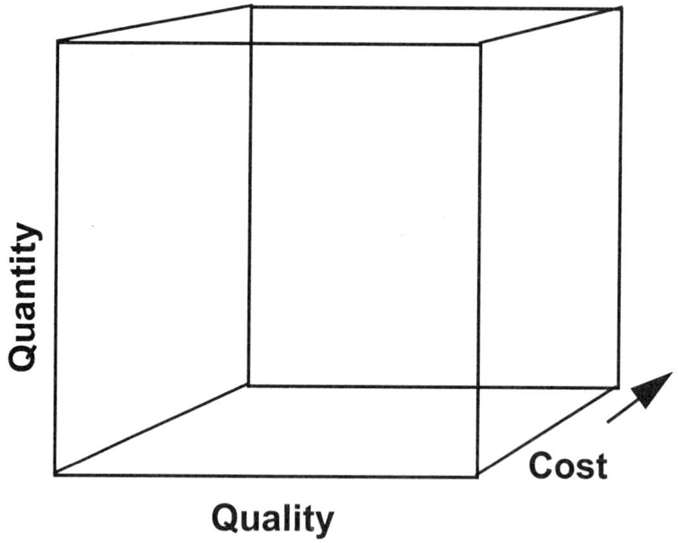

Figure 4.4 Ultimately, the value electric power consumers see in their electric power supply is based on a "value volume" of three dimensions: quantity, quality, and cost.

Electricity Pricing Has Traditionally Been Based Only On Quantity

Traditionally, the price of power in the electric utility industry was based solely upon the *quantity* used. A consumer would pay more if he or she used more power and less if their use was less. Except in rare and specially treated cases, an electric utility would offer a "one size fits all" level of quality. Customers got what was available, whether or not they needed and would have paid for higher quality, or whether they would have preferred a discount even if it meant somewhat lower quality.

The Market Comb

Numerous surveys and customer focus meetings conducted by utilities, consultants, and research firms (including the authors' firm) indicate that electric consumers differ widely in their need for, and their willingness to pay for reliability of service. Consumers also differ in exactly what "reliability," and in a broader sense, "quality," means to them, although availability of service, quick and knowledgeable response on the part of their supplier, and power of a usable nature are always key factors in their evaluation.

As a result, the electric power demand marketplace can be likened to a comb as shown in Figure 4.5. A series of small niches, that vary from the few customers who need very high levels of reliability, to those who do not need reliability, only power, and are motivated purely by lowest cost.

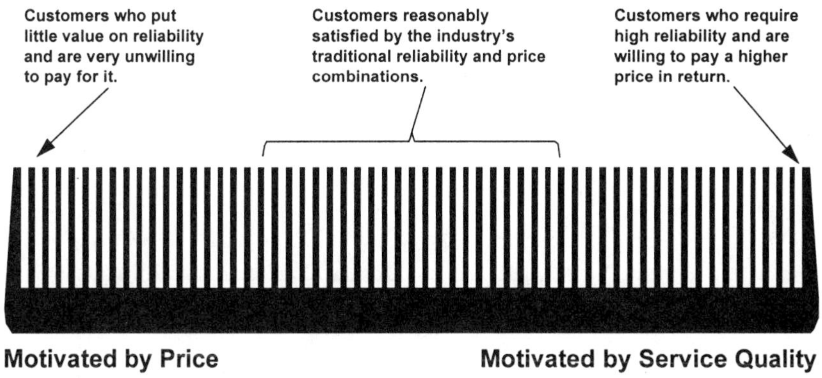
Customers who put little value on reliability and are very unwilling to pay for it.
Customers reasonably satisfied by the industry's traditional reliability and price combinations.
Customers who require high reliability and are willing to pay a higher price in return.
Motivated by Price
Motivated by Service Quality

Figure 4.5 The electric marketplace can be likened to a comb: composed of many small niches, each made up of customers who have a different need for and cost sensitivity to reliability of service. Even those who put a high value on reliability may differ greatly in how they define "good reliability," one reason why there are no broad market segments, only dozens of somewhat similar but different niches.

Offering Variable Levels of Reliability

Traditionally, electric utilities have not offered reliability as a price-able commodity. They selected equipment, and designed and engineered their systems based on engineering standards that were aimed at maintaining high levels of *power system equipment* reliability. These standards and methods, and the logic behind them, were actually aimed at minimizing utility equipment outages. The prevailing dogma maintained that this led to sufficiently high levels of reliable service for all customers and that reliable service was good. This cost of the reliability level mandated by the engineering standards was carried into the utility rate base. A utility's customers basically had no option but to purchase this level of reliability (Figure 4.10). Only large industrial customers, who could negotiate special arrangements, had any ability to make changes bypassing this "one-size-fits-all" approach to reliability that utilities provided.

The traditional way of engineering the system and pricing power has two incompatibilities with a competitive electric power marketplace. First, its price is cost-based; a central tenet of regulated operation, but contrary to the market-driven paradigm of deregulated competition.

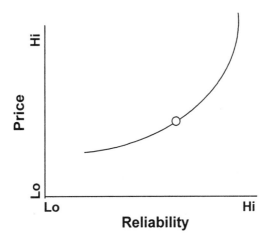

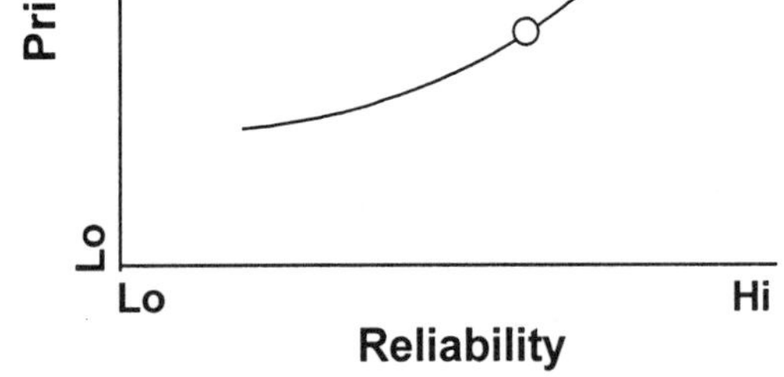

Figure 4.6 Generally, the cost of providing reliability in a power system is a non-linear function of reliability level: providing higher levels carries a premium cost, as shown by the line above. While reliability varies as a function of location, due to inevitable differences in configuration and equipment throughout the system, the concept utilities used was to design to a single level of reliability for all their customers.

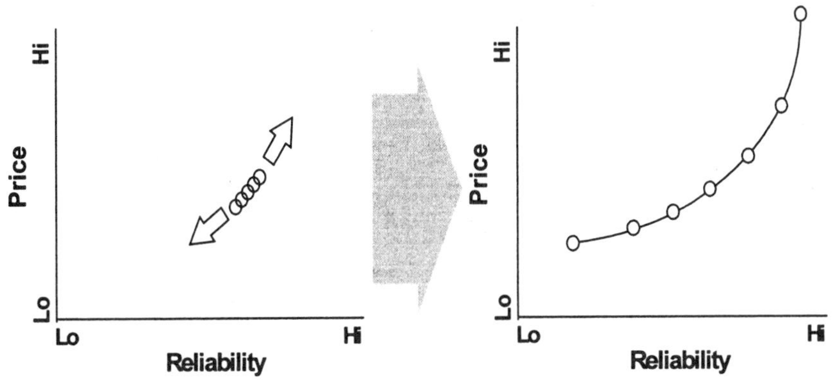

Figure 4.7 A competitive market in power will recognize the demand for various levels of reliability and providers will have to broaden the offerings they provide (left) until the range of options available to the customer covers the entire spectrum of capabilities, as shown at the right. Actual options offered will be even more complicated because they will vary in the type of reliability (e.g., perhaps frequency or duration is more important to a specific customer; seven days a week or only normal business hours. Compare to Figure 1.4.

Secondly, any "one-size-fits-all" reliability price structure will quickly evaporate in the face of competitive suppliers. Someone among the various suppliers will realize there is a profit to be made and a competitive advantage to be gained by providing high-reliability service to consumers willing to pay for it. While others will realize that there are a substantial number of customers who will buy low reliability power, as long as it has a suitably low price. The concept that will undoubtedly emerge (in fact, is already evolving at the time this book is being written) in a de-regulated market will be a range of reliability and cost options. The electric utility market will broaden its offerings, as shown in Figure 4.7.

For example, distributed generation (DG) interacts with the opportunities that the need for differentiated reliability creates in a de-regulated marketplace in two ways. First, this new marketplace will create an opportunity for DG, which can be tailored to variable reliability needs by virtue of various design tricks (such as installing more/redundant units). Secondly, DG is a "threat" that may force utilities to compete on the basis of reliability, because it offers an alternative, which can easily be engineered (tailored), to different reliability-cost combinations. Utilities may have no choice but to compete in pricing reliability.

TWO-Q ANALYSIS OF POWER SYSTEMS

Two-Q plots similar to those shown in Chapter 3, but oriented toward measurement of system capability rather than consumer needs, can be used to analyze the ability of a power system to meet projected customer demands for both quantity and quality of power. They provide a useful tool in determining both the appropriateness of the system design, and how to tailor it to best fit consumer needs.

Figure 4.8 shows a two-Q plot of the capability of a set of feeders in a utility system in the mid U.S. The area involved is part of a small town of 40,000 population and its industrial area. Over the period 1989 – 1998, the peak load in this area grew by more than 10% (from 19.5 to 21.7 MW). Planners for the utility, constrained by pressures to minimize capital cost, accommodated the load growth largely through a process of transferring switching and making small changes in the system. From a "load flow" (capacity, or quantity only) perspective the planning was successful - the utility figuring out a way to connect consumers with sufficient capacity to satisfy the peak load, at least when everything was working as expected.

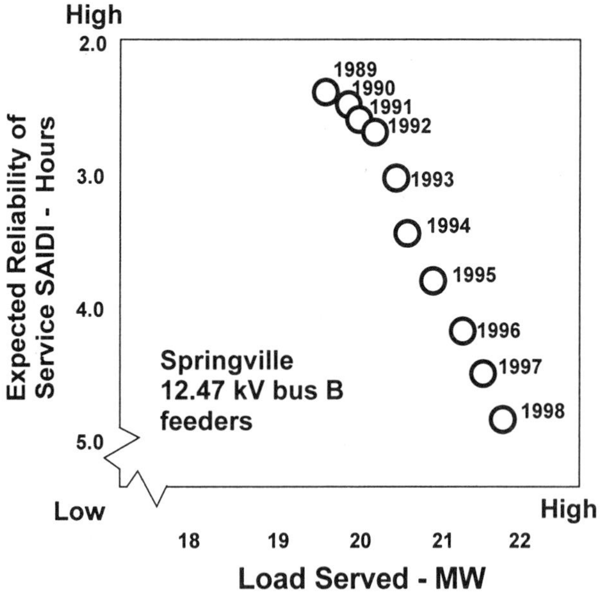

Figure 4.8 Two-Q plot of capability of a set of feeders in a large utility system in the central U.S, showing how increases made to accommodate higher load gradually eroded capability to deliver high reliability of supply. See text for details.

However, as Figure 4.8 shows, the various means taken to accommodate more quantity in power gradually eroded the system's reliability capability. This was both due to use of contingency margin for service of peak load, and use of switching and alternate paths for peak load service. Both means left no "backup' for contingencies, with the result that SAIDI in the region went from acceptable, to completely unacceptable by local consumer's standards.

Figure 4.9 shows a two-Q diagram of design options determined from detailed evaluation of the system's capability and consumer demands in the area. This study was done using a computerized method that analyzes both capacity (load flow type factors) and reliability (frequency, duration, voltage sag probabilities) simultaneously (Two Q analysis). The dashed line in Figure 4.9 shows the actual migration path for the system as modified and operated by the utility over the period 1989 – 1998. Total capital cost during the period was $41,500 (PW 1989).

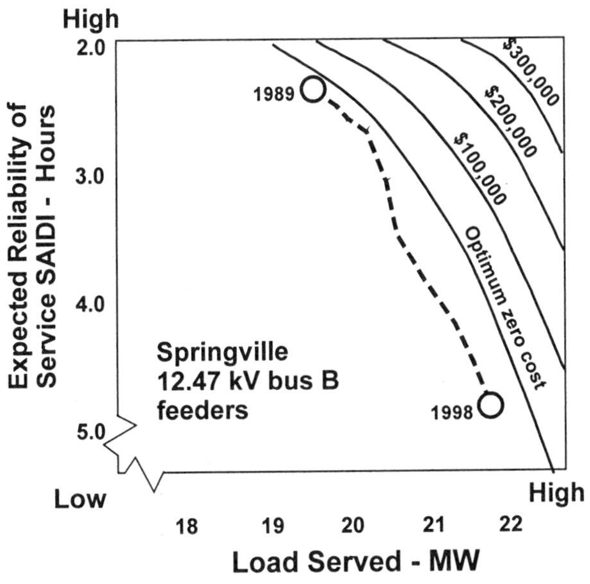

Figure 4.9 Two-Q plot of the design flexibility and options available in the region from Figure 4.8 showing results of Two-Q analysis. See text for details.

The lowest of the solid lines, labeled "optimum zero cost" shows the locus of superior performance points obtainable through re-switching and no more than $15,000 in cost. This indicates that the entire migration path for the period was inferior by roughly 40 minutes SAIDI or .66 MW quantity capability. This was due to the utility's lacking reliability-based techniques to hone the design well. Traditional single Q (quantity) design methods simply could not optimize the system to meet consumer needs at lowest cost.

The other solid lines in the drawing show the optimized capacity curves for various budget-levels (PW 1989 capital). The lines represent the locus of points for different designs or options, not the same plan. But all points on each curve are options purchasable at that budget amount. The utility's options and the trade-off between reliability of service and configuration for peak capacity is only clear.

4.7 CONCLUSION AND SUMMARY

Reliability of service is one of the major factors customers weigh in perceiving how well their electric supply system is doing its job. Whether that service is provided by an electric utility system, or a distributed generator, the planners of the supply system must set appropriate reliability goals and plan, engineer, and operate it to achieve targeted levels of customer service reliability. Various reliability indices can be used to track frequency and duration of customer interruptions over time and by location within the distribution system, in order to identify problems and ensure that goals are being met, but ultimately the only test that matters is consumer satisfaction.

Most utilities make sure that reliability targets are met, not through detailed reliability analysis in the planning stage, but implicitly through contingency and design criteria that are based on assessment of their reliability impacts. A more effective but complicated approach is to explicitly analyze reliability of planned designs, in an effort to determine if they are suitable and to engineer them to fully meet requirements at the lowest possible cost. Examples later in this book will illustrate this approach.

Electric consumers vary in their need for reliability of service as much as they vary in the amount of power they need. The electric power marketplace can be likened to a comb, a series of niches, each populated by consumers with different needs for both quality and quantity of power, and different preferences in how they would trade off cost against reliability of service, if forced to make sacrifices.

Value-based planning, although difficult, is also useful, particularly in cases where reliability is an issue to large or special industrial and commercial customers. An important element is to understand and apply how much the consumer values reliable service by analyzing their cost of interruption.

REFERENCES

P. F. Albrecht and H. E Campbell, "Reliability Analysis of Distribution Equipment Failure Data," EEI T&D Committee Meeting, January 20, 1972.

R. N. Allan, et al., "A Reliability Test System for Educational Purposes – Basic Distribution System Data and Results," IEEE *Transactions on Power Systems,* Vol. 6, No. 2, May 1991, pp. 813-821.

D. Atanackovic, D. T. McGillis, and F. D. Galiana, "The Application of Multi-Criteria Analysis to a Substation Design," paper presented at the 1997 IEEE Power Engineering Society Summer Meeting, Berlin.

R. Billinton, and J. E. Billinton, "Distribution System Reliability Indices," IEEE *Transactions on Power Delivery,* Vol. 4, No. 1, January 1989, pp. 561-568.

R Billinton R., and R. Goel, "An Analytical Approach to Evaluate Probability Distributions Associated with the Reliability Indices of Electric Distribution Systems," IEEE *Transactions on Power Delivery,* PWRD-1, No. 3, March 1986, pp. 245-251.

R. E. Brown, and J. R. Ochoa, "Distribution System Reliability: Default Data and Model Validation," paper presented at the 1997 IEEE Power Engineering Society Summer Meeting, Berlin.

J. B. Bunch, H.I Stalder, and J.T. Tengdin, "Reliability Considerations for Distribution Automation Equipment," IEEE *Transactions on Power Apparatus and Systems,* PAS-102, November 1983, pp. 2656 - 2664.

"Guide for Reliability Measurement and Data Collection," EEI Transmission and Distribution Committee, October 1971, Edison Electric Institute, New York.

W. F. Horton, et al., "A Cost-Benefit Analysis in Feeder Reliability Studies," IEEE *Transactions on Power Delivery,* Vol. 4, No. 1, January 1989, pp. 446 - 451.

Institute of Electrical and Electronics Engineers, *Recommended Practice for Design of Reliable Industrial and Commercial Power Systems,* The Institute of Electrical and Electronics Engineers, Inc., New York, 1990.

A. D. Patton, "Determination and Analysis of Data for Reliability Studies," IEEE *Transactions on Power Apparatus and Systems,* PAS-87, January 1968.

N. S. Rau, "Probabilistic Methods Applied to Value-Based Planning," *IEEE Transactions on Power Systems,* November 1994, pp. 4082 - 4088.

A. J. Walker, "The Degradation of the Reliability of Transmission and Distribution Systems During Construction Outages," Int. Conf. on Power Supply Systems. IEEE Conf. Pub. 225, January 1983, pp. 112 - 118.

H. B. White, "A Practical Approach to Reliability Design," IEEE *Transactions on Power Apparatus and Systems,* PAS-104, November 1985, pp. 2739 - 2747.

W. Zhang and R. Billinton, "Application of Adequacy Equivalent Method in Bulk Power System Reliability Evaluation," paper presented at the 1997 IEEE Power Engineering Society Summer Meeting, Berlin.

5
Cost and Economic Evaluation

5.1 INTRODUCTION

As was stated previously, reliability and service quality are very important, but "cost is king" in almost all electric distribution decisions, both those made by consumers and by utilities. Every decision contains or implies certain costs: equipment, installation labor, operating, maintenance, losses, and many others. Regardless what type of paradigm is being used, and whether traditional or new technologies are being considered, planners and decision-makers must deal with cost as a major element of their evaluation.

Many planning decisions revolve around selecting from alternatives on the basis of *when* they call for money to be spent. Two plans might be equal with respect to capacity and reliability, but different in both *when* and *how much* they require to be spent: one might eventually spend a lot, but only in the long run; the other might spend less, but call for more initial investment. At what point is the planner justified in picking the larger total expense because most of it lies far into the future?

This chapter provides a brief tutorial on engineering economics for planning; to provide a basis for the methods applied later in the book. Section 5.2 reviews the major elements of cost. Section 5.3 looks at the time value of money and future-cost discount methods. Benefit cost analysis is covered in section 5.4, and marginal benefit cost analysis, a cornerstone of aging infrastructure planning, in section 5.5. Only a quick review is given here. More comprehensive treatment is available in books listed at the end of this chapter.

5.2 COSTS

"Cost" is the total sacrifice that must be expended or traded in order to gain some desired product or end result. It can include money, labor, materials, resources, real estate, effort, lost opportunity, and anything else that is given up to gain the desired end. Usually, the costs of these many different resources and commodities are measured on a common basis – money – by converting materials, equipment, land, labor, taxes and permits, maintenance, insurance, pollution, and lost opportunity costs to dollars, pounds, marks, yen, or whatever currency is most appropriate.

In those cases where all these elements can be put on a common basis, the subsequent planning can be done in a *single-attribute* manner. The goal being to achieve the planner's goals while minimizing the total cost by trading the cost of one item against the others to find the best overall mix. However in some cases, one or more types of cost cannot be converted to money, for example, aesthetic impact or other "intangibles." In such cases, *multi-attribute* planning and cost minimization must be done, involving more complicated evaluation methods (D. Atanackovic *et al*).

Types of Cost

Initial cost is what must first be spent to obtain the facility or resource. It includes everything needed to put the facility in place, and is usually a single cost, or it can be treated as a single cost in planning studies if a series of costs over a period of time lead up to completion of the facility.

Continuing costs are those required to keep the facility operational and functioning the way its owner wants it to do. They include inspection and maintenance, fuel, supplies, replacement parts, taxes, insurance, electrical losses, and perhaps-other expenditures. They persist as long as the facility is in use. Usually, continuing costs are studied on a periodic basis – daily, monthly, or annually.

Fixed costs are those that do not vary as a function of any possible variable element of the planning analysis or engineering study being carried out. For example, the annual costs of taxes, insurance, inspection, scheduled maintenance, testing, re-certification, and so forth, required to keep a 500 kW DG unit in service, do *not* vary depending on how often or heavily it is run. They are fixed costs.

By contrast, the DG unit's fuel costs *do* vary with usage – the more it is run the higher the fuel costs. These are *variable costs,* a function in some manner of the amount of load served. Some types of maintenance and service costs may be variable, too. Highly used DG units need to be inspected and stress-caused wear and tear repaired more often than in units run for fewer hours.

Variability of Expense	Timing of Expense: Initial	Timing of Expense: Continuing
Fixed	One-time "first" costs that do not depend on usage pattern	Periodic, on-going costs that do not depend on usage pattern
Variable	One-time "first" costs that do depend on usage pattern	Periodic, on-going costs that do depend on usage pattern

Figure 5.1 Two characterizations, one based on when costs occur and the other on how they vary with usage, result in four categories of costs.

Figure 5.1 shows how these two sets of categorizations – initial, continuing, fixed, and variable - can be combined. This two by two matrix allows planners to identify fixed initial costs, variable initial costs, fixed continuing costs, and variable continuing costs. In this way, costs in studies can be characterized simultaneously by both their relationships to time and how they respond to assumptions about usage.

Sunk Cost

Once a cost has been incurred, even if not entirely paid, it is a "sunk cost". For example, once a facility has been constructed and is in operation, it is a sunk cost, even if ten years later the company still has its capital depreciation on its books as a cost.

Embedded, Marginal, and Incremental Cost

Embedded cost is that portion of the cost that exists in the current system, configuration, or level of use. Depending on the application, this can include all or portions of the initial fixed cost, and all or parts of the variable costs.

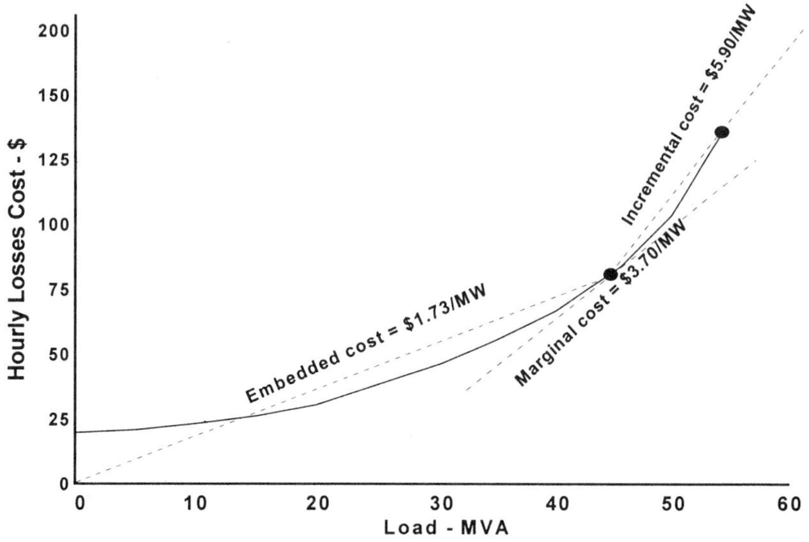

Figure 5.2 Hourly cost of losses for power shipped through a substation transformer as a function of the load. Currently loaded to 45 MVA, it has an "embedded" loss' cost of 1.73/MVA, and a marginal cost of losses at that same point of $3.70/MVA. The incremental cost of losses for an increase in load to 55 MVA load is $5.90/MVA.

Often, the "embedded" cost is treated as a fixed cost in subsequent analysis about how cost varies from the current operating point. Marginal cost is the slope (cost per unit) of the cost function at the current operating point (Figure 5.2). This point is usually (but not always) the point at which current embedded cost is defined.

Incremental cost is the cost per unit of a specific jump or increment – for example, the incremental cost of serving an additional 17 MVA from a certain substation or the incremental cost of losses when load on a feeder decreases from 5.3 to 5.0 MVA. Marginal and incremental costs both expresses *the rate of change of cost* with respect to the base variable, but they can differ because of the inconsistencies and variances in the cost relationships.

In the example shown in Figure 5.2, the marginal cost has a slope (cost per unit change) and an operating point (e.g., 45 MW in Figure 5.2). Incremental cost has a slope and both "from" and "to" operating points (e.g., 45 MVA to 55 MVA in Figure 5.2) or an operating point and an increment (e.g., 45 MW plus 10 MVA increase).

5.3 TIME VALUE OF MONEY

Expenses Over Time

Almost any electrical project has both initial and continuing costs. Figure 5.3 shows the costs for a new distribution substation, planned to serve 65 MW of residential load on the outskirts of a large city. It has a high initial cost that includes the equipment, land, labor, licenses and permits, and everything else required building the substation. Thereafter, it has taxes, inspection, maintenance, electrical losses and other continuing costs on an annual basis. Figure 5.4 shows the initial and continuing costs identified with respect to fixed and variable categories, too. Typical of most electrical planning situations, the initial cost is almost entirely fixed, and the continuing costs are mostly variable.

Decisions Based on When and How Much is Spent

Any utility planner must deal with two types of time versus money decisions. The first involves deciding whether a present expense is justified because it cancels the need for some future expense(s). For example, suppose it has been determined that a new substation must be built in an area of a utility's service territory where there is much new customer growth. Present needs can be met by completing this new substation with only one 25 MVA transformer, at a total initial cost of $1,000,000. Alternatively, it could be built with two 25 MVA transformers installed, at a cost of $1,400,000. Although not needed immediately, this second transformer will be required within four more years because of continuing growth. If added at that time, it will cost $642,000 extra - $242,000 more than if it is included at the time the substation is first built. The higher cost to add it later is a common fact of life in T&D systems' planning and operation. It's also a reflection of the additional start-up cost for a new project and of the higher labor and safety costs required for installing a transformer at an already-energized substation, rather than when the site is cold.

Which plan is best? Should planners recommend that the utility spend $400,000 now to avoid a $642,000 expense four years from now.

A second and related type of time-vs.-money decision is determining if a present expense is justified because it will reduce future operating expenses by some amount. Suppose two types of distributed generators are candidates for installation at a high-reliability customer site. The first, Type A, costs $437,000 installed. It fully satisfies all loading, voltage regulation, and other criteria. On the other hand, the higher efficiency type B unit will also meet these same requirements and criteria, and lower annual fuel costs by an estimated $27,000 per year. However, it has an initial cost of $597,000. Are the planners justified in recommending that $160,000 be spent, based on the long-term fuel savings?

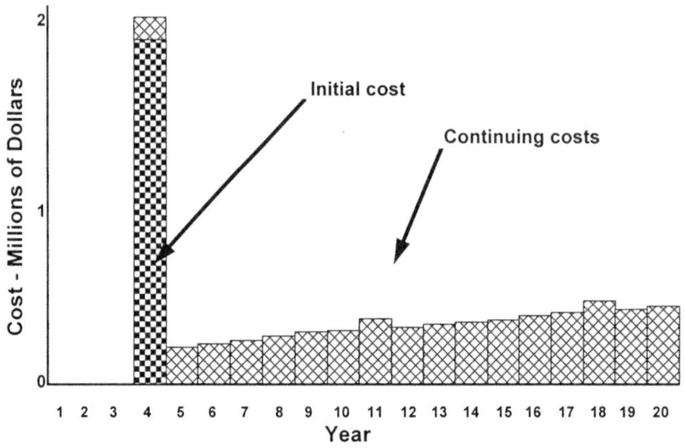

Figure 5.3 A new substation's costs are broken into two categories here, the *initial cost,* a one-time cost of creating the new substation, and its *continuing annual costs* which are required to maintain it in operation.

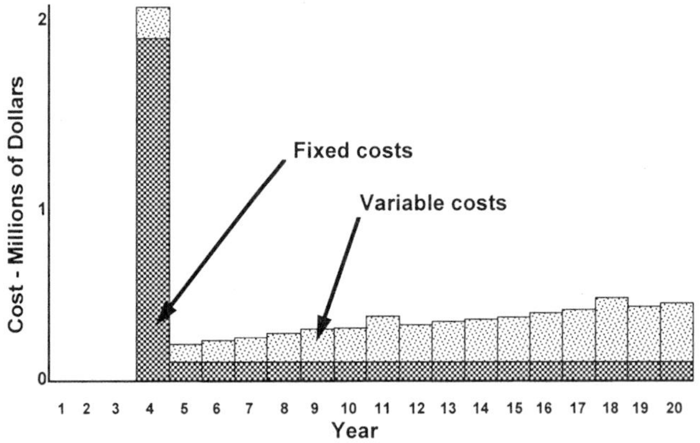

Figure 5.4 The new substation's costs can also be viewed as composed of both fixed costs – those that are constant regardless of loading and conditions of its use, and variable costs. Those change depending on load, conditions, or other factors in its application as part of the power system. In this example, the variable costs increase nearly every year, reflecting higher losses' costs as the load served gradually increases due to load growth.

These decisions, and many others in utility planning, involve comparing present costs to future costs, or comparing plans in which a major difference is *when* money is scheduled to be spent. To make the correct decision, the planner must compare these different costs in time on a sound, balanced basis, consistent with the electric utility's or DG owner's goals and financial structure.

Common Sense: Future Money Is Not Worth As Much As Present Money

Few people would happily trade \$100 today for a reliable promise of only \$100 returned to them a year from now. The fact is that \$100 a year from now is not worth as much to them as \$100 today. This "deal" would require them to give up the use of their money for a year. Even if they do not believe they will need to spend their \$100 in the next year, conditions might change and they might discover later that the do need that money. Then too, there is a risk that they will not be re-paid, no matter how reliable the debtor.

Present Worth Analysis

Present worth analysis is a method of measuring and comparing costs and savings that occur at different times on a consistent and equitable basis for decision-making. It is based on the present worth factor, P, which represents the value of money a year from now in today's terms. The value of money at any time in the future can be converted to its equivalent present worth as:

$$\text{Value today of X dollars t years ahead} = X \times P^t \tag{5.1}$$

Where P is the *present worth factor*

For example, suppose that P = .90; then \$100 a year from now is considered equal in value to today's money of

$$100 \times (.90) = \$90$$

and \$100 five years from now is worth only

$$\$100 \times (.90)^5 = \$59.05$$

Present-Worth Dollars are often indicated with the letters PW. Today's \$100 has a value of \$100 PW, \$100 a year from now is \$90 PW, \$100 five years from now is \$59.05 PW, and so forth.

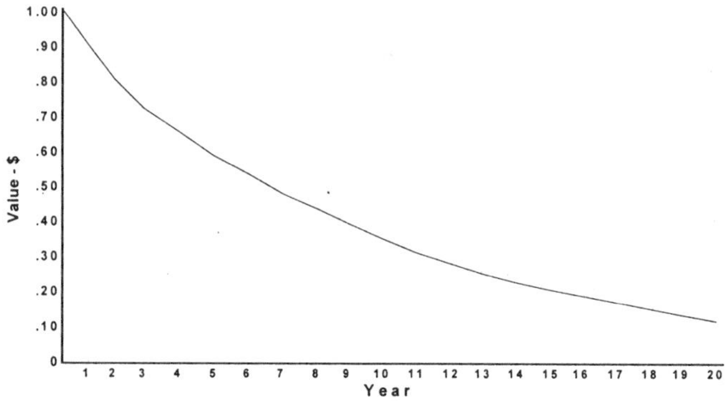

Figure 5.5 Present Worth Analysis discounts the value, or worth, of a dollar the further it lies in the future. Shown here is the value of one dollar as a function of future year, evaluated at a .90 present worth factor (11.1% discount rate).

Alternatively, the present worth factor can be used to determine how much future money equals any amount of current funds. For example, to equal \$100 present worth, one year from now a person would need

$$\$100/.90 = \$111.11$$

Two years from now, one would need $\$100/(.90)^2 = \$123.46$ to equal \$100 PW.

A continuing annual future cost (or savings) can be converted to a present worth by adding together the PW values for all future years. For example, the present worth of the \$27,000 in annual fuel savings discussed for the type B DG unit earlier in this chapter can be found by adding together the present worth of \$27,000 next year, \$27,000 the year after, and so forth

$$\text{PW of \$27,000/year} = \sum_{t=1}^{\infty} (\$27,000 \times P^t) \tag{5.2}$$

$$= \$27,000 \times (\sum_{t=1}^{\infty} P^t)$$

$$\cong \$27,000 \times (\sum_{t=1}^{30} P^t)$$

$$= \$258,554$$

Discount rate

Present Worth Analysis essentially *discounts* the value of future costs and savings *just* because they lie in the future, as shown in Figure 5.5. The discount rate used in an analysis, *d*, is the perceived rate of reduction in value of money from year to year. The present worth factor is related to this *discount rate*

$$P(t) = 1/(1+d)^t \qquad (5.3)$$

Where d = discount rate
And t = future year

If d is 11.11%, it means that a year ahead dollar is discounted 11.11% with respect to today's dollar, equivalent to a present worth factor of P = (1/1.111) = .90. Therefore, $111.11 a year from now is worth $111.11/1.1111 = $100 today. A decision-making process based on the values of d = 11.11% and PWF = .90 would conclude that spending $90 to save $100 a year from now was a break-even proposition (i.e., there is no compelling reason to do it). While if the same $100 savings can be had for only $88, the "deal" has a positive value.

A higher discount rate, corresponding to a lower present worth factor, renders the decision-making process less willing to trade today's costs for future costs. If the discount rate is doubled, to 22.2%, P drops to .82. Now, $88 spent today to obtain a $100 cost reduction a year from now is no longer viewed as a good investment: that $88 must yield at least $88/1.222 = $108 a year from now to be viewed as merely break even.

Present worth analysis does not say "no" to truly essential elements of a plan

Power system planners with a commitment to see that service is always maintained often fall into a trap of wanting to build everything as early as practical in order to have plenty of capacity margin and to allow for unexpectedly high load growth. As a result, they come to view present worth analysis, and any other aspect of planning that might say "delay," as negative.

But this is not the case. If applied correctly, present worth analysis never says "no" to essential expenses, for additions or changes that must be done *now*. For example, if a hurricane has knocked down ten miles of critically needed line; there is no question about the *timing* of the expense in putting it back up. That must be done as soon as conditions permit, and present worth analysis is not even appropriate to apply to the decision.

Present Worth Analysis should be used to evaluate and rank alternatives only when there is a difference in the *timing* of expenses (the substation could be built this year or next year), or when costs and savings that occur at different times must be balanced (more spent today will lower cost tomorrow). Present worth analysis is an essential element of keeping costs as low as possible, and it should be applied in all situations where alternatives differ with respect to *when* expenses will be incurred.

How Are Present Worth Factors and Discount Rates Determined?

There is no one rule or set of guidelines that rigorously defines what contributes to the present worth factor (PWF), or its companion, the discount rate. Likewise, there is no inviolate formula like Ohm's law or Schroedinger's equation that lies out completely and with no exceptions exactly how to compute PWF from those factors that are deemed relevant. Quite simply, PWF is merely a number, a decision-making tool that allows planners to compare future expenses to present ones. The present worth factor is often arrived at through careful analysis and computation, but sometimes it is an empirical estimate "that just works well."

> *The present worth factor is simply a value that sums up all the reasons why an organization would prefer to spend money tomorrow rather than today.*

There can be many reasons why a person may wish to limit spending today in favor of spending tomorrow. These reasons all influence the selection of a value for the present worth factor. Priorities vary greatly from one company to another and also change with time. Among them are one-reason planners often finding difficult to accept: "We just don't have the money, period." In most electric utilities, present worth factors and discount rates are determined by a utility's financial planners and based on the company's regulatory requirements for financing and use of money. The methods applied to fix these requirements quantitatively and their relation to PWF are beyond the scope of this discussion, and available in other texts (e.g., *IEEE Tutorial on Engineering Economics*). However, it is worth looking at some of the primary influences on present worth factor and how and why they influence it.

Interest rate

The fact that it takes more future dollars to equal a dollar today is often attributed to interest rate: a person who has \$100 today can invest it at the prevailing interest rate, i.e., so that a year from now it will be (1+i) times as much. Thus, \$100 invested today at an annual rate of 5% interest will be worth \$105 a year from now and \$128 five years from now. If the prevailing interest rate is 5%, then it is cost-effective to spend \$100 only if it will save or return a value a year from now that exceeds \$105. Otherwise, it would be better simply to leave the money invested and drawing interest. A PWF of .952 can be applied in present worth analysis as described earlier, and will lead to decisions that reflect this concern.

But, in practice present worth factor clearly represents something more than *just* interest rate, because present worth factor as applied by most electric utilities is nearly always greater than what corresponds to the prevailing interest rate. For example, at the time of this writing, the inflation-adjusted interest rate on safe, long-term investments is about 5%, yet most utilities are using a present worth factor of about .11, equivalent to a 12.4% interest rate.

The difference, 12.4% versus 5.5% in this case, is attributable to other factors, although it may prove difficult or impossible for a distribution planner to determine and quantify all of them.

Inflation

Inflation is *not* one of the factors normally taken into account by the present worth factor, although this is often misunderstood and new planners often assume it is a part of the analysis. Inflation means that what costs a dollar today will cost more tomorrow – 3% annual inflation would mean an item costing a dollar today will be priced at $1.03 a year from now. While inflation needs to be considered by a utility's financial planners, distribution planners can usually plan with *constant dollars,* by assuming that there is no inflation.

The reason is that inflation raises the cost of everything involved in distribution planning cost analysis. If inflation is 3%, and then on average a year from now equipment will cost 3% more, labor will cost 3% more, as will paperwork and filing fees and legal fees and taxes and replacement parts and transportation and everything else. More than that, over time inflation will call for similar adjustments in the utility's rates, in the value of its stock, its dividends, and everything associated with its expenses and financing (including hopefully the planners' salaries). It affects nearly every cost equally, at least to the extent that costs can be predicted in advance.[1] From the standpoint of planning, since inflation makes no impact on the *relative costs* of various components, it can be ignored, making the planning process just a bit easier.

Some utilities do include inflation in their planning and in their present worth analysis. In such cases, an increment to account for inflation is added to the discount rate. Given a 5% interest rate and a 3% inflation rate discussed above, this would mean that the planners' present worth factor might be

---

[1] In actuality, utility and business planners and executives all realize that inflation will not impact every cost equally, but small variations in individual economic sectors have proven impossible to predict. In cases where there is a clear indication that certain costs will rise faster or slower than the general inflation rate, that difference should be taken into account.

$$P = 1/(1+5\%+3\%) = .926$$

Present worth analysis would now compare today's investment measured in today's dollars against tomorrow's investment measured in tomorrow's inflated dollars. This type of accounting of future costs must sometimes be done for budgeting and finance estimation purposes.[2]

Inflation must be included in the analysis, but all planning can be done with "constant dollars"

However, while "inflated dollars" present worth analysis sometimes has to be done to estimate budgets, it is rarely an effective planning tool. Planning and decision-making are facilitated when expenses and benefits are computed on a common basis – by measuring costs from all years in *constant dollars.* Is plan A, which calls for $1.35 million in expenses today, worth more than plan B, which costs $1.43 million three years from now? If both plans have been calculated using their respective present worth measured in constant dollars, then it is clear that plan A is the less costly of the two. But if inflation were a factor in the present worth analysis, then one has to do further adjustment of those numbers to determine which is best. (At 3% inflation, $1.35 million in three years will equal $1.48 million in then inflated currency, making plan B at $1.43 million three-year ahead *inflated* dollars the better plan). Beyond this, errors or unreasonable cost estimates for future projects are more readily caught when expressed in non-constant dollars. For these reasons, present worth analysis in constant dollars is often used for planning, but an initial analysis, which accounts for inflation, may occasionally be needed for other purposes.

Earnings targets

The present worth factor for a dollar under the utility's control should be higher than 1/(1 + interest rate), because *the utility must be able to do better with its earnings than the prevailing interest rate.* If a utility cannot better the prevailing interest rate with its own investment, then it should liquidate its assets and invest the results in earning that interest rate. Frankly, it won't get the chance to do so: long before that, its shareholders will sell their stock and invest their money elsewhere, in companies that *can* beat the prevailing interest rate through their investment of the stockholder's money.[3] Therefore, a goal must be

---

[2] For example, when estimating a future budget, the costs often do have to be put in dollars for the years being estimated, and those are inflated dollars.

[3] Planners from municipal utilities may argue that this does not apply to them, but that is not necessarily true. If a municipal electric department cannot "earn" more from its investments than other businesses could, it is costing the city money to design and operate a system that needs to be subsidized by taxpayer money. Such subsidized

to use the company's money to earn more than it could by other safe means.

As a result, while the prevailing interest rate may be 5%, the utility's financial planners may have determined that a 12% earnings potential on all new expenditures is desirable. Rather than a PWF of .952 (5% interest) the utility would use a PWF of .892 (12%) to interject into the decision-making process its unwillingness to part with a dollar today unless it returns at least $1.12 a year from now. This is one tenet of many corporate plans, that all investments must return a suitable earning.

Risk

One hundred dollars invested with the promise of $108 payback a year from now may look good, but only if there is a very small chance that the investment will go wrong, with the loss of the interest *and* the $100 itself. Such cases are very rare, particularly if investments are made wisely, but there are other types of risks that a utility faces. For example, suppose that shortly after spending $100 to save $108 a year from now, a company gets hit by an unexpected catastrophe (e.g., an earthquake) that damages much of its equipment. It may desperately wish it had that $100 to pay for repairs, work that it simply has to have done to re-establish production. It now has no choice but to borrow the money it needs at short-term interest rates, which might be 12%. In retrospect, its expenditure of $100 to save $108 a year later would look unwise.

In practice, a present worth factor often includes a bias or margin to account for this type of "bird in the hand" value. By raising the PWF from 8% to 10%, the utility would be stating that yes; perhaps $108 is the year-ahead earnings goal. Simply breaking even with that goal is not enough to justify committing the company's resources: today's money will be committed only when there are very sound reasons to expect a better than minimum return.

Planning errors

In addition, the present worth factor often implicitly reflects a sobering reality of planning – mistakes cannot be avoided entirely. Under the very best circumstances, even the finest planning methods average about 1% "planning error" for every year the plan is extended into the future. Distribution expansion projects that must be committed a full five years ahead will turn out to spend about 5% more than if the planner could somehow go back and do things over again, knowing with hindsight exactly what the minimum expenditure is needed just to get by.

---

operation may be the policy of the city government (some city governments deliberately subsidize commercial electric rates, using the low rates as a tool to attract employers to their region). Regardless, its planners should try to get as sound a return as possible.

Adding 1% – or whatever is appropriate based on analysis of the uncertainty and the planning method being used – to the PWF biases all planning decisions so that they reflect this imperfection of planning. This makes the resulting decision-making process slower to spend money today on what *appears* to be a good investment for tomorrow, unless the predicted savings include enough margin over the element of risk to account for the fact that the planning method could be wrong.

Present Worth Factor Is a Decision-Making Tool, Not a Financial Factor

Thus, present worth factor and discount rates are merely tools. Present worth analysis is a decision-making method to determine *when* money should be spent. It can embrace some or all of the factors discussed above, as well as others. Two equally prudent planners might pick very different present worth factors, depending on their circumstances, as summarized in Table 5.1 with the inputs that determined the PWFs used in the mid-1990s by a large investor-owned utility (IOU) in the northwestern United States and a municipality in the southwestern United States.

But while the present worth factor can be affected by many factors, the simple fact is that such determinations are irrelevant to the electric service planner. Present worth factor is used to evaluate and rank alternatives based on when they call for expenditures. It is just a tool, part of the analysis, period.

A relatively low PW factor means that the planning process will be more likely to select projects and plans that spend today in order to reduce costs tomorrow. As the PWF is decreased, the planning process becomes increasingly unwilling to spend any earlier than absolutely necessary unless the potential savings are very great. A very low PW factor will select plans that wait "until the last moment" regardless of future costs.

Table 5.1 Discount Rate "Computation" for Two Utilities

Factor	IOU	Muni	Comment
Prevailing interest rate	5.7%	5.1%	Municipal bond rate is lower than prime.
Inflation factor	-	-	Both do constant dollar planning.
Earnings target	5.5%	-	Municipal has no need for "earnings."
Risk	1.5%	3.0%	Political cost more serious than financial.
Planning error	1.0%	1.5%	This IOU has the better planning method.
"We just don't have funds"	-	5.0%	This municipality is nearly "broke."
Total discount	13.7%	14.6%	
Equivalent PWF	87.9%	87.3%	

Table 5.2 Comparison of Yearly Expenses by Category for an Eight-Year Period

		Alternative A Build initially with two			Alternative B Build initially with one		
Year	Load-MW	Capital	O&M&T	Losses	Capital	O&M&T	Losses
0							
1		20			20		
2		370			290		
3	12	1010	52	54	690	44	39
4	15.5		110	119		92	99
5	18.5		110	130		92	122
6	21.3		110	142	80	92	147
7	23.8		110	155	562	101*	164*
		1400	492	600	1642	317	666
			Total = 2492			Total = 2633	

*Taxes and losses for the transformer added in year eight (alternative B) are prorated for that year, since it is assumed here that it is installed just prior to summer peak (six months into the year).

Comprehensive Present Worth Example

Table 5.2 and Figure 5.6 show more cost details on the two-versus-one substation transformer decision mentioned at the beginning of this section. In the case shown, the decision is being made now (in year zero), even though the substation will actually be built three years from now – the decision must be made now due to lead time considerations. Alternative A calls for building the new substation with two 25 MVA transformers at a total cost of $1,400,000 initially. Alternative B defers the second transformer four years (to year 7), reducing the initial cost to $1,000,000. And again, the addition of the second transformer four years later is estimated to cost $642,000 because of project start-up costs, separate filing and regulatory fees, and the fact that the work will have to be done at what will then be an energized site.

However, this decision involves other costs, which while secondary in importance, should be considered in any complete analysis. For example, no-load losses, taxes, and O&M will be greater during the first four years in Alternative A than in Alternative B. Table 5.2 and Figure 5.6 compare the expenditure streams for the eight-year period for both alternatives, including capital; operations, maintenance, and taxes (O&M&T); and losses.[4]

---

[4] The table and figure cover the only period during which there are differences that must be analyzed in the decision-making. At the end of the eight-year period, the two alternatives are the same – either way the substation exists at that point with two 25 MVA transformers installed. Present worth thereafter is essentially the same.

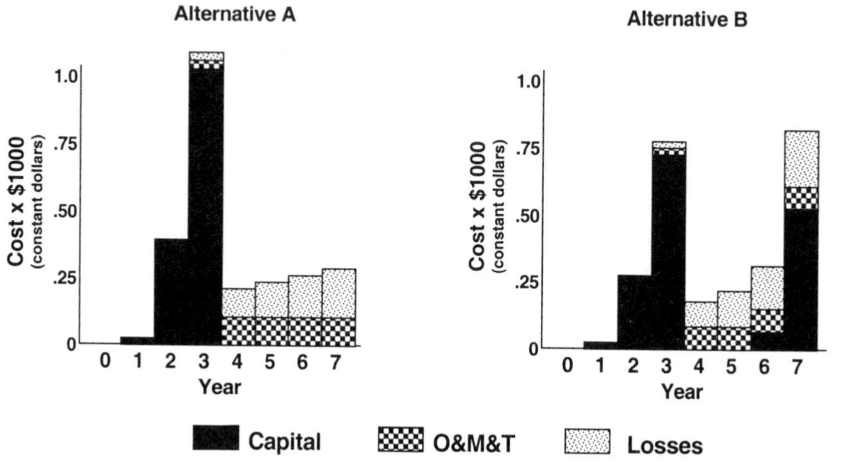

Figure 5.6 Yearly expenses for the two alternative plans for the new substation (constant dollars).

Note also, that the capital for substation construction in either alternative is not spent entirely in the year the substation is completed. A small amount is spent two years earlier (for filing and preparing for the site), and the site, itself, and some equipment purchased a year earlier so that work can begin on schedule.

In comparing differences in the two alternatives, observe also that the second transformer in Alternative A increases O&M&T costs during the first four years of operation as compared to the other alternative, because taxes must be paid on its additional value, and it must be maintained and serviced.

The difference in losses' costs is more complicated. Initially, losses are higher if two transformers are installed, because this means twice the no-load losses, and the twin transformers are initially very lightly loaded, so that there is no significant savings in load-related losses over having just one transformer. But by year six the two-transformer configuration's lower impedance (the load is split between both transformers) produces noticeable savings due to reduced losses as compared to the other alternative.

Overall, Alternative A spends fewer dollars in the period examined $142,000 less. It has the lower amount of long-term spending.

Table 5.39 and Figure 5.70 show the various cost streams for both alternatives converted to PW dollars using a present worth factor of .90 (equivalent to p = 11.11%). Note that dollar amounts have dropped substantially – discounted future dollars just aren't as valuable as the undercounted dollar figures.

More importantly, though, is the present worth evaluation reverses the order of cost preference for the two options. Even though its total spending is less, Alternative A is rated a *higher* present worth by \$51,000 PW, because many of its expenses are closer to the present.

The total difference in PW is slightly more than 3% of the average PW value of the alternative. Considering the straightforward nature of this analysis, and the high proportion of the various cost elements common to both alternatives, it is very unlikely that the analysis is in error by 3%. Therefore, this is very likely a dependable evaluation: Alternative B is the recommended "low-cost" option. Alternative A would cost more.

Looking at the sensitivity of this decision to present worth factor, the two alternatives would be considered equal if the evaluation had used a present worth factor of .937, equal to a discount rate of 6.77%. Present worth factor would have to be higher (i.e., the discount rate lower) to have reversed the determination that Alternative B is, in fact, the low-cost option.

Again, considering that nominal PWF values less than .92 are extremely uncommon, the recommendation that Alternative B is the preferred option seems quite sound. Therefore, the planner's recommendation should be to build the substation with one transformer initially, and add the other at a later time, as needed.

Leveled Value

Often, it is useful to compare projects or plans on the basis of their *leveled* annual cost, even if their actual costs change greatly from one year to the next. This involves finding a constant annual cost whose present worth equals the total present worth of the plan, as illustrated by Figure 5.7. Leveled cost analysis is a particularly useful way of comparing plans when actual costs within each plan vary greatly from year to year over a lengthy period of time, or when the lifetimes of the options being considered are far different. It also has applications in regulated rate determination and financial analysis. In general, the present worth, Q, leveled over the next "n" years is

$$\text{Leveled value of Q over n years} = Q\left(d \times (1+d)^n\right)/((1+d)^n - 1) \qquad (5.4)$$

Thus, the steps to find the leveled cost of a project are: 1) Find the present worth of the project and, 2) Apply equation 5.2 to the present worth. As an example here, using Alternative A's calculated total present worth of \$1,684,000 from

Table 5.3 Comparison of Yearly Present Worth by Category for an Eight-Year Period

Year	Present Worth Factor	Alternative A Build initially with two Capital	O&M&T	Losses	Alternative B Build initially with one Capital	O&M&T	Losses
0	1.0						
1	.90	18			18		
2	.81	300			235		
3	.73	736	38	39	503	32	28
4	.66		72	78		60	65
5	.59		65	77		54	72
6	.53		58	75	43	49	78
7	.48		53	74	269	48*	78*
		1054	286	343	1067	244	322
			Total = 1684			Total = 1633	

*Taxes and losses for the transformer added in year eight (alternative B) are prorated for that year, since it is assumed here that it is installed just prior to summer peak (six months into the year).

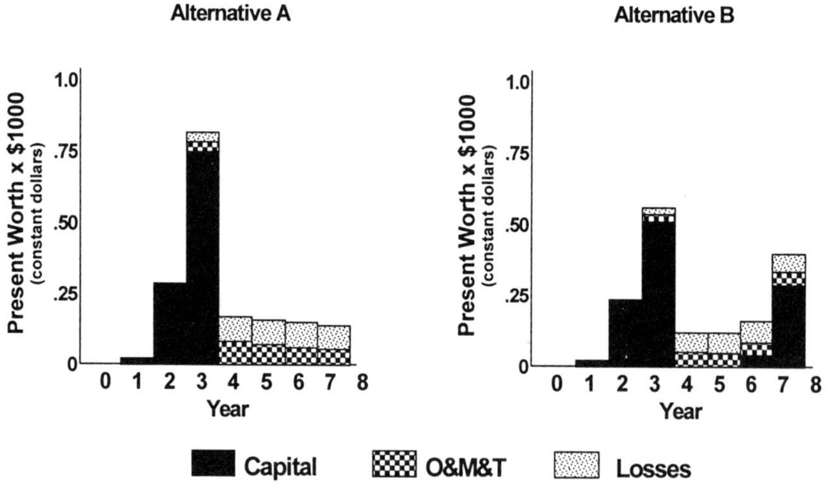

Figure 5.7 Present worth of expenses associated with the two alternative substation plans given in Table 5.2 and Figure 5.6.

Table 5.3, that value can be levelized over the seven-year period in the table as

$$= \$1{,}684{,}000\ (.1111(1.1111)^7/(1.1111^7 - 1)$$

$$= \$358{,}641$$

By contrast, Alternative B has a levelized cost over the same period of \$347,780. If the utility planned to pay for alternative A or B over the course of that seven-year period, the "payments" it would have to make would differ by a little more than \$10,800 per year.

Lifetime levelized analysis

Rather than analyze planning alternatives over a short-term construction period planners or budget administrators often wish to look at a new addition to the system over the period in which it will be in service or be financed.

For example, in the two-or-one transformer substation planning problem presented above, the substation would be planned on the basis of providing a minimum of 30 years of service, and it would be depreciated on the company books for 30 years or more. Therefore, planners should analyze cost over 30 years and compare alternative plans, A and B, on that same basis.

Figure 5.8 shows projected costs for Alternative A over a 32-year period, which includes the first through thirtieth years of service for the substation. The total present worth of the substation's costs over the period is just over \$3 million (construction plus 30 years of O&M&T plus losses), which represents a

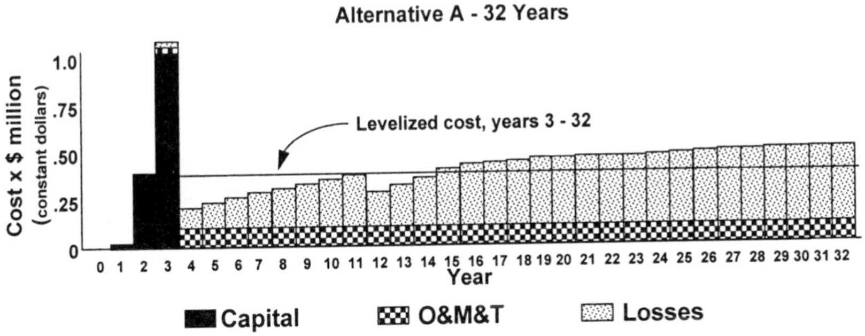

Figure 5.8 Cost of Alternative A from Figure 4.13 over a 32-year period that includes its first 30 years of service. Present worth of all costs sums to slightly more than \$3,000,000. Leveled over years three through 30, those in which the substation is planned to be in service, this represents an annual cost of nearly \$350,000 per year, which can be viewed as the annual cost of providing the substation service.

Table 5.4 Lifetime Analysis of Illumination

Incandescent

Year	PWF	Capital	Energy	Total	PW
1	.90	$2.00	$23.00	$25.00	$22.50
2	.81		$23.00	$23.00	$18.63
3	.73		$23.00	$23.00	$16.77
Total Present Worth					$57.90
		Levelized over 3 years =			$23.74

Halogen

Year	PWF	Capital	Energy	Total	PW
1	.90	$8.00	$13.00	$21.00	$18.90
2	.81		$13.00	$13.00	$10.53
3	.73		$13.00	$13.00	$9.48
4	.66		$13.00	$13.00	$8.53
5	.59		$13.00	$13.00	$7.69
Total Present Worth					$55.11
		Levelized over 6 years =			$14.95

Fluorescent

Year	PWF	Capital	Energy	Total	PW
1	.90	$17.00	$8.00	$25.00	$22.50
2	.81		$8.00	$8.00	$6.48
3	.73		$8.00	$8.00	$5.83
4	.66		$8.00	$8.00	$5.25
5	.59		$8.00	$8.00	$4.72
6	.53		$8.00	$8.00	$4.25
			Total Present Worth		$49.04
			Leveled over 6 years =		$11.62

leveled cost of $348,500/year. By contrast, Alternative B (the less expensive of the two construction schedules, as determined earlier) has a leveled cost of $342,200/year. It saves an average of $5,900/year or about $500/month in terms of a 30-year budget perspective.

Comparing cost over different equipment lifetimes

Very often the lifetimes or periods of cost being considered for two or more options are not equivalent, and yet it is desired to compare their costs on a valid basis. Leveled lifetime cost is a good way to do so. Table 5.4 compares the annual costs of providing 1000 lumens of illumination (about the equivalent of a standard 75 watt incandescent light bulb) for 2500 hours per year, using incandescent, halogen, or compact fluorescent lamps, which have expected lives of three, five, and six years respectively. The initial costs, the annual operating costs, and the lifetimes of these three options all differ substantially. In each case, initial and continuing annual energy costs are evaluated on a present worth basis over the lifetime of the particular lamp; then the leveled cost over that period is determined. This permits a comparison that identifies the compact fluorescent as having less than half the annual cost of the incandescent.

In this example, each alternative was evaluated over a different period of time – its lifetime – yet the results are directly comparable. The results are identical to those that would have been obtained from analysis of all over any much longer period, taking replacement costs and other factors into account.

For example, suppose that the analysis in Table 5.4 had compared all three options over a 30-year period. That would have required a series of ten incandescent bulbs, one every three years, six-halogen, and five-compact fluorescent lamps. Each of the sections of Table 5.4 would have extended for 30 years with appropriate PWF, computations, and sums, but this analysis would have reached the same conclusion about which of the three lamps was best in this case.

5.4 DECISION BASES AND COST-EFFECTIVENESS EVALUATION

How does a planner know that a particular project is "best"? In many cases, once present worth analysis has adjusted all costs and savings with respect to equipment lifetime, schedule, and other time-value-of-money considerations. Such decisions are straightforward – in many cases trivial. But determining what to do is not always merely a matter of *just* identifying the option with the lowest cost. Selection may be based on the relative merits of costs versus savings, the benefits versus cost or the relative gain per dollar invested. In integrated resource planning (RIP) situations, the decision may also depend on the applicable philosophy about what constitutes savings and who should benefit most from a public utility's investment in the system.

This section briefly reviews cost-effectiveness concepts and methods. More comprehensive discussions of the concepts discussed here and the methods developed to apply them can be found in the references at the end of the chapter.

Lowest Cost Among Alternatives

In many cases and from many perspectives, the goal of planning is to identify an alternative with the lowest cost. If all pertinent costs (capital, losses, O&M&T, etc.) are considered, and the appropriate type of present worth adjustment is made for timing, lifetimes, and schedules, then the alternative with the lowest PW cost *is* the preferred option.

Payback Period

The period of time required for "long-term" savings to re-pay the initial investment is often used as a measure of the desirability of making an investment in a resource or other addition to the electric supply. Returning to the "type A vs. type B DG unit" problem discussed earlier in this chapter, a fuel savings of $27,000 per year will require six years (6 × $27,000 = $162,000) to repay the $160,000 additional cost of the type B unit over type A. (If the break-even is computed based on PW dollars at a .90 PW factor, the payback of the original $160,000 takes ten years.)

"Payback period" analysis is most often used informally, to provide an additional perspective on plans and financing. Rarely is it part of the formal planning criteria or a specific cost-effectiveness test called for by regulatory authority or management. However, it does address one question that can be quite important to any investor –"When will I get my money back?"

Benefit/Cost Ratio Analysis

In many planning situations, one or all of the options being considered have both costs *and* benefits, and there are constraints or interrelated considerations with respect to both that impact the selection of the plan. In such cases, planning consists of more than simply ranking the alternatives by cost and selecting the one at the top of the list. Alternatives vary in benefit, too. This situation is particularly common in IRP situations, where options such as DSM have a cost, but also benefits accruing from their use.

Very often in such cases, the planning decisions are based on an evaluation of the expected gain versus the cost. Benefit-cost ratio is a simple concept: a ratio less than one indicates the project will lose more than it gains; the higher the benefit/cost ratio above 1.0, the greater the potential gain relative to its cost. For example, suppose that it has been determined that a particular project, with a PW cost of $160,000, would provide PW savings equal to $258,600. A planner could evaluate the benefit to cost ratio of this proposal as

$$\text{Benefit/Cost} = \$258{,}600/\$160{,}000 = 1.62$$

This alternative has more benefit than cost – 62% more.

Benefit-cost ratio is generally applied as *a ranking method,* to prioritize projects that are viewed as discretionary or optional, or in cases where there simply isn't enough funding for all the projects that present worth analysis has indicated are worthwhile. It has drawbacks however, as explained below.

5.5 BUDGET-CONSTRAINED PLANNING: MARGINAL BENEFIT VS. COST ANALYSIS

Example Problem from an Aging Infrastructure System

Table 5.5 shows the different planning options available for two feeders in a utility, which has aging infrastructure problems. Feeder 1 does not currently exist, it is being planned for a growing part of a utility service territory where new customers are developing, customers who must be served in order to meet the utility's obligation to serve (or a wireco's obligation to connection). This new distribution feeder can be built in any of several sizes and designs, as shown, using conductors and layouts of various types, all with advantages and disadvantages (including differences in cost). The details of exactly what each alternative is, and the differences in costs are unimportant to this discussion. What matters is that there *are* alternatives for the construction of this new feeder, and that they differ in both ability to handle future growth without further construction, and service quality. Note that "do nothing" is listed as an option. This is not a feasible option for the utility, but for completeness sake it is given a PW cost of $20,000,000 – the consequences, at least in part, of not serving these new customers.

Feeder number 2 is an existing feeder with an average equipment age of 45years, part of an aging infrastructure area of the system. Slow load growth (.3% per capita increase annually) has exacerbated the usual problems that aging causes, leading to below-average service quality. The table shows the capital costs for various rebuilding programs, along with their expected PW costs based on project O&M savings and restoration costs, calculated by analysis of age, equipment condition, load curves, and failure-likelihood. Again, "do nothing" is an option in this case.

Least-Cost Selection of Alternatives (Traditional Approach)

Traditionally, electric utilities have applied a planning paradigm that includes several fundamental assumptions:

1. All new customers will be connected to the system.
2. All new designs will meet standards.
3) Long-term (e.g., present worth) costs will be minimized.
4) Capital funding for any project that meets the above requirements will be forthcoming.

This paradigm can be labeled an engineering-driven paradigm, because engineering requirements (1 and 2 above) drive the decision-making process (3) and define the financial requirement (4). This is illustrated in Figure 5.14.

In this traditional approach, the two alternatives in Tables 5.14 would be evaluated for total present worth cost over a long period (usually 30 years) or alternatively on a lifetime basis, adding together present worth costs for both initial and continuing operating costs over the period. If load projections and planning criteria showed that the feeder had to be upgraded during the period, the present worth cost of the future capital cost for that upgrade would typically be included.

Table 5.5 Two Feeder Projects Alternatives Evaluated on the Basis of Least-Cost Present Worth Value System at 11% Discount Rate - \$ × 1000

Feeder Number One – New Growing Area of System

Feeder Option	Description of Approach	Meets Criteria?	Year 1 Capital	PW Cost	Comments
0	Do nothing	No	\$0	\$20,000	Do nothing is not really an option
1	4/0 large trunk	No	\$2,250	\$7,550	Must be rebuilt for growth in year 2.
2	4/0-4/0 branched	Not	\$2,350	\$7,150	High losses. Part rebuild yrs. 7 & 12.
3	336/4/0 branched	Most	\$2,625	\$6,360	Must be rebuilt for growth in year 8.
4	336 large trunk	Barely	\$2,850	\$6,100	Reinforcement needed in year 14.
5	636 large trunk	Yes	\$3,075	\$5,500	Reinforcement needed in year 17.
6	795 large trunk	Yes	\$3,450	\$4,900	Part rebuild, yr. 24. Lowest PW cost.
7	636/336 branched	Yes	\$3,550	\$4,800	Needs rebuild yr. 38.
8	795/336 branched	Yes	\$3,675	**\$4,750**	Never needs rebuild. Low losses.
9	795-636 branched	Yes	\$3,800	\$4,800	Low losses don't justify cost.
10	1113 large trunk	Yes	\$3,900	\$5,100	Too much equip. Higher O&M
11	1113/633 branched	Yes	\$4,250	\$4,950	Gold plated, not justifiable

Feeder Number Two – Aging Infrastructure Area of System

Feeder Option	Description of Approach	Meets Criteria?	Year 1 Capital	PW Cost	Comments
0	Do nothing	Yes	\$0	\$6,850	Tolerate failures and poor service
1	Fix most serious	Yes	\$550	\$5,500	Repair only most critical items.
2	Rebuild Plan A	Yes	\$900	\$4,700	Rebuild trunk
3	Rebuild plan B	Yes	\$1,400	**\$3,500**	Rebuild trunk and branches
4	Rebuild plan C	Yes	\$1,850	\$3,600	Rebuild all but <20 year old
5	Rebuild All	Yes	\$2,750	\$3,835	Rebuild all

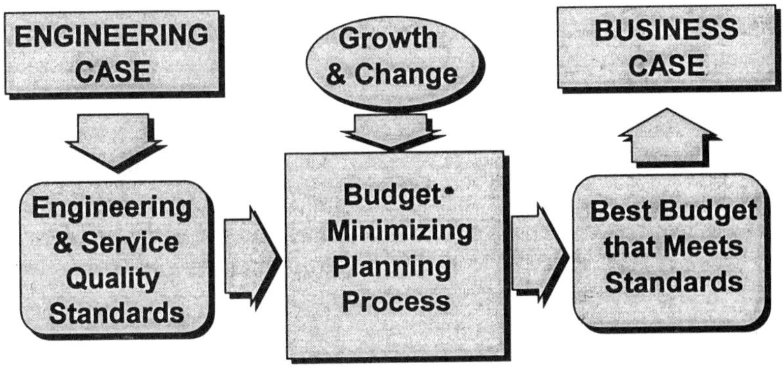

Figure 5.9 The traditional engineering-driven approach to prioritizing projects and determining utility budgets.

In each case, the alternative with the lowest PW cost would be selected. In the example shown in Table 18.1, this process would select option 8 for feeder 1 (the 795/336 branched feeder design) at a capital outlay of $3,695,000 and option 3 for feeder 2 (rebuild plan B) at a cost of $1,400,000. The utility would need a total of $5,075,000 in capital for the two projects, probably borrowing the money for this work, as is the typical financial method at most investor owned utilities.

Facing Budget Constraints

Suppose the utility that needed to do both projects listed in Table 5.13 was forced to reduce its capital budget by 30% in order to achieve financial targets for both rate and stock return reasons. This reduction would drop the $5,075,000 capital total for the two projects to an allowable $3,550,000.

One possible adjustment within the traditional paradigm to accommodate budget constraints is to change the time-value of money used in the PW analysis. The discount rate can be raised to 20%, 30%, or even higher, with the result that it "cuts" items from the approval list of projects. Each increase in discount rate reduces the perceived value of future savings from a project, and biases the decision-making process in favor of alternatives with lower initial costs, such as "doing nothing."

To achieve the desired reduction to below, $3.55 million in this example, a discount rate 38% is necessary. This will force the selection process for the two example projects to $3,550,000 by selecting only option 7 for feeder 1, and selecting nothing for feeder 2. This provides a total cost of $3.55 million (just meeting the budget) and selects "do nothing" for the refurbishment project.

Using this approach, the capital budget required by the planning process *is* reduced. However, there are four practical problems with this approach:

1. It does not provide a mechanism to force the budget below any specific limit. By trial and error, the utility can find a discount rate that "works out" to the right budget, but it is not a simple process. More important, it is not a dependable process: next year the required discount could be different. This approach does not lead to a simple rule that can be "aimed" at a specific budget target.
2. No ability to make really difficult cuts. As discussed in Chapter 6, PW analysis never says "no" to essential projects. In the face of severe budget cuts, the traditional paradigm still decrees that all new construction must be done, and work must be done to standards.
3. Using a high discount rate is inappropriate. High discount rates means future savings are unimportant, which is not true: The utility's current financial situation precludes it from *affording* long-term savings, but this is not the same as saying it does not want them. (The importance of this distinction will become clear in several more pages).
4. Customers in older areas of the system might legitimately complain that they are not getting the service and attention they deserve. New customers in outlying areas get shiny new service built to high standards. They get old equipment and lousy service.

Raising the discount rate to accommodate this budget reduction needs to change only the relative importance of the future savings against which present costs are evaluated: The planning method still assumes that the goal is to balance long and short-term costs. Truly deep budget cuts need to be addressed by looking at how necessary the various projects are *today*, not at how today's costs weigh against tomorrow's savings, in any measure. (This will be discussed at length in a few paragraphs.

Using Benefit/Cost Ratio to Make the Decisions

Ranking projects based on the benefit/cost ratio is often used to decide among project alternatives. This method can compare the present value of savings over the life of the project to today's cost to implement. Once can also evaluate projects based on their payback period. The two methods usually produce comparable results, selecting or rejecting the same sets of projects. With all proposed projects ranked in this manner, the utility can then determine how much capital it would need to finance all worthwhile projects. Alternatively, it can chose to find only those above any arbitrary benefit/cost limit.

Figure 5.10 plots the results of such analysis for the two feeder projects from Table 5.5, and others that are in the present-year budget. This particular utility needs $88 million to finance all worthwhile projects, as determined by traditional (PW least cost) planning paradigms.

The utility could achieve a budget reduction of 30%, limiting capital spending to only $61.6 million, by approving only projects with a benefit-cost ratio of 4.4:1 or better.

Table 5.6 shows the impact of using benefit cost ratio as the arbiter in prioritizing the two example feeder projects. The benefit cost ratio selected for each option for both projects is shown in the rightmost column. For feeder 1, option6 is selected (795 large trunk) at $3,450,000. For feeder 2, nothing is selected; all the options fall below the cutoff. The utility would have to have a budget of at least $77 million (only a 12.5% reduction) before any of these projects would be approved.

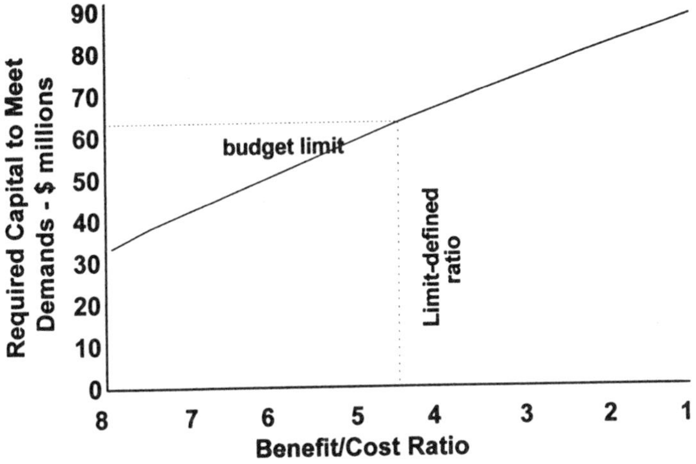

Figure 5.10 Cumulative capital requirements versus benefit/cost ratio for all "worthwhile" projects. The utility needs $88 million. Using B/C as the prioritization test, a reduction of 30% (to $61.6 million) means it can afford projects only up to 4.4:1, and only those with a B/C ratio above that limit should be approved. Benefit/cost ratio provides a way to limit spending to within a severe budget constraint, but it does not spend that budget most effectively (see text).

Table 5.6 The Two Feeder Projects From Table 5.13 Assessed By B/C Ratio

Feeder Number One – New Growing Area of System

Feeder Option	Description of Approach	Year 1 Capital	PW Cost	Benefit (PW savings)	B/C Ratio
0	Do nothing	$0	$20,000		
1	4/0 large trunk	$2,250	$7,550	$12,450	5.5
2	4/0-4/0 branched	$2,350	$7,150	$12,850	5.5
3	336/4/0 branched	$2,625	$6,360	$13,640	5.2
4	336 large trunk	$2,750	$6,100	$13,900	5.1
5	636 large trunk	$3,075	$5,500	$14,500	4.7
6	795 large trunk	$3,450	$4,900	$15,100	4.4
7	636/336 branched	$3,550	$4,800	$15,200	4.3
8	795/336 branched	$3,675	$4,750	$15,250	4.1
9	795-636 branched	$3,800	$4,800	$15,200	4.0
10	1113 large trunk	$3,900	$5,100	$14,900	3.8
11	1113/633 branched	$4,250	$4,950	$15,050	3.5

Feeder Number Two – Aging Infrastructure Area of System

Feeder Option	Description of Approach	Year 1 Capital	PW Cost	Benefit (PW savings)	B/C Ratio
1	Do nothing	0	$6,850		
2	Fix most serious	550	$5,500	$1,350	2.5
3	Rebuild Plan A	900	$4,700	$2,150	2.4
4	Rebuild plan B	1400	$3,500	$3,350	2.4
5	Rebuild plan C	1850	$3,600	$3,250	1.8
6	Rebuild All	2750	$4,435	$2,415	0.9

Results Still Not Completely Satisfactory

The use of B/C ratio as the prioritization for projects resulted in only slight real improvement over the use of a higher discount rate. Of the four objections listed earlier for that approach, there still exist numbers 1 and 4:

1. B/C ratio analysis does not provide a mechanism to force the budget below any specific limit. By trial and error, the utility can find a discount rate that "works out" to the right budget, but it is not a simple process. More importantly, it is not a dependable process: next year the required discount could be different. This approach

does not lead to a simple rule that can be "aimed" at a specific budget target.

- (Objections 2 and 3 were solved by the use of B/C analysis)

2. Customers in older areas of the system might legitimately complain that they are not getting the service and attention they deserve. New customers in outlying areas get shiny new service built to high standards. They get old equipment and lousy service.

Number 4 is a particularly damning objection in the long run, sure to get the utility into trouble with its customer base and regulators alike.

Budget Constrained Planning: Optimization Using Marginal Benefit Cost Ratio

A business rather than engineering-driven planning process

Even though many utilities may still operate within a regulatory framework, they may face limits on the amount of money they can borrow or other financial pressures that preclude as much capital spending as they would like. This constraint is completely incompatible with the fourth precept listed for the traditional paradigm earlier, which assumed that sufficient capital funding would be available for all projects that justify themselves on the basis of present worth economics.

Since the constraints facing the utility are business-related (goals related to borrowing limitations, required rate reductions or caps, and targets for satisfactory stock dividend performance), it makes sense to adopt a "business driven" decision making basis for capital spending. Many utilities are forced, or chose to, shift to the business-driven structure shown in Figure 5.11, in order to plan to meet the pressures of a competitive marketplace and industry. In this paradigm, a survival and expansion plan, based on business necessities and goals, defines the available T&D budget for the utility. Good or bad, that's all the budget there is, and unfortunately experience throughout the 1990s proved that in general the required reduction was on the order of 25%–35% less than traditional levels.

What went wrong with benefit/cost ratio prioritization?

There is, however, a problem with using the benefit/cost ratio as previously described. Consider the preceding example. The first option feed 1 had such a high benefit/cost ratio that all options that built on it also had high benefit/cost ratio. Each of the options for feeder 1 builds on the high B/C ratio of its option.

For example, option 2 costs an additional $150,000 and lowers PW cost by $400,000. Those additional dollars spent over option 1 carry a benefit cost ratio of only 4:1, yet the option is ranked as having an overall B/C ratio of 5.5 to one because its costs and benefits include those of option 1 also. By contrast, all the options for feeder 2 are much lower in B/C ratio: feeder 2's highest option is less than feeder 1's lowest option.

As a more dramatic example of the "wrong" this method applies, consider adding an option to Table 5.6. Option 1b would be a variation of option 1 that costs $500,000 more in capital, but for which that additional cost contributes nothing in terms of improved PW benefit. At a cost of $2,850 and benefit of $12,450, this combination would still have a benefit/cost ratio of 4.5:1, so it would be *still be approved* under the budget-restricted guideline and over any spending on feeder 2 - a total waste of $500,000!

Marginal Benefit Cost Analysis

Table 5.7 shows the options for the two feeders evaluated on a *marginal* benefit/cost ratio basis. In each project, the additional present worth savings delivered by moving up from one option to the next is evaluated against the that additional expense.

Given a budget constraint of $3,550, planners "buy" the most effective use of their dollars by taking the options whose marginal benefit cost ratios are shown in bold (final column) in Table 5.7. This is option 3 for feeder 1 and option 3 for feeder 2. A total of $3,525,000 in expenses for the two projects gains a total savings (over do nothing) of $15,790. No other combination of choices delivers so much "bang" for the buck.

Figure 5.12 shows the budget-requirements when all projects the utility is considering are re-ranked on the basis of their marginal present worth/marginal capital ratio. There, the total for every marginal B/C ratio is determined, a plot analogous to that of Figure 5.10 in how it is used, but using marginal rather than just benefit cost ratio. As before, slightly more than $88 million is required to meet all projects with a ratio above zero. As indicated, when limited to $61.6 million, the utility must approve only projects with a MPW/C ratio of 1.5:1 or better. But, unlike the case where it was using benefit/cost ratio as the guiding rule, it will now always "buy" an equivalent (and biggest overall) PW savings - the most it can afford.

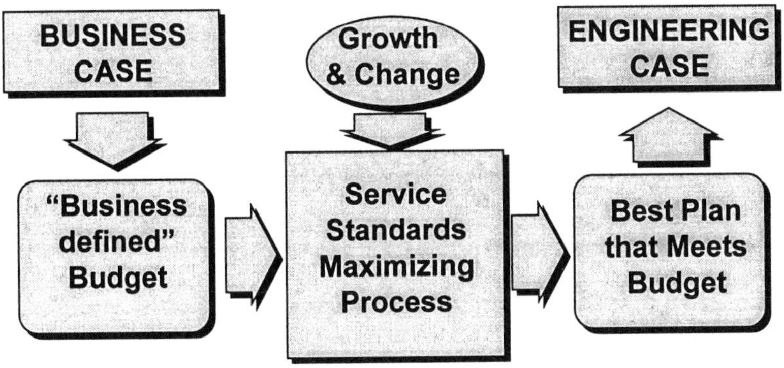

Figure 5.11 Many utilities have shifted to a business-driven structure, in which business needs define the available budget. The traditional planning paradigm has difficulty adjusting to this external change in cases where the reduction from traditional budget levels is significant. Although this paradigm stretches some traditionally rigid standards, it obtains both better reliability and better economy from the limited budget.

Application of Marginal Benefit-Cost Evaluation

Generally, the way one proceeds with the decision-making on two or more projects using this approach is shown in Table 5.8. There, the options for the two feeder projects (from Table 5.7) have been merged and sorted into order based on marginal benefit cost ratio. Planners make their selections against a budget total by moving down the table, selecting options and adding their marginal cost (delta cost) into their total, stopping when they have the appropriate budget total (equal to the limit).

Given a budget limit to work within, ($3,550,000 in this example), planning begins by selecting feeder 1 option 1 ($2,250,000), the highest item on the table. It then proceeds to add feeder 1, option 2, for an additional $100,000, as well as option 3 for that same feeder, at $275,000, for a total so far of $2,625,000 in cost. The next item on the list is feeder 2 option 1 with a marginal cost of $550,000, bringing the total spent so far to $3,175. Finally, feeder 2 option 2 can be added, at an additional $350,000, bringing the total to $3,525, the best, with the options listed, totaling under $3,550.

Table 5.7 The Example Two Feeder Projects From Table 5.16 Assessed By Marginal B/C Ratio

Feeder Number One – New Growing Area of System

Feeder Option	Description of Approach	Year 1 Capital	Delta Capital	PW Cost	Delta PW Savings	Marg. Ratio
0	Do nothing	$0		$20,000		
1	4/0 large trunk	$2,250	$2,250	$7,550	$12,450	5.5
2	4/0-4/0 branched	$2,350	$100	$7,150	$400	4.0
3	336/4/0 branched	$2,625	$275	$6,360	$790	**2.9**
4	336 large trunk	$2,750	$125	$6,100	$260	2.1
5	636 large trunk	$3,075	$325	$5,500	$600	1.8
6	795 large trunk	$3,450	$375	$4,900	$600	1.6
7	636/336 branched	$3,550	$100	$4,800	$100	1.0
8	795/336 branched	$3,675	$125	$4,750	$50	0.4
9	795-636 branched	$3,800	$125	$4,800	($50)	-0.4
10	1113 large trunk	$3,900	$100	$5,100	($300)	-3.0
11	1113/633 branched	$4,250	$350	$4,950	$150	0.4

Feeder Number Two – Aging Infrastructure Area of System

Feeder Option	Description of Approach	Year 1 Capital	Delta Capital	PW Cost	Delta PW Savings	Marg. Ratio
1	Do nothing	0		$6,850		
2	Fix most serious	550	$550	$5,500	$1,350	2.4
3	Rebuild Plan A	900	$350	$4,700	$800	**2.2**
4	Rebuild plan B	1400	$850	$3,500	$1,200	1.4
5	Rebuild plan C	1850	$450	$3,000	$500	1.1
6	Rebuild All	2750	$900	$3,835	($835)	-0.9

Table 5.8 Options for Both Feeder Projects From Table 5.7 Merged Into One List, Ranked By Marginal B/C Ratio

Feeder #	Feeder Option	Description of Approach	Year 1 Capital	Delta Capital	PW Cost	Delta PW Savings	Marg. Ratio
1	0	Do nothing	$0		$20,000		
2	1	Do nothing	0		$6,850		
1	1	4/0 large trunk	$2,250	$2,250	$7,550	$12,450	5.5
1	2	4/0-4/0 branched	$2,350	$100	$7,150	$400	4.0
1	3	336/4/0 branched	$2,625	$275	$6,360	$790	2.9
2	2	Fix most serious	550	$550	$5,500	$1,350	2.4
2	3	Rebuild Plan A	900	$350	$4,700	$800	2.2
1	4	336 large trunk	$2,750	$125	$6,100	$260	2.1
1	5	636 large trunk	$3,075	$325	$5,500	$600	1.8
1	6	795 large trunk	$3,450	$375	$4,900	$600	1.6
2	4	Rebuild plan B	1400	$850	$3,500	$1,200	1.4
1	7	636/336 branched	$3,550	$100	$4,800	$100	1.0
1	8	795/336 branched	$3,675	$125	$4,750	$50	0.4
1	9	795-636 branched	$3,800	$125	$4,800	($50)	-0.4
1	10	1113 large trunk	$3,900	$100	$5,100	($300)	-3.0
1	11	1113/633 branched	$4,250	$350	$4,950	$150	0
2	6	Rebuild All	2750	$900	$3,835	($835)	-0.9

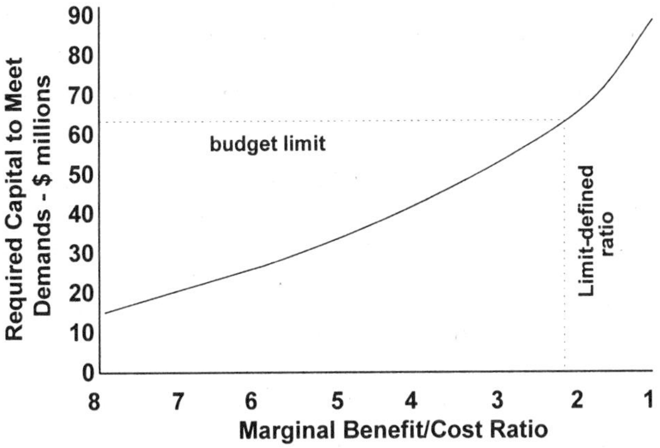

Figure 5.12 Cumulative capital requirements for the utility, versus the marginal PW/capital ratio required for that level of requirement. The utility's $61.6 million budget allows it to spend up to a 2.1:1 marginal benefit cost ratio. This approach provides a much better guiding rule for project approval and assures maximum overall "bang for the buck" using the limited capital budget.

New basic paradigm for T&D project decision-making

Generally, this approach is applied in decision making with a portion of the future costs being the performance-based rate penalties, customer paybacks, or other monetary risk measures of poor customer service quality that might be incurred by the utility. More on this concept will be given throughout this book, as prioritization based on marginal benefit-cost ratio is one or the cornerstones of the results-driven management approach and a key factor in the procedures that will be outlined to balance and optimize spending over the entire utility. This approach represents two major shifts in the thinking pattern, or paradigm applied to evaluation of projects.

1. *Zero-base planning.* No project, even a new customer extension, is absolutely necessary. All planning is done on a "zero base" basis, by beginning with "do nothing" as the base assumption.

2. *Service-quality rather than standards-driven planning.* If the cost penalty due to poor reliability is high enough to justify "doing something," then money will be spent. Otherwise nothing is approved no matter how far below standards the resulting design will be (excepting safety standards, which must be met).

Note that the method satisfies all of the four problems listed earlier with application of the higher discount rate for planning. In particular, customers in older areas of town now have the same cost per improvement of service quality directed at them as consumers in newer areas of the system.

Refurbishment alternatives receive a "marginal" advantage

The net effect of the changes made in prioritization due to use of this method is complicated. First, although this paradigm does not assume that new service extension will be done, it's impossible to argue with the service contribution that *doing something* delivers in all new construction cases (building *something* avoids nearly 8760 hours of interruption time, more "bang" than one can get with most other projects). As a result, from any practical perspective, this planning method will always decide to build *something* to serve new customers.

But there are substantial differences in *what* this process will decide to build. Taking into account the value of service, this approach will no longer build the new service extension project to a very high level meant to meet thirty-year needs, just because that is what meets standards. Instead, it adjusts both new customer construction and aging infrastructure refurbishment decisions toward equivalent levels of payback on money spent for customer service. Money is spent on making *some* refurbishment additions to raise poor service levels, and the capital is obtained by cutting what is built in new areas so that service levels there may be less than if built to traditional "lowest PW" levels.

The results indicate the types of sacrifices are often spread around the system. Restricted by budget constraints to projects with a 2.5:1 marginal ratio, the feeder 1 project is built with only a limited design not meeting traditional standards, for a capital reduction of $1,000,000 over the unconstrained (lowest PW cost) situation. For the refurbishment project, its option 2 (plan A) is selected, a reduction from the recommendation that flowed out of unconstrained PW cost minimization, of $500,000.

The reductions are spread equitably over both projects: using higher discount rates or B/C ratio made all the cusps in the aging infrastructure area. Note that with this planning method both projects get some of the limited funds, though neither gets as much as in the unconstrained budget (minimize present worth).

From the perspective of the traditional approach, the decision made for the new feeder project in this situation makes no sense. It is built with a capacity and style so that it is stressed at peak conditions from the very day it is built. However, from this new paradigm's standpoint, the service quality provided by the new feeder is no worse (in fact it is slightly better) than the service quality delivered to customers in older areas of the system.

Interestingly, the ultimate result of this planning method is not a drop in overall reliability compared to continuing with the traditional paradigm. The new paradigm results in old and new areas being more balanced in capital spending and service quality results. Due to the non-linearities in reliability-versus-cost relationships, very often the utility's service quality indices, such as SAIFI, actually improve slightly with this value system.

This planning method is recommended in situations where a distribution utility has capital budget limitations that must be addressed and there is a concern that inequities in service quality may become a real or perceived problem, particularly with allocation of funds for refurbishment of aging infrastructure areas. Spending the capital budget optimally while treating both old and new customers on an equivalent basis required the combination of several factors. First, all projects must be "zero-based" as discussed above. Secondly, service quality as defined previously is the <u>only</u> "standard" that <u>must</u> be met. Finally, all projects and alternatives are evaluated on their marginal benefit to marginal cost ratio.

5.6 CONCLUSION

Planning involves the selection of the best alternative from among those at hand. In the power industry, that selection depends to a certain extent on the values of the group making the decision. *All* cost evaluations for power delivery planning are based in part upon economic evaluation that is consistent over the range of all alternatives. Usually, this means that present and future costs must be compared. Present worth analysis, levelized costing, or some similar method of putting present and future value on a comparable basis will be used.

Intra-Project, Marginal Benefit-Cost Alternatives Comparison

The foregoing method of maximizing benefit from a limited budget by comparing alternatives within each project (i.e., for each, several alternative ways of doing the project were evaluated) against alternatives from other projects, is called *intra-project* alternatives analysis. Methods that compare only a single alternative for each project (i.e., the method used in Table 5.5) are called *inter-project comparisons.* They compare each alternative based only on its ultimate "lowest cost" alternative. By contrast, *intra-project comparison* ranks the various alternatives for each project against one another.

It is the combination of comparing multiple alternatives in on project, against those of others, and the use of marginal benefit-cost analysis, that yields the additional savings. This method – intra-project comparison based on marginal benefit cost analysis – is called *budget-constrained planning*, of BCP for short.

BCP *is* more complicated to implement than traditional prioritization methods. However, the authors have helped several utilities implement it, where it has led to bottom line savings and reliability improvements as dramatic as the improvement shown in the foregoing examples. BCP delivers results. It will be used both in capital planning, and in maintenance and operations prioritization, later in this book. When applied uniformly over an entire utility (all regions, all functions), it is the basis for Results-Driven Management.

REFERENCES

T.W. Berrie, *Electricity Economics and Planning,* Peter Peregrinus Ltd., London, 1992.

J.J. Burke, *Power Distribution Engineering: Fundamentals and Applications,* Marcel Dekker, New York, 1994.

H.L. Willis, *Power Distribution Planning Reference Book,* Marcel Dekker, New York, 1997.

H.L. Willis and R. W. Powell, "Load Forecasting for Transmission Planning," *IEEE Transactions on Power Apparatus and Systems,* August 1985, p. 2550.

H.L. Willis and G. B. Rackliffe, *Introduction to Integrated Resource T&D Planning,* ABB Guidebooks, Raleigh, NC, 1994.

World Bank, *Guidelines for Marginal Cost Analysis of Power Systems,* Energy Department paper number 18, 1984.

6
Equipment Inspection, Testing, and Diagnostics

6.1 INTRODUCTION

This chapter provides a summary and comparison of the basic procedures by which an electric distribution utility can better judge the condition of its equipment: inspection, testing, and diagnostics. These basic evaluation steps in *condition assessment* provide the data for analysis and prioritization methods of power system equipment, which will be covered in Chapter 10.

To be effective, inspection, diagnostics, and testing must be applied in a carefully coordinated program that also uses the results from system operation (trouble reports) to identify overall equipment condition and performance, and that prioritizes preventive maintenance, repair, rebuilding, and replacement. Optimum results are derived by then loading equipment to levels determined based on its assessed condition, the loss of lifetime caused by the usage, and the available economic alternatives. Procedures for organizing and managing such programs will be discussed in Chapter 10. This chapter focuses on condition assessment inspection, diagnostics, and testing methods themselves.

This chapter begins with an overview of inspection and testing and their goals, given in section 6.2. The various types of tests and diagnostics applied to equipment are covered in section 6.3. Equipment tests include a variety of methods to evaluate the electrical and mechanical capabilities of the units and to diagnose what might be wrong internally with a unit that gives Orr test results. Section 6.4 covers tests of insulating oil. Oil is used in a wide variety of power system equipment. Oil tests focus both on evaluating the condition of the oil itself, and on using oil test results to diagnosis the internal condition of the equipment. Section 6.5 concludes with a summary of key points.

6.2 INSPECTION, TESTING, AND DIAGNOSTIC EVALUATION

Inspection, testing and diagnostic evaluation is the means through which the condition of equipment is determined. These three types of condition verification all have different meanings.

Inspection involves passive evaluation of conditions based on visual or aural inspection of a device or structure. Basically, one looks, and at times listens to, the device to determine if it looks or sounds as if it is in good condition. This visual and aural inspection may be augmented by equipment to enhance the detail or better quantify the results (infrared camera, microphone and harmonic analyzer system). The best example of this is the use of infrared scanners to identify equipment that is running hotter than normal.

Olfactory inspection can also be an aid in a few cases ("Do you smell something burning? Does this oil smell sooty to you?") and is a big area for future development of inspection and test equipment. Sniffer machines, which can detect trace amounts of certain chemicals in the air or a fluid, are being developed to augment inspection and testing of equipment. (At the present time the most frequently seen examples of these devices by the general population are the explosives sniffers used in security checks at airports).

Another example of equipment-enhanced sensing for inspection is a practice at numerous utilities of leaving the radio of company cars tuned to the AM band while driving line routes for visual inspection of equipment. Cracked insulators, corroded brackets, and broken conductor strands create small amounts of radio interference, detectable as a sudden increase in "static" background noise by the radio when near them.

Testing involves checking the device and its associated support equipment (its system) to make certain it operates as it should, by checking to see that it can indeed function to specification. Thus, a breaker test involves making certain that it trips in the time and at the speed it should, and that it re-charges or re-sets itself as expected, and that its relay performs as "programmed." A test of a line switch involves cycling it once to assure that it will open and close as expected. A transformer may be tested at full load, or merely monitored in detail during peak season and its data results (maximum temperature, rate of temperature rise, losses) compared to expected values. Any deviation would trigger diagnostics to determine what was wrong.

Diagnostics involve "tests" that measure factors other than those directly-related to a device's function: tests whose results provide clues to the device's internal condition, the performance of subassemblies inside it, or the possibility of impending failure. Usually, diagnostics probe the condition of materials (e.g., insulation) and are tracked over time to determine the rate of deterioration or contamination. The amount and type of gas, and the rate of its increase over time, identify whether there are internal faults, arcing, or overheating inside the unit – problems with something other than with the oil itself. This makes them a diagnostic test, not strictly speaking, an equipment test.

"Test" has two meanings

The word "test" has a somewhat confusing dual meaning here, a fact compounded by the inexact use of the word on a day to day basis by many power systems professionals. Strictly speaking "testing" of a device always refers to actions or steps performed to measure or verify its performance against its functional specifications. Actions that check the unit to see that it has the correct turns ratio, that its losses, leakage, and thermal buildup are within spec, and that it meets any other specifications for its performance (noise) are tests of the unit itself.

However, many diagnostic procedures involve a "test" of a particular component or factor in the machine or equipment. For example, dissolved gas tests of transformer oil provide an indication of the condition of the oil itself and whether it can do its job, making them a "test" of the oil. However, dissolved gas tests of insulating oil done are done primarily because they are *diagnostic* – they provide powerful clues to the condition of a transformer's winding insulation and the causes of problems therein.

Thus a test of a component or part (transformer oil in this case) can take two forms:

1. *Testing of the oil* to determine if *it* performs its function. For example, dielectric breakdown test of transformer oil directly measures one aspect of the performance the oil is expected to provide (insulating ability).

2. *Diagnostic tests.* By contrast, the oil could be tested for acidity or by dissolved gas analysis (DGA). These tests respectively determine the content of acids in the oil (formed by oxidation) and the amount of gas dissolved in the oil. The former gives an indication of both the condition of the oil and the likely amount of sludge buildup in the transformer. The second test is widely regarded as the best possible early indicator of impending problems with the transformer itself.

Three Purposes

Inspection, diagnostics and testing fall into three different categories, with three different purposes:

1. *Acceptance, or capability testing* – new equipment is often tested to verify that it both is undamaged and ready for service, and that it meets specifications as to capability, losses, impedance, and a host of other factors important in its performance. Some of these tests may be the same as performed later during routine inspections or condition evaluation, but many are unique, one-time evaluations, such as a test for no load and load-related losses performed on new transformers, to verify what was built to specification.

2. *Routine inspection* – typically equipment in service is inspected, and perhaps some diagnostics or tests performed, on a periodic basis – annually, every three or five years, etc. This is done for what might be called "general principles" and because warranties may require it. The overall goal is to identify deterioration or damage before it becomes too serious to perform preventive maintenance, and to identify equipment that is about to fail in time to remove it from service in an orderly fashion.

3. *Condition evaluation* – on occasion, more comprehensive tests than called for on a routine basis are performed on some or all equipment to establish more precisely the condition of these units. The purpose of these tests is to identify as much as possible the condition of the equipment and any problems that might preclude its equipment being left in service with a high degree of confidence that it will perform its function dependably.

Often the line between routine service and more thorough condition evaluation is blurred. For example, routine inspection policies at some utilities call for mandatory follow-up testing of any equipment found to have possible problems. Their condition evaluation is considered a second-stage in routine testing; done only if it appears needed. Other utilities include as part of routine testing, thorough condition evaluation of randomly selected units, the goal being to test enough units to identify any trends of deterioration or deficiency endemic to all units in a certain class or category. Policy on inspection and condition evaluation is appraised and set, using methods that will be discussed in Chapter 10.

Field or Laboratory Evaluation

Inspection, diagnostic tests, and testing can be carried out in the field (where the equipment is installed) or in a laboratory or test facility. The advantage of field tests is that they can be done on-site, usually taking less time and involving less skill and expense than laboratory tests.[1] However, generally the laboratory tests

[1] At least if the cost of equipment is not included. Often field tests require specialized diagnostic equipment whose cost can be quite high, whereas laboratory tests for the same

are more accurate and thorough.

For example, dielectric testing of transformer oil can be done either as a field test using the ASTM (American Society of Testing and Materials) D-877 disc electrode method, or as a laboratory test using the ASTM D1816 "VDE cell" method. The former involves testing samples of about one pint of fluid for dielectric breakdown strength, by placing the fluid in a small container in which two small disc electrodes are spaced .1 inch apart. Higher and higher voltages are applied until the oil breaks down. The entire test takes but half an hour, and uses a portable device the size of a small suitcase that is taken to each transformer to be tested.

By contrast, the D1816 "VDE cell" dielectric breakdown test of oil is done in a laboratory. It involves testing a larger sample of fluid drawn from each transformer and taken to the laboratory, in a "VDE cell" within which the oil is circulated, with test electrodes that more closely mimic the field strength and uniformity of voltage seen inside a transformer. This test is regarded as more accurate in depicting electric strength and scatter than the disc electrode test, but it takes more time for the test itself, and for collection of materials and running back and forth from the field to the laboratory.

Non-Destructive, Destructive or Forensic Tests

Testing methods can also be categorized by the condition of the device or equipment being tested before and after the evaluation, as shown in Figure 6.1.

> *Destructive tests* ruin or damage the device or equipment. Afterwards, the entire device, or a portion of what was tested, is damaged and not suitable for service and may in fact have been completely destroyed. Examples are (for cables in this case) voltage breakdown testing of a section of cable carried out to the point of failure, to determine its dielectric strength and mode of failure, and inspection of samples of cable for treeing. The former destroys the cable section being tested. The later removes part of the cable (which must be repaired).
>
> Destructive tests are generally carried out by manufacturers on new equipment to verify strength and identify failure modes. Utilities use the tests on select samples of equipment that have been in service, in order to determine the rate of deterioration of various components. The average condition of equipment in certain age and service categories can also be determined. These tests are often applied to cables, switches, and service transformers.
>
> Usually, equipment that is destructively tested is damaged to the point of non-utility, if not destroyed outright. Regardless, this equipment should never be returned to service after the tests, unless it is part of an experiment on durability, etc.

---

or similar factors would involve only standard laboratory equipment.

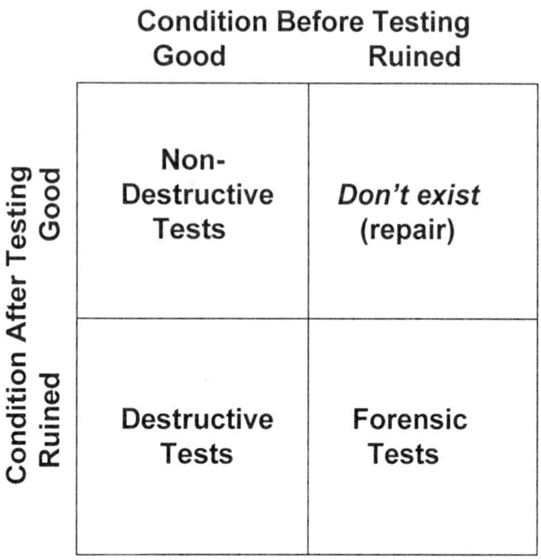

Figure 6.1 Diagnostic and testing methods can be categorized by the condition of the equipment before and after the tests.

Non-destructive tests are supposed to cause no damage or undue deterioration of the equipment. The word "supposed" is used here because there is concern in some quarters about several non-destructive testing procedures, which are suspected of putting the equipment being tested under considerable stress. While these tests do not usually fail the unit at the time of testing, some experts believe the added stress increases the subsequent failure likelihood. One example is high-energy pulse (thumper) testing for underground cable fault location. In this cable fault test, a failed cable section is pulsed from one end with a series of high frequency, high-voltage pulses. Field personnel to detect the field generated by these pulses along the cable route use portable aboveground sensors. This field will be generated only by the cable up to the location of the fault, not beyond it, making it (relatively) easy to find the fault location. The pulses from these tests are suspected of causing high stress on the cable insulation, which perhaps contributes to its premature deterioration.[2]

[2] Deterioration of insulation in cables is very roughly inversely proportional to frequency. Thus the same cable should last about 16% longer in service in a European system (50 Hz) than in an American system (60 Hz), all other things being equal. Pulses used in some cable fault testing have a high voltage and a very high frequency content – the pulse edges are of course very high frequency but the pulse repetition rate is often quite high as well. In combination with a pulse voltage raised in order to make the field more

Regardless, the concept for a non-destructive test or diagnostic is that it is benign, making no impact on the equipment's subsequent performance. Most no-destructive tests certainly fall completely into this category. Only a few are known or suspected of causing large amounts of equipment stress.

Forensic tests may be carried on failed equipment after an event in which "something went wrong," in order to determine the cause of the failure. Often, the equipment being tested was damaged or destroyed in the event, and the testing involves a type of detective work to determine how and why the failure occurred. Much of forensic evaluation involves trying to identify the exact sequence of events of the failure by examination of the failed device (what is left of it) – where did the problem begin and how did it spread?

Comprehensive forensic tests are most often done after a catastrophic failure that caused damage to public property or life, often in preparation for a lawsuit. However, forensic tests can be important to a utility trying to manage aging infrastructure because they can establish if there is a pattern in the failures of equipment. For example, if forensic evaluation of all failed transformers of a certain type shows that they fail for what appears to be a voltage breakdown in a certain part of the unit, then diagnostic testing can focus on that in the future.

Standards and Recommendations for Inspection, Diagnostics, and Testing

There are a number of standards bodies that have defined recommendations (standards) for inspections, diagnostic tests, and testing of electrical equipment. Not all of these standards groups have an applicable standard in all categories of tests or equipment. Not surprisingly, where two or more organizations have overlapping standards, they may differ in the details of their recommended testing methods. Table 6.1 lists some of these organizations.

General Categories of Inspection, Diagnostics, and Testing for Condition Assessment

Voltage withstand and dielectric strength

Voltage capability tests including dielectric strength and breakdown tests are performed to determine if the voltage withstand capability of the equipment is up to requirements. Slight degradations in insulation are detectable as leakage or by other indications given by some types of diagnostics (e.g., partial discharge classification based on field uniformity, power factor tests).

---

detectable, this often results in calculated aging rates of up to 4,000 times normal service.

Table 6.1 Standards Bodies with Guidelines or Recommended Testing Methods for Electrical Power Equipment

Acronym	Meaning	Country
AFNOR	Association Francaise de Normalisation	France
ANSI	American National Standard Institute	Mainly USA
ASTM	American Society of Testing and Materials	Mainly USA
BSI	British Standard Institute	United Kingdom
DIN	Deutsche Industrial Normal	Germany
IEC	International Electrotechnical Commission	Mainly Europe
IEEE	Institute of Electrical and Electronics Engineers	Mainly USA
JIN	Japanese Industrial Standard	Japan
VDE	Verband Deutscher Electrotechniker	Germany

Thermal testing

Thermal tests determine if the device overheats during normal service conditions. Two processes can lead to overheating. First, the unit may be generating more heat inside it than is normal – as from internal faults. Secondly, its ability to cool itself might be degraded, as from sludge covering internal components or blocking cooling ducts. Heat being the major cause of early deterioration of just about everything involved in electrical equipment, such problems can lead to short equipment lifetimes if left uncorrected.

Thermal testing of a unit such as a transformer can be done by subjecting the unit to special full-load tests in a laboratory – subjecting a "cold unit" to a step increase to full load and measuring its temperature over a 24-hour period thereafter. However, often such tests are unfeasible for equipment in the field. The same information can be determined by analysis of load-cycle and temperature monitoring for a period of several weeks, and a subsequent analysis of the results, as will be discussed in section 9.3.

Mechanical tests

Tests are often done of devices with moving parts, to determine that those parts do move with the speed, smoothness, and other properties required. Thus, line switches are checked to make certain they open and close (corrosion might prevent operation), breakers are checked for speed of operation and re-set, tap changers for proper movement, etc.

In addition, poles and structures may be tested to determine their mechanical strength. Both poles and cables can be tested using a type of penetrating probe that is pushed into the wood or insulating surface, the resistance to the penetration of the probe giving an indication of the strength of the wood or insulating material.

Physical condition determination

This is mostly from inspection. Physical condition determination aims to identify signs of past, present, or forthcoming problems. It seeks to identify any noticeable corrosion, deterioration or damage (evidence that a car hit a pole, or an insulator has been shot), or outward signs of internal malfunction (leaking oil around seals). Physical inspection of the device usually includes visual or infrared examination of the exterior of the unit while in service, but can include detailed examination of its internal parts (as for example when disassembling a circuit breaker). It may also include evaluation of noise and even odor ("Smell something burning?") using the senses, or special equipment to augment them (infrared scanners, acoustic signal analyzers, so called "sniffers" made to detect trace gases in the atmosphere).

Materials evaluation

Most diagnostic tests evaluate the chemical or mechanical properties of sub-components in an electrical device. These tests determine if the component material (e.g., XLPE cable insulation) itself has deteriorated or is defective in some way (treeing, internal voids). Performing diagnostic tests can also provide data that indicates there may be problems with other parts of the device (dissolved gas analysis of transformer oil).

6.3 EQUIPMENT TESTING AND DIAGNOSIS METHODS

This section reviews the various types of electrical test procedures used on cables, conductor, and electrical equipment. Most tests and diagnostics for electrical equipment apply a voltage of some type (DC or AC) to the equipment in order to measure current, resistance, or the rate of change in one or both over time. The methods of detection or measurement range from simple and direct (apply the voltage, measure the current, compute the resistance) to clever and complicated (use an AC voltage to excite a partial discharge at places where there are weaknesses, detect the location by timing of the return pulses)

Electrical testing methods fall into three categories: tests that use AC voltage; those that use DC voltage; and partial discharge (PD) tests, which use an AC voltage of some type but in a different manner than in "AC" tests. Generally, these testing methods require de-energization of the device, although most can be done while the equipment is in place (i.e., they are not solely laboratory tests but can be done in the field).

DC or AC?

Tests of an electrical device using an applied voltage are done for a variety of reasons:

- To determine resistance of insulation.
- To determine if internal faults exist, to identify overall condition.
- To infer amount and type of contaminants, and in some cases, to simply stress the device in the hopes that if there is a flaw, failure will occur now, not months from now while it is service.

Some tests use DC voltages, others use AC voltages of standard or non-standard frequency. Some tests (high potential voltage) can be done using either DC or AC, with slightly different diagnostics results from each.

AC and DC voltages have different effects on a device, a fact that determines which approach is suitable for some tests, and which alters the exact effect of a high potential test, which can use either approach. AC voltage tests apply a voltage similar to that seen by the device when in service. If done at normal system frequency (e.g., 60 Hz) these tests see impedances and produce stresses similar to those that the equipment is expected to endure. AC voltage tests are therefore a good proxy for the stresses and responses expected during service and can be used to verify that a machine or cable's condition matches that needed for service.

DC tests create no alternating field and "see" no effects of reactance. Thus they measure different aspects of a device's behavior than do AC tests. DC tests can be used to measure capacitance of a device, or to infer AC impedance or the "complex" condition of insulation, through measurements of time constants or rates of change in field-related factors during certain tests. One advantage is that they put less stress on insulation in some situations: voltage stress is frequency related.

It is useful to consider electrical testing in the frequency spectrum: DC testing is merely testing with voltage whose frequency is zero; various AC and PD tests use a higher frequency, one that may be lower, the same, or higher than system frequency, depending on the specific test procedure. Electrical equipment varies in its impedance as a function of frequency. So do the characteristics of flaws created in manufacturing or by the processes of deterioration and damage that the diagnostic testing may wish to identify.

Therefore, AC tests using a frequency other than standard system frequency (e.g., 7,200 Hz) excite different modes of response from the system and can be used to "tune" a test to examine only certain aspects of a device's insulation strength or operating characteristics. For example, Table 6.2 shows the

Table 6.2 Division of Stress between Paper-Cellulose and Oil Insulation in a Transformer Depending on Test Type

Insulation Type	AC Test Results	DC Test Results
Paper-cellulose	25%	75%
Oil	75%	25%

Source: Gill, Electrical Power Equipment Maintenance and Testing

difference in stress that AC or DC hi-potential testing of the windings of a transformer puts on a 15 kV transformer. By selecting either DC or AC frequency, a diagnostic test can be tailored to focus on a particular part of a device of greatest interest, or to avoid putting stress on a particular portion suspected of weakness.

Comparing DC and AC testing, qualitative differences are:

1. DC testing is considered less stressful on transformer and cable insulation than AC testing. Therefore, DC testing is less likely to do harm to a device, but AC testing is more likely to reveal flaws that might otherwise be revealed while in service.

2. Generally DC test equipment is lighter, more portable, easier and faster to use.

3. DC testing avoids problems with high charging capacitance, as for example, when testing cables.

4. Generally, DC test results are more sensitive to the temperature of the device than AC tests. They often require adjustment of their raw test results with extensive correction formulas or tables of result-vs.-temperature. These correction factors require accurate knowledge of the equipment temperature, which might not always be available (as when testing cables where one cannot measure the duct temperature along the entire length of the cable).

5. AC testing provides a more realistic test of the dispersion of voltage in the device than DC testing and tends to stress insulation more in proportion to stresses expected in service. For example, in transformer tests DC testing tends to stress paper insulation much more, and oil much less than in actual service.

Resistance Tests

Several types of resistance tests are performed on transformers. These are all variations on testing the resistance of the winding insulation. Resistance tests are essentially DC voltage tests applied with a megohmmeter.

1. *Insulation Resistance Tests* measure the resistance through the transformer insulation. They are carried out with a hand-cranked megaohmmeter or modern digital test unit that applies a DC voltage of higher-than-normal level for a brief period, and have relatively low potential to cause damage or high stress. Lower than expected values of measured resistance reveal flaws in the insulation level.
2. *Dielectric Absorption Test* is a version of the insulation resistance test in which the test is carried out over a period of up to 10 minutes, with the resistance measurement plotted over that time. Good insulation will reveal a slightly rising insulation level over time in a resistance test, poor insulation will not. Plotted as resistance vs. time on log-log paper, a good insulation system will give a straight line, a system contaminated or with deteriorated insulation will have a curved profile.
3. *Dielectric Absorption Ratio (DAR)* test is an older, field version of the dielectric absorption test, which does not plot resistance with time but only evaluates the ratio of resistance measured at 60 seconds to that obtained at 30 seconds. A value below 1.25 is considered unacceptable.

Power Factor Tests

Power factor tests evaluate the insulation strength of a material by measuring both the capacitive and dielectric components of the test current from application of an AC voltage. It then evaluates the condition of the component based on the magnitude of both, and the measured power factor of the test. As in power system analysis, power factor in the test refers to the ratio of the complex product of voltage and current to their scalar product. The voltage and current in this case are the applied voltage and the resulting current flow produced by an AC test applied to a cable, across the terminals of a device, or to a sample of oil. One advantage of measuring PF is that it is independent of sample size (e.g., the cable length). Therefore, it is accurate when one is uncertain of the size of the sample as compared to the size of the device being tested. For example, when testing cable, power factor is invariant as a function of cable length. This results from the fact that both the resistive and capacitive components increase with the length of the cable.

When applied to a cable or oil or other insulating material the power factor test assesses dielectric condition by determining if there is a lag between applied voltage in a test and the current it produces. Simplifying the concept somewhat by neglecting a few details of secondary importance, if dielectric strength is very high, the only current flow produced by the applied voltage will be a capacitive flow, completely out of phase with the voltage, producing a power factor of 0.0. By contrast, if the dielectric strength is weak, current flow through the oil will be resistive – a current will flow in phase with the applied voltage, and power factor will be far greater than 0.

In actuality, a power factor test can provide a bit more information about condition. By varying test voltage, the type of contamination affecting dielectric strength can be identified. For example, dielectric losses due to water or carbon tracking tend to vary as the square of test voltage, whereas those due to corona effects increase exponentially. Therefore, varying voltage can develop a profile that indicates whether moisture or carbon tracking damage has occurred.

PF tests are prescribed as both routine tests, and as among the first of a series of comprehensive tests done on units suspected of having problems, because in most cases the PF test causes no stress on the insulation. Thus, it is suitable for use on equipment suspected of having weaknesses that partial discharge or high-potential tests might exacerbate. Yet the PF test is reasonably dependable for indicating the level of contamination or deterioration in insulation. However, a power factor test does not identify what the *cause* of the degraded capability is, and further tests may be needed to determine what it is and to interpret that, to find out if the oil can be reconditioned or if it should be replaced.

An applicable standard for testing transformers is ASTM D-924, which specifies a type of "cell" used for PF tests of the insulating oil, and gives the results expected. Good oil (e.g., new) has a test power factor of .05. Contamination by the products of deterioration (moisture, dissolved gases, carbon, etc.) will increase the power factor observed. Values above .5% and up to 2% indicate further investigation is in order. Values above 2% indicate a need to recondition or replace the oil.

Expected results for the measured power factor depend on the type of device. Table 6.3 lists typical PF values considered to be the limits for "good" insulation for various types of power equipment insulating materials.

Table 6.3 Limits for "Good" Classification of Various Insulation and Equipment types based on PF Test

Insulation or Equipment	PF - %
Oil	< .5
Paper	< .5
Cross-link polyethylene	.05 – 1.0
Ethylene/Propylene Rubber	.5 – 1.0
Rubber	3 – 5
Varnished cambric	4 – 8
Air circuit breaker	< 1

Polarization Recovery Tests

Deterioration of cellulose insulation and insulating oil inside a transformer can reduce the time constant they exhibit for establishing a space-charge polarization across their insulation or through a sample of oil. If a DC voltage is briefly applied across insulation to build up a charge across it, and the removed, the rate at which the impressed voltage grows and/or dissipates can be tracked. The time constant of the rate of dissipation gives an indication of the quality of the insulation. It will be much shorter if the insulation has more leakage.

Thus, a test of how quickly an insulator or insulated device reacts to an applied DC (capacitive) charge applied to it (i.e., its polarization time constant), is a measure of its deteriorated or contaminated condition. Note that this test is broadly similar to the power factor test in one way: it keys on capacitive/resistive properties of the insulation to identify its strength.

Polarization recovery tests of a transformer's winding insulation are performed on a de-energized unit. The low side windings are joined and test cables are connected to the high side terminals. A known DC voltage (e.g., 2000 volts) is applied for a period of time that is varied (minimum and maximums are typically 1/50 to 5 seconds). By varying the time of applied voltage, the time constant associated with the polarization as it develops across the insulation can be determined [Gill]. Figure 6.2 shows the results of tests on transformer-type oil-impregnated paper insulation in tests of insulation with known quantities of moisture. Testing for cables of other power equipment is similar in nature.

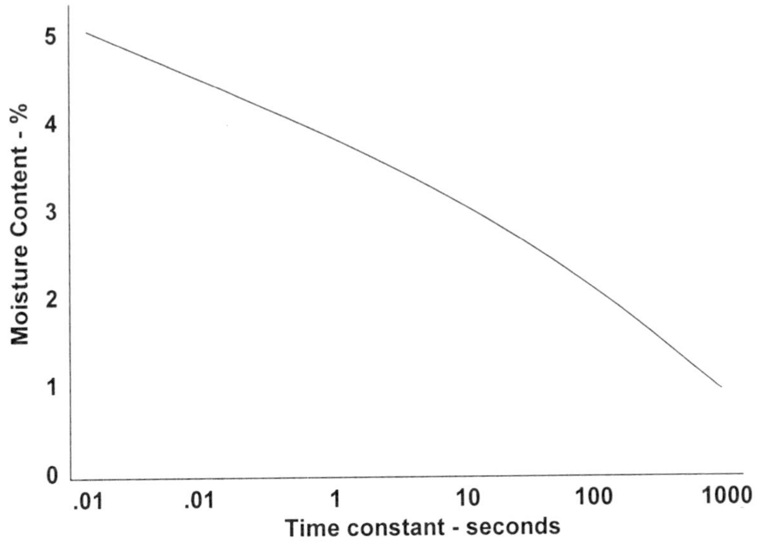

Figure 6.2 Polarization test determines a time constant of the insulation, which correlates with condition. This plot shows time constants derived from laboratory tests of oil-impregnated paper with known quantities of moisture.

Hi-Potential (Hi-Pot) Tests

Transformer winding insulation strength, both turn-to-turn and phase-to-phase-to-ground, can be evaluated for leakage and possible internal faults using an hi-potential test, in which a higher than normal voltage is applied briefly to the device in order to "stress test" its insulation strength. The test can be carried out either with DC or AC voltage.

Generally, DC rather than AC hi-potential tests are used when testing cable, as DC avoids any complications due to charging capacitance. Either, or both, AC and DC may be used on transformers, the major difference in results being that the DC test tends to stress paper-cellulose insulation more and the oil less, while AC tests do just the opposite (see Table 6.2). As a rough rule of thumb, DC voltages are preferred to test lower voltage transformers, while AC voltages are used on units rated above 25 – 35 kV because DC testing might over-stress their paper insulation. Thus, the AC hi-potential test is used on most power transformers, DC on most service transformers.

In an AC test, an AC voltage at standard operating frequency but somewhat higher than the standard operating voltage is applied briefly to the transformer terminals, putting the insulation under considerable stress to test it for flaws. Application of this test varies depending on manufacturer's recommendations and practices, the standard being followed and individual preferences that differ from utility to utility. If done as a check after maintenance, an AC voltage of about 33% (but at least 5 kV) above nominal operating voltage is applied for a period of three minutes. Slightly higher voltages might be used on new units or after a complete rebuild (acceptance tests). Higher than expected currents during this test indicate a weakness in the insulation. DC test procedures are similar in nature.

Hi-potential tests, particularly those involving AC, do put a good deal of stress on a device and fall into that category of "non-destructive" which can push a nearly failed device over the edge, discussed in section 9.2. The very high-test voltages put considerable strain on the unit and will accelerate deterioration of any incipient flaws.

Thus, high-potential voltage tests can be viewed either as an "if it can handle this it should do well in service" or as a "fail it now, not later" test. Either perspective is legitimate. For this reason, some maintenance experts (Gill) recommend against hi-potential tests as routine procedure. Instead, recommendations are usually to restrict their use to new or rebuilt equipment undergoing acceptance testing. Regardless, common sense dictates that they be applied after less stressful forms of testing a cable, device, or transformer core have been successfully passed, such as resistance and PF tests and the TTR and winding resistance tests in transformers. Those might reveal problems that hi-potential testing would worsen.

Induced Potential Test

The induced potential test can be viewed as a variation on the hi-potential test, in that it too, is a quality evaluation stress test of winding insulation. Carried out much like the hi-potential test, it is done with higher than normal operating frequencies. As with the hi-potential tests, there are slightly different forms of this test, carried out with different voltage levels and frequency. One common form involves applying twice the nominal operating voltage at 7,200 Hz for up to one minute. Others vary the voltage, frequency, and time, with lower voltages and frequencies being applied (e.g. 65% and 120 Hz).

This test puts units under considerable stress due to high voltage and high frequency (see footnote 2, earlier) and is recommended only on 5 year intervals or greater. It also must be carried out with great care to avoid creating resonance or induced voltages in other parts of the transformer and its auxiliaries.

Partial Discharge (PD) Tests

If an AC voltage is applied to a cable, it may cause a "partial discharge" inside any voids, cracks, or gas bubbles in the insulation. Generally, the volume inside the void has a lower breakdown limit than the insulation around it. Therefore, the induced voltage, if high enough, will cause this "weaker" zone to discharge without causing any discharges in "good" parts of the cable.

This partial discharge creates a small, very brief (measured in nanoseconds) electromagnetic pulse, which can be sensed by a suitable type of detector if placed on the cable. The pulse is detected by sensing it at one or both ends of the cable, and the computed time lag between pulse initiation and detection is used to determine distance from the point of measurement to the point of the defect. PD tests excite a cable using an appropriate AC voltage, and use a sensitive detector to track the pulses given off by the partial discharges. The PD test equipment analyzes the timing of both the test voltages and pulse arrival (i.e., both travel the length of the cable to the point of the flaw) to determine the location of the partial discharge. In this way, this test is similar to *time domain reflectometry*.

There are a host of complications to be overcome, however. The pulse, never very strong (these are only partial discharges, and relatively weak), suffers attenuation as it radiates away from its point of origin, so that the detector must be very sensitive to detect it. And, being of very short duration, the pulse is composed of all many frequencies.[3] The attenuation of its radiating signal is frequency dependant. Essentially, it radiates through a filter with different amounts of attenuation in different frequency bands. Attenuation in the different bands depends on the test equipment being used, the cable type, duct types, soil, and other conditions that might influence the impedance as a function of frequency in the cable. Splices and branches along the cable also affect pulse propagation and attenuation, and may create "echo" pulses. If the neutral is badly corroded or damaged, the partial discharge will not pass down the cable. If the test is measuring a branched circuit, it becomes difficult to determine which branch a particular pulse is measuring.

These problems are solved for the most part in a wide variety of commercially available sources of PD tests. While based on this overall concept, each approach has specific characteristics designed to give advantages in some types of situations and/or to work around some of the most common problems encountered. Voltages used are generally about at the normal operating voltage or slightly higher (voltage must be high enough to excite partial discharges). Some test methods use standard frequency, others very low (.1 Hz), others high frequencies. Some use DC and then switch to AC of standard or varying frequency. Other variations include a host of novel and innovative ways to detect and measure pulse characteristics and infer location from them.

The wide range of characteristics involved makes it impossible to generalize about either the results or the limitations of PD methods. The amount of stress that is put on the equipment also varies greatly depending on the exact type of test procedure used. However, some PD tests, which apply an alternating voltage up to 1.5 times nominal, can put considerable stress on cables and associated equipment.

---

[3] From a Fourier analysis standpoint: a pulse of "zero" width contains an equal amount of all frequencies. The actual pulse is so short that it has a very wide spectrum.

6.4 TESTS AND DIAGNOSIS OF INSULATING OIL

A major component in many electrical devices is insulating oil. It provides two functions. First, it provides dielectric strength, acting as an insulator, limiting current flow to where it is intended to flow. Secondly, it provides cooling capability, absorbing heat and transferring it away from where it is produced (e.g., a transformer core).

The oil in major equipment such as oil-filled transformers, circuit breakers, and voltage regulators should be checked periodically. This can include both tests to determine that it is capable of doing its jobs (insulation and cooling), and diagnostic tests for clues to the condition of components inside the device (sludge build up, indications of arcing or overheating). Analysis of the insulating oil is one of the best condition assessment tools available for devices such as transformers.

Simple, Fast, and Reasonably Dependable Diagnostics

Tests of insulating oil are among the chief means of diagnosis for power system equipment for several reasons. First, a lot of electrical equipment contains insulating oil – it cannot be ignored and in fact should be checked, meaning that most testing has to deal with it. Secondly, such tests are performed on samples that can be drawn from the device easily without de-energizing it or making any special arrangements. Most power transformers are fitted with an oil sample valve or outlet to facilitate such testing. The equipment itself is not fitted with sensors or otherwise affected and can be left in service, a big advantage. Third, the tests are then carried out away from the device, in an non-energized (except for any test equipment) situation, which is safer and faster.

Finally, and perhaps most importantly, some types of insulating oil diagnostics (e.g., dissolved gas analysis) provide dependable diagnostic information about the severity and type of internal problems in a device. For this reason alone, tests of insulating oil are the cornerstone of condition assessment, particularly for older equipment.

Of course, oil tests and diagnostics are not enough by themselves. Oil tests do nothing to assure that other aspects of a device (bushings) are in good shape. Condition of other equipment can often be determined only by detailed exterior inspection and monitoring (for leakage in bushings, etc). But any program aimed at condition assessment of transformers and similar equipment will have to rely heavily on analysis of the insulating oil, about the only means of non-intrusively diagnosing the conditions *inside* the device.

Oil Samples Required

ASTM standard D923 provides a recommended standard for collection and handling samples of transformer, breaker and other power system equipment oils and is widely followed within the United States. It recommends sampling fluid from the bottom of enclosures when the fluid (oil) has a specific gravity less than 1.0 (as for most insulating oils), and from the top of the tank for those with a specific gravity above 1.0. This standard also specifies the types of materials to be used for containers and equipment (generally glass, with no rubber in the seals) and how to clean them. It gives recommendations for how much (usually 2 quarts) and under what conditions samples should be drawn (avoid humid days), and how to handle and ship samples, if needed. A good summary of this standard is included in Gill, Chapter 5 (see references).

Freeze test

A quick, inexpensive, but inexact, sample test for insulating oil is to place the oil sample in a transparent glass bottle and put it into a freezer set to slightly below 0°C (32°F) for several hours or overnight. Water in the oil will freeze, while the oil will not. Oil with any noticeable moisture content will be milky or foggy in appearance.

Color test

This test involves passing a light of known strength and frequency content through the oil and measuring the spectrum of transmission (amount and color of light passed). The resulting spectrum is compared to a table of known colors for new oil and known changes due to various types of contamination and/or deterioration. The color test can identify quite well the degree of carbonization from switching (tap changers, breakers, oil-filled switches), which correlates with switching operations/condition. It can simultaneously evaluate oil for a number of other conditions too, but is considered inexact, so that it is often not used if more comprehensive tests for acidity, gas, and moisture are to be done.

Water Content (Karl Fischer) Test

Water can exist in the oil inside a transformer or other device in one of three ways. First, it can be dissolved into the oil. Second, it can be emulsified; a state in which it is distributed throughout the oil as minute droplets. Third, it can pool at the bottom of the tank or device. Water has several effects, all detrimental. First, it can reduce dielectric strength. Second, it can be absorbed from the oil into paper insulation and other materials in the equipment, reducing its strengths (both mechanical and dielectric). Third, it can promote corrosion (rust) of the case and other materials.

A small amount of moisture can actually be dissolved in the oil without having a highly deleterious effect on dielectric strength. However, this amount is temperature dependent – typically oil can contain up to 20 ppm at 0°C, rising to 100 ppm at 45°C and 200 ppm at 50°C. If dissolved moisture builds up when the unit is very hot and the unit is then cooled (as could happen during high loading periods). It will then "emulsify" out as micro-droplets throughout the oil – emulsified water. Even a very small amount of emulsified water in oil will reduce its dielectric strength greatly – the tiny droplets form pathways for discharge.

Thus problems caused due to moisture in the oil often occur after the load has passed peak load period and as the oil is cooling from the temperatures reached during peak conditions, but some internal elements of the windings are still warm. Emulsified oil stays that way (unless temperature rises again and it dissolves back into the oil). Typically, it will not "demulsify" (accumulate into larger drops and fall to the bottom of the tank) but will stay mixed into the oil as micro-droplets until removed.[4]

Tests for water include specific gravity tests, which measure the oil's specific gravity against that of new oil of the same type. These are non-exclusive tests, for water - deviations from a typical new value of .84 may indicate something other than water is present. Worse, the presence of other contaminants (lighter than oil) can counteract the presence of water (heavier than oil) in this test.

A water content test, often called the Karl Fischer Test, can be applied. There are three ways to apply the test, but the concept in all forms is to use iodine to absorb moisture chemically, then to detect and measure its presence in the sample. These tests are inexpensive, fast, quantitative, and dependable.

Dielectric Breakdown Tests

The oil can be directly tested for insulating strength using any of several types of test procedure. Two common types defined by the American Society for Testing and Materials (ASTM) are the D877 test, often called the "cup test" or disc electrode test, and the D1816 test, and often called the VDE test. Both identify at what voltage gradient strength the oil will break down.

The D877 test is performed in the field with portable machinery and involves putting about a pint of oil in an insulated container holding two round-faced disc electrodes separated by a .1 inch gap. Voltage is then raised at a rate of 3 kV/second until breakdown occurs. The test is usually conducted on five samples. A value of 23 kV is considered satisfactory.

The D1916 test is performed in a laboratory using about a quart of fluid in a VDE cell, a type of container within which are spherically capped electrodes with a 36 mm sector diameter and 25-mm radius with a very narrow (.08 inch) gap. The oil can be circulated in the cell, simulating oil movement inside a transformer. This test is considered more sensitive to moisture in the oil, and a better indicator of electric strength as the transformer sees it, than the D877 test.

---

[4] Emulsified water is best removed by vacuum dehydration. Cooling the oil will precipitate some but not all emulsified water as droplets large enough that they will drop to the bottom of the unit (the amount of water that can emulsify in water is a function of temperature). Typically, centrifuges or filtration will not remove emulsified water. Chemical means could also be used, but most of those methods leave traces of other chemicals which are just as detrimental, if not worse.

Acidity and Sludge

Oxidation of the oil (chemical breakdown/combination of the oil with oxygen) will take place anytime the oil comes in contact with air and water, both of which contain oxygen that will chemically combine with the oil to form acid; eventually this acid precipitates out as a sludge. Some materials commonly used in power equipment promote oxidation, acting as a catalyst. One is copper.

Sludge buildup not only can clog radiators but also effectively "removes" oil from the reservoir, reducing the net amount available for cooling. Thus, most transformers are sealed to limit the amount of oxygen (air and moisture) that can get inside the unit. Sludge will stick to internal surfaces of a device, not necessarily accumulate on the bottom of the oil container. Inside a transformer it will coat the core, cooling ducts, and radiator lines, thereby reducing thermal transfer capability and resulting in more rapid and higher heat rise during peak operation. A transformer or breaker with indications of sludge buildup should be given priority for removal of sludge. The sludge can be removed by treating/filtering the oil (slow, uncertain) or by circulating hot, new oil to remove it. Regardless, heavy sludging leads to failure.

Oxidation is accelerated by temperature. A rough rule of thumb is that the *rate of oxidation doubles for every 18°F (10°C) rise in temperature.* Therefore, overloading a transformer can greatly affect the rate of acid formation. Tests have indicated that acidity is proportional to the oxygen absorbed in the oil.

Generally, insulating oils used in power equipment have "anti-oxidation" compounds such as DBPC (2.6 di-tert-butyl persesol) added to them to prevent acid formation. These act as "sacrificial" oxidation targets, chemically combining with oxygen to remove it before it can attack the oil itself. Oxidation of the oil begins when these materials are completely consumed.

Acidity tests

There are two ASTM laboratory tests (D974 and D664) and one field test (D 1534) for acidity of oil, all of which determine a "neutralization number" which is the number of milligrams of potassium hydroxide required to neutralize the acid (return it to neutral pH) in 1 gram of fluid. A test value of .4 or less is considered acceptable. Values above 1.0 indicate severe problems. The D1534 field test involves drawing out small samples of the insulating oil and adding it to a "wash solution" already contained in a small glass cylinder. Pre-packaged, sealed small "doses" of KOH (potassium hydroxide) are then added and their number counted. The wash solution has a color-changing property (like litmus paper) – it turns pink when one too many vials of KOH have been added.

Interfacial tension tests

An interfacial tension test can determine if the acid accumulation has or is leading to accumulation of sludge. Acid buildup in an oil leads to sludging, but as shown in Figure 6.3, the latter may follow the other by a good deal of time and acid, alone, is not an indication of the presence of sludge.

Interfacial tension tests (IFT), such as ASTM D 971 and ASTM D2285 measure the sludging characteristics of the oil and give some indication of the amount of build up. These test for sludge implicitly, but measuring the tension between oil and water (i.e., akin to the surface tension between liquid and air). Interfacial tension between the two is reduced by the presence of "polar molecules" (sludge) in the oil. A reading of less than 15dynes/cm is indicative of sludge buildup (a good reading is 22 dynes/cm).

Analyzing Gases Produced by Insulation Deterioration

The different processes of deterioration or damage occurring inside a transformer or other oil-filled device create various gases, such as CO_2, acetylene, hydrogen, and methane, which dissolve into the oil or pool in the space above the oil. While dissolved gases can affect the properties of the oil, the primary reason to test for the amount and type of dissolved gases in oil is that they give an excellent indication of the type and state of deterioration taking place inside the device.

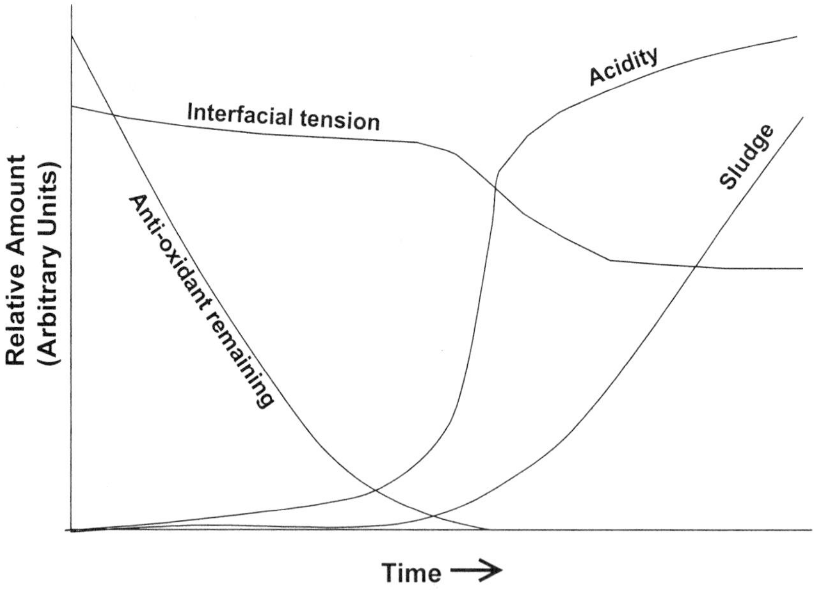

Figure 6.3 Relative rates of anti-oxidant consumption, acid accumulation, and sludge accumulation, and the resulting change in interfacial tension, in one transformer test. A drop in interfacial tension signals that acidity is already increasing in the oil, but is usually a precursor to, rather than an indicator of, sludge formation.

The chief cause of gas being generated inside a transformer is from deterioration of the paper-cellulose and oil insulation – deterioration that is greatly exacerbated by heating. Deterioration rates are an exponential function of temperature. Deterioration-to-failure times for "cold units" are often measured in centuries, but lifetime at full load might be only a few years.

Cellulose (paper) insulation overheated to 150°C or above begins a breakdown process that produces carbon monoxide (CO), carbon dioxide (CO_2), traces of hydrogen (H_2) and methane (CH_4), and water (H_2O). The presence of ethylene (C_2H_4) or ethane (C_2H_6) indicates the oil has been heated to temperatures close to its absolute physical limit of around 550-600°C. Acetylene (C_2H_2) indicates the oil has been heated to even higher temperatures, usually by electrical arcs passing through it (very bad in a transformer, not so bad in a circuit breaker). Thus, the type and amounts of gas present give an indication of both the amount and cause of deterioration or damage that have occurred inside the device.

Total combustible gas test

For the most part, the gases produced by deterioration and damage (except for water vapor and CO_2) are combustible: they will freely combine with oxygen in the presence of a flame or sufficiently high temperatures. Total combustible gas analysis (TCGA) measures the amount of these gases by drawing off some of the gas from the space above the oil in a transformer and measuring what portion of the gas sample burns. Such tests can only be performed on sealed transformers which have a nitrogen blanket (nitrogen filled space) or conservator system, but this includes most types of power transformers. Many power transformers are fitted with a gas-sampling line or an external valve to facilitate such tests. Total combustible Gas Analysis Tests can be performed in the field with a portable unit. These units use a catalytic burner and a Wheatstone Bridge circuit to measure resistivity. The burning of the gases causes changes in the resistance of one element of the circuit, which is detected and correlated with the amount of gas. These tests are quick, inexpensive and precise as to total *amount* of gas, but non-specific as to the *type* of gas, and hence the exact nature of the problem. However, large amounts of combustible gases clearly indicate a problem.

Dissolved gas tests

Dissolved Gas Analysis (DGA – Dissolved Gas Analysis) is done in a laboratory and will identify exactly what and how much of each type of gas is present. They are done not from a sample of the top-oil gas, but from a sample of the oil itself. A vacuum pump is used to extract the dissolved gas from an oil sample and pass it to a gas chromatograph. The quantity of each type of gas is then determined by gas chromatography. This gives a better indication than TCGA of both the amounts of deterioration and the type and cause of it.

Use of insulating oil tests

The presence of CO, CO_2, and water alone, determined by a DGA test, indicates operation at high but not necessarily undue temperatures, and what might be called standard deterioration – aging with time in service. Sound transformers running at peak loading typically generate small but measurable quantities of these gases. However, traces of the other gases, ethylene, ethane, and methane, indicate problems (non-standard deterioration) and give an indication that the equipment is not a sound unit.

However, the most commonly used gas test is the total combustible gas analysis (TCGA), because it is quick and can be performed in the field. It does not provide accurate data on which gases are present, and so cannot provide the information needed to determine the cause, directly. However, by doing TCGA analysis frequently (it is a very inexpensive test) the *rate* of increase of these gases can be tracked, and that rate is used to distinguish among different gassing causes.

Creation of gases from deterioration of paper insulation, even if at very high "normal temperatures" is a gradual process. A high measurement might mean only that the unit is old and has not been tested/purged in the past. By contrast, impending-failure conditions that are caused by the very high temperatures generated by short circuits or by arcing through the oil create gas at a much higher rate. Therefore the rate of increase over a short period of time (a day in the case of a unit thought to be in jeopardy) can provide a good clue as to the cause of the gassing: slow – normal deterioration; fast – an internal problem possibly indicating imminent failure.

How the Test is Carried Out is Important

It is important to bear in mind that the cost, time and resources required, as well as the sensitivity and quality of results obtained will depend greatly on both the exact nature of the equipment being used, and the procedure for its application. This means keeping in mind what is grounded and what is not, measurements of equipment temperature before and after, etc. Similarly, the level of stress put on the equipment can depend on instructions being followed precisely and how the test was applied.

Comparison of Tests and Diagnostic Methods

Table 6.4 compares the various test procedures for equipment and oil on the basis of various characteristics. From left to right, column one describes the type of test, column two the type of voltage used (AC, DC, or VF for non-standard or variable frequency), and column three identifies whether it is performed in the field or laboratory. Column four tells whether the unit can be left in service or must be de-energized while the test is being run. Column Five

Table 6.4 Power System Electrical Test Procedures

Test	Type	Where?	In serv?	No-Go	Condition	Stress level
Thermal load tests	AC	Field	Yes	No	Yes	-
Resistance test	DC	Field	No	Yes	Maybe	-
Dielectric absorp.	DC	Field	No	No	Yes	-
Power factor tests	AC	Field	No	No	Yes	-
Polarization recov.	DC	Field	No	No	Yes	x
Hi-potential test	both	Field	No	Yes	No	X
Induced potential	VF	Field	No	No	Yes	X
Partial discharge	AC	Field	No	No	Yes	x
Transf. turns ratio	AC	Field	Yes	Maybe	Maybe	-
Oil acidity tests	-	Lab/Field	Yes	No	Yes	-
Interfacial tests	-	Lab	Yes	No	Yes	-
TCGA tests	-	Field	Yes	No	Yes	-
DGA tests	-	Lab	Yes	No	Yes	-
Disassembly/inspection		Field	No	Yes	Yes	-

identifies whether the test is typically used as a "Go/No-Go test" – one that provides a definite indication that the unit is unfit for continued service. Column six (conditions) explains whether the test provides relative diagnostics on condition and deterioration that can be tracked over time. Finally, column seven gives the level of stress caused by the test, the size of the X indicating relative level of equipment stress caused by the testing.

Comparison of Tests and Diagnostic Methods

Table 6.4 compares the various test procedures for equipment and oil on the basis of various characteristics. From left to right, column one describes the type of test, column two the type of voltage used (AC, DC, or VF for non-standard or variable frequency), and column three identifies whether it is performed in the field or laboratory. Column four tells whether the unit can be left in service or must be de-energized while the test is being run. Column Five identifies whether the test is typically used as a "Go/No-Go test" – one that provides a definite indication that the unit is unfit for continued service.

Column six (conditions) explains whether the test provides relative diagnostics on condition and deterioration that can be tracked over time. Finally, column seven gives the level of stress caused by the test, the size of the X indicating relative level of equipment stress caused by the testing.

6.5 SUMMARY AND FINAL COMMENTS

This chapter summarized inspection, testing, and diagnostic procedures for power system equipment. The major goal of these activities, as it relates to aging infrastructure, is to determine the condition of power system equipment already in the field. Properly carried out, such inspections and tests can identify if that equipment is "good to go" for service and give some estimate of remaining service lifetime. They also catch some failures before they occur, permitting the utility to make repairs, saving equipment that might otherwise be ruined.

While this "ounce of prevention" function is worthwhile in reducing equipment replacement costs by saving some equipment from failure, the real value of inspection, testing, and maintenance is its ability to reduce the number of outages that occur in service. It is worth considering that exact knowledge of when a unit would fail would be worth a great deal to a utility, allowing it to schedule replacement, but at exactly the latest moment possible. While even the best of condition-assessment methods can not exactly determine expected lifetime, the information provided is a step in the right direction and worth a good deal toward improving service quality and allowing optimal management of equipment lifetimes.

The methods covered in this chapter are worthwhile only if applied in well-managed program aimed at improving and maintaining service quality at high levels while keeping costs as low as possible. Chapter 9 will discuss application of these methods to power system equipment and identify how the tests are best organized and used in an inspection and test program. Chapter 10 will discuss prioritization methods for O&M, and how utilization of units can be optimized so the utility gets the most overall "bang for the buck."

REFERENCES

J. J. Burke, *Power Distribution Engineering – Fundamentals and Applications,* Marcel Dekker, New York, 1994

P. Gill, *Electrical Power Equipment Maintenance and Testing,* Marcel Dekker, New York, 1998

H. H. Malik, A. A. AlArainy, and M. I. Qureshi, *Electrical Insulation in Power Systems,* Marcel Dekker, 1998

O. C. Seevers, *Management of Transmission and Distribution Systems,* Fairmont Press, Lilburn 1995

W. A. Thue, *Electrical Power Cable Engineering,* Marcel Dekker, New York, 1999.

7
Aging Equipment and Its Impacts

7.1 INTRODUCTION

Aging equipment is undoubtedly the primary cause and effect associated in most people's minds with the term "Aging Infrastructure." And while there are other significant factors involved as were summarized in Chapter 1, there is no doubt that aging infrastructures present an incredible challenge largely due to the effects of aging equipment. Aging equipment has high failure rates and requires, proportionally, more inspection and maintenance cost. Often parts and service equipment are hard to come by (and expensive).

This chapter looks at equipment aging – what it is, how it impacts the various types of equipment, and how those changes affect the reliability and economy of a power system's operation. Invariably failure rates increase with age, usually in an exponential manner. But characteristics vary depending on equipment type (aging impacts poles very much differently than say, cable) and conditions of use. Most important to the modern distribution utility is how an aging base of equipment changes the performance (reliability and economy of operation) of the system as a whole, and impacts the quality of customer service.

This chapter begins in section 7.2 with a qualitative look at aging and its impacts on equipment, discussing the different categories of aging impact, and the general characteristics of aging on equipment such as transformers, poles, cables, etc. Section 7.3 then looks at quantitative trends in failure rates as equipment gets older – how and why do failure rates increase? This section raises an interesting concept – what would exact information on when a particular unit of equipment be worth, and how would a distribution utility best utilize that information? Section 7.4 delves into the impact that aging and failure rates have on the installed base of equipment in a utility system and the repercussions on cost and service quality that result. Section 7.5 concludes with a summary of key points.

7.2 EQUIPMENT AGING

All electrical equipment kept in service ages. During every year of service the equipment becomes a year older. One consequence of this aging is gradual deterioration of the physical and electrical strengths of the equipment, until at some point failure occurs one way or another – the end of the equipment's useful lifetime. Table 7.1 shows several types of deterioration that affect older equipment. The "caused by" columns show which of the three primary types of aging, covered in Table 7.2, lead to the respective type of deterioration.

Table 7.1 Types of Deterioration Causing By Aging

Type of Deterioration	Caused By CA	Caused By CSS	Caused By AES	Comments
Corrosion	X	x	x	Chemical decomposition, or combination with oxygen or other ambient elements, until the material loses it's required mechanical or electrical strengths or qualities.
Dielectric loss	x	X	x	Various mechanisms (treeing, contamination) lead to the loss of electrical withstand strength.
Shrinkage/Hardening	X	x		Paper, rubber, and synthetic gaskets and seals harden and or shrink with age, losing their ability to keep out moisture or contain pressure.
Wear		X	X	Mechanical components lose tolerance and bind, or do not hold with the same bond they once did.
Moisture retention	X			Water is gradually absorbed into a material, degrading its mechanical or electric strength

Table 7.2 Categories of Equipment Aging Impact and Their Meaning

Category	Meaning and Impact
1. Chronological age (CA)	Age since construction. Certain materials deteriorate over time due to natural causes most directly associated with chronological age (paint chalking and corrosion, hardening of rubber and gasket materials, etc.).
2. Cumulative service stress (CSS)	The cumulative effect of the time the unit has been energized or in use, and the load (mechanical, electrical) it has served in that time.
3. Abnormal event stress (AES)	The cumulative impact of severe events generally not considered as "normal service." This includes through-faults for transformers, storm and auto-accident stress for poles, etc.
4. Technical obsolescence (TO)	Digital and data communications equipment, as well as other "hi-tech" equipment, can become "too old" by virtue of not being compatible with new systems and equipment.

Types of Aging Impact

Aging can having any of several particular meanings with respect to power system equipment, as shown in Table 7.2, which lists four specific categories of aging and their impacts. Generally, the terms "aged or old" and "aging" are used to refer to some non-specific combination of all four effects, with the understanding that the equipment is not as useful or dependable as similar equipment of lesser age.

Chronological age

Certain materials such as paint, paper and fabrics, rubber and synthetic gaskets and seals, and insulation, deteriorate over time. Time since manufacture alone is the direct cause of this type of deterioration. Thus, seals and gaskets begin to harden, and paint to chalk, as soon as a transformer or breaker is built, progressing at much the same rate regardless of whether the unit is installed and in service or kept in inventory in an equipment yard. A creosote pole will dry out and begin the very slow process of chemical breakdown of its preservative whether or not it is stored in an equipment yard or put in the ground. It will however, begin to rot at the ground line only after being put in the ground, particularly if the soil around it is moist and of the right pH.

As the pole example in the last sentence illustrates, very often the rate of deterioration over time is exacerbated or mitigated by ambient conditions. Transformers, poles and cables stored outside and subjected to summer heat, winter cold, direct sunlight and ambient moisture, will deteriorate faster than if kept in sealed, temperature-controlled and de-humidified warehouses. Deterioration due to chronological aging can be cut by two thirds in some cases if the equipment is keep in a controlled environment. Generally, "chronological aging" rates and average lifetimes assume that the equipment is installed in typical ambient service conditions, (i.e., poles and transformers are kept outside whether in service or not.).

In many cases, the deterioration that would occur due to time alone is accelerated by the stresses caused by putting the equipment in service. Hardening of gaskets and seals, and paper insulation, will occur at a faster rate when a transformer is operating (heat generated by electrical losses will accelerate the deterioration processes).

In some cases however (i.e., paint chalking and some types of corrosion) the rate of deterioration does not depend on whether the unit is in service. And in a few situations, putting a unit in service will reduce the rate of chronological deterioration. For example, energizing a power transformer being kept for standby service creates a low level of heating (due to no-load losses) that "cooks" any moisture build-up out of the unit, slowing that rate noticeably.

Cumulative service stress

Certain types of deterioration occur only, or primarily, due to the use or operation of the unit for its designed purpose, and are proportional to the time or the cumulative level of use in service, not chronological time. There are three primary types of processes involved:

1. *Electro-magnetic field stress* established when a cable or unit of equipment is placed in service can lead to degradation of dielectric strength and in some cases, promote corrosion. Essentially, applying voltage to the device produces a stress that will eventually break down insulation and accelerate some chemical processes (particularly if the unit is not grounded well or cathodically protected). This type of stress depends only on the voltage. It does not depend on the amount of current (power) the unit is handling.

   Every unit of electrical equipment – transformer, cable, bell insulator – is always designed to withstand a nominal voltage, that which it can withstand safely and dependably over a long period of time (decades). But, given enough time, the electromagnetic stress from that voltage will lead to various type of insulation breakdown, which can include treeing, attraction of impurities that build flashover paths, etc. Eventually, over a century (post insulators), or only a period of several decades (UG cable) the insulation will fail for this reason alone.

In most cases the deterioration rate due to this cause is very voltage sensitive. Raising voltage even slightly will increase stress a good deal. Thus, higher voltages, due to operating the unit at the high end of its nominal range, or to miss-operation, switching surges, or lightning, can accelerate this deterioration, sometimes quite dramatically. Similarly, lowering the voltage on equipment that is suspected of being weak (old cables) to the low end of the permissible scale (e.g., .95 PU) can often reduce stress, deterioration and failure rates significantly. And of course, voltage that reaches too high a level – above the tolerable range – leads to immediate failure.

2. *Wear.* Parts that move against one another, in devices such as tap changers, capacitor switches, circuit breakers, load break switches, etc., wear. The movement gradually erodes material in the moving junction between parts, loosening tolerances and often scouring smooth bearing surfaces. Such wear can lead to binding of moving parts and failure to operate, or to operate slower than necessary (breakers).

   In addition, "wear" can be applied to a type of deterioration that occurs in parts that are not technically moving against one another. The most common form of this is in overhead conductors and their associated splices, terminations, and support brackets. Conductor "moves" in the sense that it sways in the wind, and expands and shrinks, and sags more or less, with temperature (and loading). Over decades, this movement can lead to loosening of the bond between clamps and conductor, or within splices or termination fittings, and even to cracking of the conductor material itself. Eventually something breaks and the line falls.

   This gradual deterioration is accelerated greatly by *aeolian vibration.* Wind blowing across the conductor sets in motion a resonant vibration, much like blowing on a harp string might cause it to hum. This vibration stresses the bonds in splices, brackets and clamps, and can fatigue the metal strands of the conductor, over a period of only months in extreme cases. Various types of vibration dampers are available to fit to conductors to stop such vibration, but these are usually fitted only to known or suspected cases of heavy vibration. Most overhead spans experience a slight amount of this vibration, and over decades it leads to a degradation of mechanical strength in the conductor and its fittings.

3. *Heat stress.* Higher temperatures accelerate many of the physical and chemical mechanisms involved in materials deterioration, to the point that heat can be considered electrical equipment's worst enemy. In electrical equipment like transformers, regulators, and motors, heat is generated by electrical losses in the equipment. The higher temperatures created as a result cause deterioration of materials in insulation, gaskets, seals, the transformer oil itself, and in some cases, the metal in conductors and/or mechanical components. Similarly heat affects all electric equipment, "aging" or ruining the functionality of various materials or their components. Parts that should be soft and expanded, so they seal well, shrink. Parts that should insulate lose dielectric strength. Metal expands, binding and wearing, or, at very high temperatures anneals so that it becomes brittle and of low strength – essentially a different type of alloy.

   In almost all cases, the rate of deterioration of a component or material is a function of the temperature, generally increasing exponentially with temperature up to some maximum-tolerable temperature beyond which the material fails immediately. The relationship is exponential – a device might be able to operate for decades at 80°C, for several years at 90°C, for a week at 100°C, and for only a few hours at 110°C.

   In addition, temperatures that reach certain levels can cause internal physical changes in some types of electrical equipment, changes that promote miss-operation. For example, in a power transformer, extreme heating of the windings (due to losses from very high loadings) can cause hot spots on the core that are hot enough to create gassing or boiling of the oil. The gas introduced into the oil by the bubbling contaminates the oil's ability to act as an insulating medium, and a "bubble path" can provide a route for flashover, leading to immediate and catastrophic failure.

   In mechanical devices such as breakers, load tap changers, and motors, high temperatures can cause swelling of mechanical parts, leading to binding (bearing tolerances are reduced to the point that the device will not operate) and miss-operation, and/or high rates of wear. In overhead conductor, high enough temperatures (typically caused when losses generate heat sufficient to create a 100°C rise above ambient) will anneal the conductor material. As a result it hardens, becomes brittle and loses its elasticity and mechanical strength. Vibration from wind and the natural expansion of heating and cooling from diurnal temperature variations then quickly leads to minute cracking and mechanical failure and the conductor breaks and falls to the ground.

The deterioration caused by high temperature is cumulative – i.e., a period of high temperature will cause deterioration that will not be "undone" when temperatures fall. A device will suffer gradual deterioration of its components, equal to the cumulative stress of high temperature over its service life to date. Since heat (the cause of high temperatures) is due to load (the load current causes losses that create the heat), power engineers often associate deterioration rate directly with load: higher levels of load lead directly to higher rates of deterioration. The term "loss of life" refers to the rate of equipment lifetime loss associated with a particular loading (heat and temperature) level for the equipment.

Many of the chronological aging processes discussed earlier are accelerated if the device sustains high temperatures for long periods of time (as is common for many electrical devices). Thus, chronological aging and cumulative service stress is not, in practice, completely independent.

Abnormal event stress

Over its service life, any particular unit of power system equipment is likely to see a number of "events" which lie outside the normal conditions expected in service. These include electrical events such as through-faults, switching surges, and lightning strikes, and/or harsh mechanical events such as automobiles striking the device (i.e., a pole or pad mounted device), high ice loadings from freak winter storms that mechanically stress overhead conductors, and similar situations.

Auto accidents and storm damage is a prime cause of wood pole and metal structure failure, and somewhat unavoidable given that these structures are often placed on public easements and parallel roadways. However, most of these events are reported (many result in immediate failure of the pole, or equipment at the top of it) so the utility is at least aware of the event and can inspect and repair any damage done.

Ice loadings are another stressful mechanical factor, but one that is more difficult to track. During severe winter weather, ice will accumulate on overhead conductors. Given extreme conditions, several inches of ice can accumulate, weighing many hundreds of pounds per span. The cumulative weight of conductor and ice can exceeds the strength of the line, and it will part, falling to the ground. However, in the vast majority of winter storms, icing does not lead to failure – the weight is tolerated without any damage or deterioration.[1] When the ice melts after the storm, the conductor is, if not "good as new," as good as it was before the storm.

But in between these two situations – outright failure on one hand and no damage done on the other – is a narrow range of "damage-done" cases that are particularly vexing to utilities. The ice loading on a particular span of overhead conductor might reach a level that does not cause immediate failure, but is sufficient to stretch the conductor to its elastic limits, or over-stress clamps, splices, etc. This leads to accelerated deterioration and failure soon afterward. The utility has no way of knowing if any portion of its overhead lines were so affected and if so, which portions.

Through-faults are the most stressful event routinely seen by many transformers, from the standpoint of gradual deterioration leading to failure. A device downstream of the transformer experiences a fault, and for a brief time until a circuit breaker clears the fault, the transformer sees current flow through it that is from three to fifty times normal maximum. Heating caused by this high current is usually not the major impacting process – in most cases the fault current lasts but a few cycles and little heat energy is generated.

Instead, the most common damaging effect of through-faults is the magnetic field created by the very high current levels, and its shock impact on internal components. Transformers (and motors and voltage regulators as well) are designed so that the current flow through them causes an intense magnetic field – required for their operation. But a through-fault multiplies that field strength in direct proportion to its greater current level, creating a tremendous magnetic field and compressing or pulling apart nearby components, etc.

The magnetic force from a severe fault can create tremendous force. If high enough, this force can literally rip connections lose inside the transformer, leading to immediate and perhaps catastrophic failure. But typically, in a well-planned power system, fault currents are limited by design and operation to levels not so severe that they lead to such problems. Still, the mechanical impulse from a "tolerable" through-fault will make a large power transformer, weighing many tons, shake and

---

[1] Conductors, particularly those with steel reinforcement (ACSR – Aluminum Clad Steel Reinforced Conductor) are designed to have high mechanical strength to deal with periodic severe ice loadings. In some areas of the world (e.g., parts of Saskatchewan) the distribution utility uses an entirely steel wire in its OH distribution sacrificing the lower resistance of aluminum for the superior ice loading capability of pure steel wire.

"ring" as if it had been dropped several feet, and has a overall effect on it similar to such abuse. The cumulative impact of a number of such severe shocks over a transformer's lifetime can be a significant loosening of the core stack, and stretching and twisting of windings and connections – a general weakening of the mechanical integrity of the unit.

The cumulative impact of through-faults is thought by many power equipment experts to be the leading cause of power transformer failure. There are recorded cases where transformers (and motors, and voltage regulators) fail when a through-fault occurs, generally when such faults occur during peak loading conditions (when the transformer is already quite hot from high loadings and thus under a good deal of stress). Previous events have weakened the overall integrity of the unit until this one last through-fault is sufficient to cause failure.

However, more often than not, failure does not occur during or immediately after a through-fault. Given the probability of occurrence (peak loads occur less than 5% of the time, most through-faults happen during off peak times or when conditions in the transformer are of low stress level. As a result, the unit tolerates the fault, but the damage is done, and the unit subsequently fails at some later time – perhaps hours, days, or even weeks afterward when it is exposed to peak load stresses.

Few utilities have dependable, easy to access records on the through-faults experienced by major transformers during their lifetimes. As a result, one of the most useful records of cumulative stress needed to determine present condition is unavailable to engineers and planners trying to deal with aging equipment and the estimation of its remaining lifetime.

Lightning strikes are a severe-stress event that often leads to immediate failure of a device, almost a certainty in cases where the device is struck directly. Lightning is a pure current source, with the current flow being from ten to one hundred times the normal fault levels seen in power systems. A lightning strike can cause an immense "through-fault" like failure, or lead to a voltage flashover that causes other problems.

Failure modes due to lightning strikes are complex, and not completely understood, but failure occurs mainly due to heat. Although very brief, the incredible magnitude of the lightning current, often over 500,000 amps, creates heat – trees hit by lightning explode because the water in them is instantly vaporized. Similar impacts occur in electrical devices. In other cases the magnetic shock of high current does the damage.

There is considerable evidence that a good deal of damage can be done to electric equipment by indirect lightning strikes. A lightning strike to the ground near a substation can cause high voltages and or erosion of grounds. A strike in the ground may also travel to a nearby UG cable or grounding rod, burning it badly.

> Lightning strikes to overhead lines often cause immediate failure, but can instead cause a flashover, which, due to breaker operation and an interruption, leaves no outaged equipment. However, the equipment may be damaged to the extent that it will fail in the next few weeks or months. In some utility systems, there is a statistically significant increase in service transformer and line failures after the annual "lightning season."

Abnormal stress events can occur from time to time in the life of any electrical equipment. In some cases, the severity of one event leads to immediate failure, but often it is the cumulative impact of numerous events, over years or decades, that gradually weakens the device. Sometimes these events cause deterioration similar to that caused by time or service, but most often the degradation in capability is different – through-faults loosen transformer cores and twist internal fittings in a way that no amount of normal service does.

Technical obsolescence

Most types of equipment used in power systems are from "mature" technologies. Although progress continues to be made in the design of transformers, breakers, and so forth, these devices have existed as commercial equipment for over a century, and the rate of improvement in their design and performance is incremental, not revolutionary. As a result, such equipment can be installed in a utility system and be expected to "do its job" over its physical lifetime – obsolescence by newer equipment is not a significant issue. A forty-year old transformer in good condition may not be quite as efficient, or have as low maintenance costs as the best new unit, but the incremental difference is small and not nearly enough to justify replacement with a newer unit. However, in most cases, if a unit is left in service for a very long time, the issue of spare parts will become significant. Some utilities have some circuit breakers that have been in service 50 years or more. Replacement parts and fittings are no longer manufactured for the units, and have to be customized in machine shops or scavenged from salvaged equipment.

But a noticeable segment of the equipment used in utility power systems does suffer from technical obsolescence. The most often cited cases are digital equipment such as computerized control systems, data communications and similar "hi-tech" equipment associated with automation and remote monitoring and control. Typically anything of this nature is eclipsed in performance within three to five years, often providing a utility with a new level of performance that it may wish to purchase. The older equipment still does its job, but is vastly outperformed by newer equipment. Economic evaluation of the benefits of the new systems can be done to determine if the improvement in performance justifies replacement. Often it does.

But beyond performance issues alone, there are other factors that mitigate against this equipment remaining in service for decades. First, since in some cases spare parts are made for this type of equipment for only a decade or so after manufacture, there is no viable prospect of leaving the equipment in place for many decades. Secondly, even if parts are available, newer equipment requires different test and maintenance procedures. Qualified personnel to make inspections and repairs become more difficult (and expensive) to find.

While digital equipment used in control and automation is the most obvious case of "technical obsolescence," there is a much more dramatic, and significant case that is causing perhaps the most contentious issue facing the power industry as it moves toward full de-regulation. The issue is a stranded asset, and the cause is technical obsolescence of generating plants. Many utilities own generating units that are fifteen to twenty years old. These units are not fully depreciated – when the units were built, the utilities borrowed money to pay the expenses, financed over thirty year periods, and they still owe significant amounts on those loans.

But the technology of generator design has improved considerably since those units were built. Newer generators can significantly outperform those older units, producing power for as much as a penny/kWh less and requiring less frequent and costly O&M and fewer operators. In a de-regulated market, merchant generators can buy these units and can compete "unfairly" with the utilities' older units. Thus, these older units are "stranded" by the shift to de-regulation – the utility was promised cost recovery when they were purchased, but now faces competition due to technical obsolescence, and a drop in the sales that were expected to provide revenues to pay off the loans.

In the T&D arena, technical obsolescence will continue to be an important factor in control, automation, and metering areas. In addition, a high rate of technical progress will continue to impact power system equipment areas such as distributed generation, power electronics/power quality equipment, and home and building automation and control systems.

Lack Of Proper Care Can Accelerate Aging Effects

Lack of proper maintenance or inattention in setup and/or operation can, in some cases, lead to excessive deterioration due to secondary impacts (e.g., neglected seals let in moisture which leads to dielectric failure). The degree to which neglect leads to premature failure depends on the type of device and the extent to which its design depends on periodic maintenance.

Equipment with moving mechanical components tends to be most sensitive, that sensitivity proportional to the frequency of mechanical operation. Thus, tap changers and voltage regulators are most prone to early failure due to neglect. However, upon failure they can often be repaired by performing the maintenance or replacement of parts that has been deferred.

On the other extreme, most service transformers and residential power meters are essentially no-maintenance devices – they can be installed and expected to last for a normal lifetime without any maintenance. The first attention they get subsequent to installation is when they fail to do their job for some reason. At this point they are usually replaced. Equipment requiring little or no maintenance generally has only one "service mode" replacement.

Neglect or stupidity in setup or operation is another matter entirely. A neglected aspect of equipment use, often leading to rapid deterioration is poor grounding. Equipment may not be well grounded when installed, or the quality of the ground might degrade over time due to corrosion of clamps, etc. In either case, a weak ground can cause higher than normal and/or abnormal currents during "normal operation" accelerating the deterioration caused by normal service stresses. More critical yet, poor grounds often exacerbate the impact of lightning, switching surges, and through-faults to the point that the stress caused is two to five times greater. Failure is nearly assured.

Other types of poor setup include mismatch of transformer impedance in multi-phase banks composed of single-phase units, incorrect settings for regulators, and similar inattention to details in equipment utilization.

Loss of Life and Load-Related Lifetime

Overall, of all the deterioration modes mentioned above, the major contributor to aging of electrical components in a power system is high temperature, a service-stress-related cause. High enough temperatures will cut short a device's life instantly. More typically, the cumulative effect of high temperatures over years of service leads to deterioration of key elements inside the device until they fail, leading to cessation of function, and often ruining the device.

The heat generated by electrical loss causes high temperatures. Those losses are a function of the electric load on the device. Therefore, the major component of aging is related directly to the loading on the device. While other types of deterioration proceed in parallel, usually this is the one that leads to failure (unless a high incidence of abnormal events fails the unit first).

For this reason, power system equipment engineers often speak of load-related lifetimes and "loss of life" of a transformer, motor, regulator or other component, as a function of its loading. Figure 7.1 shows the expected lifetime of one particular transformer if loaded to various levels. The unit can be expected to last 40 years (350,000 hours) if loaded continuously to 80% of its rated load, or only 7.5 years if loaded to 100% on a continuous 24 hour a day basis (solid line). If serving a typical utility load (time-varying load with a 60% annual load factor and peak periods of four hours duration) the unit has an expected lifetime of 40 years when serving a peak load equal to 100% of its rating (dotted line).

Thus, operation at 150% of rated load (estimated lifetime, 350 hours) can be likened to accelerating the unit's normal 40-year (in utility service) lifetime by a factor of 1,000. The unit looses 1,000 hours of expected operational service lifetime for every hour of operation at 150%, i.e.; every hour of operation at this loading level uses about 40 days normal service lifetime. Most other types of power system equipment have qualitatively similar lifetime vs. loading relationships.

While this is a dramatic acceleration in loss of life, most utilities make it policy to operate a transformer at much higher loads during contingencies. Under emergency conditions, transformers are often operated at load levels up to 166% of rating. At this load level, expected lifetime is only 120 hours and every hour of operation uses the equivalent of four months of normal operation (see Chapter 8 for a discussion of contingency loading policies).

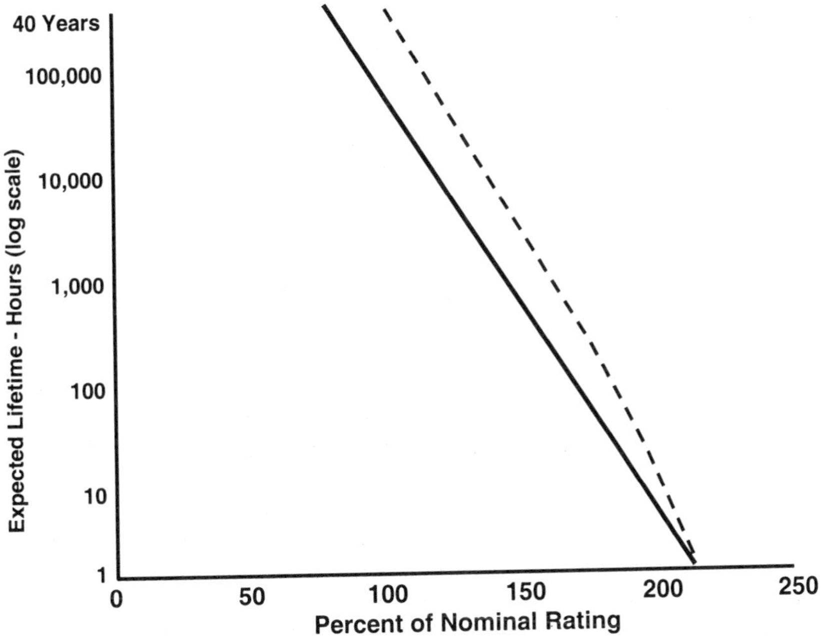

Figure 7.1 The expected lifetime (time to 50% likelihood of failure) for one particular type of power transformer, as a function of the loading in percent of nominal rating. Solid line shows lifetime for constant loadings. Dotted line shows expected lifetime if serving an annual load curve of 63% load factor with a peak load measured at 100% percent of rated capacity.

The reason is one of economics, and makes sense in almost all cases. Given that peak loads occur only occasionally, a transformer that has been installed so that it is expected to take *up to* 166% of its normal peak load during a contingency will probably see only 20 hours of such intense service during a 30 year plus service lifetime. Most contingencies will occur off peak when loading and thus the stress is far lower. Those 20 hours of loading will cumulatively lead to a loss of 7 years of the expected 40-year lifetime.

But the benefit gained (contingency coverage of other transformers) is considerable, because other options for that coverage are costly. And the savings lie entirely in the present – it has a present worth factor of 1.0. By contrast, the loss in life is realized at the end of the transformer's life, 33 years into the future (40 years minus 7). Present worth factor, in that timeframe, is minute (about 3%) so the lost service life has minimal value by comparison. Simply put, benefit exceeds cost.

Of course, a utility can get into trouble, as many have, by putting old transformers under such high levels of stress during contingencies. Many utilities have transformers that have successfully made it through forty or more years of service, during which time the units were never exposed to such high contingency loadings, due to the more relaxed loading standards that existed in the traditional utility industry. The units are partly deteriorated and weaker than new – they may have only a decade of remaining expected life. But due to financial pressures, the utilities have to switch to a policy of intense contingency loading. The problem created is that the stress of this use may absorb most of what life remains.

"Old" Depends on the Expected or Design Lifetime

Finally, it is worth noting that whether equipment is old or not depends on the expected equipment lifetime, as designed. A 20-year old wood pole is not really old (expected lifetime is at least twice that), but a 20-year old substation RTU is. A 40-year old span of ACSR is not yet really old, but a forty-year-old set of capacitor switches is.

Electrical equipment can be designed to withstand any degree of deterioration, by building it from a higher grade of materials that assures longer life (slower deterioration rates, or in some cases immunity from deterioration). Similarly, heavier material (e.g., thicker paint) can be used to allow more deterioration to occur before failure occurs, or because harsher conditions are expected. This results in a longer expected period of time, service, or abnormal events before degradation leads to failure or miss-operation.

But such measures cost money. Thus, ultimately equipment lifetime is somewhat a function of cost. In general, most electrical equipment like transformers, breakers, and regulators is designed to withstand the chronological, service, and abnormal events associated with about 40 years of service under typical ambient conditions and loading. Overhead conductors and its ancillary equipment generally have a lifetime of 60 years or more (the required mechanical strength pretty much assures immunity from electrical, thermal, and/or rapid deterioration unless abused by very harsh service). Wood poles have about a 40 – 75 year expected lifetime, depending on quality and soil conditions and steel structures typically last up to 100 years. Underground cable has perhaps the shortest expected lifetime, although lead-covered paper insulated cable has an expected lifetime of up to 60 years if well-cared for, most other types have lifetimes between 20 and 35 years.

Worn out equipment usually is really worn out

Usually, all of the types of deterioration discussed above are occurring simultaneously in any electrical device or line. Corrosion, dielectric breakdown, wear, and loss of mechanical strength all proceed at their own paces. Failure of a unit like a breaker or transformer is a type of horserace between the various failure modes that could fail it. The unit will fail when the first of these various types of deterioration reaches the point where it causes miss-operation or catastrophic failure (the device destroys itself).

One element of good design for power system equipment is a balance among the expected lifetimes of all these various components. This just makes sense: if a transformer's case is known to fail within 50 years, it is a waste of money to install a core that can easily last 100 years. Good design tries to balance the various strengths and amounts of materials in a unit so that all have roughly the same lifetime in typical service. This means that when failure occurs or is imminent due to one particular type of deterioration, the unit is most likely worn out in all its other deterioration modes, too. It may not be worth repairing because such repair essentially requires rebuilding and or replacing *everything*.

Effects of Equipment Aging

Table 7.3 lists the effects of equipment aging, in rough order of importance to most electric utilities. As units age they have a higher likelihood of failure and miss-operation (all equipment). In some cases aging results in a reduced capacity or rating (nearly all equipment, whether recognized or not – this will be discussed in the next section). Some types of devices also see increased electrical or mechanical losses with age (some motors and transformers). In other cases, the device may produce more sound pollution as it may age (motors and transformers, due to core or stator stack loosening, etc.

By far the most significant factor, and the major concern to a distribution utility, is the top effect listed in Table 7.3: increasing likelihood of failure. The next section discusses this in great detail.

Table 7.3 Effects of Power System Equipment Aging, in Order of Importance

1.	Higher likelihood of failure
2.	Increased likelihood of malfunction
3.	Reduced capacity
4.	Increased losses
5.	Greater noise

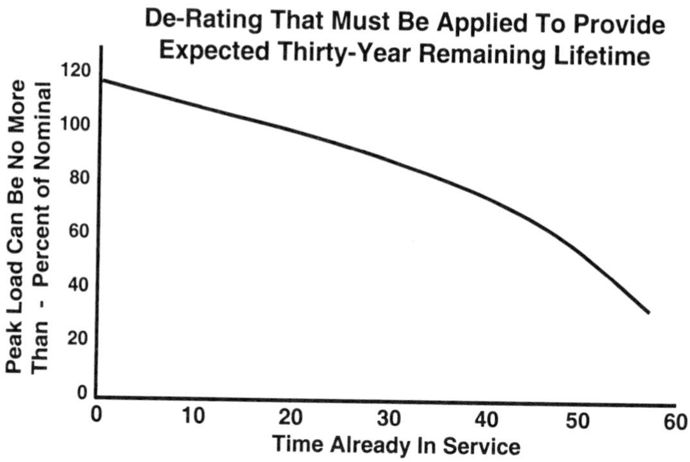

Figure 7.2 Example of the impact of "de-rating" a unit in order to increase expected lifetime. See text for details. (Note: This curve is not general; it does not accurately apply to all transformers).

De-rating to Preserve Remaining Lifetime

Table 7.3 lists de-rating as one impact of aging. Since the major modes of deterioration and failure are often heat-related or other service-stress created causes, one option available to an equipment operator is to lower the loading (i.e., cause of stress) in order to increase the expected lifetime of the equipment.

Figure 7.2 shows a de-rating curve for a power transformer that has an expected 40-year lifetime and has already seen 30 years of service at designed load levels. At this point, it has an expected 14 years of service left if kept at its present service level where peak load is 100% of its rating.[2] If the owner wishes to extend the expected time-to-failure to 40 more years of service, he must reduce loading by 29%, to only a peak of 71% of the unit's rating. Similarly, if this unit is loaded to 110%, it will very likely fail within a few years. Thus, the owner can balance the annual value to be obtained from the unit (how much power it handles) against the number of years that value is expected to be provided by the unit.

7.3 EQUIPMENT FAILURE RATE INCREASES WITH AGE

All equipment put into service will eventually fail. All of it. Some equipment will last longer than others will and often these differences in lifetimes will appear to be of a random nature. Two apparently identical units, manufactured to the same design, from the same materials, on the same day at the same plant, installed by the same utility crew on the same day in what appear to be identical situations, can nonetheless provide very different lifetimes. One might last only 11 years, while the other will still be providing service after 40 years. Minute differences in materials and construction, along with small differences in ambient conditions, in service stress seen by the unit, and in the abnormal events experienced in service, cumulatively lead to big differences in lifetime.

In fact, generally the only time one finds a very tight grouping of lifetimes– i.e., all of the units in a batch fail at about the same time – is when there is some common manufacturing defect or materials problem – a flaw – leading to premature failure.

Predicting Time to Failure

High failure rate and uncertainty make for a costly combination

As a conceptual learning exercise, it is worth considering how valuable exact knowledge of when any particular device would fail would be to a distribution utility. Suppose that it were known with certainty that a particular device would fail at 3:13 PM on July 23rd. Replacement could be scheduled in a low cost, low impact on customer manner prior to that date. There would be no unscheduled outage and no unanticipated or difficult to manage costs involved: impact on both customer service quality and utility costs could be minimized.

---

[2] It does not have only ten years of expected service left. This is a "survivor" of 30 years service. As such, statistically, it has a likelihood of lasting more than the average 40 years expected when it was new. Essentially it is making up for the few other units of its type that failed prior to 30 years of service, keeping the average at 40 years service.

It is the *uncertainty* in the failure times of power system equipment that creates the high costs, contributes to service quality problems, and makes management of equipment failure so challenging. The magnitude of this problem increases as the equipment ages, because the failure rates increase: there are more "unpredictable" failures occurring. The utility has a larger problem to try to manage.

Failure time prediction: an inexact science

With present technologies, it does not seem possible to predict time-to-failure exactly. In fact, capability in failure prediction for equipment is about the same as it is for human beings.

1. *Time-to-failure can be predicted accurately only over a large population* (set of units). Children born in 2000 in the United States have an expected lifetime of 76 years, with a standard deviation of 11 years. Service transformers put into service in 2000 have an average expected lifetime of 43 years with a standard deviation of 7 years.

2. *Assessment based on time-in-service can be done,* but still leads to information that is accurate only when applied to a large population. Thus, analysts can determine that people who have reached age 50 in year 2000 have an expected 31 years of life remaining. Service transformers that have survived 30 years in service have an average 16 years of service remaining.

3. *Condition assessment can identify different expectations* based on past or existing service conditions, but again this is only accurate for a large population. Smokers who have reached age 50 have only a remaining 22 years of expected lifetime, not 31. Service transformers that have seen 30 years service in high-lightning areas have an average of only 11 years service life remaining, not 16.

4. *Tests can narrow but not eliminate the uncertainty* in failure prediction of individual units. All the medical testing in the world cannot predict with certainty the time of death of an apparently healthy human being, although it can identify flaws that might indicate likelihood for failure. Similarly, testing of a power transformer will identify if it has a "fatal" flaw in it. But if a human being or a power system unit gets a "good bill of health," it really means that there is no clue to when the unit will fail, except that that failure does not appear to be imminent.

5. *Time to failure of an individual unit is only easy to predict when failure is imminent.* In cases where failure is due to "natural causes" (i.e., not due to abnormal events such as being in an auto accident or being hit by lightning), failure can be predicted only a short time prior to failure. At this point, failure is almost certain to advanced stages of detectable deterioration in some key component.

   Thus, when rich Uncle Jacob was in his 60s and apparently healthy, neither his relatives nor his doctors knew whether it would be another two years or two decades before he died and his will was probated. Now that he lies on his deathbed with a detectable bad heart, failure within a matter of days is nearly certain. The relatives gather.

   Similarly, in the week or two leading up to failure, a power transformer generally will give detectable signs of impending failure: an identifiable acoustic signature will develop, there will be internal gassing, and perhaps detectable changes in leakage current, etc.

6. *Failure prediction and mitigation thus depend on periodic testing as units get older.* Given the above facts, the only way to manage failure is to test older units more periodically than younger units. Men over 50 years of age are urged to have annual physical exams in order that possible failures are detected early enough to treat. Old power transformers have to be inspected periodically in order to detect signs of impending failure in time to repair them.

Table 7.4 summarizes the key points about failure time prediction.

Table 7.4 Realities of Power System Equipment Lifetime Prediction

1. Time to failure can be predicted accurately only over large populations.
2. Past and present service conditions can be used to narrow the expected uncertainty, but only slightly.
3. Testing provides accurate "time to failure" information only when it reveals flaws that mean, "failure in the near future is nearly certain."
4. The only way to assure accurate prediction of time to failure is to periodically test a unit as it gets older.

Quantitative Analysis of Equipment Failure Probabilities

Figure 7.3 shows the classic "bathtub" lifetime failure rate curve, qualitatively representative of equipment failure likelihood as a function of age for just about all types of devices, both within the power systems industry and without. The curve shows the relative likelihood of failure for a device, in each year over a period of time. Failure likelihood is not constant from year to year.

This curve and its application are probabilistic. As stated above, in a practical sense it is simply impossible to predict accurately when any particular unit will fail. However, over a large population one can predict with near certainty the overall characteristics of failure for a group as a whole. Curves such as Figure 7.4 represent the expected failure of devices as a function of age, over a large population. They can be applied deterministically to equipment only when it is in bulk, or to individual units only as expectations. Thus, based on Figure 7.3, 4 out of 1000 units that have lasted 18 years will fail in that 18th year of service. Similarly, for any unit that has survived 35 years of service, it has a probability of .1 of failing in its next year of service.

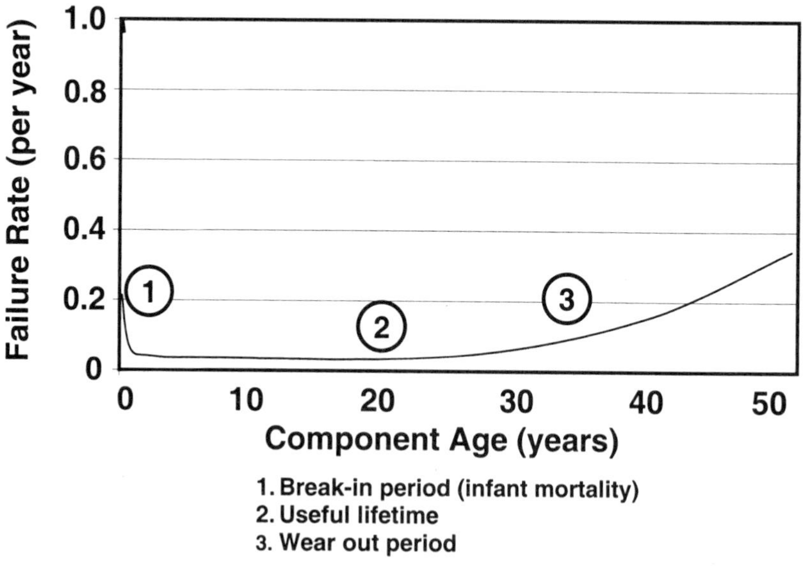

Figure 7.3 The traditional bathtub failure rate curve, as discussed in the text. Here, the device has a useful lifetime of about 30 – 35 years.

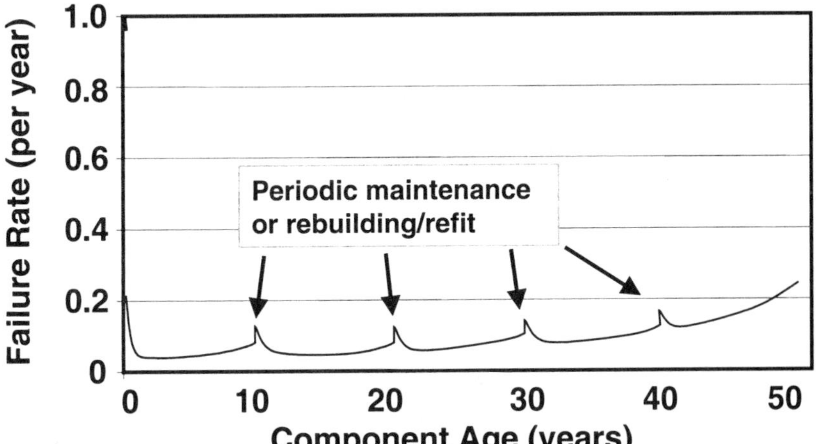

Figure 7.4 A bathtub curve showing impact of periodic rebuilding and refitting.

Figure 7.3 identifies three different failure periods of the device's lifetime. These are:

1. *Break-in period.* Failure rate for equipment is often higher for new equipment, as shown in Figure 7.3. Flaws in manufacturing, or installation, lead to quick failure of the device once placed in service.

2. *Useful lifetime.* Once the break-in period is past, there is a lengthy period when the device is performing as designed and failure rate is low, during which the design balance of all of its deterioration rates is achieving its goal. During this period, for many types of equipment, the failure rates due to "natural causes" (deterioration due to chronological aging and service stress) may be so low that the major causes of failure are abnormal events.

3. *Wear-out period.* At some point, the failure rate begins to increase, due to the cumulative effect of time, service stress, and abnormal events. From this point forward it increases with time, reaching the high double-digit percentages at some point.

Often periodic maintenance, rebuilding or refitting, can "extend" lifetime or lower expected failure rate, as shown in Figure 7.4. This concept applies only to units designed for or conducive to service (i.e., this applies to breakers, perhaps, but not overhead conductor). Here, the unit is serviced every ten years or so. Note however, that failure rate still increases over time, just at a lower rate of increase, and that these periodic rebuilds create their own temporary "infant mortality" increases in failure rate.

Failure Rate Always Increases With Age

All available data indicates that inevitably, failure rate increases with age in all types of power system equipment. Figure 7.5 shows data on the failure rate of underground equipment for a utility in the northeast United States, all of which is qualitatively similar to the failure performance seen on all equipment, in any power system.

Failure rate escalation characteristics

Figure 7.6 illustrates how the *rate of increase* of failure rate over time can exhibit different characteristics. In some cases the failure rate increases steadily over time (the plot is basically a straight line). In other cases the rate of increase increases itself – failure rate grows exponentially with a steadily increasing slope. In yet other cases, the rate climbs steeply for a while and then escalation in failure rate decreases – a so-called "S" shaped curve. Regardless, the key factor is that over time, failure rate always increases.

Eventually the failure rates become quite high

Figure 7.7 is actual data representing a large population in an actual and very representative, electric utility. Note that the failure rates for all three types of equipment shown eventually reach values that indicate failure within 5 years is likely (rates in the 15 – 20% range). In some cases failure rates reach very high levels (80% - failure in the next year or so is almost a certainty. To power system engineers and managers who have not studied failure data, these values seem startlingly high. However, these are typical for power system equipment – failure rates *do* reach values of 15%, 25% and eventually, 80%. But these dramatically high values are not significant in practice, as will be discussed below, because few units "live" long enough to become that old. What really impacts a utility is the long period at the end of useful lifetime and beginning of wear out period when failure rate rises to two to five times normal (useful lifetime) rates.

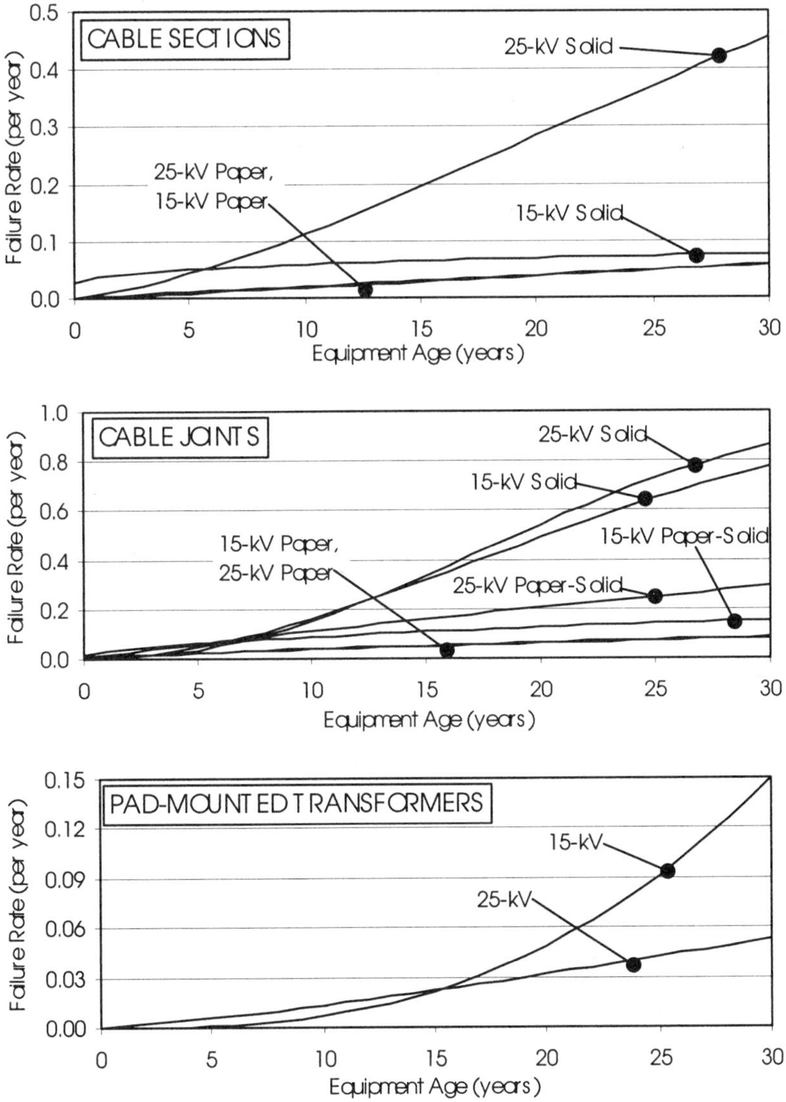

Figure 7.5 Data on failure rates as a function of age, for various types of underground equipment in a utility system in the northeastern US. Equipment age in this case means, "time in service." Equipment of different voltage classes can have radically different failure characteristics, but in all cases failure rate increases with age.

7.4 IMPACT OF ESCALATING FAILURE RATES: WHAT DO AGING UTILITY EQUIPMENT BASES REALLY LOOK LIKE?

The paragraph above stated that the very high failures rates reached after 50 years or more (see Figure 7.7) make relatively little impact on a utility's cost or service quality. The reason is that very few units survive to be so old, so there are few units that have such very high failure rates.

Example 1: A Typical Failure Rate Escalation and Its Impact on the Installed Equipment Base

Figure 7.6 illustrates a very simple example that will begin this section's quantitative examination of failure, installed base characteristics, and overall impact on the utility. In this example, the group of 100,000 service transformers is installed at one instant, a rather unrealistic assumption but one that has no impact on the conclusions that this example will draw.

As a group, this set of 100,000 units has the statistical failure rate characteristic shown in the top part of Figure 7.8. That plot gives the probability that an operating unit of any particular age will fail in the next 12 months of service. In this case there is no high break-in-period failure rate. The base rate during normal lifetime begins at 1.5% per year, rising to 2.5% by year 24, 6.6% by year 30, and to 9% annually by age 40. This curve is representative of actual service transformer failure rate curves (see bottom of Figure 7.6).

The bottom diagram in Figure 7.6 shows, as a function of age, the percent of the 100,000 units installed in year zero that can be expected to remain in service each year, as units fail according to the expectation defined by the top curve. In year 1, 1.5% of the units' fail, meaning 99% are in service at the beginning of year two. At the end of a decade, 85% are still in service. The failure rate is initially 1.5%, increasing slightly above that value each year. Despite this rise, only 15% of the units (ten times1.5-%) fail in the first decade. The reason is that the number of units left to fail decreases each year – there are only 985,000 units left at the end of the first year, etc., so 1.5% is not exactly 1,500. The number of actual failures decreases slightly each year for the first decade, to a low of only 1,440 failures in year ten, because the number of units remaining to fail drops faster than the failure rate increases.

At the end of 20 years, 71% of the units remain, and at the end of thirty, only 53% remain. The 50% mark is reached at 32 years. By the end of year 50, failure rate has escalated to more than 15%, but only a paltry 10.3% of the units remain. Only 500 (.7%) make it to year 60. Less than two are expected to be in service by year 70, when failure rate has escalated to 50%. The average unit ends up providing 43.8 years of service before failure.

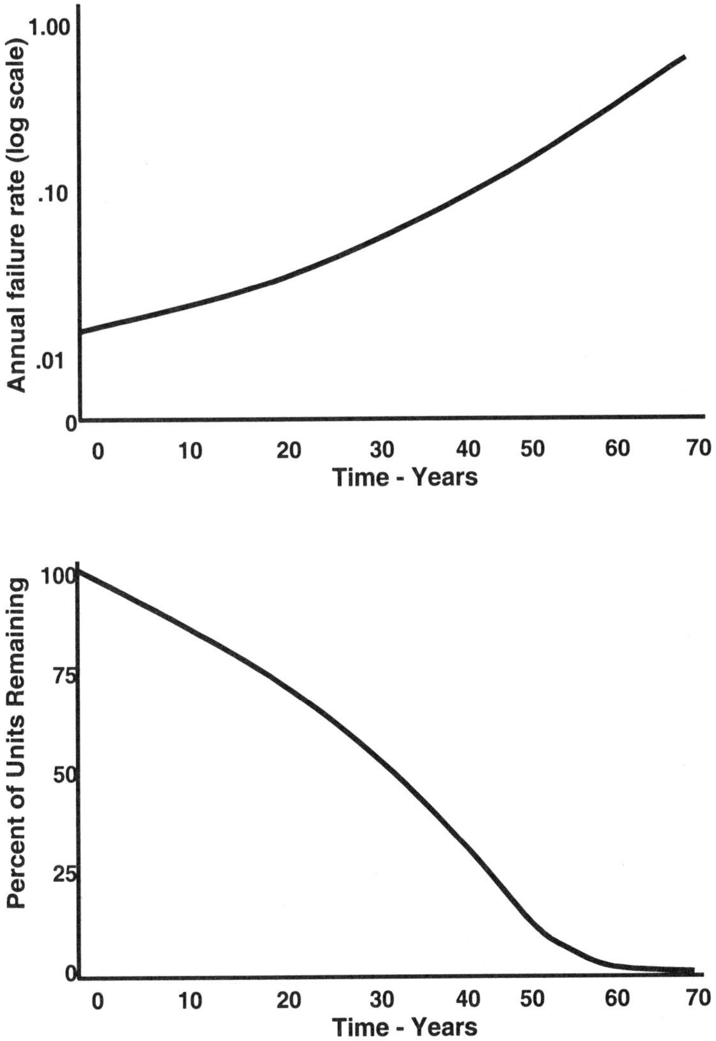

Figure 7.6 Top, failure rates as a function of age for a group of 100,000 service transformers installed in one year. These are representative of typical service transformers in typical service. Bottom, the percent remaining of the original group as a function of time. After twenty years, 60% remain, but after fifty years only 12.5% are still in service, and after 70 years, only 2 of the original 100,000 are expected to still be in service.

Failures in intermediate years are the real culprit

The very high failure rates that develop after five or six decades of service make little real impact on the utility's quality of service. Units that are 70 years old have a 50 percent likelihood of failing in the next year, but as shown above, there are only two units out of every 100,000 that make it to that age – essentially an anomaly in the system.

Instead, the high impact failure levels that plague a utility are caused by transformers of intermediate age. Figure 7.7 shows the number of units that fail each year in this example. As mentioned earlier, the number of failures is initially 1500 units per year and the count actually *drops* slightly during the first decade as the product (failure rate x number of units remaining) decreases slightly.

The number of failures peak in year 44, with 2576 expected failures. Thereafter, even though the failure rate keeps increasing every year, the number of remaining units this rate applies to is so low, that the net number of failures decreases rapidly.

Failure-Count Diagrams

Figure 7.9 is a very interesting and significant diagram, because it has another interpretation or use: this is a plot of contribution to failure by age of transformer for any utility system built with this type of unit, regardless of when they were installed. The curve's shape shows, in relative terms, how much transformers of a particular age contribute to the overall installed base's failure rate.

In the example used above, all the transformers were installed in the same year, but Figure 7.7 can be applied to cases where a utility installed them in various years. It still has the same meaning, showing the relative contribution to failures of units of various ages. The majority of the failures, nearly 2/3, will be units that are between 15 and 45 years old. Transformers installed 44 years ago contribute the most to the utility's annual failure count for service transformers. Older units fail with a higher likelihood, but there are too few to generate as high a total count. Figure 7.9 can be the foundation of studies for proposed policies that would replace units when they reach a certain age.

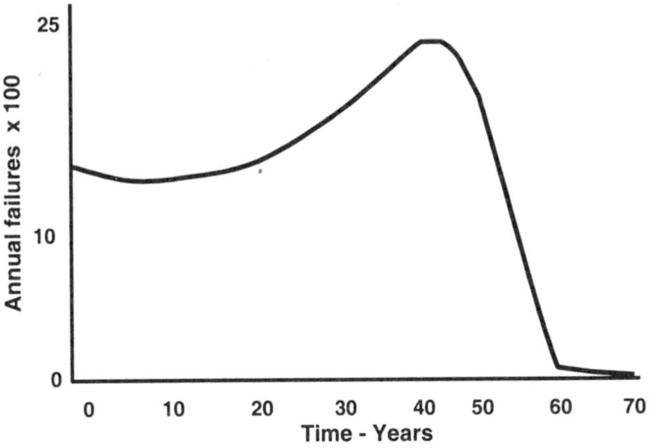

Figure 7.7 Number of failures occurring each year in the population that originally number 100,000 units. The maximum is 2,576, in year 44, the maximum combination of the escalating failure rate, applied to the decreasing number of remaining units. More than half the failures occur in the range between 20 and 45 years in service. Very old units contribute only a small portion to overall failure count.

Example 2: A More "Real World" Case

The preceding example provided useful insight into equipment failure characteristics, and led to the failure count contribution curve (Figure 7.7), a very useful analytical tool. But to see the full implications of failure rate escalation on a utility, and understand how certain replacement policies might reduce failure count, it is necessary to sacrifice example 1's simplicity for more realism. Specifically, one needs to look at a situation where failed units are replaced, which is the fact of life in the "real world."

Thus, Example 2 builds on Example 1, using the same type of service transformers. It:

1. Assumes, as before, 100,000 units installed in year 0.

2. Assumes a period of 70 years.

3. Assumes the failed units are replaced immediately with new ones, keeping the overall count at 100,000.

4. Assumes these new units follow the same failure rate curve as the original units.

This is a more realistic example, as it represents what a utility has to do – keep the same number of units in service, replacing failed units when they fail.[3] The major change in the overall equipment base and its interaction with failures in this example is that *replacement units can fail, too.*

What does this installed equipment base look like with respect to age distribution of units in service, average failure rate, and failure count? Figure 7.8 shows the distribution of ages of the 100,000 units in the system after 70 years. In year 1 (the first year) 1,500 units failed, and were replaced with new units which are now 69 years old. Those 1,500 units failed with the same characteristic trend as the original set, meaning that 1.5-%, or 23, failed and were replaced in their first year of service (year 68). In that year, 1489 of the original 100,000 units failed (1.5% of the 998,500 remaining original units). Thus, a total of 1,512 replacement units were installed in year 2. They began to fail along the same trend as the original units, so that in year 3 there were failures of units that had been installed in years 0, 1, and 2, etc.

The net result, when all of these replacements, and replacements for replacements that failed, etc., are added up, is that the utility had to install about 60,000 additional replacement units during the 70 year period. And, following the failure count contributions of Figure 7.7, most of those replaced were of "middle aged transformers" – those in the 15 –45 year old range. Thus, Figure 7.8 gives the relative age of units replaced by the utility over the course of its annual O&M in each year.

Figure 7.8 shows the resulting equipment bases distribution of transformer ages, after 70 years. It has nearly an even distribution of transformers from age 0 (new) to 30 years. Thus, despite the escalating failure rate as transformers near age 30, 30 years can be considered a reasonable definition of the "useful life" of these units. At about 35 years the count takes a rapid plunge – this is the period (about 35 years in service) during which the bulk of failure counts occur (see Figure 7.4), and thus the age when a good deal of replacements had to be made.

In this system, the average unit in service is 22 years old. However, due to the escalation of failure rates as units age, those older than the average contribute a good deal more to the average failure rate. The entire population has an average failure rate of 3.15%, or more than twice that of new units. This

---

[3] The fact that this example assumes the system is created from scratch in year zero makes no impact on the results given here. As shown earlier, by year 70 only 2 of the original units are left. The population consists of units that have been replaced, and in some cases, failed and replaced again. As a result after 70 years there is only an insignificant "start effect" involved in the data – the model's results in year 70 represent a fairly stable look at what year-to-year operation for the utility would be with respect to service transformers.

corresponds to the failure rate of a unit that is 29 years old (see Figure 7.8, top).

Replacement policy analysis

Suppose this utility decided to replace all units when they reached fifty years in service. It would have to replace only about 75 units annually – not a tremendous cost, particularly considering the units will have to be replaced pretty soon, anyway (they will most likely fail in a few years, at that age). However, the impact on the overall failure rate is insignificant, making no real difference in the average age or failure rates for the total population. The high impact on annual failure count comes from units that are aged 25 to 45 years, because there are so many of them. Replacement at age 50 gets to the units after too many have failed.

Replacement of units at age 40 has a far different effect. First, nearly 1000 units a year have to replaced, so the annual cost is roughly 12 times that of a 50-year replacement policy. But, noticeable portions of unexpected failures are avoided. The average failure rate drops to 2.6% (from 3.1%), a reduction in unexpected failures of nearly 20%, wrought by replacement on only 1% of the units in the system annually.

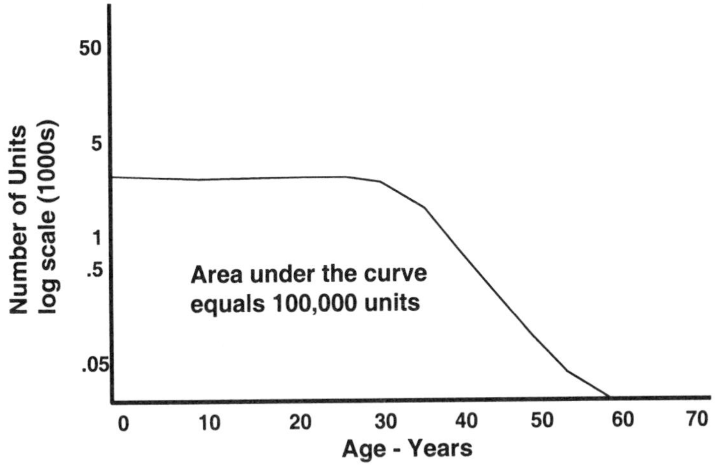

Figure 7.8 Distribution of ages of units in the example 2 system, 100,000 units that have been replaced as they fail over the last 70 years.

Average age of a unit under this policy falls to less than 18 years (if all units were replaced at age 40 and none failed until that age, the average age of units in service would be exactly 20 years). The population's average failure rate of 2.6% is equivalent to the failure rate of a 25-year-old unit. Whether this policy makes economic sense or not is something for management to weigh in its evaluation of how best to spend the (always-limited) monies it has for system improvements. That will be discussed in Chapter 12.

7.5 SUMMARY OF KEY POINTS

All equipment installed in an electric system will eventually fail, and need to be replaced. Key points brought up in this chapter are:

1. Failure is caused by deterioration, due to the effects of age, service, and infrequent but stressful abnormal events.
2. This deterioration eventually leads to failure.
3. Due to efficient design of most equipment, a unit that is near failure from one area of deterioration is probably fairly "shot" in other areas, too.
4. Regardless of equipment type, the failure increases, as a unit grows older.
5. For an installed base of equipment, the bulk of failures will come not from very old equipment but from "middle aged" equipment.
6. In all cases, due to failure rate escalation with age, the equivalent failure rate of an installed equipment base will correspond to a much older age than the average age of the equipment.
7. "Early replacement" policies can be worked out to determine if and how equipment should be replaced at some specific age (e.g., 40 years) and what that impact that would have on the net failure rate of the equipment base.

FOR FURTHER READING

P. F. Albrecht and H. E Campbell, "Reliability Analysis of Distribution Equipment Failure Data," EEI T&D Committee Meeting, January 20, 1972.

R. E. Brown, and J. R. Ochoa, "Distribution System Reliability: Default Data and Model Validation," paper presented at the 1997 IEEE Power Engineering Society Summer Meeting, Berlin.

J. B. Bunch, H.I Stalder, and J.T. Tengdin, "Reliability Considerations for Distribution Automation Equipment," IEEE *Transactions on Power Apparatus and Systems,* PAS-102, November 1983, pp. 2656 - 2664.

"Guide for Reliability Measurement and Data Collection," EEI Transmission and Distribution Committee, October 1971, Edison Electric Institute, New York.

Institute of Electrical and Electronics Engineers, *Recommended Practice for Design of Reliable Industrial and Commercial Power Systems,* The Institute of Electrical and Electronics Engineers, Inc., New York, 1990.

A. D. Patton, "Determination and Analysis of Data for Reliability Studies," IEEE *Transactions on Power Apparatus and Systems,* PAS-87, January 1968.

N. S. Rau, "Probabilistic Methods Applied to Value-Based Planning," *IEEE Transactions on Power Systems,* November 1994, pp. 4082 - 4088.

A. J. Walker, "The Degradation of the Reliability of Transmission and Distribution Systems During Construction Outages," Int. Conf. on Power Supply Systems. IEEE Conf. Publ. 225, January 1983, pp. 112 - 118.

H. B. White, "A Practical Approach to Reliability Design," IEEE *Transactions on Power Apparatus and Systems,* PAS-104, November 1985, pp. 2739 - 2747.

H. L. Willis, *Power Distribution Planning Reference Book,* Marcel Dekker, New York, 1997.

8 Obsolete System Structures

8.1 INTRODUCTION

This chapter looks at the impact that "making do" with limited numbers and sizes of substation sites and sub-transmission rights of way has on the reliability and operation of a power delivery system. As discussed in Chapter 1's introduction to aging T&D infrastructures, the second of the four interacting aspects of the aging infrastructure issue is the limited facilities which power distribution utilities invariably have in the core of thriving metropolitan areas. Difficulties in getting new sites, in expanding old sites, and in getting rights of way often mean that the utility's system can not expand in company with the demand of an ever-growing downtown area.

This chapter looks at this problem and the effects that it has on power system performance. Section 8.2 presents the basic problem with some examples. Section 8.3 then examines the impacts that "outdated" or limited system structures have on the engineering decisions that planners must make with regard to the sub-transmission and substation levels of the system, and on the operating characteristics and daily constraints faced by the people managing that part of the system. All of these effects compromise the system's reliability. Section 8.4 looks at the impacts made on the primary feeder system, which degrade its potential to contribute economy of operation and reliability to the system. Section 8.5 summarizes the various solutions and improvements recommended for the distribution utility faced with this problem. Key points are summarized in section 8.6.

8.2 OBSOLETE SYSTEM LAYOUTS

Example System

Figures 8.1 and 8.2 depict the plight of a fictional electric distribution utility, Metropolitan Power and Light, which represents the average experience of eight major metropolitan utilities on the east and west coasts and the Midwest United States.

In the period 1965 to 2000, Metropolitan Power and Light saw the peak demand in its central operating district go from 1380 MVA to 3058 MVA, an average annual growth rate of 2.3%. But while the peak load and demand for quality of service have increased greatly, the basic topology of its system did not keep pace. The original (1965) system, shown in Figure 8.1, consisted of 16 substations in the central operating district, with an average capacity of 128 MVA each, for a total installed substation capacity of 2048 MVA. This resulted in a utilization ratio of 67% (1380 MVA/2048 MVA). The majority of these substations were originally fed by 69 kV transmission loops (two, or in some cases, two sets of two, lines along the same right of way, providing a parallel and redundant (contingency backup) feed to each substations.

During the intervening 35 years, Metropolitan added one new substation site and expanded the size of four others. It also upgraded a number of 69 kV sub-transmission lines to 138 kV and completed one "loop" transmission route by adding a right of way to connect two substations (Figure 8.2). In addition, it increased the capacity of the 138 kV circuit lines in the backbone right of way through downtown by re-conductoring and bundling of conductor, and by upgrading part of the route to 345 kV.

Table 8.1 Comparison of Metropolitan Power and Light Downtown Systems in 1965 and 2000

System Measure	System in 1965	System in 2000	2000 as a % of 1965
Peak load	1380	3058	222%
Number of substations	16	17	107%
Substation capacity - MVA	2048	3400	166%
Peak load/substation - MVA	86	180	209%
Peak load/sub-trans. - MVA	52	101	194%
Number of feeders	261	386	147%
Load/feeder	5.2	7.9	150%
Substation utilization at peak	67%	90%	134%
Feeder utilization at peak	57%	74%	128%
MW-miles (x 1000)	2150	4967	230%

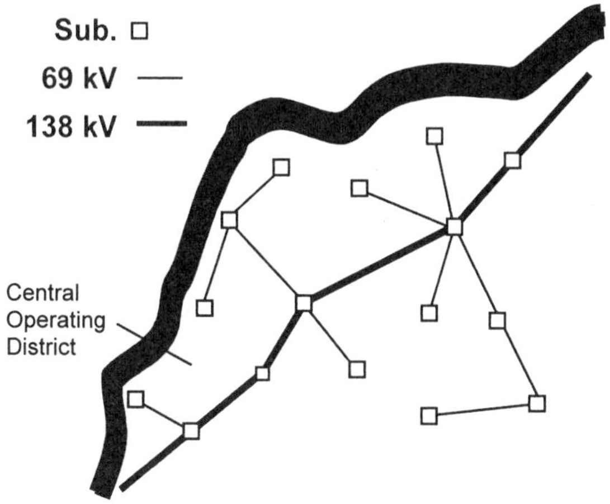

Figure 8.1 In 1965, the central operating district of Metropolitan Power and Light used 16 substations averaging 128 MVA capacity each (total 2,048 MVA capacity) to serve a peak downtown load of 1,380 MVA. This diagram shows only the overall structure of the system. Many of the apparently radial lines feeding substations are in fact loops (two circuits) or double loops (four circuits).

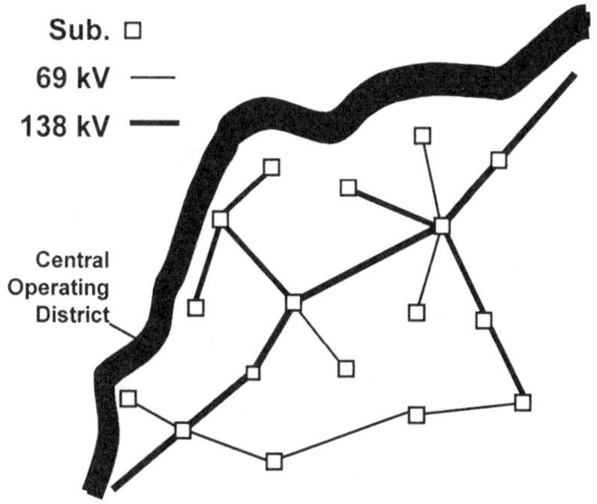

Figure 8.2 (2000) Thirty five years of slow but steady load growth result in a year-2000 peak demand of 3,058 MVA, more than double 1965's. In that time, Metropolitan Power and Light has added only one substation site and expanded the area of four others, as well as upgraded various 69 kV lines to 138 kV lines. Total substation capacity now stands at 3,400 MVA, (200 MVA each, a 56% increase), for a utilization ratio of 90%.

However, this system growth did not keep pace with peak load growth. Year 2000 capacity is 3,400 MVA, an average of exactly 200 MVA per substation, giving a utilization ratio against the 3,058 MVA peak load of 90%. Utilization ratio of the transmission lines has similarly increased. Equally indicative are the facilities ratios, load/substation and load/transmission route, as shown in Table 8.1. Metropolitan Power and Light has a lot of eggs in very few baskets.

At first glance it might appear that the system in Figure 8.2 is fully as capable of providing high reliability of service as the system in Figure 8.1. In actuality, the system is more vulnerable to certain types of outages at the sub-transmission and sub-station levels. This increased vulnerability is due to the following factors: 1) higher utilization ratio; 2) "a lot of eggs in one basket and; 3) design choices dictated by limited options for facilities.

Why Did This Happen?

Before turning to a discussion of the impacts these changes had on the cost and reliability of electric service, it is worth looking at why Metropolitan got into this situation. Over a period of three and one half decades, as its peak demand grew, its system did not keep pace in some ways, out of both choice and constraint. This did not occur because the utility planners did not forecast the growth: for the most part they saw it coming. It happened because, incrementally, the utility allowed itself to be backed into the situation. The major reasons were:

No sites available

In 1965, the downtown area served by Metropolitan Power and Light was already fully developed, in the sense that there was no vacant, never-built-upon land. However, much of it was not what would, 35 years later, be considered "downtown development." There were areas of vacant warehouses, some run down old hotels, a number of gravel-covered parking lots scattered about, an old rail yard about to be redeveloped, and other areas of downtown where land was available, if rather pricey by standards of the time.

Over the next 35 years, this price and availability situation gradually escalated and seemed to stay one step ahead of Metropolitan. As the city and the economic region around it grew the vacant warehouses, old parking lots, available rail yard, and other developable sites in the downtown area gave way to new development, and the various sites that could have become new substations disappeared. Simultaneously, the price of real estate climbed. By year 2000, very few sites suitable for construction of a large capacity (nothing else will help, given the load levels) substation built upon traditional lines exist period. Those that do are very expensive.

Political pressure

Any utility has to be cognizant of local community values and try to accommodate the wishes and preferences of local municipal leaders. Metropolitan Power and Light often found itself facing tremendous political opposition, both in city hall and at a local grass roots level, when it attempted to obtain new substation sites or rights of way for sub-transmission lines. Most of this opposition was directed at the overhead line rights of way it preferred to build to any new site it obtained, or at the inconvenience and traffic tie ups that would be caused by the construction required to build UG transmission lines in the central city. In combination with the often-perceived high costs of new sites, this political pressure blocked most moves to obtain new sites.

What really failed, in the authors' opinion, was communication between utility and municipal authorities on a key point: the utility can easily do without any one of the sites or rights of way it requests, but in the long run it cannot make do if it is denied them all. In this respect, the utility had the responsibility to communicate this issue well to municipal leaders and community alike. It failed to do so. As a result, the incremental, repetitive process of overwhelming opposition on a site-by-site basis led, after three and one half decades, to a poor-performing system.

Financial pressure

Cost was always a major element of every decision made by Metropolitan Power and Light. As a regulated utility, it had an obligation to manage its affairs in a least-cost manner. Management throughout the 35-year period was never incompetent or inattentive; it wanted to make the best decisions regarding cost and system efficiency.

But a new substation site and the new transmission line right of way and the construction required to put those facilities into service always represented a very large budget item. As such, every go/no-go decision regarding a new substation received a great deal of scrutiny. Generally, Metropolitan Power and Light's management was always pre-disposed to find some other way to bolster the system, some other change or upgrade that would avoid, or defer for a few years, the need for a very expensive new site.

In addition, in the late 1980s and throughout the 1990s, Metropolitan Power and Light, like most large investor-owned utilities, began to respond to the uncertainty created by de-regulation, and increasing pressures to raise financial performance, by reducing operating and capital budgets by up to 30%. An executive goal was to see utilization ratio on the system rise – "to get more out of existing equipment."

This led on several occasions to changes in loading policy, re-rating of transformers, cables, and overhead lines, and adoption of slightly less conservative rules with regard to available contingency margin and operating policy.

Not building substations fit right into this push for budget reduction, because it avoided both large outlays of capital, and construction of new facilities whose maintenance would have swelled the O&M budget. In addition, the re-rating of equipment mentioned in the paragraph above "created" a good deal of new capacity at many existing sites, permitting the utility to keep pace with the load growth without adding new facilities, and still meet all its engineering criteria.

No additional room available at existing sites

When it was not allowed to gain more sites, Metropolitan Power and Light had no choice but to serve the growing load from the sites it had. Load/site naturally had to go up, and it did, from 86 MVA in 1965 to 200 MVA/substations in year 2000. *But this does not imply that utilization ratio had to go up.* Metropolitan could have kept utilization ratio constant at 67% by increasing installed capacity at its 17 downtown substations to an average 270 MVA each. It did not do so for a variety of reasons, increasing it only to an average of 200 MVA, and letting peak utilization rate of its downtown substations rise to 90%.

Metropolitan Power and Light made its decision to let utilization ratios rise as a result of several mutually reinforcing trends. First, land adjacent to existing sites, for expansion, was simply not available in many cases, and its price when it was for sale was astronomical compared to what the utility was willing to pay. Secondly, the lengthy period for and uncertainty involved with negotiation and approvals for such special sites made trying to get them risky.

Thirdly, in the late 1980s and early 1990s, Metropolitan's executive management decided to cut back on capital spending. It wanted to reduce budgets and try to "get more from existing equipment." It had an inclination to let utilization ratios rise, and the situation in the central operating district played right to that goal. Perhaps Metropolitan would not have let the situation become so extreme had more expansion space been more available, but the fact is that because of its budget-reducing focus it did not fight for expansion room as hard as it otherwise might have.

Altogether, these three effects limited the number of situations where the utility added space to existing sites (it expanded only four of its sixteen original sites during the 35-year period). Thus, while the load grew by 122% during the thirty-five year period, Metropolitan increased capacity of its substations only to an average of 200 MVA/substation, up from 128 MVA, a 56% increase.

Section 8.3 will show that this lack of room was even more damaging to reliability potential than the lack of new sites.

Other Options

Every time that Metropolitan Power and Light needed a new site, there were always other options available to the utility. Among them were reinforcement re-design and re-configuration of existing facilities, and other means that could provide capacity and contingency capability. Many of these were only "stop-gap" measures to defer the need for new sites for a few years, but in every case they had a lower first cost and met the existing criteria. While a very long-term view would have shown that a few new sites were necessary, the paradigm at work led Metropolitan to take other routes in nearly every case.

A contributing factor to these decisions was the characteristics of the traditional N–1 planning criteria and tools Metropolitan's planners used to apply it, as will be discussed in Chapter 8. The utility planners' tools did not alert them to the weaknesses that were developing in their system. The limitations of the "outdated system structure" they were evolving were not completely visible to them. As a result they were expecting better performance from the system than it could actually deliver.

Incremental Mistakes and Boiled Frogs

Traditional anecdotal wisdom holds that if one throws a frog into a bowl of very hot water it will immediately hop out. But, if that same frog is placed in tepid water, which is then slowly heated over a fire, it will float in the water until it boils to death. The authors have not verified that this is true (it seems unusually cruel to the frog and the authors have no desire to have a boiled frog on their hands). However, it is a good analogy to the way many utilities, including Metropolitan, gradually backed into a situation that they never would have accepted had they foreseen it. Certainly the situation with regard to substation sites was very gradual, with no one incremental event ever so dramatic as to trigger concern.

Another example of this is the gradual erosion of feeder system strength seen by many utilities in the 1980s and 1990s. (Section 8.4 will discuss in detail the impact that the period 1965 to 2000 saw with respect to Metropolitan Power and Light's primary feeder system. As the load grew and the number of substation sites did not keep pace, the feeder system had a gradually increasing burden placed on it. From 1965 to 2000, the load it had to distribute out of every substation grew by 109% (86 MVA in 1965, 200 per substation in 2000). At times this necessitated adding new feeders – Table 8.1 shows that the number of feeders grew by 47%. But to a large extent the load growth was accommodated by allowing the loading (utilization rate on feeders) to escalate.

Table 8.1 showed that average feeder loading increased by 50% during the period; an annual compounded growth rate of 1%. Such a small annual increase is easy to lose among the annual fluctuations in loading due to weather and load transfers, and never enough to raise alarm. This gradual increase in utilization of feeders eventually degraded much of the feeder system's potential to provide high quality customer service, as will be discussed in section 8.4.

Metropolitan Power and Light's inability to see fully the impact of increased feeder loading on its system reliability was, as in the cases cited earlier in this section, partly a failure of its planning tools and engineering paradigm. Traditional distribution planing tools focused on "feeder at a time" short range analysis, techniques which cannot fully assess the feeder *system* impacts that increasing loadings and reduced contingency reach have on the emergency-backup capabilities among feeders.[1] These tools fit the Engineering and planning paradigm for distribution planning, which was to examine feeders in detail, but usually only on a one-at-at-time or small-area basis. Feeder system studies involving all the feeders in a region, modeled in company with the substations and sub-transmission, were never done. Therefore, Metropolitan's' planners never realized that the degradation of feeder-level reliability, amounting to about a 25% increase in contribution to SAIDI, was occurring. Intuition and judgement alone did not alert the planners to the problem largely because of the very slow rate of change.

Poor Planning was a Major Contributor to Metropolitan Power and Light's Problems

The discussions above twice cited the limitations of traditional planning tools as a contributing factor in Metropolitan Power and Light's getting into the outdated system structure situation without fully appreciating its implications with respect to reliability of service. Also cited were Metropolitan Power and Light's lack of effectively applying long-range planning – of using too short-term a focus and not looking at the long-term implications and costs of a de facto policy of constant deferrals and opposition to any major projects.

A longer-range view of its system, even without modern planning tools, would have revealed that problems were going to occur. Even without the best planning tools, such a perspective would have drawn attention to the issue, at least qualitatively. In addition, a longer planning horizon would have permitted Metropolitan Power and Light to identify potential substation sites farther ahead and perhaps obtain them when they were affordable (or at least, less pricey).

Thus, ineffective planning – done with tools of limited capability, and on too short a timeframe – was a contributing factor in Metropolitan Power and Light getting into the situation that in found itself in as the new century began.

---

[1] Chapter 11 summarizes the different between "feeder system planning" and "feeder planning, but for a thorough discussion of these issues see the *Power Distribution Planning Reference Book,* Chapters 16 and 17.

8.3 IMPACTS ON OPERATING CHARACTERISTICS AT THE SUB-TRANSMISSION – SUBSTATION LEVEL

Section 8.2's changes in system structure vs. load level have a number of profound impacts on the performance of the electric system, not all of which can be appreciated by reference to utilization ratios or facilities counts alone.

Site Congestion

Metropolitan Power and Light was unable to get either as many new substation sites as it needed, or to obtain expansion space it could have used at many of its existing substations. As such, many of the substations in the central operating district became congested, in the sense that space was completely used, and a lack of available space was the limiting factor on installed equipment at most sites. This lack of space caused two interrelated design compromises.

Forced to choose capacity over configuration

Figure 8.3 illustrates a common problem and its solution for many utilities. An existing substation site must be expanded in capacity. While some additional room is available, it is insufficient. Faced with the lack of room for transformers (capacity) Metropolitan's planners made the change shown at the right of Figure 8.3. Some breakers and buswork were eliminated, reducing switching flexibility, but freeing room for more transformer capacity. This permitted Metropolitan to double the installed capacity at the substation.[2]

Metropolitan traded switching flexibility during contingencies for more capacity. Lines in the system at the right are "hard wired" to each transformer (UG cable trunk connected). If *either* a transformer or a sub-transmission line fails, *both* have to be taken out of service. The failure rate of both is increased to the sum of the failure rates for both. Although this design may meet traditional N–1-planning criteria, reliability has suffered. The expectation of this outage occurring has increased, and the flexibility of switching when it occurs is more limited. (Chapter 8 will discuss this in much greater detail).

Measures like those shown in Figure 8.3 were common in the evolution of many large metropolitan utility systems throughout the last quarter of the 20th century. Often, recognizing the need for capacity, the utility built the substation originally with a compromise in favor of capacity over configuration. A situation almost exactly like that depicted in Figure 8.3 was a contributory factor in the outages the plagued ComEd of Chicago during the summer of 1999.

---

[2] Breakers at the "sending end" of the transmission lines feeding this substation now provide protection for faults on the high side at this substation.

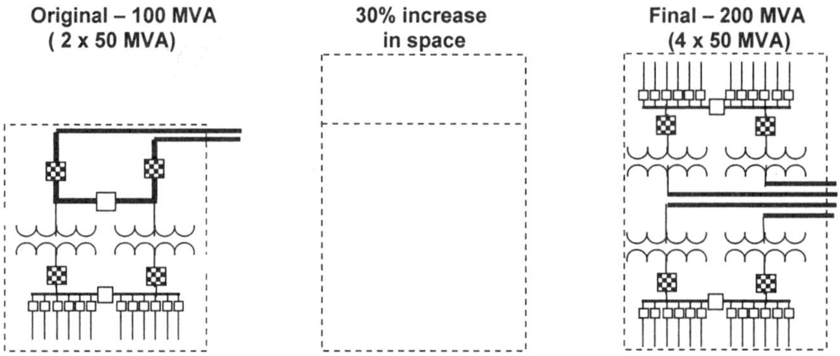

Figure 8.3 An existing site (left) needs to be doubled in capacity. Only a limited amount of expansion space (middle) can be obtained. The only viable solution is to delete high side buses and breakers and simplify the substation's configuration, so more capacity can be installed (right). The utility chose capacity over configuration.

Figure 8.3 shows only one possible way that configuration can be compromised in favor of capacity. It is not meant to represent in detail the exact way congestion problems were solved by any particular utility. But it represents fairly the type of compromises forced on utilities by crowded conditions in central-system substations and rights of way. Numerous other types of design compromises are possible, all sacrificing some or all configuration flexibility for capacity increases. The point is: lack of new sites and expansion room at existing sites forces utilities to deal with load growth by making sacrifices which always have the consequences similar to the changes shown in Figure 8.4. Capacity increases, but interconnected strength and contingency support capability decreases, and with it the reliability potential of the system.

Higher Equipment Utilization

In spite of the efforts to often choose capacity over configuration, during the last third of the 20th century capacity growth did not keep pace with peak load growth in the Metropolitan Power and Light example given here, or in most large metropolitan systems in the US. A primary effect of the changes in this case was the increase in utilization ratio, from a ratio of peak load to capacity for substations of 67% in 1965, to 90% in 2000. Again, as discussed in detail in section 8.2, this occurred primarily because the utility could not get enough new sites, or enough room at existing sites, to provide all the capacity it might like. A contributing factor was management's commitment to reduced capital spending, which favored increasing utilization of existing equipment anyway.

Regardless, an increase in utilization ratio has a tremendous negative impact on expected reliability of the system, one, which can be completely mitigated by a combination of artful engineering changes to the structure of the system (interconnection and configuration of switching) and operating policies. However, if not accommodated with such changes the increased utilization ratio will typically result in degraded reliability of service to customers.

In 1965, each substation transformer or line was loaded during peak periods to about 66% of its rating. This meant that if a substation transformer, or bus, or breaker fails, that 67% loading can be distributed over two neighbors (of equal capacity) who each take half, raising their loading during the contingency from the normal design peak of 67% to 100%. If brief overloads of equipment to 133% of rating are accepted during a contingency, the failed unit's load can be fully supported by only one neighboring unit. Peak load occurs only a few hours per year (typically about 350) so this high level of stress is expected to occur rarely (only when peak load *and* a contingency occur). Off peak, the outage of even two units could often be supported by temporarily overloading only one remaining unit of equipment.

But in a power system with 90% loading, during peak load conditions a failed transformer or line now needs *nine* neighboring units, not two, to support its outage if contingency loading is limited to 100%. (The 90% loading on the unit has to be distributed among nine neighbors, who each have only 10% capacity margin to aid the situation). If equipment overloads are accepted during contingencies, the situation is still worse than before. A 135% overload, two neighbors are required to support the outage of the failed unit, not one as before. Table 8.2 shows required "contingency support count" for various combinations of utility utilization ratio (designed peak load/capability ratio) and contingency overload limit. It plots the number of units required for contingency support as a function of utilization ratio, for systems composed of identical capacity units.

Table 8.2 Contingency Support Count – Number of Neighboring Units Required to "Pick Up" the Load of an Outaged Unit (All Units Same Capacity)

Utilization Ratio	If Neighboring Units Are Permitted to Overload To This Level (%) 100%	112%	125%	133%	150%	166%
67%	2	2	2	1	1	1
75%	3	3	2	2	1	1
83%	5	3	2	2	2	1
90%	9	5	3	2+	2	2
100%	-	9	4	4	2	2

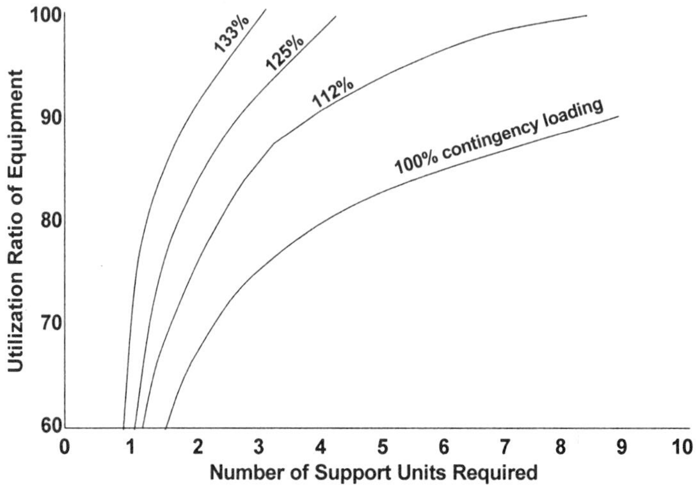

Figure 8.4 Number of contingency support units required to pick up the load of an outaged unit, as a function of designed utilization ratio for the system equipment (i.e., all units).

High utilization ratios mean more interconnection is needed

Assuming those neighbors required for contingency support are available, they will only provide the required support if they can be connected to the load of the failed unit during the contingency. Thus, in a system loaded to 67% at peak, and permitting 133% overloads during contingencies, every substation transformer, bus, and breaker needs contingency interconnection to only one neighboring substation-bus-breaker combination in order for its load to be served even if it is out of service. But in a system designed to a 90% utilization criterion, even if contingency overloading is raised slightly to 135%, interconnection to two neighbors is required.

> *A system with a high equipment utilization ratio relies on interconnection capability to provide the measure of the contingency withstand strength not being provided by a healthy capacity margin.*

Although it is a generalization and there are rare exceptions, a utility desiring to maintain reliability of service at traditional levels must increase the "strength" of its system structure. It must provide both more static and switch-able paths to neighboring equipment (as the case may warrant), and make certain there is sufficient capacity in those interconnections to handle the contingency loading. Ironically, the trends covered in Section 8.2 and the substation space problems depicted in Figure 8.3, forced utilities to do exactly the opposite. Configuration was often reduced as utilization ratio was increased, for two reasons. First, a lack of growth in overall system structure (Figures 8.1 and 8.2) occurred simultaneously with the forced rise in utilization rate. Configuration of the system layout overall did not keep pace with load growth. Secondly, capacity-over-configuration compromises often reduced the interconnection strength that had been present. Utilities often failed to recognize the seriousness of the problem due to the limitations of traditional contingency-based design methods, as will be described in Chapter 8.

More frequent and higher periods of equipment stress

Another point worth considering is the impact that the high utilization rate has on the frequency of overload for equipment in the system. In a system with a utilization ratio of 67% and contingency limits of 133%, loading one neighbor of the failed unit to 133% can cover every equipment outage.

In a system loaded to 90%, every failed unit requires two neighboring units to be loaded to 135% of rating (close enough to be considering within 133% for the sake of this example). Assuming for the moment that the number of equipment failures is the same in both systems (it would probably be higher in the 90% loaded system), this means that twice as many units are exposed to overloads in the 90% system as in the 67% system. Viewed from the standpoint of the system as a whole, twice as many units are overloaded in any given period of time.

Despite this, many utilities responded to the pressures discussed in section 8.2 by raising the contingency loading limits for their equipment. For example, one utility in the central United States raised its limits from 115% to 166%. This move was necessitated by the lack of system structure to provide the increased "contingency inter-connective-ness" of the system discussed above. The only remaining way to cover the contingencies of many units of its equipment was to accept loading to 166% during contingencies.

Chapter 6 discussed equipment loss of life as it remains in service, and the fact that overloads greatly accelerate the loss of life process, so that a unit loaded for several hours at 133% might lose the equivalent of several days or even weeks of normal service lifetime. The change from 67% to 90% loading, a 34% increase, doubles the number of "high stress events" for units like transformers and tie lines. If a utility responds to the increased utilization ratios it desires by increasing contingency loading limits so existing equipment is used at higher loading during contingencies, equipment lifetimes and failure rates suffer.

A major impact

Evidence from several large utility systems indicates that the effects covered in this section can cumulatively double the SAIDI contribution made by the substation and sub-transmission levels of the system, producing an increase of 15% - 20% in total system reliability indices.

Utilization Ratio Is Not the Culprit

High utilization ratio, per se, is not necessarily an undesirable element in power systems. Reliable and economical power systems with peak utilization ratios up to 100% can be designed and operated. However, the rise in utilization ratio must be accompanied by changes in the system's ability to isolated outaged equipment and interconnect during emergencies to available contingency margin. The requirement for interconnection capability and flexibility rises along with system utilization ratio. Configuration must in some sense provide the strength that capacity margin does not. Increasing the utilization ratio without having sufficient configuration flexibility to accommodate it, is a mistake that will lead to deteriorated reliability of service.

8.4 FEEDER SYSTEM IMPACTS

This section examines the impacts the changes discussed in section 8.2 make on the distribution utility's primary voltage feeder system. It first provides a background discussion of feeder systems and reliability, then shows how section 8.2's factors result in a general weakening of the feeder systems ability to provide reliability.

A well-designed and operated feeder system provides reliability of service to energy consumers in two ways, both of which are quite important in maintaining low SAIDI and SAIFI scores:

1. *Feeder level reliability.* It provides inherently good feeder level reliability by having immunity from outages of feeder system equipment, through switching capability combined with sufficient contingency margin to support reasonably expected outages of equipment in the feeder system itself.

2. *Substation – sub-transmission contingency support.* A strong feeder system can provide a measure of contingency support for

failures of buses, breakers, and transformers at the substations, via load transfers away from the outaged equipment to other feed sources. This materially aids good reliability of service and is often the only viable approach to provide high reliability when utilization rates are high.

Outdated system structure issues impact both areas. This section will examine in turn the impact that outdated system structures make on each of these two reliability aspects of feeder system performance.

Feeder Level Reliability

The primary voltage feeder system plays an essential role in the power system chain, taking the power from the substations to the vicinity of all customers. As such, its ability to perform its function in a reliable manner – to recover from outages of circuit elements and equipment – is essential to a good reliability record for the utility. Aging equipment, cables, and conductor in the feeder system all greatly impact reliability.

In almost all-utility systems, the majority of customer service reliability problems is caused by problems on the distribution system, either the primary feeder system or the service-level equipment it feeds. Figure 8.5 shows typical operating results – about 45% of all customer outage minutes are due to problems that occurred on the primary feeder system, nearly twice as much as on the entire system-transmission-substation level.

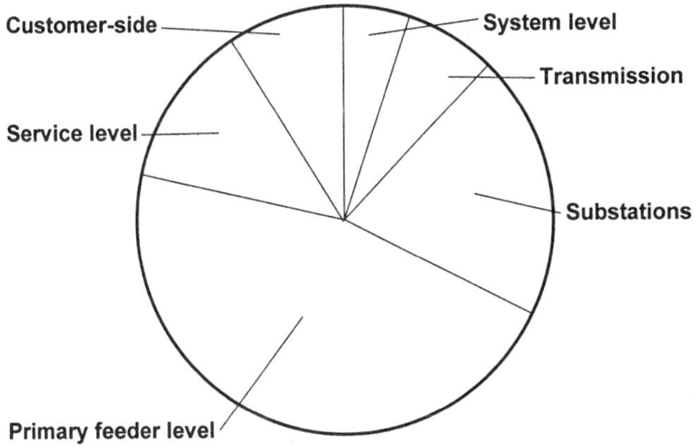

Figure 8.5 Allocation of SAIDI by level of the system for a large investor owned utility in Texas. Results here are based on the average of three years, 1996 – 1998.

The fact the feeder system is the major source of service interruptions in a power system is not the least surprising. In fact it seems inevitable. To begin, there is simply more of the primary distribution system than anything else in the power system. It consists of thousands of miles of line and tens of thousands of units of equipment.[3] There is simply more equipment to fail. Additionally, in all but very rare cases, the primary feeder is radial in design, consisting of a single serial chain of lines and equipment leading to each customer. Any failure in this type of system will lead to an interruption of service to one or more customers.

And finally, the feeder system is distributed – in fact it is the ultimate distributed resource belonging to any electric utility – a system of many, many units of equipment scattered about the system in roughly the same pattern as its customers. As such, locations of failures and repairs are scattered throughout the system, too. Unlike the substation level, the failed equipment will not be at a few well-identified sites. It often takes time – as much as an hour of searching in some cases – for repair crews to find the location of a failure.

Thus, the feeder system is the source of nearly half of a utility's service reliability problems, because of its great extent, radial characteristics, and distributed nature. As such it is particularly sensitive to the increasing failure rates of aging equipment. The aging equipment trends (Chapter 6) mean that both "clear weather" and storm-related failure rates on the feeder system tend to increase rapidly with equipment age. Attention to the feeder level is critical just because so many failures occur there during the best of circumstances. So here is the potential for the increasing failure rates of aging equipment to cause very noticeable increases in SAIDI and SAIFI – even a small increase in such a large portion of the system will make a very perceivable difference.

Outdated System Structure Impacts on the Feeder System

Basically, many of the trends discussed in Section 8.2 lead directly or indirectly to changes on the feeder system that limit its ability to transfer loads, both to restore outages that occur in the feeder system, and to provide inter-substation transfer capability to mitigate substation-level outages.

Non-optimal feeder getaways and routes

Section 8.2 alluded to the fact that one consequence of outdated system structures is that existing substations are called upon to serve more loads than

---

[3] The authors are counting service transformers as part of the primary feeder system in this example, because they are energized on the high side at the primary voltage level.

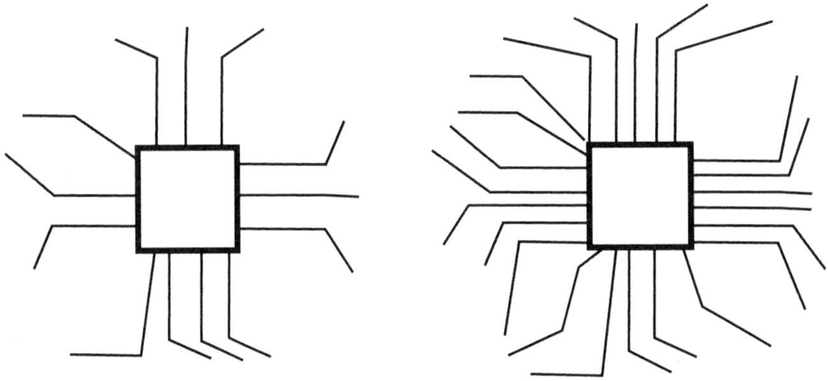

Figure 8.6 Left, as originally planned, one of Metropolitan Power and Light's substations was intended to serve 86 MVA of peak load with 13 feeders, each loaded to about 6.6 MVA. Right, as eventually operated by the utility due to outdated structure constraints and pressures, the substation served 180 MVA through 23 feeders (7.82 MVA). The additional feeder getaways often represented relatively costly designs compared to feeders planned in the initial layout of the substation, and suffered from compromised esthetics and more importantly, lower reliability than needed.

would ideally be the case. Figure 8.6 shows one consequence of this. The higher load requires more feeders (Recall that in the example in Section 8.2, the utility added 47% more feeders than originally intended to some of its substations, a typical situation).

Getting so many more feeders out of a substation than originally designed may create a lot of headaches with regard to routing, configuration, and exposure. Often, feeder getaway constraints – limits in how much space is available above and below ground to run feeder circuits out of a substation – is a limiting factor in its total load. Less than ideal (from the standpoint of economics, esthetics, and reliability) design compromises may be required simply to accommodate the sheer number of separate circuits. Reliability suffers due to a host of secondary effects.

Non-optimal routing

Similar to the getaway problem, a utility may find that additional circuits it has to "work into" an area beyond the number originally planned forces it to use routes and layouts that normally would not have been acceptable. Otherwise, more circuits and higher level of loading could have been accommodated well if the entire area had been originally planned with those in mind. Typically, however, if an area is planned to a lesser density and then upgraded, it is not possible to re-engineer it to the same level of economy and reliability after the fact. Circuit routes have been committed and constraints accepted in certain ways in the original design. Subsequent development in the area has reduced additional routes available to the planners.

The additional feeders often have to take routes that are less reliable than desired (heavy tree populations in areas difficult to trim or exposed to auto traffic along roads). Double-circuit pole routing of feeders (two feeders on one set of poles) often has to be used. Common mode failures (one event taking out more than one feeder) become much more likely. Switch locations and switching patterns often cannot be well arranged with so many feeders in such a small region.

Duct crowding/exposure

In underground distribution systems, the attempt to fit as many feeders as possible into existing duct space will reduce reliability through two secondary effects:

1. *Heating.* Putting more cables in a completely populated duct will lead to a greater amount of losses, and more heat, for all cables. Good engineering will assure heat buildup stays below limits, but nonetheless thermal stress on cables will be higher than it would be otherwise. Lifetime of cables may be noticeably shortened and failure rates increased.

Figure 8.7 Left, four feeders (lines) pass through a large cable vault (circle). Right, in the expanded feeder system, eight now populate the vault. Exposure has increased greatly: there is both more likelihood of a catastrophic failure (there is more equipment in the fault) and any problem potentially hazards twice as many feeders.

2. *Common mode failure* exposure goes up greatly, as shown in Figure 8.7. Here, more feeders means more feeders passing through each cable vault, where a fire on one can now endanger more feeders.

The net effect: a noticeable increase in failure rate

Operating experience at one large investor owned utility showed that areas where additional feeders had to be added "after the fact" had an 11% higher frequency of outages than those areas where the system corresponded to the original planning. This was an 11% increase in the frequency of outages of equipment on the feeder system, leading to an 11% increase in the contribution that feeder system makes to the overall system SAIFI. The causes seemed to be an amalgamation of all the reasons listed above.

Assuming that Figure 8.5's statistics showing that 45% of all reliability problems are feeder-related applies here, this 11% increase in problems on the feeder system means that feeder-related problems caused *just* by outdated system structure problems contribute to 45% x 11% = a 5% increase in the overall system SAIDI.[4]

Using the Feeder System to Mitigate Problems Caused by Substation-Level Outages

One of the most effective, if not the most effective, ways to support substation transformer outages in systems with high substation and sub-transmission utilization ratios is through the feeder system. When a transformer is loaded to 90% of its rating, as it is during peak in many utilities forced to high utilization rates by the issues covered in Section 8.2, avoiding lengthy customer outages when it fails means transferring a lot of load to other equipment, quickly. If the failed unit has only one neighbor that can interconnect to provide that support, that unit must go to 180% of rating – a level considered too high in all but extreme cases – or all of the load cannot be served. Concern over such high contingency loading levels is largely the reason why traditional utilization ratios for substation transformers were targeted at 66%. When one of two units at a substation failed while running at that level, the other had to go to no more than 133% of normal rating – a high but acceptable loading during a contingency.

One way to design a system so that any transformer has more "neighbors" that can support it during an outage, is to have more than two transformers at each substation. Some utilities (Houston) traditionally designed most of their

---

[4] Strictly speaking, Figure 8.5 refers to contribution to SAIDI by level of the system – duration of outages – whereas this discussion refers to frequency of outage, a difference factor. However the authors assume the two are comparable here for the purpose of demonstrating the degree of importance of this concept.

distribution substations with four transformers. Loss of any one, even if loaded to 100% at peak, meant the load could be distributed among the remaining three with only a 133% overload of each. Transformer, bus, breaker and similar contingencies could be contained within each substation, with no support needed for neighboring substations.

But in many cases, three- or four-transformer substations are not an efficient investment. In most cases, the economy of scale for transformers is greater than that for multi-transformer substation layout. Thus, for a given required capacity, two large transformers prove to be much more cost-effective than four smaller ones.[5] As a result, as much as there is any "standard" substation layout used in the United States, it is the two-transformer substation, composed of two like units (same capacity). In a power system based on these types of substations, any outaged transformer-bus-breaker combination unit would have one remaining neighbor to provide contingency support (in the substation). The utility has three choices for a viable contingency plan for these outages.

1. *Limit loading on the substation to what one transformer can take under contingency conditions.* A few utilities have shifted to permitting 166% loading during contingencies (for brief periods), permitting up to 83% loading of the substation. Most, however, limit loading to 75% under normal conditions with a 150% contingency loading, still a very aggressive contingency loading level.

2. *Provide support with a mobile transformer/substation.* The plan being that a mobile spare could be brought to the site "within a few hours." The utility accepts a brief outage during that time as the nominal result of a transformer failure. This is often not a palatable option because the average mobile substation-response time is actually 12-18 hours.

3. *Provide support for transformer outages through the feeder system.* The concept here is that the load of the failed transformer is split – the remaining transformer at the substation picks up part of the load, and the rest of the substation's load is transferred to neighboring substations via re-switching of the feeder system.

   This can be implemented via "programming" of the protection and breakers scheme in one of two ways.

   1. Upon a failure of one unit, the remaining transformer picks up

---

[5] Houston Lighting and Power (Reliant Energy) has a somewhat unusual situation in that its load density is much higher than average due to the intense amount of air-conditioning demand in the Houston area. Four transformer substations are therefore quite efficient in Houston's case.

the entire load it can within its contingency rating. For example, if loading on both transformers was at 90%, the remaining unit goes to 135%, picking up half the failed unit's load. The other half is dropped temporarily.

Feeders are now switched to transfer load from the substation to neighboring substations, so as to pick up the load that was dropped. This may be done with remote switching, but is most often accomplished by manual switching, taking up to two hours.

2. The remaining transformer immediately picks up the entire load. This means it would go to 180% of rating during peak, a level at which it is very quickly giving up lifetime to serve the load. As quickly as possible, feeders are now switched to transfer load to neighboring substations to bring loading down to a lower level for the duration of the contingency.

   This scheme puts transformers (and buses and breakers associated with them) under high stress for brief periods, but avoids customer interruptions. If the switching is well managed, as for example with a computerized restoration system at the utility dispatch center, it can be done quickly enough to avoid large stress-induced loss of life on the one remaining operating transformer.

Figure 8.8 uses a simple example to illustrate the "transfer to neighbor through the feeder system" concept.[6] At the left, a two-transformer substation with a hexagonal service area has boundaries with six (also hexagonal) neighboring substation areas. Transformers in this example are loaded to 90% of rating at peak, by design. Also shown are six feeder areas (three per transformer) alternating radially between each of the substation's two transformers.[7]

In this example, when one transformer is out of service during peak conditions, the other picks up what load it can, going to 135% of its normal rating. The remaining load, corresponding to 45% of a transformer's capacity, is transferred to neighboring substations through feeder re-switching. As shown, the concept is to transfer the peripheral load of the substation – the load closest

[6] Chapter 8, section 8.3, has a thorough discussion of switching and its planning. Chapter 11 has comments on the feeder system planning needed for such capability.

[7] This radial alternation scheme promotes feeder contingency support. See *the Power Distribution Planning Reference Book,* Chapter 8, pages 319 - 339.

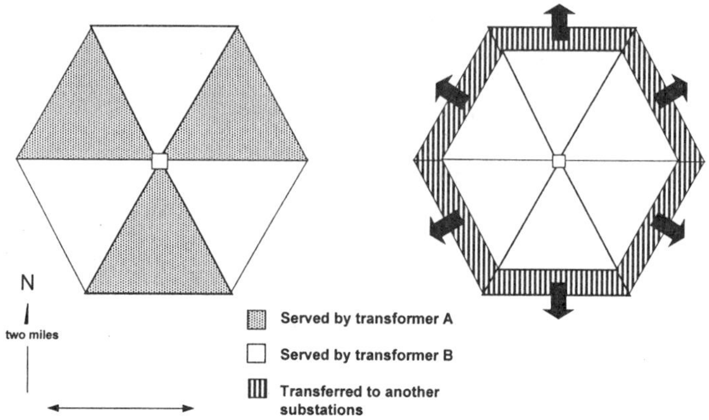

Figure 8.8 Left, hexagonal substation area with borders to six neighboring substation areas. This substation has two transformers that loaded to 90%of their rated capacity at time of peak demand. Under normal circumstances the two transformers, A and B, each serve half the load. When transformer A fails, unit B picks up half of A's load, going to 135% of its rating, and 22.5% of each of the transformers' loads is transferred to feeders from neighboring substations. In this way the utility can "get by" with loading transformers at two-transformer substations to 90% of rating at peak demand, and still have a viable contingency capability for the failure of either one. See text for details.

to the neighboring substations. This assures the shortest electrical paths for the contingency-switching feeders (for voltage drop consideration). See Chapter 8, section 3, for a fuller discussion of how switching capability like this is designed into a distribution system and what its benefits can provide. Support of transformer outages through the feeder system results in no, or very short duration, interruptions of service compared to the multi-hour or day long outages if mobile transformers or other options have to be implemented.

8.5 The Concept of Feeder System "Strength"

In order for the feeder system to work well for contingency support in the example given, the feeder system must be able to transfer 22.5% of the substation's load to neighboring substations (45% of one transformer's load equals 22.5% of the substation's). If the feeder system is not capable of transferring 22.5% of the load, the system dispatcher is forced to choose between dropping load, (interrupting service to some customers) or loading the remaining transformer to an unacceptably high, potentially damaging, level.

On the other hand, if the feeder system can transfer more than 22.5%, then that capability can be used for support transformer contingencies even if design loading of the substation is raised beyond 90%. If it can transfer 35% of each substation's peak load, then utilization ratio can be raised to 100% of transformer rating while still preserving contingency support capability.

> The contingency strength of a feeder system is the portion of the substation load that it can transfer to neighboring substations while keeping voltages and circuit loading on the feeders within the utility's standards for contingency operation.

Figure 8.9 shows the utilization ratio for a two-transformer substation that can be supported in contingencies versus feeder system strength – feeder system inter-substation transfer capability. The term "inter-substation transfers" refers to load transfers between substation service areas. The term "intra-substation transfers" refers to load transfers among feeders in the same substation area. Several different lines are plotted representing various levels of contingency loading limits. For example, the "133%" line represents the combination of

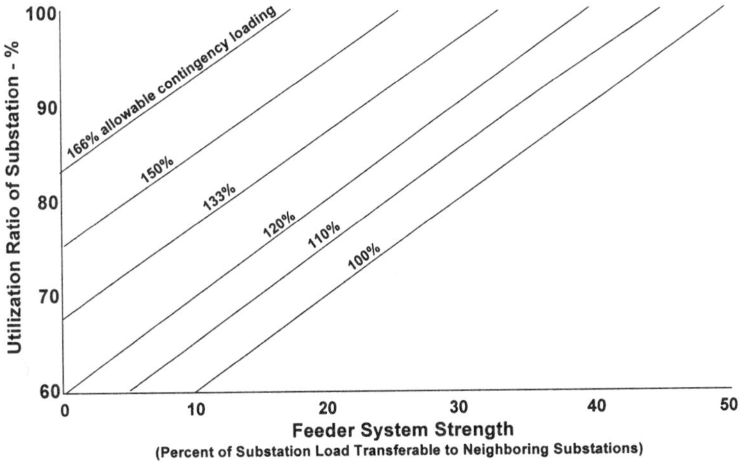

Figure 8.9 Feeder system strength (inter-substation tie capacity) vs. utilization ratio of transformers under normal conditions that can be tolerated while still preserving contingency capability for the outage of a transformer at a two-transformer substation. For example, the graph indicates that if a contingency loading of 166% is allowed by standards, then normal target loading must be limited to 83% if the feeder strength is zero, but can go to 100% if feeder strength is 17%.

utilization ratio and feeder system strengths needed to assure full contingency capability when contingency loading is limited to 133%. The Y-intercept values for each of the lines represent the maximum utilization ratio that a utility can use at a two-transformer substation given that no load will be transferred to neighboring substations in the event of a transformer outage. As can be seen, greater transfer capability through the feeder system means greater utilization ratios can be applied in the normal operation of the system.

Loss of Contingency Reach Due to Obsolete System Layout

"Reach" is the distance that a feeder system can move power while staying with the voltage standards applicable for its operating conditions. For example, a 12.47 kV, three-phase, overhead feeder of typical industry design, using 4/0 ACSR conductor, can move power corresponding to its thermal limit (the maximum amount of power possible within its current rating, 340 amps/phase, corresponding to 7.3 MVA) a distance of 1. 8 miles before voltage drop on the line reaches 7.5% - the maximum permitted on the feeder system according to many electric utility voltage standards. This distance is 4/0's *thermal reach.*

Substitution of another conductor will not materially change this distance: 336 conductor will move its thermal limit (530 amps corresponding to 11.5 MVA) 1.8 miles; 636 AA (770 amps, 16.6 MVA) will move its thermal limit 1.7 miles; #2 ACSR (130 amps, 2.8 MVA), 1.8 miles. The larger conductor has lower impedance (less voltage drop per ampere) but a higher thermal limit. The two factors compensate for one another.

Thus, the *thermal reach* of a 12.47 kV feeder system and the distance it can move power if operating at its thermal limits is 1.8 miles, regardless of conductor size.

Economic reach refers to the distance that power will travel on feeder segments before encountering the maximum allowed voltage drop, in situations conductor size and loading have been selected so that all conductor operates at or near its most *economical* loading level over the course of a year.[8] The economic factors, load characteristics, and design standards that affect determination of economic loading vary from utility to utility. But again, the economic reach of each type of conductor turns out to be about the same. Under typical conditions economic reach works out to about 3.6 miles – about twice thermal reach (economical loads are usually about ½ the thermal limit).

Contingency reach refers to the distance that power can be moved on the feeder system under contingency conditions, when higher loading levels and more liberal voltage drop limits that are permitted. Again, standards vary from

---

[8] See the *Power Distribution Planning Reference Book,* Chapter 7, Section 8.3 for a detailed explanation of how economic loading is determined for conductors.

one utility to another. But as an example the ANSI C84.1-1989 standard specifies range A voltages as design criteria for normal conditions on the system, with a 7.5% maximum voltage drop on the primary feeder system permitted (9 volts on a 120 volt scale). It states that Range B voltage ranges apply in emergency situations, permitting up to 13 volts on a 120 volt scale (10.8% voltage drop).[9]

Thus, according to this standard, during a contingency a distribution utility is permitted to operate feeders in the affected area with up to 10.8/7.5 equals 144% more voltage drop than under normal conditions. This 44% additional voltage drop can be "spent" by either moving more power (more current equals more voltage drop) or by moving power over a longer distance (longer distance equals more voltage drop), or in some combination of more power moved a greater distance which is the normal case.

Generally, due to a bit of margin built into a feeder system for a variety of secondary and tertiary reasons, there will be a bit more voltage drop margin than minimum.[10] Thus, under most circumstances it would be reasonable to expect about 50% – 60% to be the permissible additional margin, over and above normal peak load conditions, rather than just the 44%.

Consider again Figure 8.8 and its example of transfers to other substations. Note that 22.5% of each feeder's load is transferred to neighboring feeders. Roughly speaking, the additional distance for power flow is about 20% farther (assume for the moment that the other substation areas/feeder areas are the same size and shape as those in the drawing). Thus, additional loading times additional distance (122.5% x 120%) is very close to the 144% limit. If this feeder system is well designed and operated such that normal loadings are kept in the most economical range for all conductors (as it ideally would be) then no substantial reinforcement of its elements should be required to provide the required level of contingency support.

Little if any additional capacity needs to be installed to provide this contingency support. A well-designed feeder system with all conductors sized to be most economical, and with switching designed for inter-substation support, can be built at very little if any additional cost (beyond the extra planning and engineering work required) over a system that does not have these capabilities.

---

[9] See the IEEE Red Book -- *Recommended Practice for Electric Power Distribution for Industrial Plants.*

[10] There are only a limited number of different conductors available. Often in "optimally sizing" a particular segment engineers determine that one size is a bit too small, the next size a bit too large, to be exactly optimal. Necessity dictates they pick the larger size, which provides a bit lower voltage, drop than technically necessary – additional margin for contingencies, among other things. In typical systems this adds about 10% to the available voltage drop margin.

However, consider what happens when the loading on the feeder system is permitted to creep upwards, as occurred to Metropolitan Power and Light example case in Section 8.2. Over the course of three and one half decades an increased loading on the feeder system was accepted as an inevitable result of changes the utility could not stop. Although reinforcement of the feeder system was done from time to time, substituting larger conductor where necessary, many portions of the feeder system ended up running at above their most economical loading ranges. As a result, they lose reach.

The net result is that a situation like that depicted in Figure 8.10 can occur during a contingency. There, two feeders for neighboring substations are shown (each a triangular area corresponding to the triangular feeder areas in Figure 8.8). The rightmost feeder's substation transformer has failed. Its outer 22.5% load has been un-switched from the substation (opening switch B), and the feeder tie switch at point A closed, so that the feeder at the left can pick up the outer most load from feeder B. Due to higher feeder loadings, the feeder system does not have the contingency reach to provide this capability. The dotted line in Figure 8.10 shows where voltage drops below unacceptable levels.

The outdated system structure issues discussed in Section 8.2 created two effects that led to this situation:

1. *Loss of contingency capability.* A combination of higher substation utilization ratios and higher loadings on the feeder system lead to a large shift in the ratio of capability available/capability needed. Higher substation ratios raise the amount of load that needs to be transferred over the feeder system during contingencies. Also, higher loadings on the feeder system reduce the voltage drop margin available for contingency reach. As a result, the number of situations and hours per year (near peak) when contingencies cannot be supported increases. Reliability of service suffers.

2. *Contingency capability preempted by planning.* In many systems, local pockets of load growth are accommodated without reinforcement of the feeder system in the area, via load transfers. For example, suppose that in Figure 8.9, the "contingency" that occurred was a planning contingency: the load in the feeder area on the right grew to be 10% more than that feeder could handle while staying within standards, but the feeder on the left had 10% margin.

   Utility planners can avoid any reinforcement cost by performing the switching operation shown, as a *permanent solution* to the "overloading" problem. This "fixes" the normal operating problem at no cost, but leaves insufficient contingency capability in the system.

   This "solution" to local spot growth problems was common among

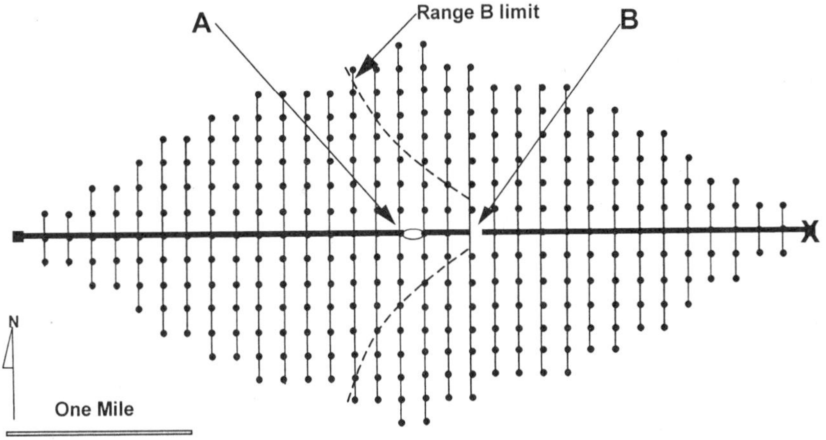

Figure 8.10 Here, the feeder on the right has experienced a failure at the substation (shown by an X). Switch B has been opened, switch A has been closed, so the feeder on the left can support both all of its load and the outlying 22.5% of the other feeder's load (the rest is supported by switching the outaged feeder truck to another feeder near the substation). The feeder on the left is now serving 122.5 percent of its normal load, over a distance of transmission that is as much as 15% more. Dotted lines show the points where voltage drop reaches range B limit.

utilities throughout the 1990s. Since few used reliability-based, feeder system evaluation tools in such planning, the full impact of these decisions on system reliability was not fully recognized.

Feeder Problems Often Don't Get the "Respect" They Are Due

While the feeder system is critical to any utility's establishing a good record of reliability, failures there never cause the widespread, "headline grabbing" problems that failures at the substation and sub-transmission level can cause. The failure of an entire feeder might drop service to between 500 and 1,500 customers. By contrast a bad problem at a substation can drop twenty feeders at once, causing a widespread outage that garners serious attention from newspapers and community leaders.

For this reason, efforts to fix aging infrastructure problems often focus on levels of the system other than the feeder system. Still, the feeder system is an important part of the system, and as shown can be an important resource in limiting the impact of substation – sub-transmission outages. Therefore, attention given to it is more important in the overall reliability equation than is often realized.

Outdated Structure Impact on the Feeder System: A Major Degradation in System Reliability

Regardless of whether one or both of the situations discussed above develop the effect on reliability is the same. A growing portion of the equipment outages, which could *potentially* lead to widespread, lengthy interruptions (i.e. outages of major equipment at the substations) now have no viable contingency support.

In several large metropolitan systems outdated system structure issues (section 8.2) eroded feeder contingency support strength due to these two reasons, to less than 40% of what was required to meet system needs, a loss of 60%. Figure 8.5 showed that substation – sub-transmission related problems cause about 25% of system SAIDI. Assuming again that Figure 8.5 is reasonably close to the distribution of frequency-of-outage causes in the system (see Footnote 6) and that degradation in results is proportional to degradation of capability (it is actually worse), this 60% reduction in capability to support substation-level problems would lead to something along the lines of a 15% increase (60% x 25).

This impact on reliability is roughly three times the impact on system reliability that issues related to feeder system reliability cause (that came to 5%). Overall then, outdated system layout problems (section 8.2) lead to a 20% net increase in reliability problems, just due to their impacts on the feeder system alone.

8.6 "FIXES" FOR OUTDATED SYSTEM STRUCTURES

There are no simple, inexpensive solutions to problems of the magnitude discussed in this chapter. However, there are solutions, which are more effective and less expensive than trying to reverse the various impacts of aging infrastructure by adding more sites, more facilities, and reducing utilization ratios to traditional levels and practices. These require innovative approaches that often break with traditional concepts and practices.

Improved Communication with Community and Regulators

Electric utilities have to do a better job of identifying to those who approve their plans that new facilities and sites are needed if long-term reliability and service quality is to be kept at high levels. The quality of results that utilities obtain from their communication with regulators and community varies greatly. All utilities would be wise to emulate the methods employed by utilities that are particularly effective in this manner, such as PSE&G.

Compact Substation Designs

The comprises of capacity over configuration as illustrated by Figure 8.4 are simply unacceptable in a system where utilization rate is kept high and reliability is a priority. A better solution is to find a way to "cram" more capacity into existing sites. This can be done to two levels of improvement:

1. *Compact designs.* Various means exist to reduce the space required for air-insulated substation equipment. Usually this entails re-building the substation upward rather than outward. Among the items used are "low footprint" transformers, and high rise buswork designed to use little horizontal real estate (which as a consequence has a much higher profile). Compact designs can typically increase the capability of a substation (as measured by additional capacity added without any sacrifice in configuration) by up to 25 – 30%. Often this is not enough to provide all the improvement needed.

2. *Gas-insulated switchgear.* The solution of favor for crowded conditions in urban areas throughout Europe is the gas-insulated switchgear (GIS) substation, the so-called substation in a bottle. Slightly more than half the metropolitan substations throughout Western Europe utilize GIS technology. GIS substations have all of their busywork and breakers contained inside a set of linked insulating capability. Voltage clearances between equipment can be much tighter than with air-insulated equipment, and the weather-protection rendered by the enclosures and gas means that much more compact designs can be used for all equipment. As a result GIS equipment requires far less space than its air-insulated equivalents.

   GIS substations can increase the capability of a substation site over that obtainable from the tightest possible use of traditional air-insulated equipment, by a factor of at least three. If need be, an old 160 MVA site could be rebuilt with close to 500 MVA of capacity. Very often full use of this capability is not needed, but usually a doubling of capability, or potential for eventual growth, is implemented when GIS is retrofitted to a site. GIS also makes sense as the construction standard for new substations in aging infrastructure areas, because it enables the utility to have a much wider choice of available sites. Its compact nature and the fact that it can be put inside medium-sized buildings, means it will fit in many places that previously had to be rejected.

   GIS substation equipment costs about 15% – 30% more than

air-insulated alternatives. However it is slightly more reliable than, and requires much less maintenance cost than, air insulated designs of equivalent capability, basically more than canceling out this high initial cost if evaluated over ten or more years of operation. In addition, its minimization of real estate needs also contributes to lower cost for land, a major consideration and often more than enough to offset the higher initial equipment cost.

Reinforcement of Feeder Trunks and Switching

Given the typical effects of outdated system structures on the feeder system, providing the capability required for support of substation-level contingencies, will mean reinforcing feeder trunks with larger conductors, and installing additional switches for improved flexibility of operation. This is usually money well spent, and if planned well and coordinated with operation planning, feeder reinforcement for improved contingency capability is among the most effective ways to spend limited funds to obtain improved reliability of service.

Improved Planning

Improved planning can contribute significantly to a utility's both avoiding the situation depicted in this chapter, and mitigating its effects if it is already in such a situation. Good planning provides foresight, rational consideration of options, and under the right conditions, innovation. The following recommendations should be taken as a whole: they provide the proper type of planning for aging infrastructure areas as well as for utilities who need to maximize both customer service and stockholder (financial) performance. Chapters 12 and 13 will discuss this in more detail.

Improved long-range planning

Section 8.2 cited ineffective long-range planning as a contributing cause to many of the causes that ultimately lead to outdated structure and poor reliability. These problems have to be fixed before lasting solutions can be put in place. Chapter 8 covers the nature of several problems with traditional planning tools. Chapters 11 and 12 discuss how improved planning methods and tools appropriate for modern high utilization-ration systems can be effectively used.

But more is needed. Section 8.2 showed that a large measure of the problems utilities created for themselves were due to simply not looking far enough ahead or not using long-range planning capability that was there. Thus, in addition to changes in method and technique, utilities must simply pay attention to and use the longer-tern perspective and evaluation of long-term consequences that flow out of long-range analysis of the consequences of growth and aging.

Two-Q approach

Part of the failings outlined in Section 8.2 stem from an inability of utilities to obtain sites, facilities, and equipment needed for purposes of reliability. Part of this stemmed from a flaw in the traditional utility-regulatory paradigm: planning, prioritization, and approval where based on capacity needs, not reliability needs. Chapter 4 introduced the concept of Two-Q planning, in which both quantity and quality of demand are viewed as burdens that set requirements for the system capability, and through which capability for both quality and quantity is arranged in an optimized fashion.

Two-Q planning explicitly identifies situations where new sites and new equipment are needed not for capacity (kVA) reasons alone, but for reliability reasons, and shows how much and why these needs arise. It makes such reliability-related requirements very clear and puts them in a context that is both communicate-able, where their justification is easier to make to executives and regulators alike. Two-Q is recommended both because it leads to a better understanding of reliability-related needs and the role elements of a long-range plan may have with regard to reliability, and because it provides a much better format and more documentation of need for their justification.

Reliability-based feeder system planning

The feeder system can be designed in a reliability-based manner to provide both better reliability itself, and more contingency support for substations, using methods discussed in Chapters 11 and 12.

8.7 SUMMARY OF KEY POINTS

"Outdated system structure" refers to a host of problems created for a utility when it cannot, or will not, expand its system's layout and number of sites in company with the increasing demand for quantity and quality of power. Key aspects of this impact are:

1. Operation of the system from fewer than an ideal number of substation feed points

2. Compromises in design that sacrifice configuration for capacity

3. Acceptance of very high utilization rates, even if not the utility's economic standard, just because there is no room for other options

4. Degradation of feeder system reliability and its ability to provide support during substation and sub-transmission level outages.

Table 8.3 Overall Impact of Outdated System Structure Issues on System Reliability

Level Of System	Typical Amount of Degradation
Substation/sub-transmission	15% - 20%
Feeder system reliability	5% - 15%
Feeder system contingency support	15% - 20%
Total	35% - 45%

The net effect of outdated system structure issues, above and beyond the impacts of aging equipment itself, is an increase in reliability indices (SAIDI and SAIFI) due to impacts on the substation – sub-transmission and feeder levels of the system. Table 8.3 lists typical levels of impact.

REFERENCES

J. J. Burke, *Power Distribution Engineering – Fundamentals and Applications,* Marcel Dekker, New York, 1994

R. E. Brown, S. S. Venkata, and R. D. Christie, "Hybrid Reliability Optimization Methods for Electric Power Distribution Systems," *International Conference on Intelligent Systems Applications to Power Systems*, Seoul, Korea, IEEE, July 1997.

R. E. Brown, S. Gupta, S. S. Venkata, R.D. Christie, and R. Fletcher, 'Automated Primary Distribution System Design: Reliability and Cost Optimization,' *IEEE Transactions on Power Delivery*, Vol. 12, No. 2, April 1997, pp. 1017-1022.

R. E. Brown, S. Gupta, S. S. Venkata, R.D. Christie, and R. Fletcher, 'Distribution System Reliability Assessment: Momentary Interruptions and Storms,' *IEEE Transactions on Power Delivery,* Vol. 12, No. 4, Oct 1997, pp. 1569-1575.

Institute of Electrical and Electronics Engineers, *Recommended Practice for Design of Reliable Industrial and Commercial Power Systems,* The Institute of Electrical and Electronics Engineers, Inc., New York, 1990.

H. L. Willis, *Power Distribution Planning Reference Book,* Marcel Dekker, New York, 1997.

H. L. Willis and R. W. Powell, "Load Forecasting for Transmission Planning," *IEEE Transactions on Power Apparatus and Systems,* August 1985, p. 2550.

9

Traditional Reliability Engineering Tools and Their Limitations

9.1 INTRODUCTION

The third of the five aspects of aging T&D infrastructure discussed in Chapter 1 is Engineering Paradigms and Approaches. Traditional power system engineering methods, and the paradigms built around them, are as much a part of existing power systems and their strengths and weaknesses, as old equipment and obsolete system layouts. Many of the key concepts and engineering approaches used in "modern" power systems were created in the mid-20th century, at a time when electric consumer demands, societal expectations, and engineering criteria were far different from today's. Like all engineering methods, the mid-twentieth century's had their limitations and advantages. Those limitations were for the most part well understood by their developers, but were far outweighed by their advantages, among the most important being that the methods could be implemented with the very limited computing resources available at the time. The power industry has carried many of these engineering methods and the concepts built around them through several generations of power system engineers. Not only are these methods in widespread use, but several of these paradigms have become dogma.

This chapter will begin by examining the traditional power system reliability engineering tool: the N-1 criterion and the contingency-based planning approach. This method was first used in the mid-20th century and had been developed into efficient computerized form by the late 1960s. It, and the conceptual approach to power system planning that goes along with it, are at the core of nearly every utility's overall power delivery planning and reliability engineering procedures. Similarly, many of the core concepts and criteria concerning the layout of distribution feeder systems were developed in the 1930s through the 1960s, and are largely unchanged in their application as the electric utility industry enters the 21st century. These will be covered in Chapter 10.

These traditional power system-planning methods proved more than adequate to meet the industry's needs through the 1950s to early 1990s. However, as experience has shown, and this chapter will explain, they are less than completely adequate for application to aging infrastructure areas in some of today's high-stress utility systems. This is due to an incompatibility between these traditional planning methods and the way that modern utilities need to plan, design, and operate their systems. As a result, systems planned and designed along proven but traditional lines, using tools that engineers could once confidently use, often provide poor customer service reliability and experience severe operating problems.

This chapter begins in section 9.2 with an examination of the basic N-1 contingency-based reliability design criterion and the typical methods of its application. Section 9.3 then explores this methodology's limitations with respect to modern needs, and discusses how they interact with the characteristics of modern utility systems, and discusses how that often results in system that does not provide the expected level of reliability. Section 9.4 looks at some other planning-related issues that have created real challenges for aging infrastructure utilities, most notably load forecasting errors and the way they interact with system reliability. Section 9.5 provides a reminder that high utilization rates and cost-reductions, per se, are not the reason that modern power systems and particularly aging systems, tend to give poor results – rather it is the inability of traditional planning tools to fully analyze the reliability implications of design in those areas. Section 9.6 rounds out the chapter by summarizing keys points and gives five recommendations for effective planning procedures to be applied to aging infrastructure areas.

9.2 CONTINGENCY-BASED PLANNING METHODS

The N-1 Criterion

The traditional power system planning method used to assure reliability of design at the sub-transmission – substation level is the N-1 criterion. In its purest form, it states that a power system must be able to operate and fully meet expectations for amount (kW) and quality of power (voltage, power factor, etc.) even if any one of its major components is out of service (a single contingency).

The system has N components, hence the name N–1.

This criterion makes a lot of sense. Unexpected equipment failures happen. Expected equipment outages (maintenance) are a fact of life. A prudent approach to design reliability should include making certain that the system can perform to its most stressful required level (i.e., serve the maximum demand, the peak load) even if a failure or maintenance outage has occurred. Just how much reliability this assures depends on a number of factors that will be addressed later in this section and the fact that it does not always assure reliability is the major topic of this chapter. However, from the outset it seems clear that this criterion sets a *necessary* requirement for any power system this is expected to provide reliable power supply.

Extension of the concept to multiple failures

The criterion can also be applied as an "N–2" criterion or "N–3" criterion in which case the system must be able to perform to peak requirements even if any two or three units of equipment are out of service, rather than one. Generalized, this becomes the N–X criterion, the power system will satisfy expectations even if any set of X and its components are out of service. Regardless, the method is generally referred to, and will be referred to here, as "the N–1" concept and criterion, even if X is greater than one.

Application of the Basic Contingency Planning Concept

Typically, this concept is applied as a criterion in the planning and engineering of the transmission/sub-transmission/substation portion of an electric distribution (T&D) system – the portion from 230 kV down to 69 or possibly 34.5 kV. At most electric distribution utilities, it is not applied to the distribution feeder system. Instead, techniques and analytical planning methods devised for application to radial power flow systems are used (see Section 9.3).

The base case

Application of the N–1 criterion begins with a *base case*, a model of the power system as designed or planned, with all equipment in place and operating as intended. An appropriate engineering description of all this equipment along with a set of expected peak loads that it will serve, forms the complete base case. In all modern utility planning procedures, this base case is a data set representing the system, to be used in a load-flow. A digital computer analysis will determine for the system represented by that data set, the power flows, voltages, power factors and equipment loadings that will result when that equipment set us asked to serve that demand. Various "minus one" or "contingency" cases are then done using this model as the base, literally deleting one element of the data set at a time and "resolving" the model to see what effect that lose had on voltages, currents, etc.

At the time the N–1 method was first developed, (which was prior to the availability of digital computers), the base case was an analog computer model built using patch-cord connections and numerical settings of rheostats and switches on a network analyzer. Essentially, an analog computer built for simulation of power system behavior. Since the mid-1960s, digital computer programs that solve the load flow computation using a set of simultaneous equations have been used. By the end of the 20th century, these programs had become very specialized, with features and analytical tricks employed to make them fast, robust, and dependable when applied to contingency analysis.

But regardless of the type of engineering computer being used, studies that build upon, or more properly "delete upon" a base case are the foundation of the contingency-analysis method. As a first step, a base case representing the system in "normal" form, i.e., with all equipment in operation and fully functional, is set up and solved, making sure that it (a system with all equipment operating) fully satisfies all loading, power quality, and operating criteria.

Contingency cases

Variations from this base case are then conducted as a series of "contingency studies." In each contingency study, one particular unit of equipment – a key transformer or line or bus, etc. is removed from the system database and the remaining system's performance studied using the engineering analysis model (load flow) applied to this "contingency case model." The analysis determines if the system can still serve all the demand, while remaining within specified operational criteria (see Table 9.1), with this one unit out of service, or "outaged." If not, additions or upgrades are made to the system model until the case does meet the criteria. Once the first contingency case is completed (the study for the first component in the system), the method proceeds to study the outage of the second. It begins with a "fresh copy" of the base case and removes the second unit in the equipment list, again performing its analysis. In this way, it proceeds through all N components, outaging each one and identifying whether performance in that situation is sub-standard, thus giving planners an indication of where problems in the system lie, and what the problems are.

Relaxation of design standards for contingencies

In most cases, electric distribution utility planners allow the contingency cases to meet less stringent requirements for loading, voltage, or other design goals, than for the "base" (no contingencies) case. For example, loading criteria may state that in the base case, no component can be loaded to beyond 100% of its normal rating. However, during any single contingency, a loading of 115% might be accepted, during a double contingency, a loading of 125% might be

Table 9.1 Transformer Loading Limits and Voltage Criteria for Various Contingency Situations Used By Four Electric Utilities in the U.S.

Utility	Base		N-1		N-2	
Number	Voltage	Loading	Voltage	Loading	Voltage	Loading
1	.97 – 1.03	83%	.96 – 1.04	125%	.95 – 1.05	133%
2	.96 – 1.05	75%	.95 – 1.07	115%	.95 – 1.05	125%
3	.97 – 1.05	90%	.96 – 1.05	135%	.94 – 1.05	166%
4	.95 – 1.04	90%	.94 – 1.05	135%	.94 – 1.06	166%

accepted. Table 9.1 lists the voltage and loading requirements for several utilities in the U.S. as a function of contingency situation.

Application of N–1 using a Computer Program

As originally conceived, prior to the existence of really powerful digital computers for power system studies, this process was done with an analog computer. Each case was set up and studied on an individual basis by the utility's power system planners, by adjusting settings and patch-cord connections on the analog computer. For this reason, initially (1950s) often only the 100 or so most important components (out of several thousand in a large power system) could be studied for contingency outage. However, beginning in the late 1960s, programs on digital computers were developed which would automatically check all single contingencies in the system [Daniels]. These programs became a staple of utility for system planning.

A modern contingency analysis program works along the lines shown in Figure 9.1. It is built around a load-flow program – a digital program that takes a data set describing the power system and the loads it is to serve, and solves a set of simultaneous equations to determine the voltages, flows, and power factors that can be expected in that system. In an automatic contingency analysis program, the basic load flow program is augmented with an outer loop, which automatically cycles through this power system data set, removing each unit of equipment and line, in turn, and solving the load flow analysis for that particularly contingency case.

For each such case, the program checks the results of that contingency case and reports any loadings or voltages that are out of acceptable range. It then

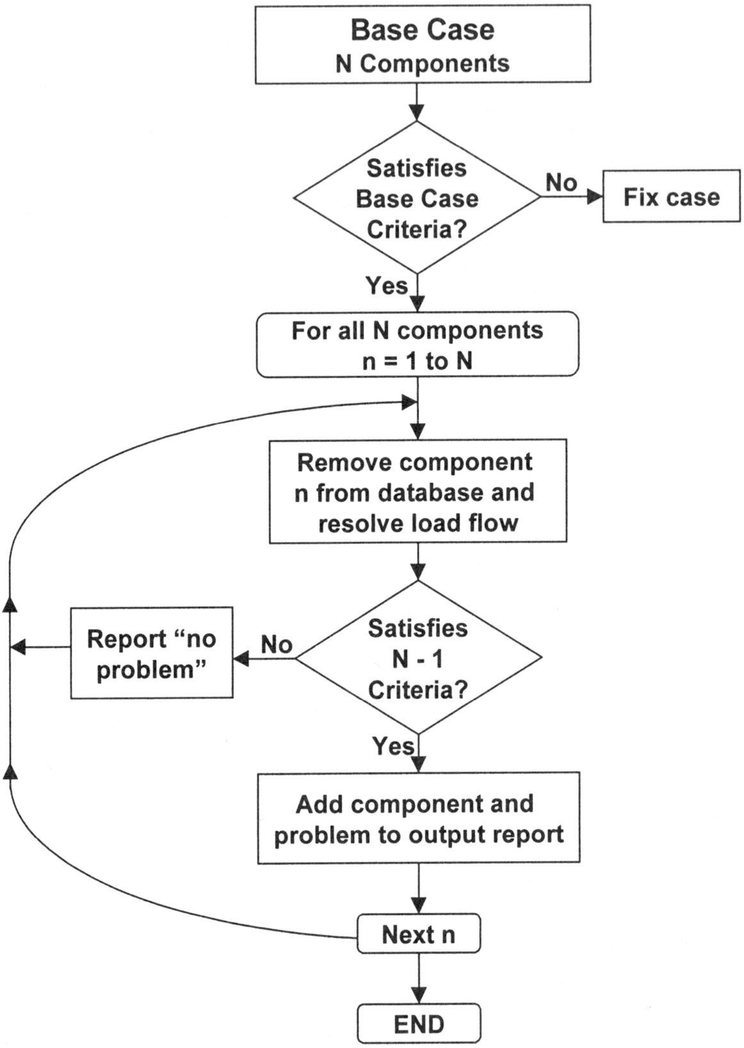

Figure 9.1 Basic approach behind the N–1 contingency planning approach. Engineering studies cycle through all components of the system and outage each one, studying what loadings and voltages would result. The system is considered to meet "N–1" criterion when all such contingency cases result in no out-of-range loadings or voltages.

"restores" that outaged component in the database, removes the next unit in turn, and runs that contingency case, cycling through all components in the system.

A system plan was considered acceptable when this type of evaluation showed that no voltage and loading standards violations would occur in every one of these instances of single contingencies. Figure 9.1 illustrates this basic approach.

A Successful Method in the Mid and Late 20th Century

From the early 1960s through the early 1990s, the vast majority of different electric utilities applied this basic approach, with a number of slight variations here and there. Typically, a utility would design its system to completely meet N–1 criterion (the system can perform despite the loss of any one component) and to meet certain N-2 criteria (a set of specific two-failure conditions, which are thought to be critical enough to engineer to tolerate). Systems designed using this approach produced satisfactory performance at costs not considered unreasonable.

Supposedly Good Systems Begin Giving Bad Results

In the late 1990s and early '00s, the operating record of large electric utilities in the United States revealed increasing problems in maintaining reliability of service to their customers. During the peak periods in the summers of 1998 and 1999 a number of power systems that fully met the traditional N–1 criterion experienced widespread outages of their power delivery systems. Table 9.2 lists only some of the more significant events in 1999, as identified by the U.S. Department of Energy. In particular, the ComEd system's (Chicago) inability to provide service was disturbing, because that system was designed to N–1

Table 9.2 Major System Outages Identified By US DOE in 1999

Area	When	Interruption Cause	Power Delivery?	Aging Infra?
New England	Jun 7-8	Wholesale generation shortage		
Chicago	Aug 12	Multiple power delivery failures	x	x
	Jul 30	Multiple power delivery failures	x	x
New York	Jul 6-7	Multiple power delivery failures	x	x
Long Island	Jul 3-8	Power delivery and grid problems	x	x
Mid Atlantic	July 6-19	Wholesale generation & grid problems	x	
N. New Jersey	Jul 5 - 8	Multiple power delivery failures	x	x
So.-Central US	Jul 23	Wholesale generation shortages		x
Delmarva	Jul 6	Wholesale grid and generation shortages		

standards and even met an N–2 and N–3 criterion in some places. There were a sufficient number of these events to make it clear ComEd was not an isolated or atypical situation [U.S. DOE, 2000].

Perhaps more significantly, the level of reliability-related problems on U.S. systems had been growing for several years prior to 1999. Based on analysis of industry survey data, the authors first noticed the trend in 1994. While at that time there were few widespread customer outage events, throughout the 1990s there was a growing "background noise level" of operating emergencies, equipment overloads, and frequent if small customer interruptions on many utility systems. Even without the major events cataloged in the DOE report (Table 9.2) the industry's record of customer service quality was falling in many regards prior to 1999.

9.3 LIMITATIONS OF N-1 METHODOLOGY

This section explains the limitations that N–1-power system planning and design techniques encounter when applied to modern (high utilization factor, lean margin) power systems. Contingency-based planning, like all engineering methods, is based on certain assumptions, uses approximations in key areas, and has limitations in its accuracy and range of application. And like other engineering methods, if applied to situations within which these assumptions, approximations, and limitations do not seriously hinder its accuracy and effectiveness, the method provides very good, dependable results. But if applied outside that range, its results may prove undependable, and the system plan may provide unsatisfactory performance.

Gradually, throughout the last quarter of the 20th century, the electric utility industry changed how it used and operated its power systems. Some of those changes meant that their systems operated in states and in situations for which the N–1 approach was not completely valid. The resulting incompatibility of method versus use of the resulting system contributed to many of the reliability problems experienced by utilities.

Of necessity, in order to fit within the space available in this book, this chapter's discussion is somewhat shorter and contains certain simplifications with respect to a "real power system." This abridging has been made in order to shorten the discussion, allowing the reader to quickly get to the heart of the matter, and to diminish distractions from the main theme due to secondary and tertiary factors. Therefore, this discussion makes the following "assumptions" or simplifications with respect to the system being discussed as an example here:

- All equipment of a specific type will be of the same capacity; e.g., all substation transformers are the same capacity.
- All equipment is loaded to the same peak level, that being the

average utilization ratio for the system at peak.

- All equipment units of any one type (e.g., all substation transformers) have the same failure rate.

The reader familiar with power systems will recognize all three as great simplifications of the many details that complicate power system planning and reliability engineering. However, real-world variations from these assumptions do not diminish the limitations and phenomena that will be discussed in the next few pages. In fact they slightly exacerbate them: these complexities generally worsen the problems explained here.

Overall Summary of the Problem

N–1 methods and the N–1 criteria assure power system planners and engineers that there is *some feasible way* to back up every unit in the system, should it fail. However, they make no assessment of the following:

- How likely is it that such backup will be needed?
- How reasonable is the feasible plan for each contingency situation or is the planner actually building a "house of cards" by expected "too many things to go right" once one thing has gone wrong?
- How much stress might the system be under during such contingency situations, and the long-term implications for both equipment life and operating policy?
- How often will conditions occur which cannot be backed up (e.g., multiple failures) and how bad could the situation become when that is the case?

As a result, systems that meet the N–1 criteria may be far less reliable than needed, even though the N–1 criteria "guarantees" there is a way to back up every unit in the system. This is much more likely to happen in modern power systems than it was in traditional, regulated power systems, due to changes in utilization and design made in the period 1990–2000, and the more volatile operating environment of de-regulation.

Utilization Ratio Sensitivity

The major culprit that led to problems that "N–1 could not see" was an increase in the typical equipment utilization ratio used throughout the industry. When it is raised, as it was during the 1980s and 1990s, an N–1 compliant system, which previously gave good service, may no longer give satisfactory reliability of service, even if it continues to meet the N–1 criterion. Similarly, power systems designed using the N–1 criterion and intended to operate at higher than traditional levels of equipment loading are very likely to have higher customer interruption rates than expected.

There is nothing inherently wrong with this trend to higher loading levels. In fact it is desirable because it seeks to make the utility financially efficient, which is potentially beneficial to both stockholders and customers. A power system that operates at 83% or 90% or even 100% utilization of equipment at peak can be designed to operate reliably, but something beyond N–1 methodology is required to assure that it will provide good customer service reliability.

N–1 is a necessary but not a sufficient criterion

Due to the success that N–1 methods had throughout the 1960s, 70s and 80s, producing power system designs that provided good reliability of service, most power system planners and most electric utilities treated the N–1 criteria as a necessary and sufficient criteria. Design a system to meet this criterion and it was, by definition, reliable. But at higher utilization factors, while N–1 criterion is still a *necessary* criterion, it alone is not *sufficient* to assure good quality of service. The reasons for this change in reliability as a function of utilization ratio are far subtler than is typically recognized. This section will review the limitations that N–1 has when applied to high-utilization-ratio systems and explain what happens, and why. The authors want to make clear that they definitely are not labeling high utilization ratios as the cause of all the industry's problems. Rather, it is the incompatibility between traditional ways of applied the N–1 criterion, and the way these systems operate, that created the problem.

Traditional utilization levels

In the 1960s through early 1980s, electric utilities typically loaded key equipment such as substation power transformers and downtown sub-transmission cables to only about 2/3 or a little more (typically about 66%) of their capacity, even during peak periods. The remaining capacity was kept as an "operating reserve" or "contingency margin." Engineers and planners at distribution utilities designed their power systems using the N–1 and other criteria, while counting on this margin.

In such systems, when a transformer or line failed, it required one neighboring transformer of equal capacity, perhaps already loaded to up to 66%, to be available to pick up its load. Depending on how close the system was to peak demand at the time of the outage, this unit might have to accept as much as 133% of its normal load (its normal 66% of rating and its neighbors 66%, too). Such overloading was tolerable for brief period. Power equipment can be run

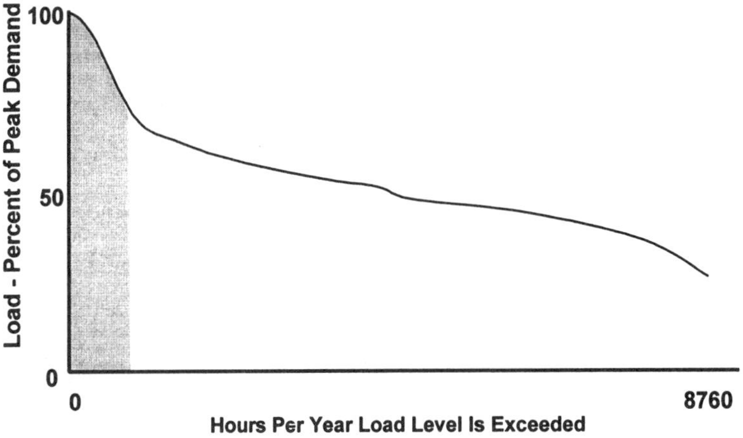

Figure 9.2 Annual load duration curve for a utility system. Risk periods for high contingency loading of a traditional power system occurs only 10% of the time (shaded area). See text for details.

above rating for brief periods without significant damage, if this is not done too often.

And in fact it was unlikely that the loading would be as high as 133%, because that would only occur if the outage occurred during a peak load period. Only when loading was above 75% that of peak demand would overloads occur (75% x 66% = 50%, so at any load level below 75% of peak, one transformer can handle the load of two without going over 100% of its rating). In the system whose load duration curve is shown in Figure 9.2 (a typical US utility system), such load levels occur only 10% of the year. As a result, it was (and still is) very likely that when equipment failures occur, they will occur at some time when loading is not near peak and hence stress on the equipment picking up the outaged units load was not unduly high.

Higher utilization rates

Beginning in the 1980s and increasingly in the 1990s, utilities pushed equipment utilization upwards to where in some systems the average substation transformer was loaded to 83% of its rating during peak periods. As was discussed in Chapter 8, in aging areas of this system, utilization rates can average close to 100% under "normal" peak conditions. Table 9.3 shows industry averages

Table 9.3 Design loading guidelines for normal and contingency loading

Average maximum peak load as planned, under normal conditions, as a percent of nameplate thermal rating.	Percent of Nameplate			
	1998	1988	1979	1970
Among utilities with > 1,000,000 customers	85	75	70	64
Among utilities with 500,000 and 1,000,000	85	75	70	65
Among utilities with < 500,000 customers	75	70	70	65
In rural areas	70	65	65	62
In suburban areas	85	80	75	67
In urban areas	90	82	75	72

Maximum planned peak load expected under contingency conditions (at least four hours duration)	Percent of Nameplate			
	1998	1988	1979	1970
Among utilities with > 1,000,000 customers	145	140	135	133
Among utilities with 500,000 and 1,000,000	140	135	130	127
Among utilities with < 500,000 customers	135	130	130	127

obtained for a comprehensive survey of loading practices across the industry.[1]

The engineers in charge of these higher-utilization rate power systems knew, and in fact, planned for, their system to accommodate these higher utilization rates in several ways. To begin, these higher utilization rates required considerable augmentation of system configuration and switching. In a system loaded to an average of only 66% at peak, each transformer and line required one neighboring unit to "stand by" to pick up its outage. This meant, for example, that in each two-transformer substations, there had to be buswork and switchgear (breakers, switches) configured so that if one of the units failed, the other could automatically pick up its load. Alternately, the load at the substation had to be partially transferred to neighboring substations (onto their transformers) through the feeder system, as will be discussed in section 9.4. Usually, some combination of stronger substation and sub-transmission-level buswork and switching flexibility, and increased reliance on feeder-level transfers was used.

---

[1] From *Electric Power Distribution Practices and Performance in North America – 1998* ("The Benchmark Report" by H. L. Willis and J. J. Burke, ABB Power T&D Company, Raleigh, NC. "Design loading" as used here, refers to the peak load on a transformer, above which it is considered so highly loaded that it should be upgraded in capacity, or load transferred elsewhere. "Emergency rating" refers to the maximum load permitted to on the substation during an equipment outage or excessive load contingency.

Those plans for these higher-utilization systems were created using N–1 methods, which "took into account" those higher utilization rates in their analysis. These N–1 Analysis applications assured that *there was a way* to back up every unit in the system, should it fail. The system plans developed as a result fully met the full N–1 criteria everywhere, and N–2 criteria in critical places, even though they were operating at these higher utilization rates. Thus, these systems did have well-engineered contingency capability. The equipment was there and it would, and did, work as intended. Any problems lay elsewhere.

Looking at N–1's Limitations

In order to understand where the problem with N–1 criterion application lies, it is important to first understand that a power system that meets the N–1 criterion *can and routinely* does operate with more than one unit of equipment out of service. Consider a power system that has 10,000 elements in it, each with an outage expectation of .16% – a value lower than one would ever expect on a real power system. One can expect that on average, about 16 elements will be out at any one time. Yet the system will usually continue to operate without problems. The reason is that the N–1 criteria has guaranteed that there is a backup for every one of these failed units, as shown in Figure 9.3.

The system will fail to serve its entire load only if two of these multiple outages occur among neighboring equipment. For example, if a transformer *and* the transformer designated to back it up both fail at the same time, as shown in Figure 9.4. If a unit, and its backup, are both out, then, and only then, will a service interruption occur.

Contingency support neighborhood

"Neighboring equipment" as used in the paragraph above means the equipment in the vicinity of a unit that is part of the contingency support for its outage. This can be more accurately described as its *contingency support neighborhood:* the portion of the system that includes all equipment that is part of the planned contingency support for the unit's outage. For a substation power transformer, this might include at least one neighboring transformer (usually at the same substation) which would provide capacity margin during its outage, along with portions of the high-side and low-side buswork and switchgear, which would operate in a non-standard configuration during its outage.

Figure 9.5 illustrates this concept, showing several "contingency support neighborhoods" as in the example system used in Figures 9.3 and 9.4. As stated in the introduction to this section, this discussion simplifies the real world somewhat. Here, every unit is the same size and contingency support is always grouped exclusively in sets of neighbors. Actual design is more complicated, but the complications do not make any substantial net impact this discussion.

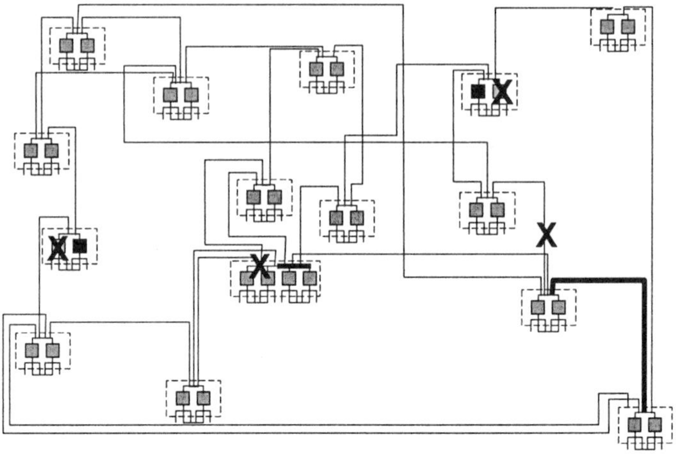

Figure 9.3 One-line diagram for a small part of a large power system. Four equipment outages are shown, indicated by an X, two transformers, one high-side bus, and one sub-transmission line. Each outaged unit has a neighboring unit (shaded) that has picked up its load: the N–1 criteria assured that this was the case. The system continues to operate smoothly because no two of the outages occur close enough to one another.

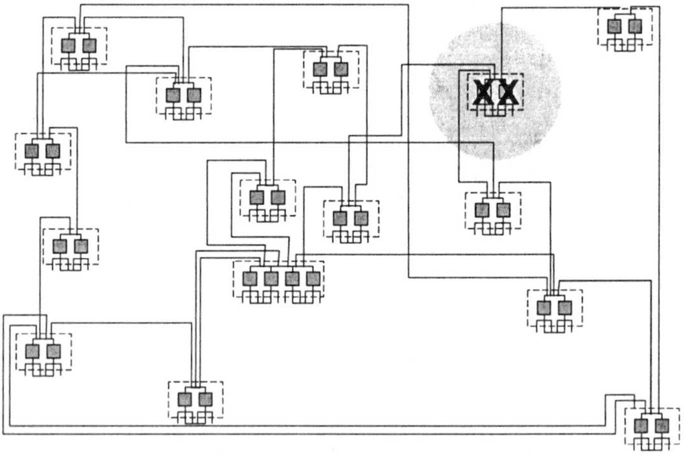

Figure 9.4 One set of dual failures in the same contingency support neighborhood, as illustrated here with the failure of two neighboring transformers (each was the designated backup for the other), will lead to interruption of service to consumers. Here, the shaded circle indicates the rough area of the system that would be without power.

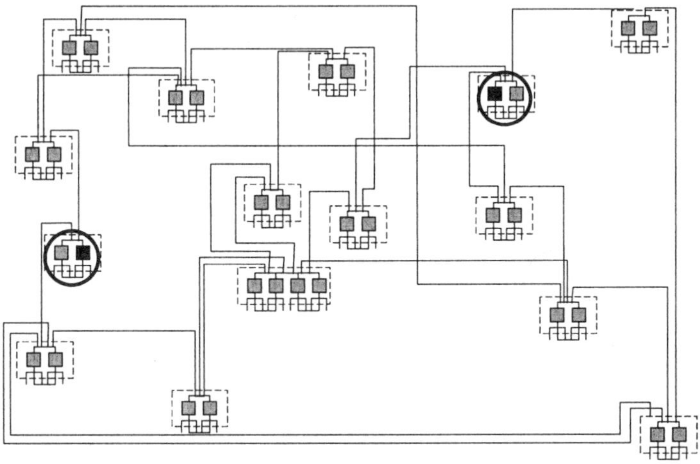

Figure 9.5 Every unit in the system has a "contingency support neighborhood" that includes all the equipment that provides contingency support for the unit. Shown here are two transformers (shaded) along with their neighborhoods (circled). Equipment in a neighborhood provides contingency margin (capacity) as well as connectivity flexibility (switching, flow capability) during the outage of that unit.

Problems in this N–1 system that lead to customer outages occur only when *two or more* equipment units fail simultaneously within *one* contingency support neighborhood. Such a "double failure" does not have to be among just the unit and its like-type of support unit, i.e., the failure of both transformers at a two-transformer substation. Failures of one transformer and a line, a breaker, or a bus needed to support the contingency re-configuration of the system to support its outage can also lead to a failure to maintain service. Still, such occurrences are very rare. While there are perhaps 10 to 15 units out of service in a system of 10,000 elements, it is most likely that they are scattered singly throughout the system. The likelihood that two are concentrated in one-minute neighborhood is remote.

Traditional power systems had "small" contingency support neighborhoods

In traditional power delivery systems, those whose utilization ratio for power transformers and sub-transmission lines was nominally targeted to be about 66% of equipment rating during normal (design) peak conditions, the contingency support neighborhood for any unit of equipment was small. As discussed earlier, every unit in the system needed one backup unit of like size. A 32 MVA transformer would be loaded to 22.5 MVA (66%) at peak. If it failed, its partner at a substation, also already serving 21 MVA, would pick up its load too, briefly running at 133% (45 MVA) loading so that all demand was served. Therefore, the contingency support neighborhood for both units was a small locality that included the other transformer and the various switchgear and buswork needed to connect the two to the other's load during a contingency.

Systems with high utilization rates have larger contingency support neighborhoods

Suppose that the area of the power system being considered has 88.5% loadings on all transformers, instead of 66%. In that case, when any transformer fails, and if the utility is to keep within a 133% overload limit, a failed unit's load has to be spread over *two* neighboring transformers, not just one. The size of the "contingency support neighborhood" for each unit in the system has increased by a factor of fifty percent. Previously it included one neighboring transformer, now it includes two.

More importantly, the probability that an outage will occur among the designated support units for each transformer is double what it was in the system loaded to only 66%. Previously, whenever a transformer failed, there was only one unit whose failure stood in the way of good service. Now, if *either* of its two designated support units fails, an interruption of service to the utility's customers will occur. Two possible failures, each as likely to occur as the one failure that could have taken out the 66% loading system

Thus, in a system where utilization rate has been pushed upward, every contingency support neighborhood is proportionally larger, and thus a greater target for trouble to occur: There is more exposure to "simultaneous outages." In a system loaded to 66%, there is only one major target. Failure to serve the load occurs only if a unit of equipment and *one specific neighbor* designated as its contingency support are both out of service. In a system or area of a system loaded to 88.5%, it occurs if a unit and *either one of two neighbors* is out. In an area of a system loaded to over 100%, (as some aging areas are) it occurs whenever the unit and *any one of three* designated neighbors is out (Figure 9.6).

Basically, the whole problem boils down to this: the contingency support neighborhoods are larger. But there are still "N–1" neighborhoods: each can tolerate only *one* equipment outage and still fully meet their required ability to serve demand. A second outage will very likely lead to interruption of service to some customers. In these larger neighborhoods, there are more targets for that second outage to hit. "Trouble" that leads to an inability to serve customer demand is more likely to occur. The analysis below estimates the relative likelihood that this occurs in example systems loaded to different levels.

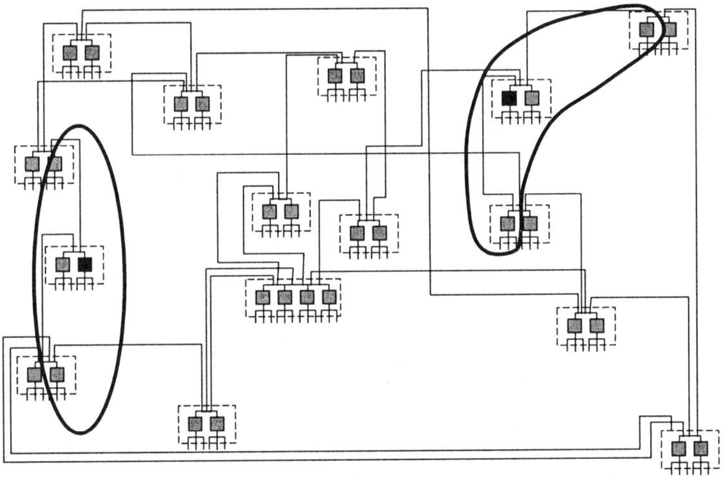

Figure 9.6 At 100% loading, each transformer from Figure 9.5 needs three nearby units to cover its load, expanding the "contingency support neighborhood" involved. See text for details.

A system as discussed earlier, with "10,000 major elements." might contain 1,200 substation transformers. Assuming that the outage rate for them is .25%, this means:

1. In a 66% utilization system, there are 600 two-transformer contingency support neighborhoods. Failure to serve the load occurs only if both transformers of this pair fail. That is:

   - Failure probability = $.0025^2 = .000625$
   - Hours per year = .000625 x 8760 hours/year x 600 pairs = 32.9 hours/year.

2. In an 88.5% utilization system, with three-transformer contingency support neighborhoods, failure to serve the load occurs only if all three or any two transformers of this triplet fail. Over the whole system, annually, that is

   - Failure probability = $.0025^3 + 3x(.0025^2x(1 - .0025)) = .00001871872$
   - Hours per year = .00001871872 x 8760 hours x 400 triplets = 65.6 hours/year.

3. In a 100% utilization system, with four-transformer contingency support neighborhoods, failure to serve the load occurs if all four, any three, or any two transformers of this triplet fail. Over the whole system, annually, that is

   Failure probability = $.0025^4 + 4x(.0025^3x(1 - .0025)+6x(.0025^2)x(1 - .0025)^2 = .000037375$

   Hours per year = .000037375 x 8760 hours x 300 quadrals = 98.22 hours/year

By comparison to a traditionally loaded system, a power system at a higher utilization rate is two to three times as likely to experience a situation where a pattern of equipment outages fall outside of the N–1 criterion. For example, one of the "N-2" situations that might lead to an interruption of load. Systems run at higher equipment utilization rates are more likely to experience events that could put them in jeopardy of being unable to serve all customer loads. N–1 analysis does not measure or evaluate this in any manner.

> N–1 criterion assures planners and engineers that a feasible way to handle every equipment outage has been provided. It does nothing to address how often situations outside of that context – i.e. those that will lead to unacceptable service quality, might occur.

High Utilization Coupled With Aging System Equipment Leads to Greatly Increased Service Problems

Chapter 7 discussed the effect that aging has on equipment failure rates. In aging areas of a power system, the failure rate for equipment is three to five times that of normal areas of the system. Coupled with the high utilization rates common in these aging areas and the result is a ten to one or slightly worse incidence of customer service interruptions due to equipment outages.[2]

[2] Here the authors will bring in a real world factor. Chapter 8 discussed why utilization rates are usually above system average in aging infrastructure areas of the system. A system where the average has gone from 66% to 88.5% may have seen only a modest increase in the utilization rate for equipment in newer areas of the system, while increases in aging areas make up for those below-average statistics. Thus, the aging part of the system has much higher utilization than other parts. Its customer service problems stand out both because of the higher failure rate in this area, and the higher likelihood that outages lead to customer interruptions. As a result aging areas often have a customer interruption rate up to twelve times that of newer areas of the system.

Increased High-Stress Levels and Periods

The situation is slightly worse than the perspective developed above when one looks at the stress put on the system's equipment, and the portion of the year that the system is likely to see: high- and medium stress events due to equipment outages.

In a 66% utilization system, every transformer is paired with one other: whenever the unit it is backing up fails, it must support the load of two transformers. Given the transformer failure rate of .0025%, this means each transformer can expect to have its partner our of service and thus be in this "contingency support mode" about 22 hours (.0025 x 8760 hours) per year. Given that the system is at peak demand about 10% of the time, this means that a transformer can expect about two hours of severe loading time per year.

When utilization ratio is 88.5%, each transformer is partnered with two other units: failure of either one will put it in a contingency support mode. Thus, neglecting the slight amount of time when both its partners are out of service, (and thus customer service in interrupted), the amount of time it can expect to be in this mode is twice what it was in the 66.5% system: or about 44 hours, 4.4 of that will be high stress. Similarly, in 100% utilization, each transformer will see about 66 and 6.6 contingency support and high-stress hours period year. *Stress put on system equipment is much higher in high utilization systems.*

"Low standards" Operating Hours are Increased

It is also worth considering that standards on loading, voltage regulation and other operating factors are relaxed during contingency situations (see section 9.2). Since the amount of time that the system spends in these "medium stress times" is greater in high utilization systems, which means that the distribution system spends more time in "sub-standard" situations – twice is too much if the utilization rate is 88.5%, three times as much if utilization is 100%.

The Result: Lack of Dependability as Sole Planning Tools

The limitations discussed above can be partly accommodated by modifications to the traditional approaches and changes in N–1 criteria application. But the overall result is that resource requirements (both human and computer) rise dramatically, and the methods become both unwieldy and more sensitive to assumptions and other limitations not covered here. The bottom line is that N–1 and N–2 contingency-enumeration methods were, and still are, sound engineering methods, but ones with a high sensitivity to planning and operating conditions that are more common today than in the mid-1960s when these methods came into prominence as design tools. These limitations reduce the

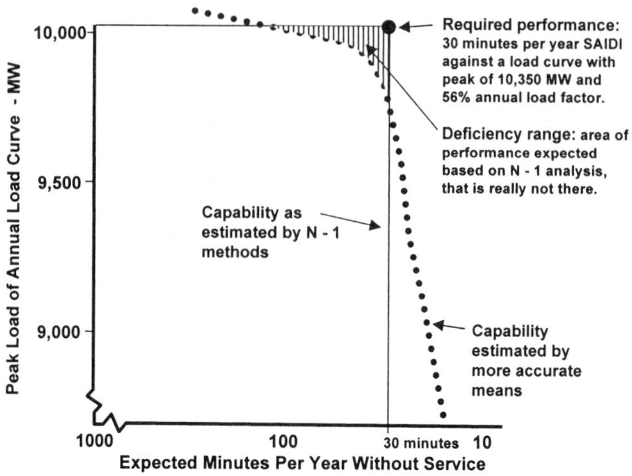

Figure 9.7 Contingency-based analysis (solid lines) determined that a particular power system could sustain the required peak load (10,350 MW) while meeting N–1 everywhere and N–2 criteria in selected places. Meaning it met roughly a 30 minute-SAIDI capability at the low side bus level of the power delivery system.

dependability of N–1 analysis, and the use of N–1 as a reliability criterion, as a definition of design sufficiency in power system reliability engineering.

Figure 9.3 illustrates this with an example taken from a large utility system in the Midwestern U.S. Traditional N–1 analysis determines that a power system operating at an average 83% utilization factor can serve a certain peak load level while meeting N–1 criteria (defined as sufficient reliability). Basically, the "rectangular" profile on the diagram given by N–1 analysis indicated that the system passed N–1 criteria everywhere, and N–2 criteria at a set of selected critical points, while the demand was set to that of projected design levels for system peak load.[3]

Actual capability of the system shows a rounded corner to the load vs. reliability of service capabilities of the system. It is capable of delivering only 9,850 MW with the required 30 minute SAIDI. If expected to serve a peak load of 10,350 MW, it has an expected SAIDI of four times the target, 120 minutes per year.

---

[3] N–1 analysis does not determine an actual estimated reliability value, but in this case subsequent analysis showed that a valid N–1 criterion was equivalent to about 30 minutes SAIDI, and that value is used here as the target reliability figure.

By contrast, an analysis of the system's capability using a reliability computation method that does not start out with an assumed "everything in service" normalcy base, which accommodates analysis of partial failures of tap changers, and which accommodated some (but not all) uncertainties in loads and operating conditions, determined the profile shown by the dotted line. At high loading levels (those near peak load), the system is incapable of providing the reliability required) – the rectangular profile is actually rounded off. The system can provide the peak load but with much less reliability than expected.

9.4 OTHER PLANNING RELATED CONCERNS

Partial Failures

Traditional N–1 contingency planning methods use "zero-one" enumeration of failures. In the contingency case analysis method (Figure 9.1), every unit of equipment and every line in the system is modeled as completely in service. In each contingency case, a unit is modeled as completely out of service. But modern power systems often encounter partial failures:

- A transformer may be in service but its tap changer has been diagnosed as problematic and is locked in one position, limiting system operation.
- An oil-filled UG cable's pumps are disabled and the cable has been de-rated, but is still in service.
- Concerns about a ground that failed tests have dictated opening a bus tiebreaker to balance fault duties.

At the traditional loading levels that existed when contingency analysis was developed, such partial equipment failures seldom led to serious operating limitations and were safely ignored, while the contingency analysis still remained valid. In systems operating at higher loading levels, partial failures cause problems under far less extreme situations, and often cannot be ignored in reliability planning. For example, a power transformer loaded to 85% at peak, whose tap changer is locked into one position, is subject to voltage regulation problems that can easily reduce its ability to handle load by close to 15% in some situations.[4] The contingency margin (100% - 85%) that the typical N–1

---

[4] Loss of its load tap changer does not cause any lowering of the transformer's capability to carry load. However, flow through it is now subject to variation in voltage drop – higher flows result in higher voltage drops. It may be unable to do its job, within the electrical confines of its interconnection to other parts of the network, due to this variable voltage drop, which may limit it to partial loading only. Ignoring this and accepting the higher voltage drop that goes with full capability during a contingency would lead to

method assumes is there may, in fact, be mostly nonexistent.

Connectivity Sensitivity

As mentioned earlier, success in handling a contingency depends on the power system in the "neighborhood of contingency support" around the failed unit being connected in such a way that the neighboring units can provide the support while remaining meeting all electrical standards and satisfying all operational requirements. At higher equipment utilization rates, this neighborhood is generally larger everywhere within the power system.

This greater size is not, *per se,* the cause of problems for traditional N–1 analysis. Standard contingency-based analysis and the engineering methods that accompany it *can* fully accommodate the detailed electrical and capacity analysis of any and all contingency support neighborhoods, regardless of their sizes.

But each of these wider contingency neighborhoods involves more equipment and interconnections. Thus, accurate modeling is sensitive to more assumptions about the exact amount and location of loads in the surrounding areas of the system, and the way the system operator has chosen to run the system at that moment, and myriad other details about operating status. There are *more assumptions* involved in accurately depicting the status of each of the N components' contingency support neighborhood.

And the analysis of each of these N contingency support neighborhoods is more *sensitive* to these assumptions. The range of uncertainty in many of these factors about future area loads and operating conditions is ± 5% to ±10%. Such ranges of uncertainty are not critical in the N–1 contingency analysis of a system operating at 66% utilization. The available contingency margin (33%) is considerably larger than the range. But when operating at 90% utilization, the uncertainty ranges of various factors involved often equal the assumed available contingency support capacity, and there is a larger neighborhood, within which it is more likely something will be different than assumed in the N–1 analysis.

Aging and High Utilization Systems Are More Sensitive to Load Forecasting Errors

A projection of future need is the first step in power delivery planning. The forecast of future peak load defines requirements for the capability of the system, and starts the process of evaluating alternatives for feasibility, value, and cost. It defines the constraints for selecting the alternative which best meets those requirements. Poor load forecasting has been a contributor to a significant

---

serious problems of another type (unacceptably low service voltages to customers, or higher demands on other transformers).

number of aging infrastructure system problems around the nature – in the authors' experience roughly half. Two areas of forecasting deserve special attention.

Weather normalization

Peak demand levels depend very much on the peak seasonal weather. In summer, the hotter the weather, the higher the demand. In winter, colder weather increases demand levels. Therefore, a projection of electric demand can, and should, include an assessment of the impact of temperature on demand. For example, Figure 9.8 shows the peak demand vs. peak temperature relationship for a small municipal electric system. The peak demand is:

- Summer Peak Load (MW) = 495 MW + (T – 57) x 12.5 MW (9.1)
- Winter Peak Load (MW) = 495 MW + (57-T) x 8.2 MW (9.2)

Recommended practice for electric load forecasting is to adjust historical weather data to a *standard set of weather conditions* to which the system design is targeted, then to project future demands under this same constant weather criteria for all planning purposes. In this way all weather data and forecasts are based on comparable situations: increases or decreases due to "real" reasons versus those due to variations in weather are distinguishable. Similarly, all planning should target a specific "design weather standard." The forecast load, which defines requirements for the system plan, should be adjusted to this weather criterion. Essentially the system is being designed to serve peak-demand for weather this extreme, but not worse. Weather conditions (and peak loads) that exceed those conditions are treated as contingencies, just like other contingencies.

What temperature should planners select for this standard weather condition? Temperatures vary from year to year. Setting the design conditions at the mean, or most expected temperature, means that the forecast loads will be exceeded, and the system's capability exceeded by the demand, roughly every other year. On the other hand, it isn't cost-effective to install equipment to handle the worst possible weather conditions: "the heat storm of the century,"

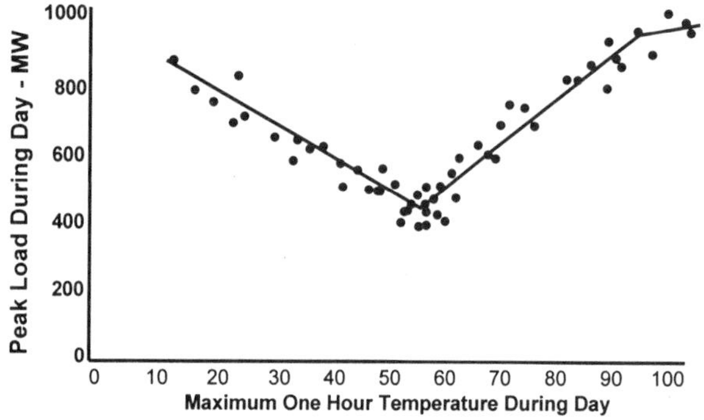

Figure 9.8 Peak daily loads and temperatures are related with a "jackknife" function (solid line). Shown is peak daily-load versus peak daily temperature for all Tuesdays in a year (several Tuesdays thought non-representative because they were holidays or similar special events were left out of the analysis). Only Tuesdays are used in order to reduce the effects that different weekday activity patterns may have on load variation. From *Spatial Electric Load Forecasting,* published by Marcel Dekker, Chapter 3.

etc. Generally, the recommended practice is to define a set of "design weather conditions" "extreme enough" to be rare but not so extreme as to be totally unexpected. Situations and needs vary, but a reasonable criterion is "design weather conditions are defined so that they *will be exceeded no more than once every ten years."* See the *Spatial Electric Load Forecasting*, pages 76-80 for a discussion of both the techniques used to determine such adjustments and for recommendations on what constitutes "extreme-enough weather."

Impact of Mistakes in Weather Normalization on Reliability

The weather normalization method used in planning and the design weather targets set for the system are among the easiest matters for "rationalization" when efforts are being made to cut costs. For example, if planners re-set their design weather conditions from a criterion of once in ten years to once in five, or lower the forecast target in some other manner, the budget requirements that flow out of their system planning process will fall. As an example, the utility whose load is diagrammed in Figure 9.8 has an annual growth rate of nearly 1.0%. Summer peak demand sensitivity is 1.25%/degree F. Reducing the design weather target by one degree Fahrenheit, about the equivalent of going from once-in-ten to once-in-five years, will reduce the load growth forecast for a four year period to that forecast over five years. Assuming for the sake of this analysis that budget directly corresponds to amount of growth, that means it results in an annual budget reduction of 25% over the next four years. For this reason, a number of utilities succumbed to the temptation to change weather normalization too much.

- *Weather normalization* that targets a "too average" weather condition puts the power system in high-stress situations too often.

A low load forecast results in several detrimental impacts. First, it generally leads to a situation where the system is serving more load than intended. Usually, this does not create severe problems when all equipment is functioning, although it does age equipment somewhat faster than expected (accelerated loss of life). One can view operation at loads above design conditions as a contingency. Poor load forecasts used in the planning or operation of a power delivery system effectively "use up" its contingency capability [Willis and Powell, 1985]. Poor normalization of weather data for forecasting or poor spatial forecasting (poor correlation of loads with areas and equipment) results in deterioration of a system's contingency withstands capability. This greatly exacerbates the reliability-of-service problems discussed up to this point in this book.

Equally important, and far less frequently recognized as a key impact of poor forecasting, the system serving a load above that for which it was designed will operate for many more hours of the year in a state where service quality is at jeopardy if complete or partial failures occur, or if "things don't go exactly right." Figure 9.9 compares the annual load duration curves for a "average year" as used in design of the system and 1999 (a one-in-ten year), for a large investor-owned utility in the central U.S. The difference in peak demand between an average year and an extreme year is 4.4%. However, as shown, not only peak load changes but, annual load factor. The period of time when the system is above 75% of peak (defined as "high stress" earlier in this chapter) increases by 28%. As a result SAIDI increases significantly.

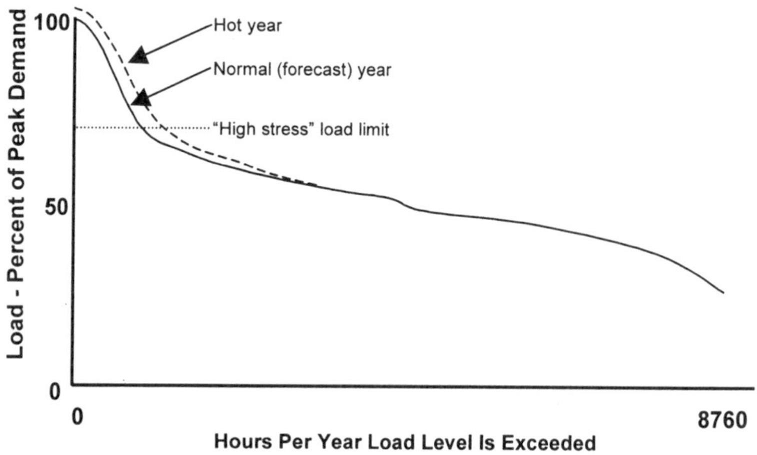

Figure 9.9 When load exceeds forecast, it usually does so no only during peak periods but for an entire season (summer, winter). Shown above are forecast vs. actual annual load duration curves for a utility in the mid – United States, for 1998. As a result of the higher temperature summer weather, peak load was about 3.3% higher, but the period of high stress for the system was 30% more for the entire year. The system operated in high-stress modes for more than twice as many hours during the summer as expected.

Spatial forecasting

A great deal of power delivery planning is about *where* equipment and facilities should be placed. It does little good to add substation capacity in the wrong substation area, or to bolster feeder capacity in the wrong feeder area. A spatial forecast, which associates electric load and growth with location, is typically used to associate electric load growth with location, so that planners know both where and how much load growth to anticipate. Figure 9.10 displays a long-term spatial forecast.

A number of methods are in use for spatial forecasting, from simple trending methods (extrapolation of weather-adjusted substation and feeder peak load histories) to quite comprehensive simulations involving analysis and projection of changes in zoning, economic development, land-use, and customer end usage of electricity. Results vary greatly depending on method and resources used, but engineering methods exist to both determine the most appropriate methods and forecast characteristics needed for any utility application, and to evaluate the efficacy of a forecast. The most important factor is that a utility employs some

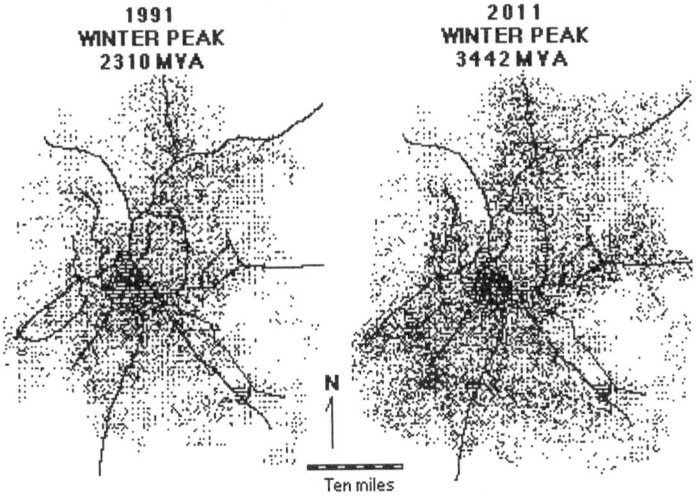

Figure 9.10 Maps of peak annual demand for electricity in a major American city, showing the expected growth in demand during a 20 year period. Growth in some parts of the urban core increases considerably, but in addition, electric load spreads into currently vacant areas as new suburbs are built to accommodate an expanded population. Forecasts like this, done for two-, four- and similar periods out to twenty years, set the requirements for both short- and long-range power delivery system planning.

Table 9.4 Percent of Utilities in North America Using Some Type of Formally Recognized Spatial or Small Area Load Forecasting Method

	1998	1992	1988	1979
Among utilities with > 1,000,000 customers	45	75	70	35
Among utilities with 500,000 - 1,000,000 customers	50	70	65	30
Among utilities < 500,000 customers	12	15	15	10

legitimate means of studying and projecting load on a detailed enough location-basis to support its planning needs.

Traditionally, good spatial forecasting required both considerable labor and above-average engineering skills and was considered a "high-expertise" function within state of the art distribution planning methods. The best traditional methods worked very well but had rather high labor and skill costs methods

[Willis and Northcote Green – 1983, Engel et al – 1996]. Many utilities cut back on both the quality of the technique used and the effort devoted to data collection and forecasting study, when they downsized professional staffs during the 1990s. Table 9.4 illustrates the reduction the number of utilities using the best class (simulation-based) spatial forecast methodology, but does not reflect reductions in the data or time put into the forecasting effort. As a result, at a time when spatial forecasting needs are at an all time high (see below), the quality of local-area forecasting done at many utilities deteriorated sharply.[5]

In the very late 1990s, new forecasting methods were developed that reduce labor and skill requirements considerably, but these have limited availability and are not widely used [Brown et al, 1999]. However, this means methods that can provide the information needed within reasonable labor and skill limits are available to the industry.

Impact of spatial forecast errors on reliability

In some manner, every T&D plan includes a spatial forecast, for the total load growth allocated in some manner among the various parts of the system. Classically, the viewpoint on forecast sensitivity of T&D systems has been that if the spatial element of the forecast is done poorly, the result is a very poor use of capital. A projection of the wrong locations for future load growth identifies incorrectly those portions of the system that need to be reinforced. Capital additions are made less effectively than possible.

But in addition, a large effect of poor spatial forecasting is a loss of contingency capability. Normally a power system designed based upon a mildly incorrect spatial forecast (i.e., one pattern of ":where load is") will suffer from less than the planned contingency withstanding capability (i.e., provide less reliability in service) than expected. It will operate well enough during times when "everything is going well" but suffer from problems that are both more serious and take longer to fix than expected during contingencies. Essentially the poor forecast "uses up" the contingency capability built into the system [Willis and Tram, 1984].

Systems with high utilization ratios are more sensitive to this degradation of contingency planning due to spatial forecasting errors. Although subtle, the effect is best described this way: The contingency neighborhoods described earlier increase in size as a result of the higher utilization ratios used (Figures 9.5 and 9.6. While it may seem that this makes the planning less in need of detailed spatial forecasts (there are fewer "units" – contingency support neighborhoods – and they are on average far larger), the opposite is true. Contingency capability is very sensitive to the allocation of load within each

---

5 Load forecasting problems related to local area forecasting were identified as a major contributing problem in six of events (Table 9.2) investigated by DOE's P.O.S.T. report.

support neighborhood. The analysis given earlier assumed the load in each neighborhood was evenly split among units in that group: if it is even slightly unbalanced, the system's contingency capability is greatly degraded. The forecast of where load is within each contingency support neighborhood is critical, so that proper, operationally efficient allocation of loads to equipment can be arranged. One sure sign of poor spatial forecasting (and planning) and a system with less contingency withstand capability than it could have, is that considerable operational adjustment to loading patterns (load balancing) has been done, by altering switching and feeder loading. Again, as with poor weather normalization, this "uses up" the contingency capability of the system, something in very short supply in aging areas of the system. SAIDI increases.

Interconnection Complexity

In the slightly simplified power systems used as examples earlier in this chapter, the small contingency support neighborhoods needed at 66% loading required interconnection with only two neighboring units assured success without overload during N–1 conditions. But at higher utilization ratios, contingency support neighborhoods grew in size and number of mutually supporting components. Interconnection of more equipment, into a scheme where it could support one another during contingencies, was necessary for success of the contingency plans.

In aging areas or where for other reasons planners and engineers have accepted near 100% utilization of equipment, there is a requirement for an even stronger and more widespread interconnection. Everywhere in a high-utilization system, each of its N units must have a strong-enough electrical tie to a wider neighborhood of equipment around it, to support its outage.

> At higher equipment utilization rates, the importance of configuration and operating flexibility in the design of the system becomes more critical to reliability.

Traditional contingency-based study methods *can* deal with the *analysis* of these issues relating to the wider neighborhood of support around every one of the N units in the system, and its great complexity. They can determine if the required number of neighbors are there, if they have enough margin of capacity to accept the load without overloads, and if the system's electrical configuration makes it possible for them to pick up the demand that was being served by the failed unit. Basically, modern N–1 methods can and will determine if the failure of each unit in the system is "covered" by some plausible means to handle its failure and still provide service.

Again, N-1 methods do not work with probabilities nor determine the system's sensitivity to multiple failures, so they cannot determine the failure sensitivity or failure likelihood of these complicated interconnected schemes.

At higher utilization ratios, complexity of the contingency backup cases has increased. Determining if a particular contingency backup plan is really feasible, if it is really connected with sufficient strength to survive the likely failure states, or if it depends on too much equipment operating in exactly the wrong way, is something N–1 methods do *not* fully address. Configuration needs to be studied from a probabilistic basis – is this entire scheme of rollover and re-switching likely to really solve the problem?

9.5 THE PROBLEM IS NOT HIGH UTILIZATION RATES

A point the authors want to stress again is that high equipment utilization is *not* the cause of poor reliability in aging infrastructure areas of a power system. It is possible to design and operate power systems that have very high (e.g., 100%) utilization factors and provide very high levels of service reliability. This is accomplished by designing a system with the configuration to spread contingency burden among multiple equipment units, with the flexibility to react to multiple contingencies, and that can apply capacity well in all the situations most likely to develop. Such designs require detailed analysis of capacity, configuration, configuration flexibility, failure probabilities, and the interaction of all these variables.

This does not mean that every power system should be built with high equipment utilization. The authors are not "taking sides" on the issue of utilization rate, because it has no single answer. Equipment utilization is only one factor in the design of a power system. In some cases, the best way to "buy" reliability along with satisfactory electrical (power flow) performance is to use capacity – to build a system with low utilization ratios. But in other cases, particularly where the cost of capacity is very high (as it is in many aging infrastructure areas), good performance comes from using the equipment to its utmost. Achieving high reliability of service even in these situations where equipment is highly stressed and contingency margins are small or non-existent requires using configuration and interconnection flexibility in an artful manner.

The key point is that traditional planning tools are incompatible with these needs. They cannot provide dependable analysis of, nor serve as good guides for such engineering. They lead to designs with necessary and sufficient qualities for good reliability only if the capacity margin is used to purchase reliability. They do not identify weak points in a system nor provide an indication of how and where solutions to these problems may be pursued through changes in the design of the system. *Traditional planning tools are partly undependable for planning reliability in aging infrastructure and other "high stress" areas of a power system.*

9.6 SUMMARY AND CONCLUSION

Traditional Tools Have Shortcomings With Respect to Modern Needs

Many of the problems faced by a utility owner/operator of an aging power T&D infrastructure are compounded by the fact that the tools being used by its planners and engineers cannot directly address one vital aspect of the required performance: reliability. Traditionally, reliability was addressed in power system design by engineering contingency backup capability into the system: every major unit of equipment could be completely backed up should it fail. This was termed the "N–1" criterion and methods that engineer a system based on this criterion were often referred to as "N–1" or "N–X" methods.

Such methods addressed a key and *necessary* quality for reliability: there must be a feasible way to do without every unit in the system, some way of switching around it during its outage and/or picking up the burden it was serving during its outage. Realistically, no power system can be expected to provide reliable service to its energy consumers unless it possesses this necessary qualification of having complete N–1 contingency capability.

But through many years of use during periods when equipment loading levels of equipment were lower than is typical in the 1990s and 2000s, the power industry came to view N–1 criterion as a necessary and sufficient. For power systems with roughly a 33% redundancy (contingency margin), the criterion is effectively a necessary and sufficient criterion to assure reasonable levels of reliability. However, when a power system is pushed to higher levels of equipment utilization efficiency, N–1 is still a necessary criterion, but no longer *sufficient* to assure satisfactory levels of reliability.

Basically, the shortcoming of the N–1 criterion, as well as engineering methods based upon it, is that they do not "see" (respond to, identify problems with, or measure) anything with respect to reliability of service except the "yes/no" satisfaction of this one criterion. Therefore, they cannot alert engineers to a flaw in the design of a power system with respect to its reliability, due to either the likelihood that events may get worse than the system can stand. "Yes, you have a feasible backup plan, but for this system it is very likely that while this unit of equipment is outaged, something else will go wrong, too." Or, because some parts of the system are key elements for reliability of the system around them, to the extent that they are "important enough" that they need more reliability built into their design. "This unit is more important than you realized: A second backup is really needed for the reliability level you want to achieve."

Furthermore, they cannot provide quantitative guidance to planners and engineers on what to fix and how to fix it in a satisfactorily effective but economical manner.

A Deeper Problem, Too

Finally, there is a more subtle but perhaps more fundamental problem associated with the use of N–1, one that is difficult to fully demonstrate in a textbook, but nonetheless real. N–1 engineering procedures solve reliability problems by using *capacity*: It is a criterion and method that uses contingency margin as the means to achieve reliability. This is partly the nature of the criterion and the tool, but also the fault of the paradigm – the way of thinking – built for the engineers around the N–1 tools.

By contrast, experience with planning and engineering tools that directly address reliability will quickly show any engineer that very often *configuration* is the key to reliability. In fact, capacity and configuration are both key factors in achieving reliability and artful engineering of power system reliability requires combining both in a synergistic manner.

Traditional N–1 analysis tools (Figure 9.1) can analyze the functional and electrical characteristics of configuration. However, they cannot do so on a probabilistic basis, analyzing how the likelihood that the interconnection will be there and determining how that interacts with the probabilities that the capacity will have failed. Table 9.3 summarizes the limitations of N–1 analysis that need augmentation in order to meet modern power system reliability planning needs. Such methods will be discussed in Chapter 14.

Explicit Reliability-Based Engineering Methods

What is needed to assure the *sufficient* requirement in power system reliability engineering along with the necessary, is a method that address capacity, configuration with an explicit, quantitative evaluation of the reliability of service they provide. Ideally, this must be a procedure that computes the reliability of service delivered to every element of the system, in much the same manner that a load flow computes the current flow and voltage delivered to every point in the system. It should be a procedure that can then be used by planning engineers to explore the reliability performance of different candidate designs, and that, in the same manner that load flows identify equipment that is heavily loaded (i.e., key to electrical performance) would identify equipment that is heavily loaded from a reliability standpoint (i.e., key to reliable performance).

Table 9.5 Seven Desired Traits of the Ideal Power System Reliability Planning Method (that N – 1 Methods Do Not Have)

Trait	Comments
1. Analysis of probabilities	Computes actual likelihood of all equipment in any area of the system being in service at one time.
2. Partial Failures	Can model the de-rating of equipment in stages, rather than just zero-one outage
3. Connectivity Sensitivity	Can model impact of variations in local loads, possible changes in feeder switching, etc., on expected performance.
4. Quantitative reliability	Computes actual expected reliability of service at points throughout the network in a quantitative manner.
5. Contingency Prioritization	Quantitative assessment includes both likelihood and importance against requirements in a way that permits prioritization of contingency results.
6. Cost sensitive	Relates or can relate results directly to money and risk.
7. Solutions focused	Can determine or recommend possible fixes based on both effectiveness and economy.

Modern planning needs are best met using power system design and planning techniques that directly address reliability performance: Explicit, rather than implicit, methods. Such techniques have been used for decades in the design of systems where reliability is of paramount importance – nuclear power plants for commercial and shipboard use, spacecraft design, and throughout the aircraft industry. Adaptation of these methods to power systems provided a much more dependable design method to achieve operating reliability.

Such analysis begins with the same "base case" model of the power system as traditional techniques did. But probabilistic analysis starts out with a true normalcy base – a recognition that the natural condition of the system is "some equipment out" and that that condition will always be in a state of flux, with equipment being repaired and put back in service, and other equipment out of service. Like contingency enumeration methods, probabilistic analysis determines the consequences of every failure or combination of failures – can the system continue to operate and how close to the edge will it be during that time? But unlike traditional methods, probabilistic analysis determines if and how every combination of two or more simultaneous failures could interrelate to create problems, whether they are nearby or not in the immediate neighborhood, and it determines if that combination of failures is likely enough to be of concern. By tracing reliability through the configuration of the system while analyzing expectation of failure-to-operate, it effectively analyzes configuration and its relationship to reliability, too.

Depending on the exact method used (Chapter 14 will cover three basic types), partial failures can be accommodated using conditional capacity levels, partial failure states, or by recognizing sub-units within each main unit. Assumptions about loads, operating conditions, and other aspects of system operation can be modeled using probability distributions with respect to those variables.

Evaluations determine and prioritize problems in the system as they will occur in the actual operation. Areas of the system, or operating conditions, that are especially at risk are identified. A big plus is that the method can be fitted with optimization engines to solve the key identified cases – "Find the lowest cost way to make sure this potential problem won't be a real problem."

Such methodology for reliability-based design of a power system can be created using a highly modified form of N–1 (Figure 9.1) analysis in which probability analysis is used at every stage, or by combining N–1 with reliability-analysis methods. What is best for a particular utility or situation depends on a number of factors specific to each case. But the important point is that there are proven, widely used methods to perform this type of work available to the power industry, and in use by selected utilities. Such methods are a key aspect of solving aging infrastructure problems: the reliability of such system areas must be well analyzed and solutions to it well engineering. These types of methods will be covered in Chapter 14. Table 9.6 summarizes the overall recommendations on planning method improvement needed to meet modern reliability engineering needs.

REFERENCES

H. A. Daniels, *Automatic Contingency-Based Analysis of Power Systems,* PhD Dissertation, University of Texas at Arlington, 1967.

M. V. Engel, editor, *Power Distribution Planning Tutorial,* IEEE Power Engineering Society, 1992

H. L. Willis, *Spatial Electric Load Forecasting,* Marcel Dekker, New York, 1996.

H. L. Willis and J. E. D. Northcote-Green, "Spatial Electric Load Forecasting – A Tutorial Review," *Proceedings of the IEEE,* February 1983.

Table 9.6 Overall Recommendations for Planning Methods

Method	Comments
1. N–1 contingency analysis	No reason to not use it, as it tests for a necessary criterion of reliability and it explicitly demonstrates for each contingency how the system will react and handle the demand during that contingency. However, some of the better explicit reliability analysis methods implicitly merge N–1 into their analysis, so an explicit N–1 analysis is no longer needed.
2. Explicit reliability analysis	Probabilistic analysis of both capacity and configuration, including contingency interconnection, with computation of the expected reliability of service at all customer and intermediate points in the system. Can be based on a very modified form of N–1 analysis (dynamic analytical simulation) which combines (1) above with this step into one comprehensive method.
3. Economic cost comparison	All reliability engineering is done on the basis of "bang for the buck."
4. Good load forecast	All the reliability engineering possible is wasted if the system is designed to a different demand level than required. Good weather normalization and spatial allocation methods are necessary.
5. Coordinate feeder plans	Chapter 10 will explore engineering of reliability on the primary distribution system. Effective, economical design of reliability requires coordinating sub-transmission – substation level reliability with primary distribution reliability, as explained there.

H. L. Willis and T. D. Vismor, and R. W. Powell, "Some Aspects of Sampling Load Curves on Distribution Systems," *IEEE Transactions on Power Apparatus and Systems,* November 1985, p. 3221.

United States Department of Energy, *Power Outages and System Trouble (POST) Report,* March 2000, Washington

10
Primary Distribution Planning and Engineering Interactions

10.1 INTRODUCTION

This chapter is in essence a continuation of Chapter 9's discussion about engineering paradigms and methods. Here, the focus is on the distribution system. The power distribution system (primary feeder system) accomplishes more of the power delivery function in most electric utility systems than any other level of the system. Power is routed into a few locations in fairly large chunks and the distribution system then routes and sub-divides that power into very small allotments delivered to many different locations. No other layer of the power system accomplishes so much dispersion over area or so much sub-division of power from large to small.

The primary distribution level also has more impact on customer power quality than any other layer. It is both immediately downstream of the lowest-level voltage regulation equipment in the system (substation voltage regulators or load-tap changers), and by the nature of its topography, the source of the majority of voltage drop that the customer sees. That topography also means that the majority of outages that the customer sees are caused by the distribution system – there is simply more miles of exposed line at jeopardy to weather, accident, and random failure. Thus, in most systems, the distribution system is the source of the majority of both the voltage-related and availability-related power quality problems seen by energy consumers.

This alone would guarantee that distribution should receive a good deal of attention anytime a utility needs to improve reliability, as for example when hard pressed by aging infrastructure issues.

However, the distribution system is also a resource that can be used to bolster reliability of the entire system: if combined artfully with the sub-transmission – substation level designs, can result in improved performance in terms of "bang for the buck." Such planning requires, as was the case in Chapter 9, engineering methods different from traditional approaches, but methods that are proven, and widely available.

This chapter will begin with a look at distribution, the approach to its planning and operation that is typically taken in the industry, and what can be accomplished by making a shift to a different way of looking at and engineering the distribution system, all in Section 10.2. This will introduce two key concepts. First, using the distribution system to bolster reliability at the substation area through a built-in ability to transfer large loads between substations. Two, the use of marginal benefit-cost ratio balancing as an optimization tool in power system planning. Together these lead to a much improved cost effectiveness of the entire system. Section 10.3 will then summarize the engineering of distribution layout – how the spatial or geographic configuration of trunks, branches, and switching can be used to build a distribution system that has the characteristics needed. Many distribution engineers are unaware of the extent to which this aspect of distribution systems can be engineered. Section 10.4 concludes with a summary of key points.

10.2 DISTRIBUTION PLANNING AND THE PERCEIVED ROLE OF DISTRIBUTION

At its most basic, a power distribution system must be designed to deliver power over wide areas while maintaining reasonable voltage drop, and with some means of providing redundancy (contingency switching) should any one flow pathway fail. Naturally, it is desirable to perform this function at the lowest cost compatible with providing satisfactory power quality within standards. This defines the traditional goal of power distribution planning, and reflects the traditional viewpoint on the role of power distribution system.

During the early to the mid-20th century, every utility developed its own distribution planning method to satisfy these needs. At each utility these evolved during subsequent decades into very tight institutionalized standards for "the way we do things." In many cases, these were nearly inflexible "template" or "cookie-cutter" approaches to layout style, conductor size set, and switching zone and conductor sizing. In all cases, they resulted in a distribution system that adequately fulfilled the traditional needs of distribution well enough to get by. The system routes power from substations to customers without violation of voltage, loading or power factor standards. It does not cost too much. It is serviceable and expandable as needed in the future.

Remarkably, these approaches to distribution differ substantially among utilities, although all those distribution systems used the same components and equipment and all fulfill much the same role. At many utilities, power distribution planning is very parochial, built upon principles developed in-house decades ago, lacking any external influence, and never comprehensively re-examined and revised with respect to modern needs. The reasons have a lot to do with the distribution aspect of the utility business. Compared to utility's transmission planners, distribution planners seldom have to interact with neighboring utilities through intertie meetings or power pool planning groups. They are seldom exposed to the different ideas and their particular design habits and guidelines can be challenged as that could be improved. "We do it differently."

As a result, nothing is quite as tightly institutionalized, nor varies quite as much, throughout the power industry, as the quality of distribution planning. As an example, one large IOU in Florida has standards for conductor policy and feeder layout that are the exact opposite in several regards of the practices at a large IOU in the western U.S. What is required at one utility is not permitted at the other. In fact, neither utility is correct – both paradigms are appropriate in some cases, but not all. A more effective and lower cost system results when flexibility is introduced into the design: when layout, switching, and conductor sizing are *engineered*, rather than designed by guideline. Section 10.3, later in this chapter; will summarize several concepts related to such engineering.

Frankly, engineering of a distribution system so that it performs the basic electrical functions that were traditionally required of it (i.e., electrical flow within standards, feasible manual switching capability for contingencies) is rather straightforward. At many utilities, this was done for decades with a standardized layout rule, a set of tailored nomographs and tables and some rough calculations (slide rules at best), all guided by a series of rote-memorized rules of thumb. When digital computers became available, load flow computations and spreadsheet-level tools were adapted by each utility to its particular distribution paradigm. The basic concepts underlying each utility's method -- how things are done and the fact that the engineering is largely based on a set of inflexible rules – remained in place.

Both because adequate distribution planning is so straightforward to do, and because it is done at many utilities with the fairly straightforward template-guideline-type of approach, one often hears the terms "it isn't rocket science" applied to distribution planning. This belief that distribution is easy to plan, and that significant and high-expertise effort is wasted in its engineering, is exacerbated by the facts that distribution is of relatively low voltage compared to other parts of the power system, that the equipment involved is small and inexpensive (compared to other levels,) and, that the system is normally laid out in radial form (which makes power flow easy to calculate). All of this tends to make distribution seem simplistic, and in fact it is, if viewed only from the traditional perspective, that the task is just to "get it so it works."

Distribution Can Be Asked To Do A Lot More

However, that traditional interpretation of the distribution system's job is obsolete and incomplete with respect to modern societal needs and technical capabilities. When reliability of supply is a major performance metric, and when cost must be minimized, not just kept within traditional bounds, distribution planning becomes very challenging. The important point to realize, however, is that the effort required meeting this challenge more than pays for itself: the gap between the performance of really good distribution, and merely "adequate" ones, is immense. This gap manifests itself in two ways:

- *Economics:* Advanced engineering and flexibility of approach can yield considerable reduction in cost.
- *Reliability*: The potential of the primary system to maximally contribute to improved reliability is partially untapped in the vast majority of cases.

A key factor in both is that the distribution system is both engineered well (optimized) and its design is coordinated carefully with that of the sub-transmission and substation levels in ways not traditionally done.

Table 10.1 shows the results generated by this approach at six utilities with which the authors have recently worked. These utilities are representative of all the utilities the authors have worked with over the past decade – they are an average set, not one picked to show outstandingly good results. Together they demonstrate both the margin of improvement that can be expected by updating and modernizing primary distribution planning and design paradigms, the variation in improvement and type of improvement that are seen among different utilities.

Table 10.1 lists four statistics for each system. The first, "Electrical - $," is the reduction in overall (lifetime) cost of distribution that could be wrought with respect to the traditional primary distribution paradigm of moving power. This is the saving in *new* capital additions that results from improving how the distribution system is designed to do its traditional job – moving power.

The second statistic, "Existing - $" shows how much the useful utilization of an *existing* distribution system can be improved with respect to that traditional paradigm. Existing systems encompass both the good and bad points of their past engineering and design, and those systems cannot be thrown away and re-designed based on new and improved rules. However, as shown, some

Table 10.1 Improvements in Distribution System Cost and Performance Derived from Revision of Traditional Design Guidelines and Engineering Methods

Case	Utility System Type	Electrical $	Existing $	Reliability $	System Reli.
1	Urban UG, Atlantic coast	8%	2%	79%	3.4:1
2	Rural, small towns, central US	12%	0%	25%	.8:1
3	Metropolitan, Midwest US	6%	3%	40%	1.8:1
4	Entire system, IOU, central US	2%	3%	36%	1.5:1
5	Rural, small towns, southern US	12%	11%	31%	1.6:1
6	Entire system, N.E. US	22%	16%	33%	.92:1
Averages		10%	6%	41%	1.7:1

improvement can be wrought in the MW/$ capability of these systems. However, improvement is only useful where useful (if one improves a 5.0 MVA feeder so it can carry 5.4 MVA, the increase is useless unless it is in an area where one *needs* a .4 MVA increase in capability). The values shown in Table 10.1 reflect both what is possible and what was found useful in each system.

The third statistic, "Reliability - $," represents the reduction in the cost of improving reliability that was attained by revising design guidelines and applying the most effective reliability-based planning and engineering methods to these systems. These improvements while rather dramatic, are typical. As stated earlier, distribution systems are simply not designed from a reliability standpoint. When this is done and the engineering is optimized, the results are a very considerable improvement.

The final statistic "System Rel.(Reliability)" represents the margin in "bang for the buck" that the revised distribution reliability improvement made over the existing cost of improving reliability that was in effect before the revision. This statistic goes to the heart of the use of distribution as a resource in aging infrastructure areas, and of the authors' overall theme of using all of the power system chain optimally to obtain the greatest "bang for the buck." In four of the six cases listed, reliability improvement was between 20% and 40% less expensive to buy at the primary level than at other levels of the system. In these cases, spending on the distribution system could deliver reliability improvement for less cost than spending on other levels of the system.

In two other cases, this was not the case, and the value shown is negative. Of note here is that the utilities were unaware of this entire issue, because the traditional tools they were using for the design of all three levels were not capable of directly engineering reliability or of even measuring such things as reliability improvement per dollar. Thus, in all cases, improvements between 9% (1/.92 – see final column) and 70% were possible by adopting reliability-based engineering methods and using them well at the distribution system and for coordination of design among the distribution and sub-transmission-substation levels.

A Quantum Difference in "Bang for the Buck"

Table 10.1 made it clear than revision of distribution system guidelines and adoption of reliability-based engineering methods often (actually in all cases the authors know) provides a good improvement. The increases in cost effectiveness of 10% and 6% versus the traditional distribution function are noticeable and quite valuable. However, it is the improvement from the "modern perspective" – that which views distribution's role as one of both performing the electrical duties and fitting into the optimum reliability vs. cost structure of the entire system – where the improvement represents a quantum leap. By looking at the distribution system as a reliability resource, and optimizing its use for that purpose, an average 40% increase in cost effectiveness of reliability improvement is obtained at the distribution level. By coordination with the planning and design of other levels of the system, a 70% increase in overall effectiveness of money spent on reliability improvement is achieved.

The authors are aware than many readers, particularly some with long experience as distribution planners at power delivery utilities, will dispute that the improvements shown are real (they were), or will argue that that their system is different (not likely) and that "it won't work here" (it will). The six cases shown are all real utilities implemented long enough in the past that time has proven the expected savings and performance increases to be valid.

The 6% and 10% savings against normal needs come about because modern techniques can find better ways to reduce costs than the formalistic methods of the past. But the quantum gains in reliability and reliability improvement per dollar come from the fact that because of its dispersed nature, the feeder system can often be reinforced at reasonable cost to cover weaknesses in reliability (contingencies) at the substation level.

The concept is easy to understand, if not to engineer (at least with traditional tools). Suppose that for whatever reason, reliability at one substation is a particularly degraded level of expected reliability. Perhaps the utility has no choice but to accept very high equipment utilization at the substation and there is a less than ideal configuration of buswork and switchgear there to provide contingency flexibility. The feeder system in this substation area could be reinforced to provide "more than the normal amount of primary feeder level reliability strength" to make up for this deficiency. Specifically, since this substation will be out of service more than usual, the feeder system can be designed to transfer its loads to other neighboring substations, maintaining service during those contingencies. This is really what is necessary to provide good service in high-utilization systems, particular in the aging areas where substations are congested and highly flexible switching and buswork is not an option.

Good Distribution Planning *Is* Rocket Science

Despite the fact that what might be called "commodity" design can create a power distribution system that will work "well enough," the power distribution system is by far the most difficult of all layers in the power system to plan and engineer to a truly high standard. Meeting this challenge is difficult, because distribution systems are like piranha. A single piranha is small, and not of great concern, but in aggregate, a group of piranha become something entirely different.

Like the piranha, almost every unit in a distribution system is small, of low voltage, commodity design, and low price. But in aggregate a distribution system becomes large, expensive, and most importantly, *very* complicated. This is due to both the sheer number of components (perhaps millions) and the their interconnectedness or interaction with one another (particularly if viewed from the reliability perspective). A distribution system presents a tremendous challenge in *combinatorial decision-making* – how to select equipment, route lines, site key equipment and switches, taper conductor, arrange normal and contingency flow patterns, and makes a myriad of other decisions needed. The *combinatorial* aspect is tremendous, resulting from the fact that not only are there dozens or even hundreds of individual decisions to make in even a small distribution layout problem, but also because these individual decisions all interact with one another. No decision can be made in isolation: they all influence one another.

It is possible to sidestep this combinatorial challenge by using a set of standardized layout rules, templates on sizing and design, and inflexible guidelines on equipment and engineering. This was how utilities approached distribution in the 1930s. By sacrificings the gains in cost and performance that might be obtainable, in exchange for a tremendous simplification in the process of finding a workable plan for every situation. This template approach does assure that minimum electric requirements are met, and that cost, while perhaps not optimum, is not unduly high (perhaps no more than 10% above the minimum possible).

However, maximizing benefit, and particularly, tailoring distribution in each area of the system to fit the specific needs and the characteristics of other levels of the system there, means facing up to that combinatorial challenge. The possibilities within all that flexibility have to be exploited. That means dealing with all those interrelated decisions and obtaining plans that are close to optimality. Distribution planning, done to the most effective manner possible, *is* rocket science. Section 10.3 will summarize design and planning methods that accomplish these goals in a practical sense - they deliver results in the real world. However, they require new ideas and slightly more work than traditional methods, which brings up a final point about modern distribution planning:

Feeder System Strength

The "strength" of a feeder system will be defined here as the percent of the substation load that can be transferred to neighboring substations during contingencies. This is a characteristic of a distribution system that can be designed to have a value of anywhere from zero to one hundred percent. The authors have worked with systems at both ends of the scale and everywhere in between.

There is no single, recommended target for the feeder system strength. It depends on the voltages, capacities, demands, configurations and constraints on the sub-transmission – substation level, the capabilities of the type of distribution system being built, and the need for reliable service as interpreted by the consumers and the utility alike. Figure 10.1 shows the reliability versus

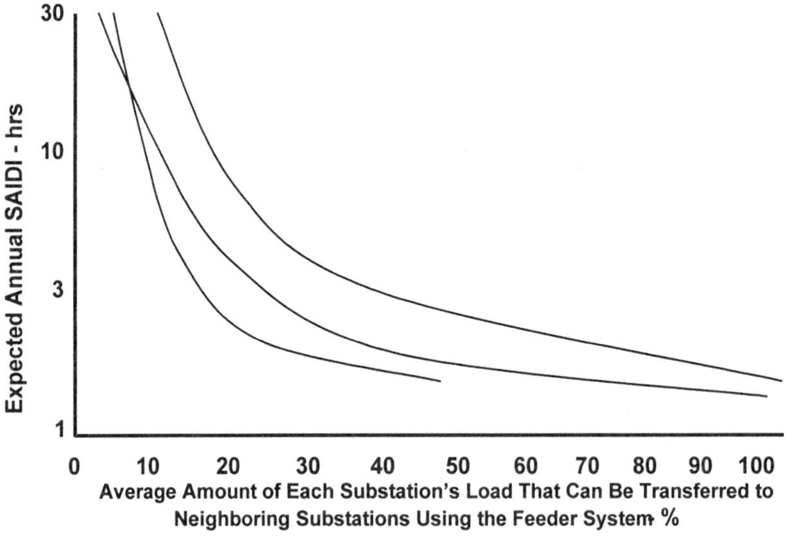

Figure 10.1 Feeder strength (% of the average substation's load that can be transferred to three large metropolitan central systems in the U.S. The "strength" of the feeder system, in terms of its ability to transfer a portion of each substation's load to neighboring substations, is a key factor in reliability of most power systems.

feeder transfer strength for the central (metropolitan) division of the electric utilities in three American cities. As can be seen, in all three systems the transfer capability of the system is a key variable in determining the overall capability of the system. Further, feeder system strength required to achieve a specific overall level of service reliability for the entire system (e.g., a SAIDI of 100 minutes) is far different in each of these systems.

In fact, usually there is a need for different levels of feeder strength in different areas of the same utility system. Very often in older (aging infrastructure) areas of the system, a stronger feeder system is the optimal (lowest cost, most "bang for the buck") solution to problems of reliability due to combinations of aged equipment and limited or congested substation sites.

So, while there is no recommended value for feeder system strength, what is always recommended is that the characteristics of the feeder system be tailored to, and its planning both on an electrical and reliability basis be coordinated with, that of the rest of the system.

The Systems Approach

It is possible through the use of an appropriate type of reliability analysis (Chapter 14) to determine the impact of various levels of feeder strength on the overall customer-reliability delivered by the power system. For example, determining curves like those shown in Figure 10.1, and engineering the system's reliability to achieve the desired reliability target at the total lowest cost. This means balancing how much is spent on reliability and exactly how it is designed into the sub-transmission, substation, and feeder levels, realizing that through appropriate arrangement of configuration, each layer can to some extent provide contingency capability for reliability problems encountering in the other layers.

A key aspect of applying the systems approach to reliability engineering is the use of *marginal benefit cost analysis* – computing for each level of the system how much a small increment of reliability *increases* costs, based on the present configuration and the options available. Comparison of the values among levels and areas of the system leads to optimization.

Figure 10.2 illustrates a simple example. In this diagram, one sees the reliability vs. cost curves for each level of a candidate system design – how much it costs to build the system that will provide various levels of reliability (so many minutes of outage annually due to events or failures on that level). In every case there is an exponentially increasing cost to buy increasing levels of reliability (decreasing levels of outage time).

The top diagram has large dots indicating the cost vs. reliability point of each level of the system for one particular candidate design, a plan which is expected to provide a total of 119 minutes SAIDI, at an annualized cost of $172/kW peak

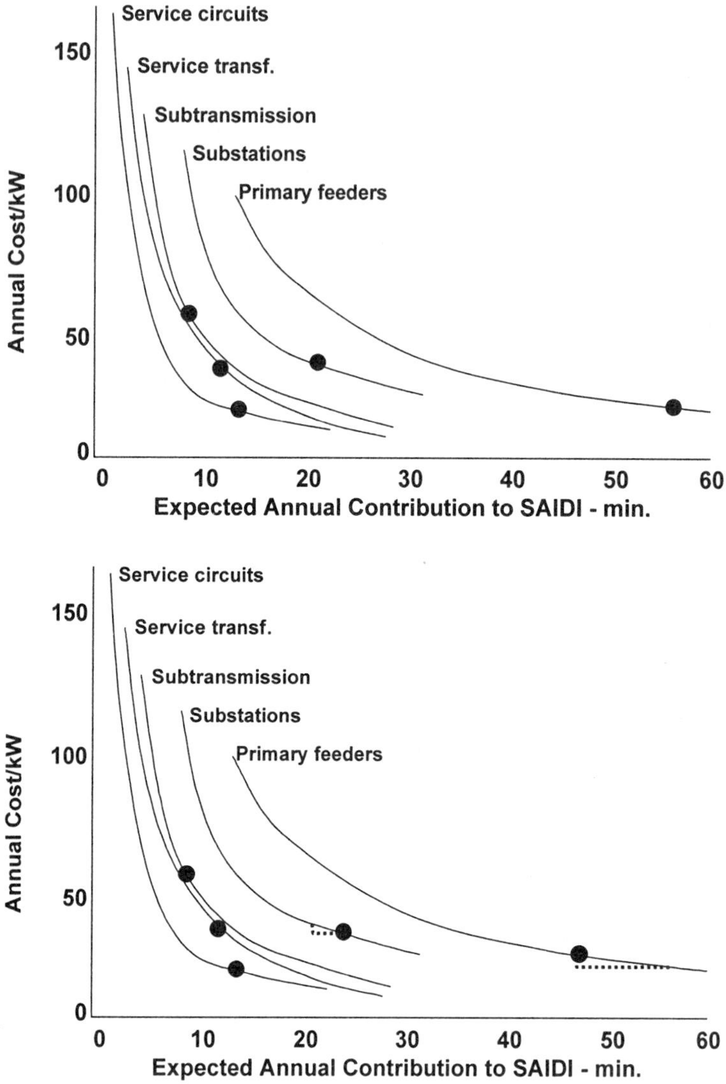

Figure 10.2 Top, cost vs. SAIDI curves for the different layers of a power system. Dots show the status of each layer in a candidate design, which in aggregate provide 119 minutes of SAIDI annually at an annualized cost of $172. Bottom, by adjusting the feeder system design to "buy" ten minutes of SAIDI improvement for $5.00, and saving $5.00 by cutting at the substation level, a net improvement of five minutes SAIDI is obtained at no cost: SAIDI is now 114 minutes and cost is still $172.

demand (the sums of the reliabilities and costs of all the dots, respectively, equals these two values). The important items for improving reliability are not the actual value (i.e. position of the dot) of reliability vs. cost for any of these points. Instead, it is the *slope* of the cost vs. reliability curve at those points. For example, the primary feeder level design is at 57 minutes and $24/kW annualized cost. However, the slope at that point indicates that spending an additional five dollars will improve reliability by ten minutes (moving the dot along the curve to the left by ten minutes raises cost by five dollars). By contrast, buying ten minutes at the substation level (see curve and dot for that level) would cost $32.00. Clearly, the way to improve reliability at a bargain price is to buy improvements in the feeder system.

But beyond this, in this candidate design, one can begin improving reliability without spending any additional budget by "trading" reliability and cost between the feeder and substation levels. This can be done by "buying" ten minutes of reliability at the feeder level for five dollars, and "selling off" five dollars worth of reliability at the substation level, which because of the slope of that curve, means giving up only five minutes of SAIDI. Overall performance is improved by five minutes at no change in overall cost.

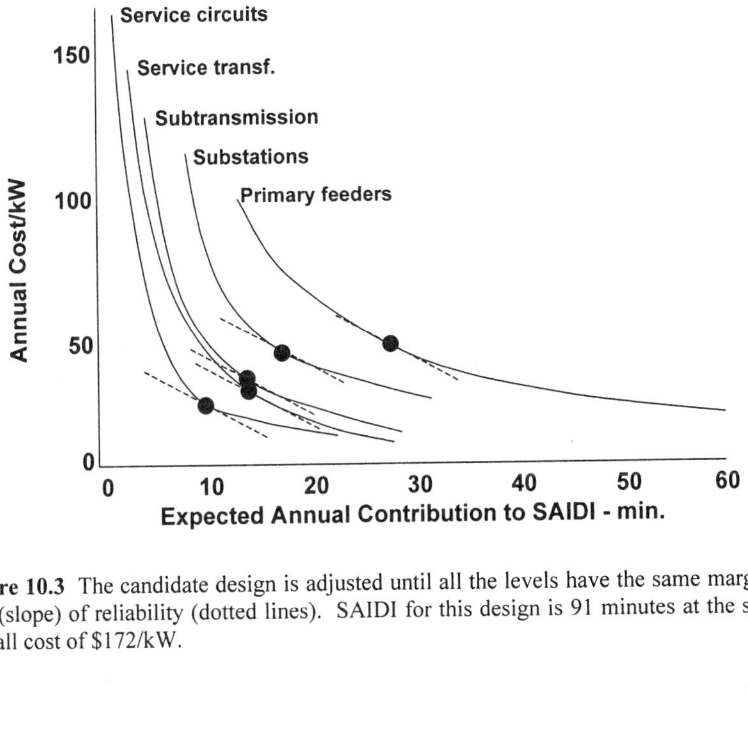

Figure 10.3 The candidate design is adjusted until all the levels have the same marginal cost (slope) of reliability (dotted lines). SAIDI for this design is 91 minutes at the same overall cost of $172/kW.

Thus modified (bottom plot), the slopes of the feeder and substation levels are still not equal, so this "trick" can be applied again, buying 13 minutes at the feeder level in exchange for eight minutes at the substation. This brings SAIDI down to 110 minutes at no increase in cost – annualized cost is still $172. Far beyond this, the points in all the curves can be adjusted in this way, with the result shown in Figure 10.3. The final result is a SAIDI of 91 minutes at the same $172/kW annualized cost.

The example shown is a slightly simplified example based on case 5 in Table 10.1. In the real world, the curves for each level have discontinuities and small but frequent discrete jumps, and there is some interaction and interdependence among the curves. But the concept shown is a valid and workable approach, and it leads to viable, noticeable improvements in reliability, at no net increase in cost. In this case, to accomplish these improvements, more was spent on the distribution system and less on substation configuration and redundancy: distribution was both designed to a different style (see large trunk vs. multi-branch in section 10.3) and its strength was augmented in selected areas.

Changing the Distribution Paradigm

Chapter 1 (pages 28–30) discussed a number of "cultural barriers" that often stand in the way of improving power delivery system performance using "non traditional ways. In the authors' experience, the technical methodology required to successfully implement the distribution system performance improvements discussed above are far less challenging to most utilities than the cultural changes required to fit it into their organization. Considerable commitment from executive management, and steady effort to accomplish the change, are required.

10.3 FLEXIBILITY AND EFFECTIVENESS IN FEEDER LEVEL PLANNING

Generally, N–1 criteria and planning methods were traditionally applied to only to the sub-transmission – substation portion of the power delivery system, everything from the low side (feeder) breakers upward toward the wholesale grid (Figure 10.4). It was and still is uncommon to see that method applied to the distribution feeder system, although concepts allied with N–1 criteria often are. The primary reasons for this difference in methods on each side of the substation feeder breaker are:

Tradition – in vertically integrated electric utilities of the 1930s through 1990s, the Transmission Engineering Department most typically engineered all of the system above the feeder breaker, the Distribution Engineering Department engineered the feeder system and everything downstream of it. At most utilities they evolved different methods.

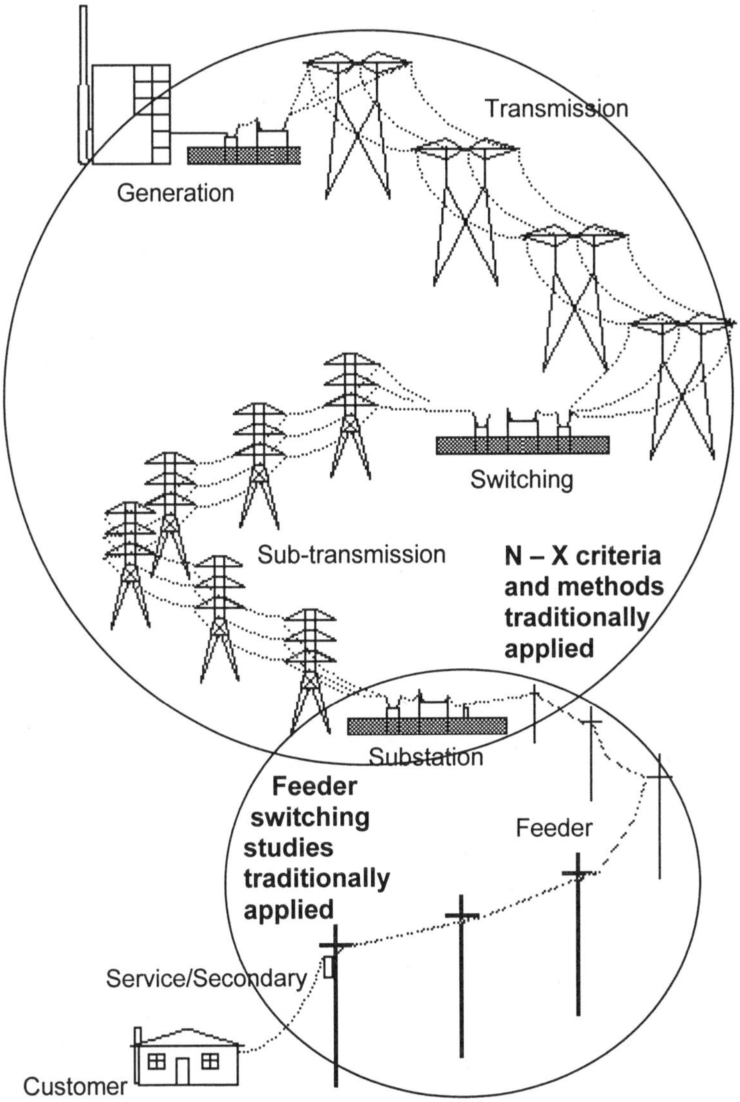

Figure 10.4 N-1 criterion (Chapter 9) is applied to the power system at the transmission, sub-transmission, and substation level analysis. Generally, feeder switching studies are done to assure "reliability" at the feeder level.

Applicability – The N–X study method (Figure 9.1) can not be applied to distribution feeder systems without significant modification of the traditional "modified load flow" approach. Distribution level application would involve representation of manual and/or automated switching operations, in precedence, which requires a much more involved algorithm, both in terms of internal complexity and user effort.

Different paradigm – Generally, at the distribution feeder level, reliability is addressed through switching redundancy design, as will be discussed below.

Feeder Contingency Switching

Traditionally, reliability-assurance methods for feeder level engineering varied greatly throughout the power industry, there being no standard or norm to which most utilities adhered. However, reliability engineering of feeders recognized certain characteristics at the feeder level:

- Feeder circuits are radial, as such, an equipment failure leads to an interruption of flow of power to customers.

- Switching can restore service in advance of repair. Switches to isolate failed sections and to switch feed of power to alternate routes are provided at points throughout the feeder system, so that line crews can restore service prior to making repairs. Figure 10.5 illustrates this basic concept.

Multiple Switches Divide Each Feeder into "Switch Zones"

Although practice varied greatly among electric utilities both within the U.S. and internationally, the traditional concept for "building reliability" into a feeder system design was to place several switches at points on each feeder to segment it, with portions transferred to neighboring feeders. Figure 10.6 illustrates this point. The tie switch near the feeder permits the feeder to be fed from an adjacent feeder leaving the same substation. (If the design of the substation-feeder system is sound, the adjacent feeder will emanate from a low-side bus different than this feeder's, permitting service to be restored with this type of switching, even if the outaged feeder's low-side bus is out of service).

Alternatively, a second feed option is usually available on most feeders. The tie switch at the end of the feeder permits it to be supported from a feeder from another substation, as was shown in Figure 10.5.

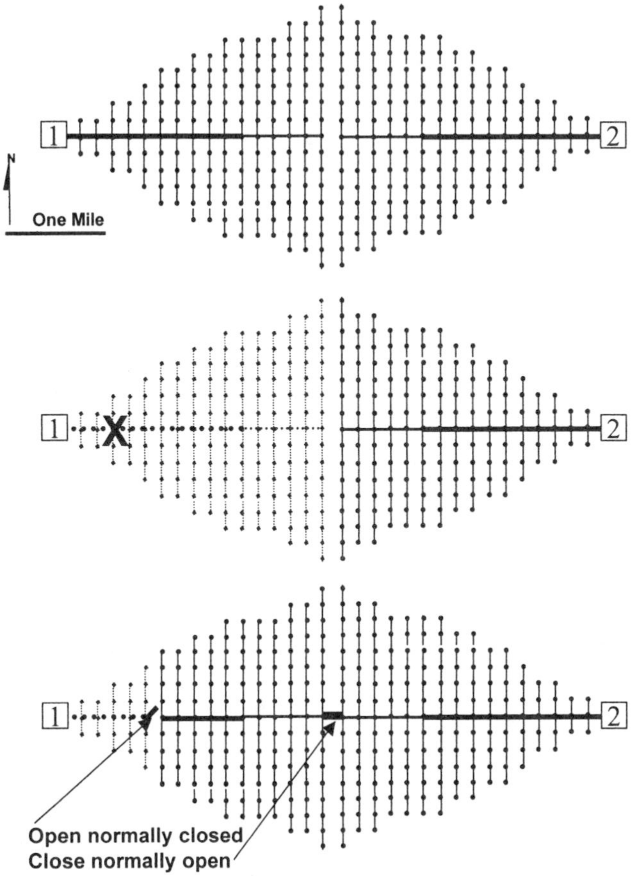

Figure 10.5 Example of feeder switching. Top, feeder from substation 1 and feeder from substation 2 are both in service. In the middle, a line failure at the location shown interrupts service to all customers on line now shown as dotted (the entire feeder). Bottom, a normally closed feeder switch is opened to isolate the failed section, and a normally open feeder tie switch between the two feeder end points is closed, restoring service to a majority of the customers on the outaged feeder. Only those portions shown as dotted lines are now without service (they must await repair of the damaged line). After repairs are made, the switching sequence is reserved to go back to normal configuration.

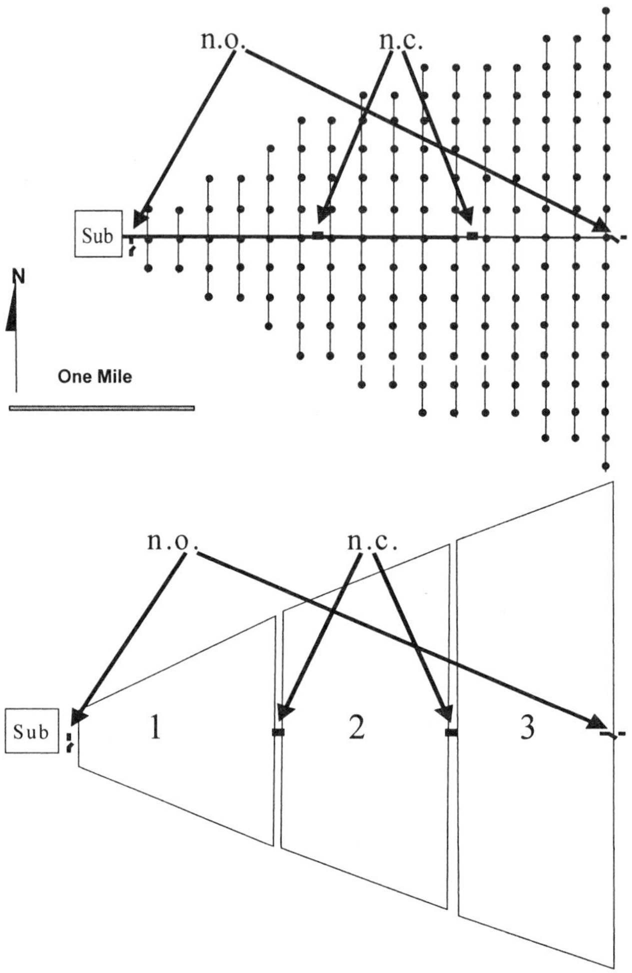

Figure 10.6 Top, a single feeder in a feeder system, in this case one of four identical (symmetrical) feeders from a substation. It has normally closed (N.C.) and normally open, (N.O.) switches which tie it to other feeders. Bottom, these four switches segment it into "switch zones" in which each is isolated (if a failed line section is in that zone) or transferred to adjacent feeders during a contingency, so service is maintained despite the failure in one zone.

Voltage, Loading Criteria, and Line Upgrades for Feeder Contingencies

During a contingency, when loads have been transferred, the feeder picking up additional load will experience higher than normal loading levels and voltage drops on the re-switched portion of its system. Referring to Figure 10.5, bottom, in that switched contingency configuration, the feeder from substation 2 is supporting roughly about 180% of its normal load (all of its customer demands and about 80% of feeder 1's). Current flow on its trunk is thus 80% higher than under normal conditions, and voltage drop correspondingly 80% higher. In addition, the power flow to the portion of feeder 1 that has been picked up during the contingency is over a far longer route than under normal conditions – voltage drop will be even more than 180% of normal: it might reach over two times the normal voltage drop. Figure 10.7 illustrates this.

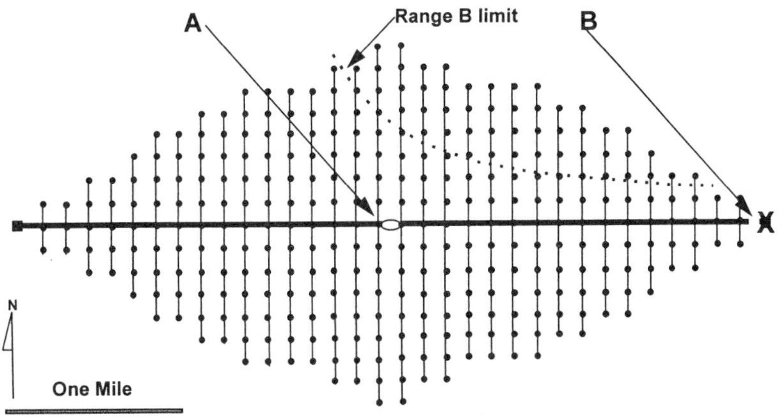

Figure 10.7 Two large-trunk feeders tied together for contingency support. The feeder on the right has experienced a failure of equipment at the substation (shown by an X). Switch B has been opened, switch at A has been closed, so that the feeder on the left supports both sets of feeder loads. The feeder on the left is now serving twice its normal load. Even though its conductor has some margin over peak load, the 200% load pushes it far beyond acceptable loading limits. Dotted lines show the points where voltage drop reaches range B limit using normally sized conductor. Voltages are symmetrical about the trunk, but the profiles are shown only for the northern side. This means voltages beyond the dotted line are unacceptably low, even when compared to the already relaxed standards applied during contingencies. Thus, to avoid unacceptable loading levels and voltage drops during contingencies, the conductor size on both feeders must be increased to a larger wire size, far beyond that needed for normal service.

Again, although standards and practices vary greatly throughout the power industry, often utilities will reinforce the trunk of each feeder (use larger conductor or cable than needed to serve the peak demand), so that capacity and voltage drop capability are available for contingencies. This reinforced capacity is not needed during normal operations; it exists purely to augment contingency capability. Its cost is justified on the basis of reliability improvement

Generally, utilities also apply different standards for what is acceptable loading and voltage drop during contingencies, than they do for what is acceptable under normal circumstances. While practices vary, many utilities use ANSI C84.1-1989 range A-voltages as design criteria for normal conditions on their system, and range B for contingencies scenarios. Range A specifies a 7.5% maximum voltage drop on the primary feeder system (9 volts on a 120 volt scale, maximum) during normal conditions (no failures). Range B, which applies during contingencies, permits voltage drops of up to 13 volts on a 120 volt scale (10.8% voltage drop).[1]

Similarly, loading criteria during emergencies typically permit excursions past the limitations set on expected loading during normal service. A particular conductor might be rated at a maximum of 500 amps, a substation transformer at 24 MVA for normal service – whenever everything is operating as planned. Under emergency or contingency operation, however, the conductor might be permitted a loading of 666 amps (133%) for six hours, and the transformer up to 30 MVA (125%) for four hours. While not all utilities follow these specific standards, most follow something similar. The establishment and application of emergency loading and voltage standards is a recommended practice, so that while the same rigid requirements of normal service need not be met during contingencies, some minimally-acceptable standards do have to be met, to provide a uniform target for contingency planning and operations.

Traditional Engineering and Planning Feeder Systems for Reliability

Reasonable levels of feeder system reliability can be assured engineering each feeder into an appropriate number and location of switchable zones, and arranging the line pathway of neighboring feeders so that they have the capacity to support these switchable zone loads [Willis, 1998].

During the latter fourth of the 20th century, practices at the feeder system level varied tremendously amount traditional regulated, vertically integrated electric utilities. The quality of engineering and planning, and the amount of scrutiny and detail paid to reliability design at the feeder level, varied greatly over the industry. Some utilities had tough standards and rigorous engineering

---

[1] See the IEEE Red Book -- *Recommended Practice for Electric Power Distribution for Industrial Plants.*

methods. Most, however, applied simple formula methods based on topology and standardized layouts.[2]

Weaknesses in Traditional Systems

As was the case at the sub-transmission - substation level discussed in section 9.2, as budget pressures increased throughout the 1980s and 1990s, utilities responding by cutting back on reinforcement of their feeder system as their loads grew, and accepted higher utilization ratios (peak load/capability ratio) as a consequence. This had an impact at the feeder level that is qualitatively similar to that at the sub-transmission – substation level discussed earlier. Higher utilization ratios eroded contingency margin. Situations where contingency currents were too high, and contingency voltage drops too low, became not uncommon in many utilities, limiting their ability to restore service quickly during storms and when equipment failed, and generally leading to degraded customer availability.

Furthermore, at many utilities where budgets were cut back significantly, planners were forced to use the contingency switching capability of the existing system to accommodate load growth. When the customer demand in one feeder area grew to where it exceeded the feeder's capability, a zone would be transferred to neighboring feeders. At several utilities (e.g., Commonwealth Edison of Chicago) widespread application of this stopgap approach during the 1990s "used up" much of the contingency withstand capability of the feeder system.

Optimizing Engineering and Planning of Feeder Systems for Reliability

As was observed earlier, adequate and even superior customer service level reliability can be assured by properly engineering the switching and capacity margin of the feeder system. This level of reliability can be accomplished by engineering each feeder into an appropriate number and location of switchable zones. A key factor of success is the artful arrangement of the circuit pathways for each feeder and its neighbors so that all have the capacity to support switchable zone loads, and so that there are a sufficient number of strategically placed intersections at which to put switches. This type of engineering can be involved and can require specialized techniques and approaches to work best, but there are proven methods in well- documented sources. See the *Power*

---

[2] For example, switches were not located so they would segment the feeder into zones designed for improved reliability, but located based purely on topology, at branches points and intermediate locations along long trunks.

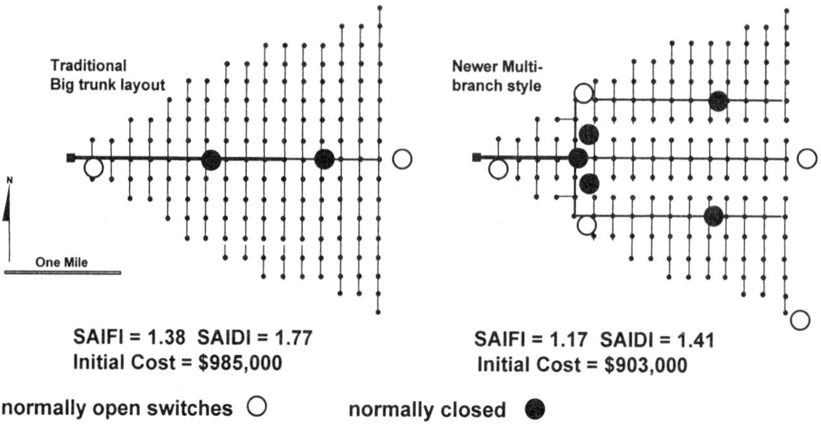

Figure 10.8 Left, the traditional large trunk feeder design, a simple to engineer and operate feeder layout based on 1930s engineering technologies. It has tie points to feeders at both ends of the trunk (Figure 10.7). Right, a multi-branch feeder serving the same area, designed to the same standards, has more switches and switching zones, with ties to similar multi-branch feeders on both sides of it, and at its end points. It is more difficult to engineer. But it but costs less and delivers better reliability. The point is that switching and layout can be engineered to deliver maximum reliability per dollar.

Distribution Planning Reference Book (Willis, 1997), Chapters 8 and 9, for a detailed discussion of switch zone engineering methods and design trade-offs. As an example of the difference such methods can make, Figure 10.8 shows a feeder layout called a short-contingency path, multi-branch feeder. This feeder has a more "dendrillic structure" (repeated hierarchical branching into more and smaller flow paths) as compared to the large trunk feeders shown in Figures 10.5 and 10.6. Large trunk feeders are the most traditional types of feeder design, having only one path that is used in contingency switching (their trunk). Large trunk designs date from the 1930s and are used, mainly due to tradition, as the layout concept for feeder engineering at about half of all utilities in the U.S.

By contrast, the multi branch feeder shown has a more complicated layout consisting of branches and sub-branches. These segment it into more and smaller zones. Basically, the feeder is broken into smaller "chunks" for contingency operation, and those chunks are transferred among several neighboring feeders during a contingency. This means:

> *A smaller portion of the feeder customer base will be unrestorable due to the failure.* The switched zone containing failed equipment can never be transferred. Service in that one

zone can be restored only by repair. Smaller zones mean a smaller portion of the customer base falls into this category during every outage. Consequently, a higher proportion of the customers can be restored through re-switching, cutting average interruption time.

Less contingency margin is needed. Loading levels and voltage drops during contingencies are far lower than with large trunk designs. Such a system is more challenging to design, although with modern electronic feeder CAD systems, this is quite within the grasp of any reasonably trained distribution-planning engineer.

Short pathways mean lower voltage drop. Although not obvious from visual inspection, the increase in MW-mile of flow during contingencies in the multi-branch shown is much less than in the large trunk designs. This means less money must be spent on conductor upgrades for contingency capacity and voltage drop capability (Figure 10.7).

High engineering and operating complexity. A distribution system composed of this type of feeder does involve slightly more work to design, and to operate well during contingencies. However, the widespread use of CAD systems for feeder planning and CAM systems for distribution operations makes such capabilities commonplace.

The net impact of the above is that the multi-branch feeder has a 12% better SAIDI and 8% lower capital cost to serve the same load, while requiring roughly 10% more engineering effort and the same overall O&M cost.

The point of this discussion is not to promote the use of multi-branch feeders (there are other types of layout types besides large-trunk and multi-branch, and all have advantages and disadvantages that need to be carefully weighed in each situation). Rather, what is important here is that feeder layout, switching, and reliability can be engineered well, using proven methods. How this is done will be summarized in Chapters 13 and 14.

Strategic Planning – Analysis and Optimization of the Utility Feeder System's "Strength"

Feeder systems can be engineered to provide only contingency support capability for feeder-level failures or to provide support through the feeder system for substation outages as well. Providing support for feeder outages – designing the feeder system so that it has switching and contingency capability to provide backup during failures on the feeder system, is certainly a best practice, and practically mandatory for, urban and suburban systems.[3] Layout, conductor size, and switching can be engineered so that all, or only a portion, of each feeder's load can be transferred to neighboring feeders.

The primary means of supporting a feeder's outage is to transfer its switching zones to adjacent feeders out of the same substation. This keeps the substation load as close as possible to actual, and is generally easier to accomplish in practice (line crews have to drive shorter distances to reach the various switches involved). It is also often considered a more prudent policy to transfer loads only between feeders of one substation, rather than between feeders served by different substations.[4]

Designing a Feeder System to Partly or Fully Support Substation Outages

If a particular feeder can be transferred so that it is completely served through a tie to a feeder emanating from a neighboring substation, then its electric consumers can remain in service during the failure of its substation (or the portion serving it). The ability to transfer the entire feeder load requires great strength among neighboring feeders in terms of feeder trunk capacity and voltage (factors such as discussed in Figures 10.7 – 10.8). A general capability to do this throughout a system – a feeder system designed so all feeders for any one substation can be transferred among its neighbors – results in what is termed a "strong" feeder system."

---

[3] Given that most feeder systems are radial, there is no other way to assure quick termination of customer interruptions caused by feeder failures. In the design of rural distribution systems, there generally is no switching capability, due to the distances and scarcity of the system. But in suburban and urban systems where there are neighboring feeders in all areas, the design of the system to switch around feeder-level outages is a given in almost all utility systems.

[4] The issue here is mostly a desire, when switching back to the original configuration after the outage is repaired, of avoiding a short-term interruption of service to any customers. To do so, the utility has to use "hot" or "make-before-break" switching. The "to" tie switch to the permanent feed feeder is closed before the "from" switch connecting it to the temporary contingency feeder is opened). For a moment during this process the transferable section has two feeds. This avoids a short interruption to the customer, which occurs if the "from" switch is opened prior to closing the switch to the other source. Hot switching between feeders served by the same substation is generally "safe" in the sense that it will cause no loop flows or overloads. But hot switching on feeders of different substations can create high loop flows, if phase angle differences between the substations, caused by circuit flows on the transmission system is large. Some utilities have a firm policy of not doing such switching between substations, and as a result, prefer to do only switching only between feeders served by the same substation.

Clearly, to permit this to work well, in addition to a strong feeder system, the substations themselves must also have contingency withstand capability to accept the additional transferred loads. However this usually is not a design problem. In fact, use of a strong feeder system permits increased utilization ratios (peak load/capability ratio) at the substation level, i.e., lower contingency margin in substation designs. If a substation transformer fails, the remaining transformer(s) at the station do not have to pick up that transformer's load – it can be distributed over many transformers by transferring some of the feeders to neighboring substations.

Often, a utility will design its feeder system to support a limited amount of feeder load transfers to neighboring substations. For example, suppose that the outermost 1/3 of load on every feeder can be transferred to neighboring feeders. This is not an undue amount of load (1/3 that of transferring the full feeders) and more importantly, is the portion of load closest to the neighboring feeders, meaning flow paths and voltage drops are lower, so that it is not typically difficult to arrange.

With this feeder system strength, the loss of one of two identical transformers at a substation can be supported, even if they are both loaded to 100% at peak, with an overload of only 33%. During the contingency, load amounting to 66% of a transformer's capacity is transferred to neighboring substations (the outermost 1/3 portion of feeders served by *both* transformers), leaving the remaining 133% to be served by the one unit in place. Overloading is equivalent to what occurs without feeder transfers, if the transformers were loaded to only 66%. The ability to transfer 33% of feeder loads results in an ability to increase transformer loading by 50% (from 55% to 100% or rating) with no increase in substation contingency stress levels.

Optimizing the amount of strength in a feeder system can be accomplished by balancing the marginal cost of feeder strength against the marginal savings in transformer capacity. Again, as throughout this book, the keys to success in gaining reliability/dollar are:

- *Systems approach* – in this case the feeder and substation levels need to be considered simultaneously, trading strength at one level against the other, in order to achieve reliability targets at minimum cost.

- *Marginal cost optimization* – as described elsewhere, the planners "buy" and "sell" reliability between levels until there is no bargain at either level.

In any real system, application of this design approach results in areas of the system with stronger and weaker feeders, and weaker and stronger substations

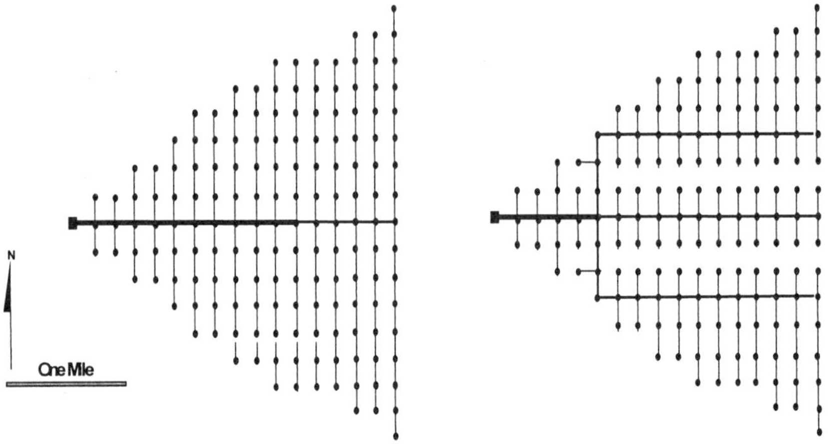

Figure 10.9 Feeder configuration evolution. By selectively reinforcing segments and changing configuration, an existing large-trunk feeder (left) can be modified, or through gradual changes, evolved, into a multi-branch feeder with greater inter-substation tie capability (right). Engineering such changes so that they both work well, and fit within constraints of route and duct size, etc., and fit budget and work management schedules, generally requires a modern feeder system CAD software package with optimization capabilities. But, this is well within the capability of several existing commercial, feeder design software systems.

respectively. These differences are due to constraints and factors unique to locations throughout the system. At substations where there is a limitation in capacity, a high utilization factor already, and very high marginal cost for any additions, reinforcement of reliability by use of a strong feeder system is generally the lowest cost option. Similarly there are areas where the reliability can be bought at the substation level less expensively than at the feeder level. But the fact is that the system if designed like this overall, delivers the performance required at minimum cost.

Increasing Feeder Strength to Augment Aging Infrastructures

Generally, in aging infrastructure situations it is impossible, or very expensive, to augment the substations – sites are small relative to need and tightly packed with equipment. Reinforcement of the feeder system to add strength is a good means of improving reliability overall, as it can usually be achieved within existing constraints and at lower cost. Such design, called configuration evolution, is not practiced by most utilities but is feasible and long proven in numerous cases (Figure 10.9).

Details of configuration evolution can be found in the *Power Distribution Planning Reference Book.* The important point here is that the reliability of a system, aged infrastructure or not, can be managed through artful use of modern reliability-engineering methods.

10.4 CONCLUSION

Distribution systems often provide fertile ground for cost-effective improvement of power system reliability. This is not only due to the fact that many power quality problems start at the distribution level and are therefore most effectively solved there. It is also because the distribution level can be used to provide load transfer capability between substations, thus reinforcing weaknesses in the overall customer service reliability chain due to weaknesses at the sub-transmission and substation levels. A distribution system's strength – its ability to transfer load between substations – can be tailored as needed at locations throughout the power system.

In aging infrastructure areas, reliability-based augmentation of the distribution system is a particularly attractive option for improvement of overall customer service quality. Basically, the cost of "buying" reliability improvements at the primary feeder level is often far less than at the sub-transmission and substation levels, which are far more constrained. The distribution has a small granularity – it can be upgraded in stages – and essentially evolved into a system designed to bolster sub-transmission/substation level weaknesses. This is not to say that distribution can provide all, or even most, of the "cure" needed in aging infrastructure areas, but it is one of the most effective among the many options which a utility can assemble into a viable overall plan.

Explicit Distribution Reliability Methods

Distribution planning and methods traditionally used at most utilities applied standardized concepts, often not completely articulated, but instead institutionalized as tacit knowledge or culture, for how the system should be laid out, tables used, and guideline-driven design methods to simplify effort and engineering. While adequate to meet the traditional expectations for performance of the distribution system, these did not address reliability directly nor get the maximum possible from the distribution system in terms of reliability interaction with other levels, or economy.

For these reasons, modern planning needs are best met using power system design and planning techniques that directly address reliability performance.

The most important points are:

1. *Systems Approach* – planning of the distribution system must be completely coordinated with the sub-transmission and substation levels. The goal is not to design an optimal distribution level, or substation level, but to design an optimum system. Any one portion of that whole might be decidedly "non-optimum" until its context within the entire system is taken into account.
2. *Balancing of cost vs. performance* among multiple levels of the system using adjustment of design based on achieving equivalent marginal cost of reliability. This results in near-optimum use of limited funds to "buy" reliability improvements.
3. *Engineering of configuration and switching* – a wide variety of possibilities in the overall layout of a feeder exist and should be reviewed in order to pick the best approach for the needs in each area. Very often standards at a utility will be so tightly institutionalized that engineers are not aware of neither other options nor the advantages they could obtain with flexibility of configuration and layout.
4. *Reliability-based engineering* – explicit rather than implicit methods for reliability engineering of the distribution system. Applications of these methods to power systems provided a much more dependable design method to achieve operating reliability.

Dependable, practical, methods for applying these concepts to aging infrastructure (and others types of) power systems have been proven in numerous applications. While these are different than traditional methods, and require more skills and somewhat more labor, they are often essential in order to achieve any type of meaningful performance improvement from the distribution system within a reasonable budget limit. This chapter served only as a summary of modern distribution planning methods. Additional details are found in the references below, particularly in those by Engel and Willis

REFERENCES

ABB Power T&D Company Inc., *Electric Power Transmission and Distribution Reference Book, Fifth Edition,* Raleigh, 1998.

M. V. Engel, editor, *Power Distribution Planning Tutorial,* IEEE Power Engineering Society, 1992.

Institute of Electrical and Electronics Engineers, *Recommended Practice for Electric Power Distribution for Industrial Plants* (the Red Book), Institute of Electrical and Electronics Engineers, New York, 1994.

Institute of Electrical and Electronics Engineers, *Recommended Practice for Design of Reliable Industrial and Commercial Power Systems* (the Gold Book), Institute of Electrical and Electronics Engineers, New York, 1991.

H. L. Willis, *Power Distribution Planning Reference Book,* Marcel Dekker, New York, 1997.

United States Department of Energy, *Power Outages and System Trouble (POST) Report,* March 2000, Washington DC.

11
Equipment Condition Assessment

11.1 INTRODUCTION

Assessment of the condition of power system equipment, and decisions about what service and utilization is appropriate for each unit, begins with the inspection, testing, and diagnostic methods covered in Chapter 6. These basic tools for *condition assessment* provide the data, which when interpreted, leads to decisions regarding:

1. Whether the unit is suitable for service or must be withdrawn and repaired,
2. Whether more comprehensive, invasive, and expensive tests are justified before making such a decision.
3. The condition of the unit – if and how it deteriorates with age and service and its viability for future service.

This chapter discusses the testing needs and approaches for each major class of power system equipment, and how they are organized into a program of condition assessment. Sections 11.2 through 11.6 discuss in turn the testing and diagnostic needs of various categories of equipment, in sections for transformers, breakers and switchgear, underground cables and equipment, overhead lines and equipment, and service transformers. Section 11.7 discusses assessment analysis methods – using test results to reach conclusions about equipment condition – and draws some conclusions about what testing and diagnostics can and cannot do. Section 11.8 concludes with a comparison of needs among the various classes of equipment, and a summary of key points.

To be effective, inspection, diagnostics, and testing must be applied in a carefully coordinated program that also uses the results from system operation (trouble reports) to identify overall equipment condition and performance, and that prioritizes preventive maintenance, repair, rebuilding, and replacement. Optimum results are derived by then loading equipment to levels determined based on its assessed condition, the loss of lifetime caused by the usage, and the available economic alternatives. Procedures for organizing and managing such programs will be discussed in chapters 12 and 15. This chapter focuses on condition assessment inspection, diagnostics, and testing methods themselves.

11.2 POWER TRANSFORMERS

Transformers are one of the basic building blocks of power systems. They alter the voltage-current constitution of alternating current power passing through them, essentially changing the economy of scale of transmission of the power from one side of the transformer to the other.[1]

Power transformers at distribution substations constitute a major category of distribution equipment. Generally, their capacity is summed throughout a system, or at a substation, in order to determine the capability of that system. This is only one of many indications that they are considered *the* key aspect of a distribution system. Regardless, large power transformers are a major concern to any electric utility when it comes to reliability evaluation, because each one feeds large numbers of customers and its replacement would involve a considerable amount of time and expense. Much of the contingency planning of power systems relates to concerns about transformer loading limits, and arrangements for contingency operation when one fails (see Chapters 8, 9, and 13).

Transformer Aging and Deterioration

Transformers "age" in the sense that the strengths of their components deteriorate as a function of chronological time, time in service and loading, and due to severe abnormal events like through-faults and switching surges. The amount of load served has a great deal to do with the rate of deterioration. Chapter 7 discusses in detail aging and loss of life with service time.

Many aspects of a transformer deteriorate with time: 1) the insulation level of its windings; 2) its oil and its bushings; 3) the mechanical strength of its core stack and internal bracing and electrical connections; 4) the desired chemical and physical properties of its materials (anti-oxidants in the oil, corrosion resistance of paints, flexibility of insulation, etc.). However, deterioration of the insulation (of both winding and the oil itself) is the major area of concern, and the predominant area evaluated, to determine a transformer's "condition."

There are several reasons for the focus on internal insulation strength. First,

---

[1] See the *Power Distribution Planning Reference Book,* Chapter 7.

a core winding insulation failure will not only prevent the transformer from doing its job, but probably lead to a large fault causing very severe damage if not catastrophic failure. Bushing failures, corrosion, and other types of deterioration can lead to failure of the device, but rarely cause such high levels of damage.

Secondly, many other defects, when found, can be repaired or replaced rather quickly and inexpensively. Bushings can be replaced, tap changers repaired, and most non-core/winding related equipment serviced while the unit is left in place but de-energized. However, the winding insulation is both the most difficult and the most expensive item to repair in a transformer. The unit must be withdrawn from service for a long period of time and returned to a factory or refurbishment center for what is essentially a complete rebuild.

The insulating oil in a transformer, if significantly degraded, can be reconditioned or replaced. This will improve the oil and have a positive effect on the condition of the winding cellulose insulation. But in some cases the winding insulation is sufficiently deteriorated so that replacing the oil alone will not necessarily make a satisfactory improvement in the transformer as a whole.

For all these reasons, the main focus of power transformer condition assessment is on insulation strength, the various tests focusing on direct (measures of dielectric strength) and indirect (measures of contaminates, dissolved gas analysis) means of determining the winding and oil insulation strength. However, other portions and aspects of the unit should be checked regularly and maintained as needed.

Acceptance Tests

New units are usually tested for insulation resistance, dielectric strength, turns ratio, amount of losses, and run through AC hi-pot and power factor tests. They may be subjected to load cycle (thermal tests). Acceptance tests are not a factor in condition assessment of older equipment and will not be discussed here.

Routine Transformer Inspection

Routine physical inspection of transformers includes : Examining the exterior of the unit for signs of leakage or corrosion of the case; examining radiator joints and grounding straps, etc., for cracked or dirty bushings, for damage (vandalism, weather), deterioration (paint chalking), and loosening of brackets or seals for attached sub-components (radiator fans, pressure sensors). It should also include testing the operation of ancillary devices such as oil pumps, fans, pressure relief valves, and the tap changer mechanism and its control system. This level of inspection should be done annually, if not on a more frequent basis.

Overall, inspection tends to focus on aspects of the transformer other than insulation strength and winding condition. Inspection may include viewing the unit through an infra-red scanner to enhance detection of hot-spots caused b loose terminations of incoming lines, leakage current, or other simil developing flaws. Similarly, it can include enhanced "audible noise" analy using a spectrum analyzer and pattern recognizer programmed to identify acoustic signatures indicative of trouble.

Routine inspection should include a thorough review of all operating records (temperature, pressure, load records) since the last routine inspection, with particular attention paid to abnormal events such as nearby lightning strikes, through faults, and storm damage.

Depending on the results of these inspections, follow up tests or diagnostics aimed at determining more precisely the condition and any necessary maintenance may be required.

Routine or Periodic Tests

Along with routine inspection, certain tests are done on a periodic basis. These tests have two uses:

1. "Good to go" evaluation. The tests determine if any measured values deviate from their permissible or recommended ranges. This indicates a need for further testing or for maintenance, and caution in using the device at high loadings. Values above a certain event lead to recommendations that the unit be withdrawn from service.
2. Condition monitoring. Periodic values are compared to identify long term trends – a particular value (e.g., dissolved methane in the oil) could still be below the maximum permitted – but significantly higher than its value in the last test, indicating a possible problem just beginning to develop.

Many of the most common periodic tests and diagnostics for a transformer focus on measuring some aspect of insulation strength, or look for the products or signs of insulation and/or oil deterioration. Table 11.1 lists the periodic inspections and tests that are done on power transformers. The frequency of tests shown are the authors' impressions of typical industry practice (not necessarily recommended practice) with regard to units that are not thought to be high risk. Units that are suspected of having problems or a high potential for failure should be checked much more often – in extreme cases on a daily basis if not continuously (with real time monitoring).

Four Categories of Risk/Recommended Evaluation

IEEE Standard C57-1004-1991classifies oil-filled power transformers into four categories or conditions of risk based on the results of dissolved combustible gases tests in their oil. Within each of the four categories, the standard then recommends specific inspection and test intervals that vary depending on the *rate* of change of the gas content as measured from interval to interval. It also contains recommendations for use of units and the time when they should be

Table 11.1 Inspections, Tests, and Diagnostics Performed on Power Transformers

Monitor for . . .	
Liquid level	Continuously
Load current	Continuously
Temperature	Continuously
Voltage	Continuously
Inspections and Tests	
Exterior for signs of damage, deterioration	Routinely, quarterly
Interior for signs of damage, deterioration	5 – 10 years
Ground connections	Semi-annually
Lightning arresters	Semi-annually
Protective devices and alarms	Semi-annually
Radiators, pumps, valves, and fans	Semi-annually
Tap changer function	Semi-annually
Other exterior ancillary devices	Annually
Solid Insulation	
Hi-pot (AC)	5 years
Induced voltage	5 years
Insulation resistance	1 - 3 years
Power factor	1 - 3 years
Polarization index and recovery voltage	1 – 3 years
Insulating Oil	
Acidity	Annually
Color analysis	Annually
Dielectric strength	Annually
Interfacial tension	Annually
Power factor	1 – 3 years
TCGA	Annually
When Condition Is Suspect	
All inspections and tests above	Immediately
DGA (gas chromatography)	Problem identification
Insulation resistance	If winding faults suspected
TTR	If winding faults suspected

Table 11.2 IEEE Standard C57.104-1991 Power Transformer Risk Conditions

Condition	Meaning
1	Transformer is operating properly.
2	Greater than normal gassing. Suspected higher deterioration rate. Additional monitoring and testing is warranted.
3	High level of decomposition has almost certainly occurred. Internal faults are likely. Periodic gas analysis should be used to establish a trend. The unit may need to be scheduled for repair.
4	Internal faults are very likely. Continued operation may result in failure.

Table 11.3 Summary of Risk Categories and Recommended Actions Based on TCGA Tracking, from IEEE Standard C57.104-1991

Risk Condition	Defined by TDCG (ppm)	Rate of Increase (ppm/day) . . .	Means Operating Recommendations of . . .
1	< 720	< 10	Good for normal operation, and test again annually
		> 10	Good for normal operation, but test again quarterly
2	721 – 1920	< 30	Exercise caution in use and do further tests soon
		> 30	Exercise caution in use of unit and do DGA tests to determine causes.
3	1920 – 4630	< 10	Schedule an outage to check unit, and test monthly until then. DGA tests.
		> 30	Exercise extreme caution in use of unit. and test weekly
4	> 4630	< 10	Exercise extreme caution and test weekly, schedule repair soon.
		>30	Consider removal from service now, and test daily until then

Table 11.4 Limits to Amounts of Individual Gases in Each Category

Risk cond.		H_2	CO	CO_2	C_2H_2	CH_4	C_2H_4	C_2H_6	TDCG*
1	below	100	350	2,500	35	120	50	65	720
2	up to	700	570	4,000	50	400	100	100	1,920
3	up to	1,800	1,400	10,000	80	1,000	200	150	4,630
4					above level 3				

The total dissolved combustible gases total does not include CO_2, which is non-combustible.

removed from service, based on the gas test results. Tables 11.2 – 11.4 summarize that standard.

The full standard gives more detailed breakdowns of recommended testing frequency versus TDCG totals and increase rates than shown here. Table 11.2 summarizes the meaning of the four categories, which are defined by the totals shown in column two of Table 11.3. The table also shows how rate of increase data is interpreted in each of the four categories. Table 11.4 gives the concentrations of various gases (which must be determined by DGA) that are considered tolerable within each category.

Thermal Load Tests

The electrical losses created inside a transformer produce heat, which leads to elevated temperatures unless the heat is moved away from the core with some means of cooling. If the unit has internal problems that create higher than expected amounts of heat, the unit might operate at temperatures noticeably higher than it should for a given load. Sources of excess heat include short circuits, severe sludging problems, damaged radiators or damaged pumps. This could shorten lifetime significantly and bring about premature failure.

Units can be "laboratory tested" to determine their temperature versus load characteristic by running test currents of known amounts through them for known periods of time, while measuring their internal temperature rise. However, such tests are expensive (the unit has to be de-energized, taken to the lab, and left there for a considerable period).

Nearly the same accuracy in determining thermal behavior of a transformer can be carried out in the field. This is done by monitoring load, ambient temperature, and temperature inside the unit on a periodic basis (every 15 minutes) and then calculating the thermal time constant and the maximum rise expected at full load from the resulting data (Figure 11.1). The actual computation is straightforward, although it is generally best to gather much more data than theoretically sufficient – several weeks of readings. Use of an analytical method based on signal processing, rather than time-series analysis methods, is also recommended.

This computed "thermal response" signature of the transformer will reveal whether the unit is performing up to its design specifications or if thermal capability is degraded for some reason. Typically, such field tests will not indicate the cause of the problem, only that a problem exists. For example, Figure 11.1 gives no indication whether the deviation from expected performance is due to a degradation in cooling capability (due perhaps to sludge buildup or blocked cooling ducts), or to creation of greater amounts of heat inside the unit due to a defect such as an internal fault.

The causes of thermal problems have to be traced through other inspections and tests. These begin with an inspection and test of the entire cooling system –

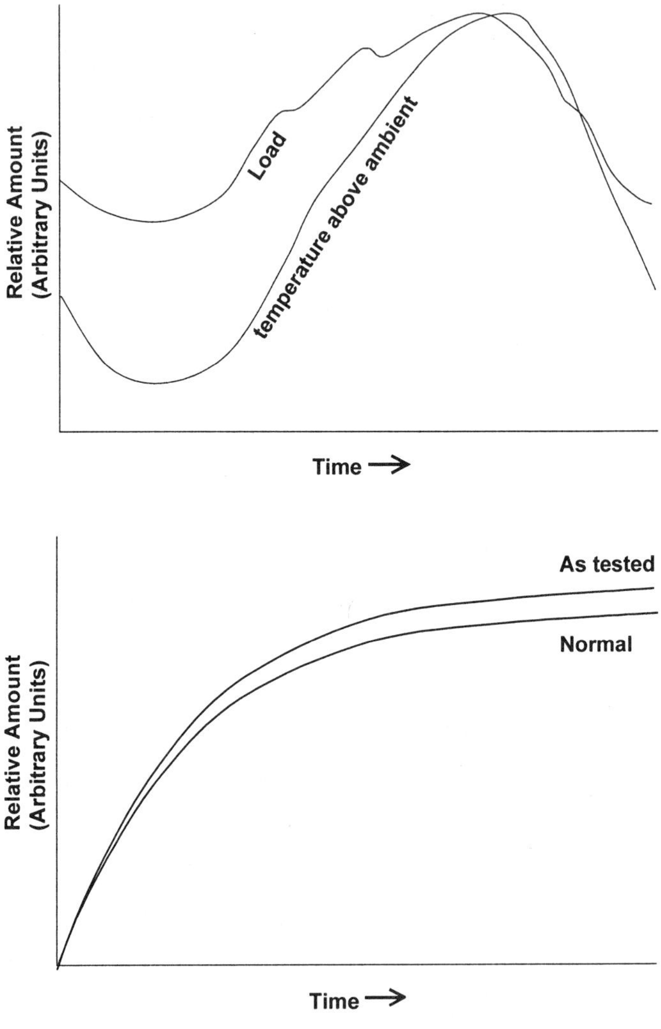

Figure 11.1 A thermal load "test" of a power transformer. Top, electrical load and temperature rise above ambient are monitored over a 24-hour or longer period. Bottom, de-convolution of the two time-series produces a calculated impulse response function (temperature rise and time for an arbitrarily short pulse of loading) for the unit. This can then be used to compute the step response shown, which is the rate of rise and final temperature that would result from having the unit serve a load equal to its rating on a continuous basis. This example reveals an as-tested signature that both have a higher rise time and asymptote (high eventual temperature), indicating some type of internal problem.

pumps, fans, radiators. If these are fine, then tests for sludging, such as acidity and interfacial tension tests of the transformer oil, are recommended.

11.3 SWITCHGEAR AND CIRCUIT BREAKERS

Circuit Breakers

Circuit breakers are electromechanical devices. They are tested in both mechanical and electrical performance and for signs of deterioration.

Routine inspection

Physical inspection aims to identify signs of past, present, or forthcoming problems, and is carried out routinely (every quarter or semi-annually). The unit is visibly checked to identify any noticeable corrosion, deterioration or damage, or outward signs of internal malfunction (leaking oil around seals). Physical inspection of the device usually includes visual or infrared examination of the exterior of the unit while in service. Inspection of internal parts of a circuit breaker (condition of contacts, etc.) is a big part of that maintenance, but not normally considered part of "inspection" because it is done only after de-energization and disassembly.

Records should be kept on routine inspections, so that gradually deteriorating paint, corrosion problems on a particular unit, etc., and other similar problems can be tracked over time.

Temperature rise test

This can be viewed as a form of inspection and must be done while the unit is in service. The temperature of the breaker is measured accurately and correlated with load. Significant rises in temperature can identify internal electrical problems. Use of an infrared scanner can aid in this test, particularly for determining where in the unit the additional heat is coming from.

Electrical tests

All electrical tests done on breakers require de-energization, but can be carried out in the field if recommended safety precautions are taken, and if instructions for the test are followed properly. If they reveal problems, the unit will have to be disassembled and visually checked, or its components tested, to determine the problem and its fix.

Standard electrical tests are performed on circuit breakers. An insulation resistance test applied twice, once with the contacts open and once with them closed. An AC high potential test (or a DC version if AC equipment is not available) can be performed, but generally to only 75% of factory test voltage, so that it does not unduly stress the unit. Manufacturers' recommended procedures for such tests differ substantially depending on the type of unit. Power factor tests may be done – they are usually done on air circuit breakers.

Contact resistance tests are the only special type of electrical test for circuit breakers, in the sense that these tests are not done on other types of electrical devices like transformers. This is simply a measurement of the resistance from terminal to terminal across each phase, made with the breaker closed, and is a way of inferring the condition of the breaker contacts without disassembly of the unit. A standard ohmmeter is not used. Instead, a current source (typically about 100 A) is used in order to put some magnitude of current close to normal operating levels through the breaker to simulate actual load. Expected values for medium voltage breakers are in the range 100 – 250 $\mu\Omega$, but each manufacturer provides specific target data for each type of unit that it manufacturers. Signs of possible deterioration from this test include either a higher resistance than expected or significant deviations in the measured resistance from one phase to another.

Time and travel tests

Mechanical tests of the circuit breaker are called time-travel analysis. They are carried out with the unit de-energized, but are done on site (in the field). These are tests of the breaker mechanism to determine if it is opening quickly enough, or binding in some part of its analysis or otherwise not performing as expected. Modern test procedures use electronic time-travel analyzers rather than the older style mechanical devices, which were used for testing into the late 1980s. These new test units provide not only an accurate reading of opening and closing times for the unit, but also detailed analysis of mechanical performance. They provide plots of opening and closed speed during each phase of the cycle, and numerical tables that identify quantitatively the breaker's performance in each of several categories. Time-travel tests of a breaker include:

1. The opening and closing time
2. Contact speed and gap distance as a function of time during the cycle.
3. Contact bounce
4. Trip operation
5. Time to re-charge or re-energize the trip mechanism

Tests of ancillary and associated equipment

Tests of breakers should include, either as part of the breaker inspection and testing, or as a separate program, inspection, testing, and appropriate diagnosis of the rest of the protection system that includes the breaker. Instrument transformers, protective relays, panels, and wiring should all be routinely checked for signs of deterioration and to see that they function properly.

- *Instrument transformers* in service are tested periodically to determine that their breakdown voltage remains sufficiently high, and for ratio and phase angle performance. Other than that, they are "left alone." However, if de-energized for any lengthy period (as when the surrounding equipment goes through a long period of maintenance), and particularly if left outdoors, they should be removed and dried.

- *Protective relays* controlling the units need to be tested to determine that their entire system is functioning correctly. Testing of relays involves methods often tailored specifically to each type of relay. However, a highly recommended general procedure is to record the condition and settings of the relay system in "as found" condition at the start of each inspection/test. Many of the problems with relaying are not due to deterioration or degradation of units, but mistakes in setup, calibration, or installation. Some of these can be systematic (due to training errors or poor habits of personnel) which will be found only if accurate tracking of problems of this nature is maintained.
- *Wiring and panel equipment,* including any recorders and indicators, should be checked and assured to be in operating condition. Indicators and records should be checked periodically (e.g. every 5 years) for accuracy, and re-calibrated if need be.
- *A complete system test* (of the relay-breaker combination) should be performed at the end of the inspection/testing. The fact that all of the equipment is individually in good condition does not absolutely assure that it will function as a whole. A final test of the complete system is the only sure way to verify this.

Switchgear

Switchgear other than breakers is generally tested and inspected on the same periodic basis as breakers. Switches and other mechanical devices should be routinely (semi-annually or annually) operated. Unlike the situation with breakers, in many cases visual inspection of contacts and mechanisms is possible with much of this equipment.

- *Routine electrical tests* include only the insulation resistance tests. They are relatively easy and quick to apply in the field and put no particularly high stress on the equipment. If they indicate problems other tests might be called for. These include hi-potential tests and power factor tests.
- *Radio interference* is a very good way to identify problems due to dirty or cracked bushings, post insulators, and corrosion-caused weaknesses in connections. As mentioned earlier, a small AM

radio can detect the interference given off by some types of electrical problems.

A detailed discussion of breaker and switchgear inspection and testing procedures, including comprehensive lists of recommended procedures on a step-by-step basis and a method for a detailed RIV (radio interference voltage) test of substations is included in *Electrical Power Equipment Maintenance and Testing*, Chapters 7 and 8 [Gill].

11.4 UNDERGROUND EQUIPMENT AND CABLES

Underground cables

Underground cable is available in a wide variety of types, including three-phase, single phase, either carrying or not carrying a concentric neutral, and with various types and degrees of "armor" or protective sheathes. Cables are made from a diverse range of materials, including copper, aluminum, and alloys for the conductor. Insulation and construction may be composed of any large set of materials with somewhat differing properties, and different construction methods, including paper, cotton derivatives (varnished cambric), natural or synthetic rubbers, and other materials. Cables also vary in the type or material and construction of their outer sheath.

With all of these types of materials and construction, there is no one type of inspection or testing program that is applicable to all types of cable. Each manufacturer provides information or recommendations. Table 11.5 lists general categories of inspection and testing and the typical periods of their application.

Acceptance tests

Acceptance tests are usually performed on cable when received from the manufacturer. This includes insulation resistance measurement and DC high potential tests. Acceptance tests often include a series of destructive tests on samples of the cable, carried out in a laboratory, to verify it's withstanding – capabilities.

Routine inspection

It is impossible to visually inspect the vast majority of installed underground cable simply because it is underground and inaccessible, inside ducts or directly buried. Despite this, routine annual inspection should be carried out on those portions (the ends of cable sections) that can be inspected, and on all associated

Table 11.5 Cable Inspection and Testing

Activity	Purpose	Frequency
Visual inspection	Check for visible deterioration, leaks, corrosion	6 mo.
Indentor test	Track material deterioration	1 – 5 yr.
Insulation resistance	Non-destructive test of insulation quality	1 – 5 yr.
PD test	Detects flaws/incipient failures. High stress test.	Infrequent
Hi-potential (DC)	Detects flaws/incipient failures. Very high stress	Infrequent
Fault location	Identify location of failure	As Needed

terminations and ancillary equipment. This last group of equipment is particularly important, because a good deal of cable problems occurs at terminations and are due to failures or deterioration of equipment other than the cable itself.

Inspection involves passive evaluation of condition based on visual (including infrared) examination of cable section ends, termination, conduit, and other equipment. Corrosion and deterioration of ground connections can be a particularly important factor for cables and deserves special attention. Cables should be checked for oil leaks, swelling of insulation. Oil circulation systems for oil-filled cables should be checked fully, including full evaluation of the condition and capability of the pumps.

Inspection may include a type of thermal analysis based on comparison of "temperature as measured" versus "temperature expected" using data obtained from monitoring of load and temperature at points on the cable. This is similar in concept to that shown earlier for transformers (Figure 11.1). However, it cannot be done quite as accurately as with transformers because the thermal factors for cables are not known quite as accurately as they are for transformers. Factors such as the thermal transfer capability deep inside the duct, etc. can only be estimated. Therefore these types of tests are useful only for tracking over time so that deterioration – deviations from a previously verified "good" performance level can an be identified.

Condition evaluation of installed cable

Underground cable failures are one of the most vexing trouble areas for modern electric utilities. Cables cannot be inspected visually; deterioration rates can be exacerbated by many factors not visible because the equipment is buried. Repair times are lengthy, and often very costly. Many types of cable have very long service lifetimes (e.g. paper insulated, lead covered), but some types of extruded cable, particularly bad batches of it, gained reputations for lifetimes far shorter than needed (i.e. less than 20 years).

Therefore, indirect means of testing cable to determine the condition or state of deterioration of its insulation and hidden or difficult-to-access splices, is a significant part of a utility system reliability improvement program. Electrical tests require de-energizing the cable, but can be done in place. The most frequently performed tests that can be carried out include:

1. *Insulation resistance tests* as described in Section 9.3 are electrical tests that can be done routinely on cable. They are relatively quick and inexpensive and cause no undue stress on the cable, so frequent testing will not degrade cable quality.

   Since the measured resistance of a good cable will depend on its length, which is different for nearly every section, these tests are useful only if they are done on a periodic basis (including ideally, when initially installed) with good records kept, so that results can be plotted over time. Unfortunately, small flaws or early signs of deterioration may not show up on a resistance test. Therefore a partial discharge or DC hi-potential test may be required.

2. *Partial discharge (PD) tests,* discussed in Chapter 6, are available from a variety of commercial suppliers of PD test equipment and services. Interpretation of the results of PD test is still something of a "black art," but a number of very clever means of analyzing timing and spatial pattern of results have been developed to infer a tremendous amount about cable condition. A particularly good discussion of partial discharge methods including diagrams of different types of partial discharge display patterns and their interpretation, is included in *Electrical Insulation in Power Systems* (Malik et al), Chapter 12.

3. *Indentor tests,* while not strictly an electrical test (they are a type of mechanical procedure) can be done while the cable is de-energized and being otherwise tested for electrical properties. These tests involve pushing a small indentor against the insulation of the cable and measuring the material's resistance to its penetration. Material will harden during service, so that this test provides a quantitative measure of material deterioration.

4. *High potential tests* can be done to stress test the cable and reveal any developing flaws. As mentioned in Chapter 6, hi-pot tests of UG cable usually use a DC rather than AC voltage approach, to avoid capacitance problems that might occur with AC hi-pot tests. As also discussed in Chapter 6, these tests are well-known for the high level of stress they can put on cable, which can worsen existing weaknesses to the point that the cable fails much sooner than it otherwise would. Therefore, these tests are usually done on a non-routine basis.

5. *Destructive or forensic laboratory* testing is often carried out on

samples of cable taken out of service. While this type of examination is frequently done only on the "good portion" of failed cable sections removed after a fault, it is often performed as part of a random sampling process for UG cable. Portions of cable are destructively tested (to the point of insulation failure) in controlled laboratory conditions. Other portions are sliced and analyzed under a microscope for evidence of treeing or other deterioration in the insulation.

Fault Location

When an underground cable fails due to a fault or open circuit, various test procedures can be used to try to locate the fault. The traditional method is a manual sectionalization to isolate the fault, section by section, which is tedious and time consuming, and often abuses equipment.[2] Other methods include the radar method and arc-reflection method (two forms of time reflectometry) and a resonance method in which electrical signals or pulses are tracked down the cable, pulse return timing or resonant frequency indicating distance to the fault. Several methods inject a signal or pulse that excites some type of electromagnetic signal from the fault location, which can then be found using a portable location (this only works if it is possible to walk above the cable route).

Fault location is not part of "condition assessment" of cable operating in normal fashion, and will not be discussed further here. A very comprehensive discussion is included in Malik et al.

Repair and replacement

Flaws and problems found by routine inspection can generally be repaired: such things as corroded casings, poor grounds, leaking seals, non operative cathodic protection, and oil pumps not up to specification can and should be repaired immediately. Similarly, poor terminations or equipment at terminal sites that is judged to be bad can be repaired or replaced.

If and when a cable fails, the fault can be repaired with a splice, but it is often advisable to replace the entire section if post-fault analysis indicates the cable had a severe amount of deterioration. Splices have a failure rate higher than equivalent lengths of un-spliced cable. They represent a potential weak-spot in the already weak system (the remaining cable is aged and potentially at risk of failure itself).

But when PD or hi-pot tests indicate that a cable section insulation level is poor, it *must* be replaced. There is no repair for badly deteriorated cable.

---

[2] An expedient for finding faults in URD systems at many utilities involves "opening and closing" URD primary sections by pulling elbow connectors with a hot-stick. This is not recommended and most elbow connectors are not designed or rated for such duty. The stress put on the elbows is unknown and cumulatively could lead to early failure.

11.5 OVERHEAD LINES AND ASSOCIATED EQUIPMENT

Damage, Not Deterioration, Is the Major Concern

A majority of power distribution in most utility systems consists of overhead primary (MV) and overhead service, or secondary (LV) lines and equipment. The basic materials for overhead lines include poles, conductor, insulators, and their associated hardware, all of which are incredibly robust equipment. Their expected lifetimes in normal service are 50 years or greater. While deterioration is a concern, the very slow rate of change under most conditions and the very great times to failure for this equipment generally mean that routine monitoring to track the rate of deterioration is not an effective or justifiable O&M procedure.

This is not to say that condition assessment is not an issue in overhead distribution. Far from that, assessment and tracking of condition should be a high priority in many distribution systems throughout the United States, particularly in those where large portions of the system exceed 50 years of age, an age where deterioration of wood, metal, and bell insulators may have reached a point of concern. Condition assessment of overhead equipment in these areas is a critical part of good reliability management.

But one definite fact that colors all approaches to inspection and maintenance of overhead lines is that by its very nature, overhead equipment is quite exposed both to the elements and to external sources of damage such as errant automobile traffic, nearby vegetation and fires, etc. *A majority of overhead outages occur due to damage,* not deterioration, from severe weather, contact with trees, automobiles and construction accidents, forest and building fires, and vandalism. For these reasons, identification of damage, not deterioration, is a large part of inspection and testing for overhead lines.

Inspection of Overhead Lines

Generally, elements of the primary and service voltage systems are not tested with electrical tests like insulation resistance, power factor, hi-potential etc., as are substation equipment and UG cable systems. This is not to say that they could not be, and that such tests have not been done on a special or experimental basis. Such tests aim to identify deterioration of insulation, and as mentioned above, deterioration is typically not a big concern on overhead lines; damage from external causes is. And while these tests could perhaps identify some types of damage, visual inspection of overhead lines can reveal a great deal more. Visual inspection, done to a high standard and organized with good record keeping will identify damage and in fact can detect signs of deterioration in many elements of the system.

Therefore, inspection is the chief means of determining overhead system condition. The only electrical test routinely done on overhead distribution involves checking the grounding of poles and neutrals in grounded and multi-grounded Y-connected distribution systems.

Poles and Pole Assemblies

Poles are made of natural wood (treated, straight tree trunks), laminates of wood, composite wood (wood epoxy mixes), concrete, fiberglass or steel. Concrete, fiberglass, and steel poles have very long lifetimes, if their materials are properly manufactured (cured, etc.).

The vast majority of power distribution poles in use are wood. Wood poles have an expected lifetime of from 35 to 85 years, depending on the type of wood, chemical preservative and treatment method, installation, soil and climate conditions, and periodic inspection and treatment given them. Cross-arms have a similar lifetime. Deterioration takes two forms: 1) a general weakening of mechanical strength of the entire pole, due to aging, cracking, etc., and, 2) rotting of the pole near ground level due to trapped moisture. They are also subject to damage from automobile accidents, trees falling, fire and all manner of unusual situations (a property owner nearby may use the pole as one corner of a garage or barn being built, etc. Deterioration, like damage, tends to be very pole-specific. One pole can be near failure due to deterioration; those nearby might have a considerable lifetime left.

Visual inspection is notoriously unreliable in determining the state of deterioration of a wood, although it is very good at identifying damage to poles and their assemblies. Testing for mechanical strength and rotting takes several forms. The simplest, but least exact, is the hammer test. The pole is struck with a large hammer near ground level. The resulting "thump" has a very different sound depending on whether the pole is sound or rotted. This test is very quick (many poles can be checked per person-hour, and inexpensive (the equipment, a standard small sledge, costs very little). However it is not particularly quantitative. It tends to identify poles with severe stages of deterioration, but does not measure interim levels of deterioration. In addition, dependability of results has a good deal to do with the skill of the person doing the test.

A type of sonogram approach uses a device that sonically scans a cross-section of the pole (usually near ground level) to determine its density and detect any voids within. The sonogram device is reasonably portable (about the size of a small rug vacuum cleaner) and is taken from pole to pole. This test is more expensive than the hammer test, and requires more operator skill, and a good deal more time for each pole than the hammer test, but it is much more accurate in determining condition on a pole. Coring to obtain a sample of interior wood for analysis is also often used.

A type of indention-mechanical/electric resistance test can also be used. A portable device about the size of a small suitcase is attached to the pole near ground level, using a web belt to tighten it to the pole. The device then pushes two thin electrical probes into the pole, measuring the mechanical resistance to their insertion, one measure of the condition of the pole. The electrical resistance through the pole is then measured with the probes, using an ammeter. The combined mechanical-electrical values can be correlated with the pole's mechanical strength (different tables are needed for each type of wood, e.g., fir).

Deteriorated poles can be repaired by injection of strengthening and/or drying materials into them near the base or by reinforcement with sheathes or stub-poles. The most effective long-term approach however, is replacement.

Inspection of Cross-arms, Insulators, Conductors and Hardware

Overhead line equipment includes cross-arms, mostly wooden, which can deteriorate much like poles can, insulators and attachment hardware, and conductor and associated hardware like splices and vibration dampers.

- *Careful visual inspection* of a distribution line, often called "driving the line," can identify damage of cross-arms, hardware, conductor, and deterioration if it has reached advanced stages (near failure). One factor in favor of such inspection is that it is inexpensive, requiring no special equipment. However, it fails to find some incipient problems (in fact, a majority may go undetected), and it is not easy to carry out on sub-transmission or distribution lines that do not parallel roads or highways.

- *Radio interference tests.* As mentioned at the beginning of this chapter, a simple and inexpensive enhancement to visual inspection of electrical equipment exposed to the elements is radio interference monitoring. Routine policy at several utilities is to leave a radio on the AM band but not tuned to any station while driving line routes for visual inspection of equipment. Cracked insulators, corroded brackets, and broken conductor strands create low levels of radio interference, so a sudden increase in radio "static" and noise is a sign of a nearby problem. Generally, inexpensive analog radios work best for this purpose: expensive stereo systems and digital radios have filters and signal enhancement circuits that eliminate all or most static and interference.

- *Infrared scanning.* Two improvements in overhead line inspection are infrared scanning and inspection by helicopter, often combined into a much more effective and rapid means of inspection lines. Infrared scanners are essentially infrared video camera-recorders that display on a CRT an image of equipment as seen in the thermal infrared band. Intensity of items in the image is proportional to temperature, so brackets splices or conductor sections that are overheating for whatever reason show up clearly.

- *Inspection by helicopter* provides several advantages. First, a number of important deterioration modes are visible from the air better than from the ground. Cross-arms crack and rot from above where standing water tends to collect after rains. They may look fine from below but can have signs of serious deterioration if viewed from the air. Loose attachments of conductor to insulators are easier to see when the insulator does not block view of the tie point, as it does if viewed from the ground.

Second, in all but very heavily congested urban areas, a helicopter can "fly" line routes while staying close enough to visually inspect in detail. Often, it is possible to approach closely enough that service-level equipment and lines can also be checked.

Third, the helicopter can cover much more line per hour than inspection by foot or automobile. Helicopter surveys often use a video recorder in both visual and infrared bands to record the inspection for later review in detail as needed.

Replacement and Repair of Basic Line Equipment

Damage to or deficiencies in basic overhead line equipment found by inspection can be repaired on an as-identified basis by replacement of the failed or failing components. Little of this equipment is "repairable" in the sense that it can be taken down, reworked or refurbished, and put back up. Instead, replacement with new is the only alternative. The exception is conductor. Birdcaged conductor, broken strands, and similar damage can be repaired by replacement (splicing) of only the damaged part of the conductor.

One recommended exception is annealing. Experience has shown that failures or severe deterioration due to annealing at one part of an overhead line will generally indicate severe deterioration in other parts nearby. Ideally, if a conductor falls down or other problems from annealing are suspected, the entire segment should be replaced or carefully inspected at close range (from a bucket truck) for signs of annealing.

Switches and Cutouts

A portion of manually operating switches in many electric distribution systems are suspect simply because they have not been operated in years. Over time, switches will become inoperable for a number of reasons. Even a minor amount of corrosion on the contacts of closed switches, or in the operating mechanism of open or closed switches, will freeze the switch so it cannot be operated. Water working its way into mechanisms can freeze, warping rotating surfaces and jamming devices. Closing a switch into a current higher than its rating can "weld" its contacts closed, so it will not open next time an attempt is made to operate it. Overloading can reduce the temper in the switch springs, causing it to malfunction. In addition, connections sometimes just weaken or loosen.

Cutouts and fuse connections also suffer from the same types of problems as line switches. However, they are almost never test-operated. There is some likelihood that each will go bad, but given their number, the difficulty in reaching them, and that fact that any quick test would assuredly cause a customer interruption,[3] few utilities test anything but a handful of cutouts, and

[3] To test without an outage a jumper has to be placed in parallel with the cutout prior to the test, then removed afterward. While certainly within the capabilities of any well-trained line crews, this increases the time, and hence the cost, of such tests.

only at special sites, each year.

Infrared scanning of both switches and cutouts can reveal a large portion of mechanical and electrical problems. Loose and corroded connections or a partially broken conductor near them shows up as a hot spot. The corrosion that causes frozen contacts often creates higher impedance in the switch contacts, creating a hot spot at the top of the blades.

Fuses

Fuses can be damaged by overheat, nearby lightning strikes, or long periods of high loads. However such problems are very rare, and fuses have very long lifetimes in service, assuming no fault occurs that requires them to blow.

One of the problems with fusing experienced to varying degrees by some utilities is a mismatch between designed and actual protection size of fuses on overhead lines. Mismatches occur because of mistakes made by line crews, expediency (a trouble-man has no 30 amp fuses so installs a 40 amp), and "coppered out" fuses.

The last problem arises due to either laziness on the part of line crews (rare) or to good intentions, overwork, and poor coordination of line repair records (more common than is appreciated). After long outages, particularly in very cold or very hot weather, "cold load pickup" can be a nuisance when restoring service. Starting current and loss of diversity due to the lengthy outage may mean that the "cold loads" are up to three times higher than normal. As a result, on primary feeders with tight fuse coordination, the initial load seen on a branch or lateral that is closed into service after a long outage may exceed the specified fuse rating. The fuse links keep blowing when the line crews try to close the cutouts to put the customers back into service.

The proper way to "fix" this problem is to jumper around the fuse for a period of thirty minutes to one hour, until load diversity returns. However, one expedient solution frequently used by line crews working against a tight deadline (common during storms) is to replace the fuse link with something that will not blow. This might simply be a larger fuse link than specified, but even a larger fuse size (i.e., a 40 amp instead of a 30 amp) may not provide enough capacity for the initial cold load surge. However, a length of #6 solid copper, or a large strand of aluminum from AA conductor, will fit in a fuse barrel, and "cures the problem." Usually, the intent of the line crews is to return after one or two hours and change the fuse link back to the proper fuse. But due to difficult schedules and forgetfulness, a noticeable portion of such coppered-out fuses probably remain in service. Accumulated over many years, it can be a noticeable problem – engineering records and visual inspection show that fusing is installed on the system, but protection doesn't work properly.

It is impossible without pulling the fuse element to verify the size of fuse actually installed in a fuse barrel. The "solution" to this inspection problem is a detailed assessment of trouble reports from the field. Statistical analysis of outages in which protection coordination is suspected of not functioning properly can reveal the degree of the problem – the "advanced TLM" discussed in Chapter 15 is somewhat successful at this. However, nothing short of site by site inspection, will identify all problems.

Regulators, Capacitor Banks, Reclosers and Sectionalizers

These types of power system equipment are considerably more complicated than basic line equipment, and should be inspected, tested, and maintained in a manner similar to the transformers, tap changers, breakers and station capacitor banks it resembles. Routine inspection of this equipment, including visual and infrared inspection from the ground or helicopter, is typically included with inspection of the entire line.

Unlike the case with the basic overhead line equipment, electrical tests are typically applied to this equipment. Line regulators are tested in the same manner as small medium-voltage transformers/tap changers. Reclosers and sectionalizers fall into a category similar to breakers and relay sets. All of this equipment is routinely tested on something like a 2-5 year cycle for proper operation, and usually with instrumentation to assure proper turns ratios or operating speed and synchronization, as the case may be. Insulation resistance is also measured as a routine course. Reclosers and sectionalizers will have specific inspection, test and maintenance procedures recommended by their manufacturers.

Shunt and series capacitor banks have unique characteristics (high capacitance) which require special test procedures, generally involving DC potential testing and induced potential tests. Capacitor switches in some systems are particularly failure prone and are tested annually, generally just prior to peak period.

11.6 SERVICE TRANSFORMERS AND SERVICE CIRCUITS

Because they are so numerous and there is no contingency backup for each one, service transformers can be an issue in the reliability of any system. In most cases, restoration of service can be achieved only by replacement since no re-switching option exists. Service transformers and service-level circuits are predominantly "fit it and forget it" devices. At most utilities, they are not inspected, except as part of general inspection of overhead lines. This inspection does catch a good number of problematic service transformers and weak connections of service-level wiring. Many utilities also depend on meter readers and other personnel in the field to "keep their eyes open" for problems. But despite this, deterioration and damage can go undetected. Fortunately, service transformers and low-voltage circuits, whether overhead or underground, are very robust, so lifetimes even in this light-inspection environment are quite good (see Chapter 7).

One very effective "inspection" method for service transformers is a transformer load management (TLM) program. "Program" here means both a computer program and all the procedures and internal organization required collecting the data and using the information produced to manage the service transformer asset base. TLM involves gathering customer-billing data on a monthly or quarterly basis. This is input into the TLM program, and correlated with transformers (a TLM requires data telling it which customers be served through which transformers). The TLM program then estimates the peak load and peak duration, and hence the total stress on the unit, for each service transformer. Its estimate is based on the number and type of customers and the total kWh sold through each transformer during the analysis period.

The TLM does this calculation for all service transformers in the system. It then produces a sorted list of units for which the calculated stress is higher than recommended – a so called "jeopardy list." This is sent to the field so that each unit can be checked by line crews (leads can be amped to determine true loading level, the unit can be visually checked for signs of leakage caused by overheating, etc.). Overloaded units are replaced with larger capacity transformers.

A good TLM has among the highest payback ratios of any activity related to maintenance and asset management of distribution equipment. Often, it pays for itself in a matter of months, by permitting overloaded units to be changed out before high loading levels lead to premature failure. Advanced approaches like this can apply much the same concepts to evaluation of service quality and lifetime problems with the service-level circuits.

11.7 EVALUATING AND PRIORITIZING EQUIPMENT CONDITION

Theory of Condition and Correlation with Test Results

The theory of equipment aging and deterioration is well understood by the power industry. Three major factors "age" electrical equipment: voltage-frequency stress, thermal stress, and water damage.

1. *Voltage aging.* Given enough time, equipment that is simply energized (e.g. a cable kept in service but carrying no load) will fail due to the accumulated stress of voltage – somewhere, somehow, its insulation will break down. The level of stress, and hence the rate of deterioration in insulation strength, is a function of both voltage level and frequency. Insulation lifetime is exponentially related to voltage level by something between a 5th and 25th order relationship depending on type of material and design. This means doubling voltage will cut lifetime by anywhere between a factor of 2^5 (32) and 2^{25} (33 ½ million) an incredibly wide range, indicating that insulation type and design are key factors in equipment quality.

   The voltage-lifetime relationships for typical electrical insulation is in the range of 5th to 10th order. For example, XLPE insulation has a relationship that is ninth order. Doubling voltage means lifetime is cut by a factor of 2^9 (512), meaning the unit loses the equivalent of about 3 weeks of lifetime for every hour it is energized to twice its nominal voltage. Its lifetime will be cut by 35% if it is operated at 1.05 PU rather than 1.0 voltage.[4]

2. *Frequency* also plays an important role in aging. Time to failure is related to frequency, roughly on a linear basis (doubling frequency will roughly half lifetime, halving it will roughly double lifetime. Frequency is regulated so tightly – within .02% even in "sloppy" systems – that the equipment installed in any power system will see very nearly the same operating frequency all its lifetime. Basically one can assume that the equipment sees nominal system frequency during all hours of operation.

   However, voltage spikes due to lightning strikes, switching surges or other unusual transients have a very high frequency component due to the rapid rise and fall times of their transient pulse. By their very nature these events are higher than normal voltage, and that higher voltage causes added stress in and of itself. But even though brief, these disturbances can cause considerable stress on insulation due to their high frequency content. For example: assuming that lifetime and voltage are inversely related by a 9th order relationship (XLP cable), one can compute that a spike to 1.5 PU voltage, lasting one half millisecond (.0005 sec), would cause roughly 50% more stress than normal due to its higher voltage, multiplied by a factor of 16.6^9 (almost 150 billion) due to the higher frequency component of the spike. This means that this event lasting 1/2 of a millisecond ages the cable by the equivalent of about 2 years. In fact the impact may be somewhat different than this computation predicts, due to a host of secondary effects that can only be computed with an EMTP analysis.[5] But the stress will still be

---

[4] A caveat here. This does not include the effects of any current increase due to the raised voltage. Generally, raising the voltage on a cable or unit of power system equipment will also raise the current passing through it (since it is a constant or very close to constant impedance). That increases the losses and hence the rate of thermal aging. The effects discussed here are due only to voltage – increasing the applied voltage on an *unloaded* section of XLPE cable by 5%, for example, is most likely to cut time until voltage breakdown occurs due to voltage stress alone, by a factor of about 35%. If that also leads to an increase in load, the additional thermal loading could contribute further acceleration of the cable aging.

[5] This very simple computation neglects a host of electromagnetic transient (EMT) related factors that alter the induced stress, including changes in: 1) the impedance of the cable or insulators as a function of frequency, 2) the non-uniform distribution of stress that occurs as a function of high frequency, and 3) the fact that the exponential order of the relationship is itself a function of frequency.

great. Brief voltage events put undue stress on insulation.

Secondly, many diagnostic procedures such as hi-potential and partial discharge tests, and some cable fault location methods, use pulses (again, a high frequency content) or variable frequencies higher than normal. They can produce noticeably high stress on insulation.

3. *High temperature* has two effects on electrical insulation. First, some types of insulation (XLPE) soften at high temperatures to the point that plastic deformation occurs in the cable. The concentric conductor falls or "flows" through the insulation; mechanical stress on the cable may result in deformation. Minimum insulation distances can decrease as a result and failure can occur.

   Secondly, and more important for the vast majority of equipment which does not suffer plastic deformation, higher temperatures accelerate *all* types of deterioration. The different modes of deterioration and the different deterioration rates of different materials are all affected in varying degrees, but qualitatively all forms of power system equipment lifetime loss are accelerated by higher temperature. A rough rule of thumb is that lifetime of cables is halved for every 9°C increase in temperature. Transformer insulation behaves in a roughly similar manner.

4. *Combined stress of all three factors* is what all-electrical equipment in service sees, the simultaneous effects of voltage, frequency and temperature, over time. Basically, the relative stress levels from these three factors are multiplicative, meaning that each exacerbates any problems due to the others:

   Total amount of stress = C x (voltage stress factor)
   times
   (frequency stress factor)
   times
   (thermal stress factor)

   Thus, if voltage and frequency at the moment are causing high levels of stress, and the insulation is at high temperature – say 9°C higher than normal (which can be expected to roughly halve lifetime), then the accelerated rates of lifetime loss due to the voltage and frequency are doubled.

Condition Assessment from Inspection and Test Data

Go/No-Go vs. Condition Measurement

Table 11.6 reprises data from Chapter 6's Table 6.4 on the types of test and diagnostic methods available for power system equipment. Some of the tests are basically "go/no-go" tests of equipment, such as the high potential test and a transformer turns ratio test. A unit will either pass, or fail this test. Units that pass are put in service. Units that fail should be withdrawn from service.

Other tests, such as the total combustible gas test (TCGA) provide a numerical measure that is useful in assessing condition, and by tracking over time, provides clues as to the gradual deterioration of the unit and its continued capability to provide dependable service. These diagnostic tests lead to classification and fuel decision-making on future inspection, testing, and/or replacement of units. For example, TCGA is used to determine classification of transformers by risk category and to recommend inspection and testing intervals (IEEE Standard C57.104-1991).

Some tests can be viewed as both go/no-go and condition assessment. Insulation resistance can be used both as a certification test and diagnostic. Below a certain value of insulation resistance, the unit is not put into service, and the value can be tracked over time so that changes in the unit can be detected. Like it, most of the test procedures discussed in Chapter 6 provide quantitative results – numerical values that can vary over some range. For example, insulation resistance tests measure resistance in megaohms, and can return a value of from 1100 to 1500 ohms when tested on the windings of high voltage power transformer. A power factor test for cross-linked cable can give a value of from just above 0.0 to over 2.0%. Transformer turns ratio tests for a 138/12.47 kV transformer should give 11.06, but can give values below or above that ratio depending on internal faults in the high or low side windings.

Generally, go/no go tests are dependable indicators of the viability of power equipment for service during the next year. If a unit "passes" the appropriate tests, it will usually provide good service in the next year. Leave a unit that has failed such a test in service and it will very likely fail, often catastrophically but nearly always to a point of at least destroying any residual future value, in the next year.

Condition assessment' predictive capabilities

Given the wide ranges of numerical values that these tests return, it might be expected that "condition assessment" analysis would be able to translate these test results into a dependable evaluation of the condition of each transformer, cable, or other power system equipment. This would be something akin to an accurate ranking of "equivalent life" – this unit has the equivalent of 35 years ($35/40^{th}$ of normal service lifetime) left, that unit only 11 years, etc. But despite over a century of commercial power system application, electrical engineering technology is still not at this level. This is not to say that many attempts haven't been made to develop such methods of translating test results to accurate status estimation. But at best they result in "fuzzy" or inexact evaluations. The use of inspection and test data to determine the overall condition of equipment is part science but still part "black" art.

With very few exceptions, the results from any one type of test are relatively straightforward to analyze and apply. The real challenge is combining results from multiple tests into a weighted or "single valued" evaluation that is both easy to use and accurate. For example, exactly what do the results mean when a cable passes an insulation resistance test with perfect scores (equivalent to when new), but PD tests indicate it has a number of weaknesses? Certainly, a qualitative interpretation is that the cable is currently operable but that it has imperfections that will develop into flaws over time. But how does the concern for this particular problem rank with respect to concerns about other cables that had other patterns of test scores? If budget is limited (and when isn't it?) which cable will get the most attention?

Consider the data shown in Table 11.7, which lists total combustible gas accumulation rates, and insulation resistance test results, for a dozen 25 MVA 138-12.47 kV transformers tested by a large utility in the southern U.S. in the spring of 1984. Individually, each of these two tests provides a numerical value that is a measure of the "condition" of the unit. Higher gassing rates indicate faster rates of deterioration or internal damage. Lower resistance indicates deteriorated insulation condition.

Table 11.6 Power System Electrical Test Procedures

Test	Type	Where?	In Service?	No-Go	Condition	Stress Lvl
Thermal load tests	AC	Field	Yes	No	Yes	-
Resistance test	DC	Field	No	Yes	Maybe	-
Dielectric absorp.	DC	Field	No	No	Yes	-
Power factor tests	AC	Field	No	No	Yes	-
Polarization recov.	DC	Field	No	No	Yes	x
Hi-potential test	Both	Field	No	Yes	No	X
Induced potential	VF	field	No	No	Yes	X
Partial discharge	AC	Field	No	No	Yes	x
Transf. turns ratio	AC	Field	Yes	Maybe	Maybe	-
Oil acidity tests	-	Lab/Field	Yes	No	Yes	-
Interfacial tests	-	Lab	Yes	No	Yes	-
TCGA tests	-	Field	Yes	No	Yes	-
DGA tests	-	Lab	Yes	No	Yes	-
Disassembly/inspectionField		No	Yes	Yes	-	

But what do the data mean when taken together, as when plotted in Figure 11.2? What *additional* information on condition can be inferred by using the combination of the two in some analytical manner? Note that, except for unit numbers 8 and 4, there is an identifiable if rather wide inverse relationship exhibited between the two tests – gassing rate goes up as resistance goes down. Roughly speaking then, either one of these tests provides roughly the same assessment as the other (ranking of units by either factor is not that much different), all except for the two outliers, 8 and 4. Of those, unit 4 apparently has considerably degraded insulation, but a low gassing rate. It may be an old unit, with deteriorated insulation due to age, but in otherwise very good condition (inferred because it is producing very little gas). Unit 8, on the other hand, is very likely a failure waiting to happen. Although its insulation tests quite well, the unit is producing gas at a high rate. If deterioration is not the cause (unlikely given the high resistance of the insulation), then the gas is being produced by some other mechanism (e.g. internal arcing) all of which are very undesirable, such as due to internal faults or arcing.

Table 11.7 Test Data for Twelve Power Transformers

Unit	Resistance - MΩ	TCGA – ppm/day
1	1520	5
2	1455	21
3	1430	12
4	1405	4
5	1360	26
6	1510	15
7	1480	14
8	1470	25
9	1475	8
10	1480	13
11	1490	3
12	1520	<1

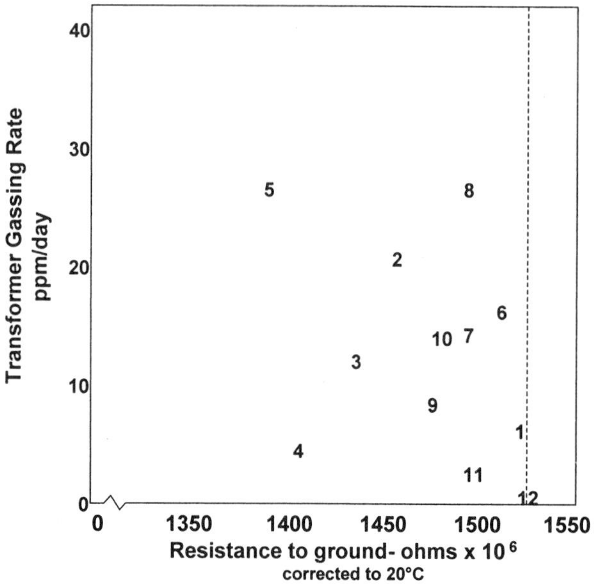

Figure 11.2 Position of transformers 1 – 12 based on their results in two tests.

Table 11.8 provides additional data on these twelve transformers, including age at the time of the tests (years in service up to 1984) and status of the unit fifteen years later (in 1999), a "result" not known at the time of the test. Note that Table 11.8's data verifies that unit 4 is indeed an older unit as was surmised from the data interpretation discussed above. Note also that that unit provided good service for (at least) another 15 years. Similarly bearing out the test results, Table 11.8 shows that unit 8, although only 7 years old at the time, was withdrawn from service within a year of these tests and completely rebuilt.[6]

Figure 11.3 plots the years of remaining service given by these transformers at their test score's position on the diagram.[7] Note that either of the tests

---

[6] The high gasing rate on a relatively good unit with high insulation scores led the utility to do a DGA test (see Chapter 6) on the unit which indicated high levels of acetylene – an indicator of internal arcing. The unit was withdrawn from service, torn down and rebuilt. During that process traces of carbonization were found.

[7] These units are part of larger group which included units moved to another location (their positions were upgraded with larger units) and units that failed due to suspected lightning fails and/or vandalism. These units have been removed from the test data, leaving, as far as is known, only a group including units that failed due to "normal" causes.

Table 11.8 Service Lifetime Data for Twelve Power Transformers

Unit	Age at Time of Test	Status 1999
1	5	S.I.S.
2	17	Failed 1990
3	22	Failed 1986
4	31	S.I.S.
5	34	Failed 1994
6	4	Failed 1993
7	11	Rebuilt '86 Failed 97
8	6	Rebuilt 1985
9	13	S.I.S.
10	9	Rebuilt 1996
11	9	S.I.S.
12	New	S.I.S.

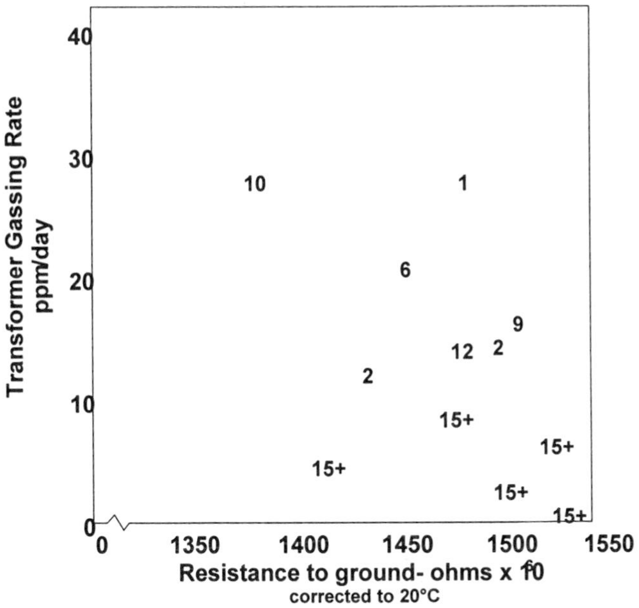

Figure 11.3 Remaining lifetime in service of the twelve units, plotted for each unit in Figure 11.2, at its position in the plot. Thus, unit 8 (highest data point vertically) had only 1 year of service remaining. Clearly, units near the lower right hand corner were "in good condition," but other lifetimes were not completely proportional to test results.

provide a fair ranking of remaining lifetime versus test results, in that remaining lifetime at the time of the plot clearly correlates, for the most part, with vertical or horizontal position on the plot. However, the correlation in either direction is only fair, not excellent. A combination of both tests helped identify problems somewhat better than either would alone. Certainly insulation resistance tests along would not have caught unit 8 as a critical unit to investigate further (it was the combination of high resistance and high gassing that indicated a real problem there).

Yet the tests, even combined, did a poor job of getting all the details right. A combination of poor scores in both tests clearly indicated transformer 5 was the "worst scoring" unit of the lot, but it lasted longer in service than units 2, 3 and 7, all of which tested much better in both tests. Unit 7 was a particular failure for the predictive capabilities of both tests. These tests gave no indication of any significant reason to think it did not have a good number of years of life remaining, like those other units nearby on the plot. It might have failed due to something other than age and deterioration (a direct lightning strike) but there was no indication of that.

Greater amounts of data improve condition assessment

The utility that owned these twelve units and actually did the tests shown above in 1984 had more test data than just the two measurements given in Table 11.8 and shown in Figures 11.2 and 11.3. It also had the total combustible gas counts, not just the rate of increase. Plotted in three dimensions (not shown), this additional data did not help improve the analytical results for these twelve units, except to verify that unit 4 had a low gas total, that thus despite its age really was in good condition.

Thus, a third set of measurements helped resolve one of two anomalies in the test results (units for which the other two test results differed as to evaluation of relative condition to the set as a whole). But it did little to improve the long-term predictive capability of the data set as a whole – it changed the prognosis for units 2, 3, 5 and 7 in no substantive way.

Periodic and Frequency Testing is Required for Efficacy

The data shown here and the results they give in application are typical. Despite careful measurement, estimation of remaining lifetime – long term-condition assessment – is inexact. This test (and others) bears out several lessons:

1. *Multiple tests* help, but mostly because it gives more opportunities to "flunk" a unit on the basis of a single test. Note that unit 8-looked fine from the standpoint of insulation resistance. But it scored so poorly on the TCGA test, that it received almost immediate attention.

2. *Explanations based on other data* help make decisions. Unit 4 was an old unit, something known to the test personnel from their property and maintenance records. That age data provided an explanation for why unit 4 had a lower resistance reading – age means deterioration means lower resistance. Therefore unit 4's reading was basically expected. Had unit 4 been three years old, the interpretation of that test data would have been far different – there is a problem.

3. *The best of tests are imperfect predictors.* As mentioned above, roughly 4 out of 12 units (2, 3, 5 and 7) in this set ultimately had service lifetimes different than might have been expected based on their test results. These strike the authors as typical results – long-term condition analysis is dependable only about 50% of the time.

Thus, the overall conclusion one reaches about using test data for condition assessment is that:

1. *Multiple tests* have value because they test for different things or test the same thing (insulation strength) in different ways, and therefore increase the number of ways that flaws or weaknesses can be detected.

2. *Good records* are a critical part of inspection and testing and can be used in conjunction with test data to improve evaluation.

3. *Periodic testing* must be done, because long-term predictive capability of the tests is far from perfect.

4. *Models of equipment behavior* and function are essential, as they permit the test results to be interpreted both for consistency and importance.[8]

This mirrors what was said in Chapters 6 and 7 when comparing modern equipment diagnostic technology to medical diagnostic capabilities. In both fields, the tests are fairly good at identifying a reason (if there is one) for immediate concern due to an impending failure, but poor at predicting how condition will deteriorate, or not, over the long term.

Basically, both the results shown here and experience gained elsewhere all indicate the same thing: that testing and diagnostics, while giving some indication of long-term condition prognosis and remaining lifetime, are basically

---

[8] Models contribute value by relating test results to how the device functions or reacts and helping identify if the results are merely important or critical. "Model" here does not necessarily mean a computer model. It could be simply the conceptual understanding that test personnel have of how a device like a transformer ages and behaves in use.

accurate only at identifying flaws and weaknesses that *currently* or soon will limit capability. Thus, the only viable way for a utility to maintain a good picture of its equipment bases' condition, is to perform frequent "check ups" on that equipment.

11.8 SUMMARY AND FINAL COMMENTS

Understand the Difference Between Condition Assessment and Failure Prediction

As mentioned in Chapter 6, exact knowledge of when a unit would fail would be worth a great deal to a utility. But even the best condition-assessment methods can not predict exactly the expected remaining lifetime, other than in extreme cases where the unit's failure is imminent. In many such cases monitoring and testing can usually determine the unit is about to fail.

In all but those "near to failure" cases, the best that good condition assessment can do is to rate the *relative likelihood* that a unit will fail. One hundred transformers judged to be in "poor condition" will see far more failures in the next twelve months than one hundred that have been assessed to be in "very good" condition. Just how poor and just how good the units have been assessed will give some indication of failure rate. The information provided is a step in the right direction and worth a good deal toward improving service quality and allowing optimal management of equipment lifetimes. But again, this condition assessment is not a prediction of expected lifetime. Instead, it can and should be used to set priorities and general operating policy.

Good Record Keeping

Good records on the service and maintenance history of major equipment should be a part of any comprehensive condition assessment and improvement program. Records of the annual peak load and duration of peak served, and kWh served through the unit should be maintained. This equipment operating record library should also contain information on the number, type, and time of all through-faults and other types of abnormal events that occurred to the unit. Similarly, records of all inspections and maintenance should be maintained along with any comments on findings during inspections and repair ("Lower northwest corner of case dented").

Record keeping has a good deal of value. Recall Chapter 6's discussion of the value that exact knowledge of future time-of-failure would be worth. Accurate prediction of the exact failure of time is probably not feasible and is certainly not within the capabilities of existing technologies. Nevertheless, an old transformer is worth considerably more to a utility, in terms of being able to estimate its likelihood of failure, if records are maintained so its operating history is available to help determine its condition. Therefore, good record keeping has real value.

The most common failing among utilities in terms of record-keeping is a lack of historical data on through-faults, which many power systems engineers (including the authors) believe is a leading cause of transformer failure.

Inspection, Test, and Diagnosis Program

The methods covered in this chapter are worthwhile only if applied in well-managed program aimed at improving and maintaining service quality at high levels while keeping costs as low as possible. The recommended method of determining priority and of setting the intervals for inspection, testing, and diagnostic evaluation is to use the appropriate approach as described in that chapter.

REFERENCES

J. J. Burke, *Power Distribution Engineering – Fundamentals and Applications,* Marcel Dekker, New York, 1994

P. Gill, *Electrical Power Equipment Maintenance and Testing,* Marcel Dekker, New York, 1998

H. H. Malik, A. A. AlArainy, and M. I. Qureshi, *Electrical Insulation in Power Systems,* Marcel Dekker, 1998

O. C. Seevers, *Management of Transmission and Distribution Systems,* Fairmont Press, Lilburn, GA, 1995

W. A. Thue, *Electrical Power Cable Engineering,* Marcel Dekker, New York, 1999.

12
Prioritization Methods for O&M

12.1 INTRODUCTION

The inspection, testing, and diagnostic techniques, and conditional assessment methods covered in Chapters 6 and 11 provide raw information on the capabilities of power system equipment and their measured symptoms of deterioration. From that, a distribution utility must make decisions about how it allocates its maintenance, refurbishment and replacement resources to best achieve its goals of providing good customer service at a low cost. This chapter examines the analytical and decision-making methods used in prioritizing these *maintenance resources,* which include inspection and testing, preventive maintenance, refurbishment and upgrading, repair, and replacement.

Three Activities, But One Decision-Making System

Any utility that wishes to provide good customer service in spite of the challenges created by an aging infrastructure must prioritize its equipment needs and utilize both its physical assets and its O&M resources as effectively as possible. Doing this involves managing three interrelated activities:

1. *Prioritization of O&M* resources including those for inspection, testing, and diagnostics, along with preventive maintenance, repair, and replacement.

2. *Prioritization capital spending,* to meet new expansion needs and for replacement of equipment that is too old or costs too much to repair.

3. *Optimum utilization* of its existing and any new equipment, including the loading level for each, which equipment will be put or left at critical locations, and what use can be made of marginal equipment.

This chapter focuses on the first of these three functions. Chapters 5, 13, and 15 discuss elements of the other two, with Chapter 15 discussing management methods for integration of all three into an overall strategy for the utility.

The three activities listed above *must* be viewed as part of the same overall managerial process if a distribution utility expects to obtain the greatest performance possible from its aging system. All three affect how the distribution company utilizes its assets. The first two activities spend money. While that money is spent on different types of expenses (O&M vs. acquisition of new equipment) in each activity, money is spent. The utility should look at how its total budget should be allocated, beginning with whether O&M or capital buys it more performance.

The third activity listed above – decisions on equipment utilization and where to use what also "spends." Equipment utilization "uses up," or spends, the remaining lifetimes of existing equipment. High loading levels get more value from equipment in the short term, but increase the loss of life rate, "spending" that remaining lifetime faster. Sooner or later, that means increased O&M costs (as equipment ages, inspection periods must be more frequent – see Chapters 6 and 11) and non-deferrable capital spending (higher utilization cuts the time until a replacement is needed).

The distribution utility's goal is to balance these three aspects of usage: to utilize the equipment so that the total value of spending, viewed from the macro level as using up its assets (equipment); using up its O&M budget, and; using up its capital budget, is maximized. "Value" in this case is essentially customer service quality.

This chapter presents both the ideas of, and a sound implementation concept for, reliability-centered maintenance (RCM). In the authors' opinion it is the best overall guiding principle for getting the most possible from maintenance resources. This is accomplished largely through the discussion of a series of example cases that become increasingly comprehensive as one works through the chapter. Section 12.2 begins this discussion with some general concepts on maintenance and reliability focus. Sections 12.3 and 12.4 then examine a large (for a textbook) RCM example in a series of evolutions from simplistic to fairly detailed evaluations. Sections 12.5 and 12.6 follow this with a discussion of several issues that are key to practical implementation. The chapter concludes with a summary of key recommendations in section 12.7.

12.2 OVERVIEW OF RELIABILITY-CENTERED MAINTENANCE

The recommended approach to allocation of O&M resources (money, people) for electric utilities faced with aging infrastructure problems is to adopt a form of Reliability-Centered Maintenance (RCM).[1] What form of RCM depends on details of the utility's particularly situation. The recommended method may not be precisely what the utility once thought of as "reliability-centered maintenance." But the basic concept – allocate maintenance resources based on how they contribute to reliability – is a sound mechanism if applied correctly.

Maintenance and Maintenance Management

The traditional view of maintenance's purpose is that it is used:

- To make certain equipment performs its intended function in a satisfactory manner
- To reduce long-term costs by servicing equipment before deterioration causes any avoidable damage
- To avoid unexpected outages by detecting failure in advance

Resources can be allocated by a utility to activities aimed at achieving any and all of these goals (inspection, testing, service, repair). There are several "philosophies" or perspectives on how maintenance can be managed.

1. *By the book* – do maintenance on all equipment on a periodic basis. The period between service and the maintenance to be done are exactly as prescribed by manufacturer's recommendations or applicable standards.
2. *By the book* (modified) – do maintenance on all equipment on a periodic basis. To reduce cost, the period between service work and the particular maintenance to be done each time are respectively the lengthiest period and leanest amount of service the manufacturer/standards will permit.
3. *Only as needed* – do not schedule maintenance, period. Only equipment that gives operational signs of needing service (e.g., tap changer non-operative) will be serviced, and only when something does not work.
4. *Reliability-Centered Maintenance.* Allocate maintenance resources among equipment in a manner that will maximize the improvement in "reliability."

Among utilities in the United States, most traditionally adopted a maintenance program that was somewhere between 1 and 2, performing all

[1] The authors will use "RDM, to denote *results*-driven management, which encompasses the intra-project, marginal benefit-cost methods described here (see page 179-180) applied to both maintenance and capital spending, and the more commonly seen all-caps designation, RCM, for "Reliability-Centered Management, which will mean here the expanded version of RCM described in this chapter.

the prescribed maintenance on equipment, but often on a slightly less frequent basis than the manufacturers would recommend.

During the 1990s, as electric utilities throughout the U.S. faced both increasing competition and uncertainty as to cost-recovery rules in the future, many made every effort to cut spending, including cutting back on maintenance budgets. As a result, there was a shift from (1) above toward (2), with some utilities adopting procedures that were very close to (3), essentially performing maintenance only when equipment failed or gave signs of imminent failure.

Always: A Budget-Constrained Situation

A point worth reiterating is that nearly every modern distribution utility, and certainly every aging infrastructure utility, is permanently in a "budget constrained" situation – there will always be more worthy, and *justifiable*, uses for resources than there are resources. A modern electric utility simply cannot afford to replace all of its aging equipment just because that equipment is old. It does not have enough funds. Therefore, the number of potential projects (capital) or programs (O&M) that are worth funding will always exceed the number that can be approved. The utility will permanently find itself in a "budget-constrained" situation, where it must make decisions between worthy projects on the basis of "bang for the buck."

Prioritization for utilities in this situation is about much more than deciding on the order in which projects will be done. It is about deciding which projects will be done, and which won't be done at all, even though those projects may be worthy and, in some sense (perhaps the traditional sense) justifiable. Prioritization in a budget-constrained situation means determining which projects and programs are the *most* worthy – which will do the *most* good – and funding only those with the limited budget.

Techniques to apply that concept to capital projects and to non-maintenance related operating costs are covered in parts of Chapters 5 and 15. Sections 5.4 and 15.3 – 15.5 emphasize how such prioritization can be made using "budget-constrained" planning methods which use *intra-project marginal benefit-cost comparisons* as a way of maximizing the "bang from the buck" obtained from the limited budget.

That same basic concept is used in this chapter with respect to maintenance. All projects, and all alternatives for all projects, are evaluated with respect to how well they help improve customer service quality. Those that provide enough "bang for the buck" are approved. Those that don't are not.

Reliability-Centered Maintenance

As mentioned above, the real question in allocating resources when faced with service quality challenges *is how much will each activity contribute to the utility's goals, and at what cost?*

In RCM, the costs of various inspection, testing, service, repair and replacement activities that the utility could do, are all assessed on the basis of

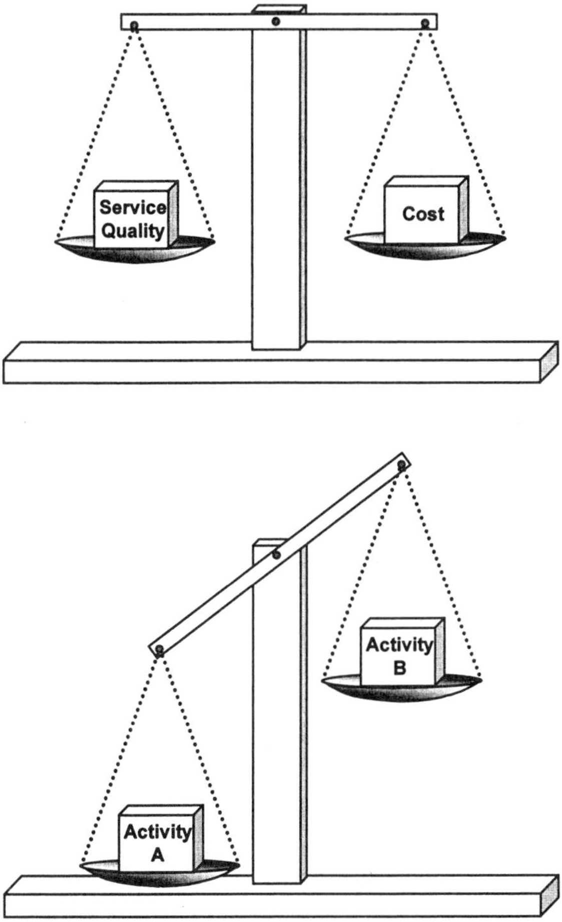

Figure 12.1 Reliability-centered maintenance does not seek to balance service quality and cost (top), although management practices used to administer it and similar marginal benefit-cost based programs may (see Chapter 15). Instead, RCM seeks to maximize bang for the buck from O&M dollars, which it does by identifying inequalities in what various potential projects and programs will provide for the dollar (bottom).

how much they will contribute to the utility's goal (good customer service) versus their cost. Essentially, the merits of the numerous possible maintenance activities are judged against one another. Various *possible* O&M activities are *in competition* with one another for funding. Winners are those that score particularly high on a "bang for the buck" basis – those that provide more customer service quality improvement for less money. Figure 12.1 illustrates this concept. RCM is depicted at the bottom of the figure. Basically, RCM involves weighing the merits of various projects and picking those with the greatest benefit at the least cost. Figure 12.1 also illustrates what RCM isn't. It is not, as shown at the top, an attempt to "balance" or optimize the total amount of maintenance budget against the value of the resulting service – so called Value-Based Maintenance (VBM). VBM has a place in the planning and decision-making at an aging infrastructure utility, but that place is as a strategic planning tool to determine overall budget, and to prioritize the total maintenance budget versus capital and other operating budgets. It will be discussed later in this chapter (section 12.4).

RCM Prioritizes Programs and Projects

As applied in the utility industry, RCM prioritizes a group of projects or programs based on some "bang for the buck" measure of effectiveness, such as reduction in customer minutes out per year, or kVA minutes not served, or SAIFI, etc. The two important points are that:

1. Effectiveness is determined by both the number of customers served and by the improvement in reliability that is expected.
2. Cost is the total cost of the service or maintenance program.

Projects are then ranked on the basis of bang for the buck and the most effective one picked. Working down such a list, a utility chooses all it can afford until it reaches a point where its maintenance budget is exhausted. The remaining projects on the list are not funded.

12.3 BASIC RELIABILITY-CENTERED PRIORITIZATION

Table 12.1 lists 35 distribution system maintenance projects or programs that cover a range of typical maintenance activities.[2] These include both major maintenance projects on "large equipment" and smaller projects, as well as a major program (pole inspection/replacement). The entire list of 35 projects has a cost of $3,174,500. However, for this example, the utility has only about 40% of this, or $1,375,000, to spend. It must select the best $1,375,000 worth of projects from out of that list.

These projects could be prioritized in any number of ways, but in this example they will be ranked on the basis of impact on customer service quality

---

[2] This is literally a "textbook example" created by the authors to illustrate well certain key concepts in RCM. However, all of the data given is both consistent with the authors' experience in actual utility systems, and with the failures rates, repair costs, and expected improvements given in the various equipment-specific appendices in the IEEE Gold Book (*IEEE Recommended Practices for the Design of Reliability Industrial and Commercial Power Systems*) and its references.

per dollar. Impact is the decrease in customer-minutes out of service that is expected from each project during the next few years, defined as

$$\text{Impact} = \Sigma R_t \times P_t \tag{12.1}$$

Where R_t = reduction in customer minutes out of service per year, in year t, expected due to the activity
And P_t = present worth factor for year t, t = 0 (this year), and years 1, 2, 3 and further into the future

The impact must be evaluated over multiple years, because maintenance projects have a positive impact on equipment reliability that lasts more than one year. This sum usually needs to be evaluated over only the short term, for the next three to seven years. Three years is used in this example, mainly because the tables would become unwieldy if more years of impact improvement were displayed. However, in actual implementation, decisions comparing replacement vs. maintenance often need to take a longer perspective, as will be demonstrated with examples given later in this chapter. The concept and method of implementation are identical, however.

Present worth factor is applied to customer outage reductions in future years, in order to adjust all factors onto a present-time basis for the current year. In this case a factor of .9 is used.

Cost in this example is total cost of the proposed activity. In all cases in this example (at least for now), the costs are operating budget items within the present year (year 0), and so no present worth adjustment or consideration of costs beyond the present year needs to be done. Programs or projects that include: a continuing cost for several years or a commitment to future expenses, or; make partial payment today with future payment of the remainder for some service or replacement done next year or thereafter, need to be reevaluated with present worth evaluation of cost. Cost being defined on the same present worth basis as benefits, as

$$\text{Cost of project } i = \text{Cost}_i = \Sigma C_{i,t} \times P_t \tag{12.2}$$

Where $C_{i,t}$ = cost of project i in year t
And P_t = present worth factor for year t

Reduction in failure rate due to maintenance

Table 12.2 shows the 35 projects listed with their new failure rates, and present and expected outage rates for the next three years with and without the proposed project performed. The failure rates used here are within the typical ranges seen for distribution equipment in various stages of aging.

Table 12.1 Example List Of Projects As Evaluated By RCM Ranking

Project Number	Equipment Type	Work Proposed	Cost of of Work	Number of Customers	Peak Load (kVA)
1	40 MVA transf. EC03	Service/Rebuild	$65,000	10,899	38,000
2	40 MVA transf. EC01	Recond. Oil/Servc.	$22,000	7,800	37,000
3	40 MVA transf. JT235	Inspect, Test, Follow up	$11,000	842	31,000
4	40 MVA transf. DT03	Inspect, Test, Follow up	$11,000	4,612	41,000
5	HV breaker # EC03	Complete Service	$22,000	10,899	38,000
6	HV breaker # EC01	Complete Service	$22,000	7,800	37,000
7	HV breaker # 221	Complete Service	$22,000	634	26,000
8	HV breaker # A0121	Complete Service	$22,000	11,300	31,500
9	LV breaker #ES06	Complete Service	$6,500	1,681	6,891
10	LV breaker #EC06	Complete Service	$6,500	2,200	7,100
11	LV breaker #NB01	Complete Service	$6,500	580	6,535
12	LV breaker #NB02	Complete Service	$6,500	650	7,800
13	LV breaker #NB03	Complete Service	$6,500	1,191	4,883
14	LV breaker #CN1A	Complete Service	$6,500	1,058	4,338
15	LV breaker #CN2A	Complete Service	$6,500	891	3,653
16	LV breaker #DT3N	Complete Service	$6,500	2,091	8,573
17	LV breaker #DT2N	Complete Service	$6,500	112	10,000
18	UG cables, feeder EC06	Replace All	$27,000	2,200	9,020
19	UG cables, feeder EC07	Test, Replace	$18,000	592	4,805
20	UG cables, netw. DT3B	Replace Complete	$43,000	245	12,600
21	UG cables, netw. DT3A	Test, Replace	$31,000	340	11,300
22	OH feeder BN06	Test, Replace	$58,000	950	4,503
23	69 kV circuits 34 & 37	Test, Replace	$170,000	22,005	103,800
24	OH feeder EC17	Test, Replace	$58,000	1,578	7,350
25	Poles, central east dist.	Test, Replace	$290,000	80,134	328,000
26	OH feeder BN04	Inspection, Replace	$43,000	1,345	5,780
27	OH feeder BN05	Inspection, Replace	$17,000	1,609	6,108
28	OH feeder CT14	Inspection, Replace	$56,000	1,671	6,875
29	OH feeder CT05	Inspection, Replace	$43,000	1,389	7,324
30	OH feeder JM07	Inspection, Replace	$86,000	1,555	4,120
31	Old Town sub feeders	Test, Replace	$210,000	22,789	72,007
32	Riverside sub feeders	Test, Replace	$335,000	28,093	85,000
33	Hillside sub feeders	Test, Replace	$219,000	14,004	45,070
34	Central sub. feeders	Test, Replace	$310,000	28,345	110,000
35	Farmton substation	Test, Replace	$905,000	43,000	124,000
Total			$3,174, 500		

Table 12.2 Example List Of Projects With Change in Failure Rates for the Next Three Years (%/yr.) Without and With the Proposed Work Being Done

Project Number	Equipment Type	Work Proposed	New	Failure Rate Now	Yr +1	Yr +2	Failure Rate if Worked Now	Yr +1	Yr +2
1	40 MVA transf. EC03	Service/Rebuild	1.00%	6.0%	6.3%	6.6%	1.8%	1.8%	1.9%
2	40 MVA transf. EC01	Recond. Oil/servc.	1.00%	5.0%	5.3%	5.5%	3.0%	3.1%	3.2%
3	40 MVA transf. JT235	Inspect, Test, Follow up	1.00%	4.7%	4.9%	5.2%	3.2%	3.3%	3.4%
4	40 MVA transf. DT03	Inspect, Test, Follow up	1.00%	6.0%	6.3%	6.6%	4.0%	4.1%	4.2%
5	HV breaker # EC03	Complete Service	0.60%	2.0%	2.1%	2.2%	1.1%	1.1%	1.1%
6	HV breaker # EC01	Complete Service	0.60%	2.0%	2.1%	2.2%	1.1%	1.1%	1.1%
7	HV breaker # 221	Complete Service	0.60%	2.0%	2.1%	2.2%	1.1%	1.1%	1.1%
8	HV breaker # A0121	Complete Service	0.60%	2.0%	2.1%	2.2%	1.1%	1.1%	1.1%
9	LV breaker #ES06	Complete Service	0.92%	2.5%	2.6%	2.8%	1.5%	1.5%	1.5%
10	LV breaker #EC06	Complete Service	0.92%	2.5%	2.6%	2.8%	1.5%	1.5%	1.5%
11	LV breaker #NB01	Complete Service	0.92%	2.5%	2.6%	2.8%	1.5%	1.5%	1.5%
12	LV breaker #NB02	Complete Service	0.92%	2.5%	2.6%	2.8%	1.5%	1.5%	1.5%
13	LV breaker #NB03	Complete Service	0.92%	2.5%	2.6%	2.8%	1.5%	1.5%	1.5%
14	LV breaker #CN1A	Complete Service	0.92%	2.5%	2.6%	2.8%	1.5%	1.5%	1.5%
15	LV breaker #CN2A	Complete Service	0.92%	1.8%	1.9%	2.0%	1.2%	1.2%	1.3%
16	LV breaker #DT3N	Complete Service	0.92%	1.9%	2.0%	2.1%	1.3%	1.3%	1.3%
17	LV breaker #DT2N	Complete Service	0.92%	3.7%	3.9%	4.1%	1.9%	1.9%	2.0%
18	UG cables, feeder EC06	Replace all	3.50%	15.0%	15.8%	16.5%	3.5%	3.6%	3.7%
19	UG cables, feeder EC07	Test, Replace	4.10%	9.0%	9.5%	9.9%	7.0%	7.2%	7.5%
20	UG cables, net DT3B	Replace Complete	2.80%	8.5%	8.9%	9.4%	2.8%	2.9%	3.0%
21	UG cables, net DT3A	Test, Replace	2.60%	6.0%	6.3%	6.6%	4.6%	4.8%	4.9%
22	OH feeder BN06	Test, Replace	4.50%	10.0%	10.5%	11.0%	7.8%	8.0%	8.3%
23	69 kV circuits 34 & 37	Test, Replace	0.09%	0.3%	0.3%	0.3%	0.2%	0.2%	0.2%
24	OH feeder EC17	Test, Replace	2.90%	17.0%	17.9%	18.7%	11.4%	11.6%	12.0%
25	Poles, CE dist.	Test, Replace	0.03%	10.0%	10.5%	11.0%	6.0%	6.2%	6.4%
26	OH feeder BN04	Inspection, Replace	3.40%	10.2%	10.7%	11.2%	7.5%	7.7%	7.9%
27	OH feeder BN05	Inspection, Replace	3.80%	11.4%	12.0%	12.6%	8.4%	8.6%	8.9%
28	OH feeder CT14	Inspection, Replace	1.70%	5.1%	5.4%	5.6%	3.7%	3.8%	4.0%
29	OH feeder CT05	Inspection, Replace	3.40%	10.2%	10.7%	11.2%	7.5%	7.7%	7.9%
30	OH feeder JM07	Inspection, Replace	3.20%	9.60%	10.1%	10.6%	7.0%	7.2%	7.45%
31	Old Town sub feeders	Test, Replace	4.50%	13.5%	14.2%	14.9%	9.9%	10.2%	10.5%
32	Riverside sub feeders	Test, Replace	6.60%	19.8%	20.8%	21.8%	14.5%	14.9%	15.4%
33	Hillside sub feeders	Test, Replace	8.10%	24.3%	25.5%	26.8%	17.8%	18.3%	18.9%
34	Central sub. feeders	Test, Replace	2.80%	7.1%	7.5%	7.8%	4.5%	4.6%	4.8%
35	Farmton substation	Test, Replace	1.70%	3.9%	4.1%	4.3%	2.0%	2.08%	2.15%

This example illustrates one highly recommended procedure. The improvements made in failure rate for each project are defined using a *formalized method* of estimating improvement (in this case a probabilistic method discussed later in this chapter). It involves classifying maintenance into categories, with each category modeled as having a certain expected impact.[3] Such an institutionalized set of standardized expectations for improvement assures uniformity of purpose and that no one "cheats" the system by using their own maintenance efficacy figures. Categories used here are the following:

1. *Complete rebuild* returns failure rate to a failure rate equal to "as new condition" plus 5% of the degradation expected over thirty years. Thus, if a new transformer (but one past its infant mortality times) has an annual failure rate of .05%, and a thirty-year old one has a failure rate of 2.7%, then a recently rebuilt one has an expected failure rate of .05% + .5 x 2.2% = .61%/year.

2. *Full service done on the unit"* cuts the amount of degradation of the equipment by two-thirds. As an example, suppose that a HV breaker has a new failure rate of .92%/year, whereas one badly in need of full servicing has an annual failure rate of 3%. Then, a recently serviced unit has a failure rate of .92 + .33 x (3.00-.92) = 1.61%.

3. *"Inspection and all repairs found to be needed were made"* cuts failure rate by 40%. This, if a group of old low-voltage breakers, all in need of servicing, has an average expected annual failure rate of 3.5%, but only .056% when new, then inspection and needed repairs reduces the value to an expected annual failure rate of .56 + .60 x (3.5 -.56) = 2.3%.

The above are rough rules of thumb used for this example only, and not recommended as hard and fast rules the reader should apply to any utility system. They illustrate that repair activities have been *classified* by expected improvement on a category or equipment class basis. No attempt is made to determine what the improvement made by a specific type of service on a specific device would be. Such categorization is highly recommended. Repair and improvements, like failure rates themselves, are probabilistic in nature. Both the concepts of failure rate itself, and failure rate improvement, are probabilistic, and their application works well only when used as probabilistic expectations applied to large sets of equipment.

Obtaining Failure-Rate Reduction Data

A concern for any utility maintenance management team will be "where do we find the data on failure rate reductions for various maintenance functions?" Ideally, these data should be developed by the utility specifically for its situation, or at least based on studies done elsewhere with judgmental

---

[3] "Expected" here means in a probabilistic sense. Failure is a random process that is best modeled probabilistically. So is improvement.

adjustment for any unique factors for the utility's situation.

This is not as difficult as it seems to many. Data is available at most utilities, and if not, then "generic" data based on published and available public domain data, such as that included in various IEEE books and publications, can be used as a starting point. The authors have never seen a situation where the appropriate data cannot be developed to the point where the RCM approach can be applied in a dependable manner.

One strong recommendation is to never use a lack of data as an excuse not to take this approach. In the author's experience, that is most often used an excuse by people who simply want to continue to do things as they have always been done. If a utility cannot develop a set of reasonable estimates of the impact maintenance makes on its equipment, on a broad category basis as required here, then it is basically admitting it has no idea of the value of the maintenance it is performing. Doing anything whose value cannot be estimated, in the present industry environment, is as extremely imprudent. A utility that does not develop failure rate reduction data has only a very weak basis upon which to justify its maintenance expenses.

Prioritization Based on Ranking

Table 12.3 shows the result of applying the formulas given earlier to the data on the 35 projects shown in Tables 12.1 and 12.2. There, the customer minutes of improvement per year for years zero, one and two is computed and a present worth sum formed. This is then divided by cost to obtain reduction in customer service minutes expected per dollar – "bang for the buck."

The projects are listed in Table 12.3 in descending order according to their merit score (improvement per dollar), given in the leftmost column. The "bang/buck ratio" for projects on this list varies from a high of 12.84 to a low of 0.01. Figure 12.2 plots the merit score (bang/buck ratio) versus total budget required to buy all projects of that score or better, for this set of 30 projects. For example, the top project on the list (merit score 12.84, has a cost of $27,000). The utility can buy all projects with scores of 12.84 for $27,000. The next project on the list has a score of 11.89 and costs $11,000. The utility can buy all projects with scores of 11.89 and better for $38,000. And so forth. The first eight projects together total only $168,000 – just 5% of the total list's cost – but provide over 25% of the total improvement in reliability from the entire list.

The plot has a "Scree" bend, a sharp change in slope between a few very worthy projects and a large number of others.[4] Here it occurs at a budget value of about $250,000. This is typical. Most evaluations of this type are dominated by a few very high-scoring projects, amidst a plethora of others.

[4] Scree is a geologic term for the pile of rocks and debris at the foot of a volcanic cinder cone or other tall "monument" rock feature (e.g. Shiprock). As the hard rock cinder cone weathers, bits of rock break away and tumble down to ground level, often rolling some distance from the cone's near vertical face, and creating the scree – a wide, nearly flat mound surrounding the near-vertical rock monument.

Table 12.3 Projects Evaluated By RCM Ranking, Listed in Descending Order of Merit

Project Number	Equipment Type	Work Proposed	# Cust	Reduced Cust-Min Out Yr. 0	Next	+2 yr.	PW Sum Cust-min	Cost (yr. 0)	Bang/ Buck
18	UG cables, feeder EC06	Replace all	2,200	121440	128436	135502	346789	$27,000	**12.84**
4	40 MVA transf. DT03	Test, Follow up	4,612	44275	48703	52681	130779	$11,000	**11.89**
2	40 MVA transf. EC01	Recond. Oil/servc.	7,800	74880	81432	87461	219012	$22,000	**9.96**
1	40 MVA transf. EC03	Rebuild Complete	10,899	222343	235749	249132	636315	$65,000	**9.79**
20	UG cables, net DT3B	Replace Complete	245	67032	71207	75341	192144	$43,000	**4.47**
25	Poles, CE district	Test, Replace	80,134	575234	625675	672069	1682718	$610,000	**2.76**
21	UG cables, net DT3A	Test, Replace	340	22195	25198	27777	67373	$31,000	**2.17**
3	40 MVA transf. JT235	Test, Follow up	842	5982	6606	7163	17729	$11,000	**1.61**
32	Riverside sub feeders	Test, Replace	28,093	177997	199134	217621	533491	$335,000	**1.59**
33	Hillside sub feeders	Test, Replace	14,004	108895	121826	133136	326379	$219,000	**1.49**
		Approval Limit Based on Budget							
31	Oldtown sub feeders	Test, Replace	22,789	98448	110139	120364	295068	$210,000	1.41
34	Central subs. feeders	Selected Replace	28,345	131634	143981	155199	386928	$310,000	1.25
35	Farmton substation	Complete Rebuild	43,000	385968	415741	443829	1119637	$905,000	1.24
27	OH feeder BN05	Inspection, Replace	1,609	5870	6567	7176	17592	$17,000	1.03
19	UG cables, feeder EC07	Test, Replace	592	5570	6348	7014	16964	$18,000	0.94
8	HV breaker # A0121	Complete Service	11,300	6265	6760	7225	18202	$22,000	0.83
5	HV breaker # EC03	Complete Service	10,899	6043	6521	6969	17556	$22,000	0.80
10	LV breaker #EC06	Complete Service	2,200	1376	1493	1602	4018	$6,500	0.62
6	HV breaker # EC01	Complete Service	7,800	4324	4666	4987	12564	$22,000	0.57
24	OH feeder EC17	Inspect, Replace	1,578	10680	11752	12714	31555	$58,000	0.54
9	LV breaker #ES06	Complete Service	1,681	1052	1141	1224	3069	$6,500	0.47
16	LV breaker #DT3N	Complete Service	2,091	811	891	963	2394	$6,500	0.37
13	LV breaker #NB03	Complete Service	1,191	745	808	867	2175	$6,500	0.33
29	OH feeder CT05	Inspection, Replace	1,389	4534	5072	5543	13588	$43,000	0.32
26	OH feeder BN04	Inspection, Replace	1,345	4390	4911	5367	13158	$43,000	0.31
14	LV breaker #CN1A	Complete Service	1,058	662	718	770	1932	$6,500	0.30
23	69 kV circuits 34 & 37	Test, Replace	42,300	5198	5708	6169	15332	$70,000	0.22
12	LV breaker #NB02	Complete Service	650	407	441	473	1187	$6,500	0.18
30	OH feeder JM07	Inspection, Replace	1,555	4777	5344	5840	14317	$86,000	0.17
11	LV breaker #NB01	Complete Service	580	363	394	422	1059	$6,500	0.16
28	OH feeder CT14	Inspection, Replace	1,671	2727	3051	3334	8174	$56,000	0.15
15	LV breaker #CN2A	Complete Service	891	310	342	371	919	$6,500	0.14
22	OH feeder BN06	Inspect, Replace	950	2508	2856	3153	7632	$58,000	0.13
7	HV breaker # 221	Complete Service	634	351	379	405	1021	$22,000	0.05
17	LV breaker #DT2N	Complete Service	112	30	34	37	90	$6,500	0.01
Totals and averages							6,158,862	$3,174,500	1.81

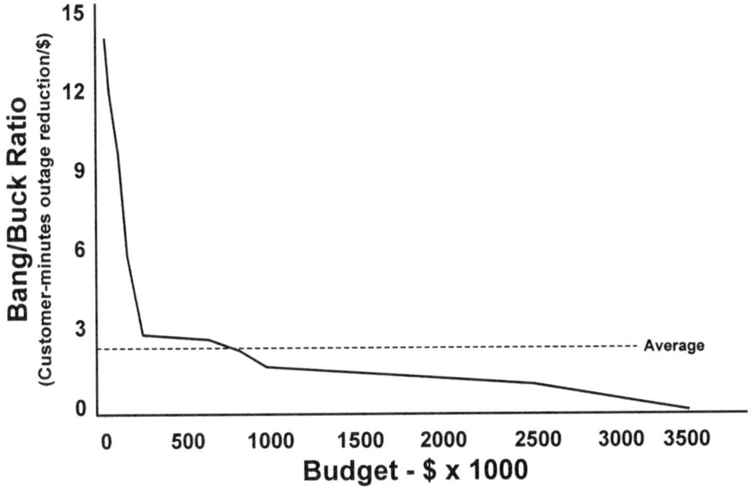

Figure 12.2 Top, merit scores of projects versus cumulative budget required to buy down to that level. This set of projects exhibits the "Scree pattern" typical of many RCM prioritization results. There are a very few extremely high scoring projects (above 3), a number that "pretty good" (3 – 1.5) projects, and a host of projects with lesser merit.

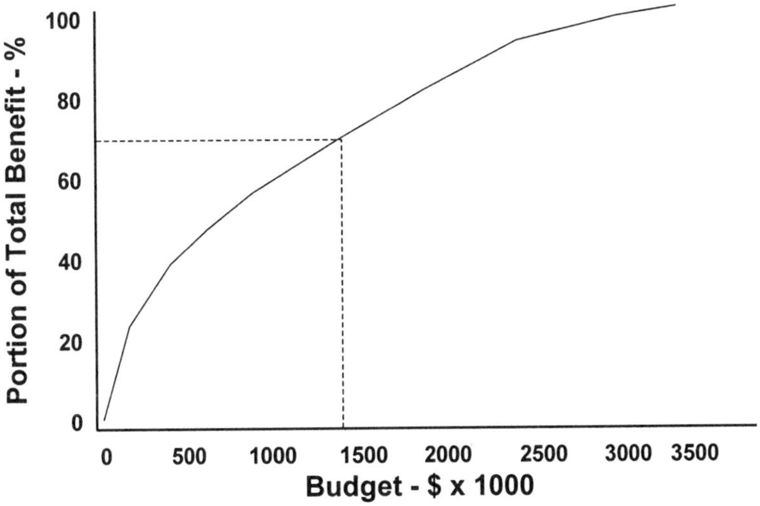

Figure 12.3 Total "bang" that can be purchased versus budget. By selecting the top ranked projects, the utility can buy over two-thirds of the total benefit available for its budget limit of only 40% of the total cost (dotted line).

Figure 12.4 plots total benefit purchased versus budget. It shows how much money is required to buy what total level of improvement. The cost of the first ten projects in Table 12.3 (those with bang/buck values shown in bold) is $1,374,000, exactly $1,000 less than the allowed budget total in this example. The utility can spend its budget on these projects – only 43% of that needed for the entire list – and receive 68% of the total benefit in customer-minute reductions of the entire list (4,152,729 customer minutes avoided out of 6,158,862 customer minutes possible from the entire list). The "cutoff value" for funding at the $1,375,000 budget limit is 1.49-customer service minutes per dollar. No other set of projects it can afford will attain as much improvement.

What This Prioritization Accomplished

RCM prioritized the proposed projects in a way that permitted the utility to rank and select projects that would use budget-dollars most effectively. Among the projects listed, no other subset will provide more value, assuming the utility's goal is to reduce customer minutes of outage, than those ranked highest on the list. For any budget limit, the utility merely has to begin selected projects from the top, working its way down the list until it exhausts it budget.

RCM, then, is basically a tool to optimize reliability-related operating activities against a budget limit.

Changing the Figure of Merit

RCM does create one conundrum for a utility. It must choose a figure of merit – a particular performance factor to minimize – in this case we chose the customer-minutes of outage. Clearly, that choice will greatly affect just what projects are selected, and what benefits the utility "buys" with its maintenance budget.

Continuing with this same example, what if the utility had selected minimization of kVA-minutes of outage, not customer-minutes of outage, as its goal for prioritization? There is considerable merit in using kVA-minutes as the decision-making criteria in RCM. Often, a circuit or transformer will serve only a small number of customers, but those customers each buy a lot of power, and spend a good deal more on power than customers who buy less. One can argue that every kilowatt these larger customers buy should be treated the same with respect to attention and quality control, as those bought by smaller customers. Further, some of these large consumers may be "more important" in terms of societal outage impact.[5]

---

[5] Opinion in many parts of the industry holds that interruption of electric service to the central core of cities and key municipal facilities is much more unfavorable than outages to most other loads, partly due to the impact such outages have on the function of society as a whole. Regardless, the criticism for poor service leveled at many utilities bears this out. ComEd was criticized severely for dropping downtown Chicago, with its large buildings, etc., areas with relatively small numbers of customers compared to their load.

Table 12.4 Projects Evaluated By kVA-Minutes, Listed in Descending Order of Merit

Project Number	Equipment Type	Work Proposed	Load kVA	Reduced Cust-Min Out Yr. 0	Next	+2 yr.	PW Sum Cust-min	Cost (yr. 0)	Bang/ Buck
20	UG cables, net DT3B	Replace Complete	12,600	3447360	3662064	3874675	9881705	$43,000	**229.8**
4	40 MVA transf. DT03	Test, Follow up	41,000	393600	432960	468325	1162607	$11,000	**105.7**
21	UG cables, net DT3A	Test, Replace	11,300	737664	837466	923193	2239169	$31,000	**72.2**
3	40 MVA transf. JT235	Test, Follow up	31,000	220224	243214	263723	652732	$11,000	**59.3**
18	UG cables, feeder EC06	Replace all	9,020	497904	526588	555557	1421834	$27,000	**52.7**
2	40 MVA transf. EC01	Recond. Oil/serv.	37,000	355200	386280	414878	1038903	$22,000	**47.2**
1	40 MVA transf. EC03	Rebuild Complete	38,000	775200	821940	868599	2218511	$65,000	**34.1**
25	Poles, CE district	Test, Replace	328,000	2354515	2560978	2750877	6887606	$610,000	**11.3**
19	UG cables, feeder EC07	Test, Replace	4,805	45205	51525	56931	137692	$18,000	**7.65**
34	Central subs. feeders	Selected Replace	110,000	510840	558756	602288	1501574	$310,000	**4.84**
32	Riverside sub feeders	Test, Replace	85,000	538560	602514	658447	1614165	$335,000	**4.82**
		Approval Limit Based on Budget							
33	Hillside sub feeders	Test, Replace	45,070	350464	392082	428480	1050407	$219,000	4.80
31	Old Town sub feeders	Test, Replace	72,007	311070	348010	380316	932335	$210,000	4.44
27	OH feeder BN05	Inspection, Replace	6,108	22282	24928	27242	66783	$17,000	3.93
35	Farmton substation	Complete Rebld	124,000	1113024	1198882	1279880	3228720	$905,000	3.57
5	HV breaker # EC03	Complete Service	38,000	21067	22734	24298	61209	$22,000	2.78
6	HV breaker # EC01	Complete Service	37,000	20513	22136	23659	59598	$22,000	2.71
24	OH feeder EC17	Replace as needed	7,350	49745	54737	59220	146976	$58,000	2.53
8	HV breaker # A0121	Complete Service	31,500	17464	18845	20142	50739	$22,000	2.31
12	LV breaker #NB02	Complete Service	7,800	4880	5295	5678	14245	$6,500	2.19
10	LV breaker #EC06	Complete Service	7,100	4442	4820	5169	12967	$6,500	1.99
9	LV breaker #ES06	Complete Service	6,891	4311	4678	5016	12585	$6,500	1.94
7	HV breaker # 221	Complete Service	26,000	14414	15555	16625	41880	$22,000	1.90
11	LV breaker #NB01	Complete Service	6,535	4089	4436	4757	11935	$6,500	1.84
29	OH feeder CT05	Inspection, Replace	7,324	23906	26744	29227	71649	$43,000	1.67
16	LV breaker #DT3N	Complete Service	8,573	3327	3655	3950	9815	$6,500	1.51
13	LV breaker #NB03	Complete Service	4,883	3055	3315	3555	8918	$6,500	1.37
26	OH feeder BN04	Inspection, Replace	5,780	18866	21106	23066	56545	$43,000	1.31
17	LV breaker #DT2N	Complete Service	10,000	2693	3000	3270	8042	$6,500	1.24
14	LV breaker #CN1A	Complete Service	4,338	2714	2945	3158	7922	$6,500	1.22
22	OH feeder BN06	Inspect, Replace	4,503	11888	13536	14947	36178	$58,000	0.62
28	OH feeder CT14	Inspection, Replace	6,875	11220	12552	13718	33628	$56,000	0.60
15	LV breaker #CN2A	Complete Service	3,653	1273	1403	1520	3768	$6,500	0.58
23	69 kV circuits 34 & 37	Test, Replace	103,800	12755	14008	15137	37623	$70,000	0.54
30	OH feeder JM07	Test, Replace	4,120	12657	14160	15474	37934	$86,000	0.44
1b	40 MVA transf. EC03	Complete Service	38,000	601920	644328	685126	1736767	$38,000	

For example, Table 12.1 showed that project 19 (UG cables, feeder EC07 – test and replace as found necessary) had only 592 customers, about ¼ of the customer count on most feeders, but served 4805 kVA peak load, rather typical for a feeder. Clearly, it serves customers who have a higher demand than average. As a result of the relatively low customer cost, the evaluation shown in Table 12.3 put it mid-pack in the table – it is not to be funded this year. Service to those larger customers will suffer, and for that matter, service to smaller customers on that feeder will suffer in comparison to other similar customers on other feeders.[6]

[6] An actual fact feeder EC07 (#7 out of East City substation) service three customers of

Table 12.4 lists the 35 projects re-evaluated and prioritized on the basis of kVA-minutes of outage reduction expected per dollar, rather than customer minutes as done earlier. The values are much different since they are based on a different measure. They vary from a high of 229.8 to a low of .44. Again, the ranking exhibits a scree-like shape (not shown). The utility's available budget is reached somewhere in the 12th project (approval of 11 leaves it with about $225,000 left, approval of project 12 spends about $100,000 too much. Assuming the first 11 projects are approved and the remaining budget is spent on part of project 12 with proportionate results (it involves many feeders so it would be feasible to only do a portion of it), the total budget of $1,375,000 buys 22,273,339 MVA minutes of reduction, or 64% of the possible reduction. Project 19, the feeder with few but larger-than-average customers, just barely makes the cut, for although it moves up greatly in relative score, a number of other projects do, too. Other changes are approval of project 34, which did not make the approval list in Table 12.3, and non-approval of project 31, which falls off the approval list.

The top projects will be at the top of just about any list

While the change in figure-of-merit altered the composition of the approval list, eight of the top eleven projects stayed the same regardless of the prioritization scheme. Looked at another way, in this example 84% of the budget is spent on the projects that make it into both lists.

This is often the case - the very top rated projects stay top rated even if the exact definition of "merit" is changed. The reason is that projects that tend to score extremely high generally do so because they provide a tremendous improvement in *equipment reliability* per dollar. Whether serving many customers or a high load, that large increase in equipment reliability pays off with a handsome increase in merit score. Careful examination of Tables 12.1 through 12.4 will show that the winners here are all projects where a correction from "very poor" to "very good" reliability is made at a good price. Another way to look at it is that prioritizing on the basis of kVA minutes gains 90% of the optimum result from the standpoint of optimizing for customer-minutes.

Both merit indices used above were duration-related. Had a frequency of outage index been used, such as customer interruption count, or kVA-interruptions, the results would have been somewhat less similar to those in Tables 12.3 and 12.4. But even so, over 2/3 of the projects at the top of the list

---

roughly 1 MVA peak load each and 499 typical small residential and commerical consumer sites. Had this feeder, which has very severe reliability problems, served only residential and small commercial customers (about 1700 of them would be typical) the higher customer count would have led to a figure or merit of 2.71, and would have been on the approval list. As a result, one can argue that these 499 customers were "screwed" only because they happened to be on a feeder which served several large consumers and therefore had a low customer count.

Table 12.5 Comparison of Top Projects

Ranking Order	As Prioritized by Customer-minutes	As Prioritized by kVA-minutes
1	**18**	**20**
2	4	4
3	**2**	**21**
4	1	3
5	**20**	**18**
6	25	2
7	**21**	**1**
8	3	25
9	**32**	**19**
10	33	34
11	-	**32**

would remain the same even if a frequency-related merit score (e.g., SAIFI) were used.

In general, the use of any reasonable reliability definition seems to gain at least 2/3 of "optimality" as judged by any other figure of merit. Qualitatively, this is a general result:

> The use of any reasonable definition of "merit" as the prioritizing index in RCM will provide qualitatively similar overall results.

What "Reliability Definition" Should A Utility Select?

The answer is quite simple: whatever index its regulatory authority has indicated will be used to evaluate performance of the utility. Generally, that is some combination of SAIDI, SAIFI, and a measure of "outliers" – customers who have much worse than average reliability.[7] Regardless, if law or regulation defines "what is important," the utility would be imprudent to pick anything else, particularly if performance incentives or rate penalties are linked to its performance as assessed by that measure.

This is where performance based rates (PBR) provide a great advantage to the utility. PBR not only provides a completely unambiguous definition of *what* "good performance" means, but it provides a quantitative measure of *how much* it is worth. This permits not only prioritization of projects as shown here, but optimization of the total budget as will be discussed in Chapter 15.

[7] Most utility regulatory commissions have some provision in their regulations requiring attention to customers/feeder who have much worse (e.g., three times) than average duration or frequency of outages, or who are perennially on the list of worse performance every year.

With very few exceptions, there are no significant problems applying the methods discussed above to any type of regulatory definition of "good performance" or PBR rate schedule. The examples given here are a bit simpler to execute, and display on textbook pages, than what is generally required in the most actual utility cases, but the concepts are the identical and the execution requires nothing not shown here. Non-linearities in PBR functions (see Chapter 14, Figures 14.6 and 14.8) create some challenges in analyzing individual projects and prioritizing a list to a budget constraint, but the authors have never encountered any system that cannot be optimized in the manner discussed above.

12.4 PRIORITIZATION OF THE *TYPE* OF MAINTENANCE

Section 12.3's RCM prioritized projects by picking from a list of proposed maintenance projects. Projects were prioritized based upon:

- The expected reliability improvement,
- The number of customers, load or other measure of importance of the equipment's service and,
- Cost.

Projects were ranked so that the best can be selected against either a goal of using a constrained monetary budget most effectively (as in the example) or against a goal of achieving some target reliability improvement at the lowest cost possible. Applied with good data, focus on objectives, consistency and common sense, that basic method provides good results. However, results can be improved.

Table 12.6 Project 8 and Two Alternatives

Project Number	Equipment Type	Work Proposed	# Cust	Reduced Cust-Min Out Yr. 0	Next	+2 yr.	PW Sum Cust-min	Cost (yr. 0)	"Bang/ Buck"
8	HV breaker # A0121	Complete Service	11,300	6265	6760	7225	18202	$22,000	0.83
8b	HV breaker # A0121	Partial Service	11,300	4746	5204	5617	13979	$9,000	1.55
8c	HV breaker # A0121	Test, Replace	11,300	3797	4231	4612	11340	$6,750	1.68

Table 12.7 Project 8 and Two Alternatives Listed As Marginal Project Decisions

Project Number	Equipment Type	Work Proposed	# Cust	Reduced Cust-Min Out Yr. 0	Next	+2 yr.	PW Sum Cust-min	Cost (yr. 0	"Bang/ Buck"
-	HV breaker # A0121	Do Nothing	11300	0	0	0	0	0	-
8c	HV breaker # A0121	Test, Replace	11300	3797	4231	4612	11340	6750	1.68
8b	HV breaker # A0121	Partial Service	11300	949	973	1005	2639	2250	1.17
8	HV breaker # A0121	Complete Service	11300	1519	1557	1608	4222	13000	0.32

The improvement is based on letting RCM determine what and how much maintenance should be done to each unit. In section 12.3, each item on the list had a specific maintenance activity (e.g., full rebuild, inspect and replace as needed, etc.) already assigned to it. "Bang-for-the-buck" can be improved if the RCM method is used to also decide *what* should be done to each unit of equipment.

Half Measures Sometimes Deliver More than One-Half the Results

Application of "intra-project" alternatives evaluation

Table 12.6 lists project 8, from Table 12.1, which was the complete service on a HV breaker #121, along with two other alternative categories of service that are possible for that unit – *different types of service* on that same breaker. These are option 8b, "partial service," and option 8c, "test and replace as found needed." The table shows data equivalent to that shown in Table 12.3 for all three alternatives (i.e., the evaluation done against customer outage minutes). Both options 8b and 8c provide less reliability improvement than the original project proposed, but they also have a lower cost. They provide more "bang for the buck." Their figures of merit (right-most column in Table 12.6) indicate that unlike the original project (complete service) both 8b and 8c score high enough to have gained approval in Table 12.3 (the selection limit was 1.49).

Table 12.7 lists these two projects in a slightly different way. First, they are listed in order of increasing cost. Secondly for completeness, "do nothing" is listed as an option. Thus, do nothing (the least expensive) is listed at the top. The next line gives project 8c, along with its costs and benefits, then 8b is listed, and finally 8.

Most importantly, the costs and benefits of each option have been computed as *marginal differences.* The least cost project (8c) is listed with its costs and reliability improvements as it was in Table 12.6. But project 8b, the next least costly, is listed with marginal costs and benefits. The cost shown is its cost over and above 8c's. The improvements shown are its improvements minus those that 8c would deliver. Similarly, the original project's reliability improvements and cost have been converted to the marginal differences that choosing that alternative over 8b.

This *marginal benefit and cost table* has converted the list of possible projects into a *decision chain.* It tells the utility what it gets for what it has to pay when it picks any option over a less expensive option. Table 12.7 lists the three projects and "do nothing" as a serial decision chain. This is illustrated in Figure 12.4. These projects are now decisions among *different categories of service* for the same breaker, a chain of increasingly expensive projects, each providing more reliability improvement than the preceding, but at higher cost.

While viewing a set of possible alternative projects in this manner may seem obtuse, and of no practical value, when dealing with an optimization method as

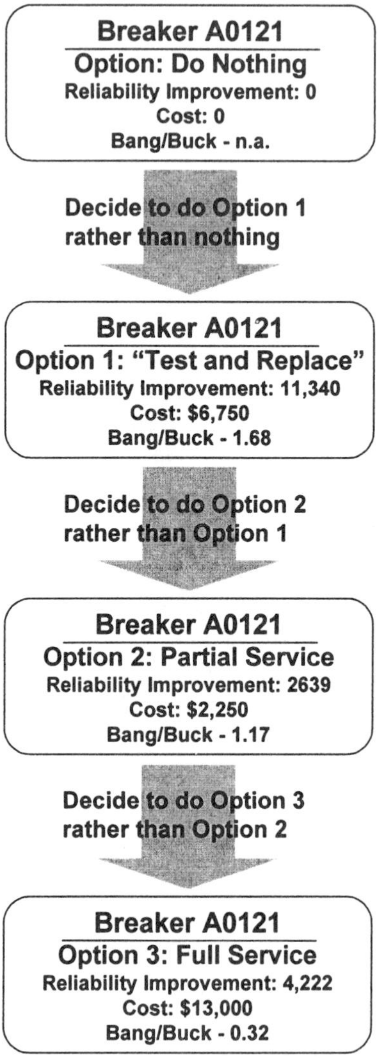

Figure 12.4 The three possible projects for breaker A0121, viewed as marginal decisions. This is *intra-project evaluation* – comparison of project alternatives that all fall within one equipment unit or set of equipment – as opposed to *inter-project comparison* – comparing projects applied to one unit of equipment to those on others.

described here, it provides a big improvement. The "chain" of options shown in Figure 12.4 basically reduces a decision with multiple outcomes (which of four alternatives do I choose?) to a series of binary (yes-no) decisions. It simplifies performing the decision-making process with a formal (i.e., rigorous, repeatable, documentable, defendable) procedure that essentially performs a series of binary evaluations (this is better than that). It also illustrates a key point. Doing nothing should *always* be the basis for comparison: all activities should justify themselves against a "zero base."

Looking at Table 12.7, note that the scores for these three options (last column), the lowest cost option, 8c has the same as it did in Table 12.6 – a score of 1.68 ($11,340 customer minutes expected reduction over $6,750). But option 8b has a score of only 1.17, rather than 1.55 as evaluated in Table 12.6. While project 8b looks good enough when viewed on its own, viewed as a decision to spend $2,250 more than option 8c to gain the improvement it gives over 8c, it does not look quite as good – definitely not good enough to gain approval in Table 12.3. That $2,250 is not nearly as effective in improving reliability as the $6,750 spent on only buying option 8c. Similarly, when viewed as a marginal option, option 8 itself looks very inefficient, with a sore of only .32.

Table 12.8 shows the RCM evaluation and ranking of all projects from Table 12.3 with these three projects substituted for project 8. Note that 8c, with its figure of merit score of 1.68, makes it into the "winners" category. Projects 8b and 8 fall far below the approval limit – their *additional cost* over 8c is definitely not justifiable based on their *additional benefit.* The recommended action, then, is to select to perform only "testing and replacement as found needed" on this breaker.

What got bumped off the list to make room for project 8b? In this case, nothing. Instead, whereas originally the utility could just afford to buy the top ten projects, it can now not quite afford to buy all of the final projects on the approval list (#32, Hillside substation feeders – testing and replacement as found needed). By the time it gets down to that item in Table 12.8, it is $6,7500 short of having enough to fund this project completely. Prorating that project by assuming that some small portion could be deleted with proportional impact on results, indicates that this change in using alternatives for project 8 results in a .03% improvement overall.

Using the approval limit's marginal cost of reliability ($1.41/customer minute) the additional reliability wrought by this improvement over the results of Table 12.3 would cost the utility another $859 to buy if it stuck to the method used in Table 12.3.

This small improvement may hardly seem worth the effort of going through this analysis. However, the point of the discussion above was to introduce the concept of extending the RCM evaluation to deciding how to maintain each unit of equipment, not just what gets maintained. By evaluating options in this manner – letting the RCM decided what is done, not just what equipment is

Table 12.8 Projects Evaluated By RCM Ranking, Including Two Options for Project 8

Project Number	Equipment Type	Work Proposed	# Cust	Reduced Cust-Min Out Yr. 0	Next	+2 yr.	PW Sum Cust-min	Cost (yr. 0)	"Bang/ Buck"
18	UG cables, feeder EC06	Replace All	2,200	121440	128436	135502	346789	$27,000	12.84
4	40 MVA transf. DT03	Test, Follow up	4,612	44275	48703	52681	130779	$11,000	11.89
2	40 MVA transf. EC01	Recond. Oil/Servc.	7,800	74880	81432	87461	219012	$22,000	9.96
1	40 MVA transf. EC03	Rebuild Complete	10,899	222343	235749	249132	636315	$65,000	9.79
20	UG cables, net DT3B	Replace Complete	245	67032	71207	75341	192144	$43,000	4.47
25	Poles, CE district	Test, Replace	80,134	575234	625675	672069	1682718	$610,000	2.76
21	UG cables, net DT3A	Test, Replace	340	22195	25198	27777	67373	$31,000	2.17
8c	*HV breaker # A0121*	*Test, Replace*	*11300*	*3797*	*4231*	*4612*	*11340*	*$6,750*	*1.68*
3	40 MVA transf. JT235	Test, Follow up	842	5982	6606	7163	17729	$11,000	1.61
32	Riverside sub feeders	Test, Replace	28,093	177997	199134	217621	533491	$335,000	1.59
		Approval Limit Based on Budget							
33	Hillside sub feeders	Test, Replace	14,004	108895	121826	133136	326379	$219,000	1.49
31	Oldtown sub feeders	Test, Replace	22,789	98448	110139	120364	295068	$210,000	1.41
34	Central sub feeders	Selected Replace	28,345	131634	143981	155199	386928	$310,000	1.25
35	Farmton substation	Complete Rebuild	43,000	385968	415741	443829	1119637	$905,000	1.24
8b	*HV breaker # A0121*	*Partial Service*	*11300*	*949*	*973*	*1005*	*2639*	*$2,250*	*1.17*
27	OH feeder BN05	Inspection, Replace	1,609	5870	6567	7176	17592	$17,000	1.03
19	UG cables, feeder EC07	Test, Replace	592	5570	6348	7014	16964	$18,000	0.94
5	HV breaker # EC03	Complete Service	10,899	6043	6521	6969	17556	$22,000	0.80
10	LV breaker #EC06	Complete Service	2,200	1376	1493	1602	4018	$6,500	0.62
6	HV breaker # EC01	Complete Service	7,800	4324	4666	4987	12564	$22,000	0.57
24	OH feeder EC17	Inspect, Replace	1,578	10680	11752	12714	31555	$58,000	0.54
9	LV breaker #ES06	Complete Service	1,681	1052	1141	1224	3069	$6,500	0.47
16	LV breaker #DT3N	Complete Service	2,091	811	891	963	2394	$6,500	0.37
13	LV breaker #NB03	Complete Service	1,191	745	808	867	2175	$6,500	0.33
8	*HV breaker # A0121*	*Complete Service*	*11,300*	*1519*	*1557*	*1608*	*4222*	*$13,000*	*0.32*
29	OH feeder CT05	Inspection, Replace	1,389	4534	5072	5543	13588	$43,000	0.32
26	OH feeder BN04	Inspection, Replace	1,345	4390	4911	5367	13158	$43,000	0.31
14	LV breaker #CN1A	Complete Service	1,058	662	718	770	1932	$6,500	0.30
23	69 kV circuits 34 & 37	Test, Replace	42,300	5198	5708	6169	15332	$70,000	0.22
12	LV breaker #NB02	Complete Service	650	407	441	473	1187	$6,500	0.18
30	OH feeder JM07	Inspection, Replace	1,555	4777	5344	5840	14317	$86,000	0.17
11	LV breaker #NB01	Complete Service	580	363	394	422	1059	$6,500	0.16
28	OH feeder CT14	Inspection, Replace	1,671	2727	3051	3334	8174	$56,000	0.15
15	LV breaker #CN2A	Complete Service	891	310	342	371	919	$6,500	0.14
22	OH feeder BN06	Inspect, Replace	950	2508	2856	3153	7632	$58,000	0.13
7	HV breaker # 221	Complete Service	634	351	379	405	1021	$22,000	0.05
17	LV breaker #DT2N	Complete Service	112	30	34	37	90	$6,500	0.01

Table 12.9 Projects 1 and 35 with Multiple Alternatives

Project Number	Equipment Type	Work Proposed	# Cust	Reduced Cust-Min Out Yr. 0	Next	+2 yr.	PW Sum Cust-min	Cost (yr. 0)	Bang/ Buck
For Project 1									
1d	40 MVA transf. EC03	Test, Replace	10899	104632	115095	124497	309060	$11,000	28.10
1c	40 MVA transf. EC03	Complete Service	10899	41853	42899	44315	116357	$10,000	11.64
1b	40 MVA transf. EC03	Recond. Oil/Servc.	10899	26158	26812	27697	72723	$17,000	4.28
1	40 MVA transf. EC03	Rebuild Complete	10899	49700	50943	52624	138174	$27,000	5.12
For Project 35									
35d	Farmton substation	Test, Replace	43000	181632	206297	227473	551552	$195,000	2.83
35c	Farmton substation	Serv. Transf. Only	43000	36326	37235	38463	100993	$65,000	1.55
35b	Farmton substation	Full Service	43000	81734	83778	86542	227234	$300,000	0.76
35	Farmton substation	Complete Rebuild	43000	86275	88432	91350	239858	$345,000	0.70
For Project 9 (example for 9 – 17)									
9b	LV breaker #ES06	Test, Replace	1681	637	716	785	1918	$1,100	1.74
9	LV breaker #ES06	Complete Service	1681	414	425	439	1152	$5,400	0.21

maintained, meaning an improvement in the total *value* obtained by the utility can be made. As will be shown below when the concept is extended to all projects, the resulting savings are significant.

Intra- vs. inter- project alternatives evaluation

Tables 12.6 – 8 showed one example of letting RCM determine *what should be done* to a unit of equipment. The earlier "optimizations" – those carried out in section 12.3 – basically decided only among projects where it had already been decided what would be done to each unit. Comparison like that, where decisions about what maintenance will be done has already been made, is called *Inter-Project Alternatives Evaluation.* By contrast, evaluating the possible maintenance options on one unit of equipment to the possible maintenance options on others unit is termed *Intra-Project Alternatives Evaluation.* It applies the optimization to a broader range of options and thus provides an improvement.

Comprehensive Example Using Intra-Project Optimization

Table 12.9 lists three alternatives to project 1 and four for project 35 – the original and two and three alternatives respectively, for each. Also shown is one example of splitting the nine LV breaker projects (numbers 9 through 17 in Table 12.1) into two alternatives. In the original list (Table 12.1, optimized in Table 12.3) nine LV breakers were listed for "full service" at $6,500 each. Here, an option – inspect and replace (as found needed) - is an alternative at only $1,100 in cost but with less effective reliability improvement. All of these options are shown figured as marginal costs and marginal improvements and grouped together as sets of ordered, marginal decisions.

Table 12.10 Projects Evaluated By RCM Ranking Using "Intra-Project" Alternative Evaluation for Twelve Projects (#s 1, 8, 9-17, and 35).

Project Number	Equipment Type	Work Proposed	# Cust	Reduced Cust-Min Out Yr. 0	Next	+2 yr.	PW Sum Cust-min	Cost (yr. 0)	Bang/ Buck
1d	40 MVA transf. EC03	Test, Replace	10899	104632	115095	124497	309060	$11,000	**28.10**
18	UG cables, feeder EC06	Replace All	2200	121440	128436	135502	346789	$27,000	**12.84**
4	40 MVA transf. DT03	Test, Follow up	4612	44275	48703	52681	130779	$11,000	**11.89**
1c	40 MVA transf. EC03	Complete Service	10899	41853	42899	44315	116357	$10,000	**11.64**
2	40 MVA transf. EC01	Recond. Oil/Servc.	7800	74880	81432	87461	219012	$22,000	**9.96**
1	40 MVA transf. EC03	Rebuild complete	10899	49700	50943	52624	138174	$27,000	**5.12**
20	UG cables, net DT3B	Replace complete	245	67032	71207	75341	192144	$43,000	**4.47**
1b	40 MVA transf. EC03	Recond. Oil/servc.	10899	26158	26812	27697	72723	$17,000	**4.28**
35d	Farmton substation	Test, Replace	43000	181632	206297	227473	551552	$195,000	**2.83**
25	Poles, CE district	Test, Replace	80134	575234	625675	672069	1682718	$610,000	**2.76**
10b	LV breaker #EC06	Test, Replace	2200	834	938	1027	2510	$1,100	**2.28**
21	UG cables, net DT3A	Test, Replace	340	22195	25198	27777	67373	$31,000	**2.17**
9b	LV breaker #ES06	Test, Replace	1681	637	716	785	1918	$1,100	**1.74**
8c	HV breaker # A0121	Test, Replace	11300	3797	4231	4612	11340	$6,750	**1.68**
3	40 MVA transf. JT235	Test, Follow up	842	5982	6606	7163	17729	$11,000	**1.61**
32	Riverside sub feeders	Test, Replace	28093	177997	199134	217621	533491	$335,000	**1.59**
35c	Farmton substation	Serv. Transf. Only	43000	36326	37235	38463	100993	$65,000	**1.55**
		Approval Limit Based on Budget							
33	Hillside sub feeders	Test, Replace	14004	108895	121826	133136	326379	$219,000	1.49
31	Oldtown sub feeders	Test, Replace	22789	98448	110139	120364	295068	$210,000	1.41
16b	LV breaker #DT3N	Test, Replace	2091	492	564	625	1505	$1,100	1.37
34	Central subst. feeders	Selected Replace	28345	131634	143981	155199	386928	$310,000	1.25
13b	LV breaker #NB03	Test, Replace	1191	452	508	556	1359	$1,100	1.24
8b	HV breaker # A0121	Partial Service	11300	949	973	1005	2639	$2,250	1.17
14b	LV breaker #CN1A	Test, Replace	1058	401	451	494	1207	$1,100	1.10
27	OH feeder BN05	Inspection, Replace	1609	5870	6567	7176	17592	$17,000	1.03
19	UG cables, feeder EC07	Test, Replace	592	5570	6348	7014	16964	$18,000	0.94
5	HV breaker # EC03	Complete Service	10899	6043	6521	6969	17556	$22,000	0.80
35d	Farmton substation	Full Service	43000	81734	83778	86542	227234	$300,000	0.76
13	LV breaker #NB03	Complete Service	1191	294	301	311	816	$5,400	0.74
35	Farmton substation	Complete Rebuild	43000	86275	88432	91350	239858	$345,000	0.70
12b	LV breaker #NB02	Test, Replace	650	246	277	304	742	$1,100	0.67
11b	LV breaker #NB01	Test, Replace	580	220	247	271	662	$1,100	0.60
6	HV breaker # EC01	Complete Service	7800	4324	4666	4987	12564	$22,000	0.57
24	OH feeder EC17	Inspect, Replace	1578	10680	11752	12714	31555	$58,000	0.54
8	HV breaker # A0121	Complete Service	11300	1519	1557	1608	4222	$13,000	0.32
29	OH feeder CT05	Inspection, Replace	1389	4534	5072	5543	13588	$43,000	0.32
26	OH feeder BN04	Inspection, Replace	1345	4390	4911	5367	13158	$43,000	0.31
10	LV breaker #EC06	Complete Service	2200	542	556	574	1508	$5,400	0.28
23	69 kV circuits 34 & 37	Test, Replace	42300	5198	5708	6169	15332	$70,000	0.22
9	LV breaker #ES06	Complete Service	1681	414	425	439	1152	$5,400	0.21
30	OH feeder JM07	Inspection, Replace	1555	4777	5344	5840	14317	$86,000	0.17
16	LV breaker #DT3N	Complete Service	2091	320	328	338	889	$5,400	0.16
28	OH feeder CT14	Inspection, Replace	1671	2727	3051	3334	8174	$56,000	0.15
14	LV breaker #CN1A	Complete Service	1058	261	267	276	725	$5,400	0.13
22	OH feeder BN06	Inspect, Replace	950	2508	2856	3153	7632	$58,000	0.13
15b	LV breaker #CN2A	Test, Replace	891	188	217	241	579	$1,100	0.11
12	LV breaker #NB02	Complete Service	650	160	164	170	445	$5,400	0.08
11	LV breaker #NB01	Complete Service	580	143	147	151	397	$5,400	0.07
15	LV breaker #CN2A	Complete Service	891	122	125	130	340	$5,400	0.06
17b	LV breaker #DT2N	Test, Replace	112	18	21	24	57	$1,100	0.05
7	HV breaker # 221	Complete Service	634	351	379	405	1021	$22,000	0.05
17	LV breaker #DT2N	Complete Service	112	12	12	13	33	$5,400	0.01

A different set of decisions

Table 12.10 shows the result of an RCM ranking based on customer-minutes that includes these alternatives added to the projects and alternatives used in Table 12.8 (including project 8's options). Among the first eight projects at the top of the list are all four of the alternatives to project 1. Although the RCM method had a list of partial measures it could consider for this unit, it chose to "fully upgrade" project number 1 through selection of its alternatives 1d, 1c, 1b to 1. Thus, for that transformer it picked, in a serial, four-decision fashion, the full service as was originally selected in Table 12.3.

But on other projects the prioritization often selected only an interim level of permitted maintenance. For project 35, alternatives 35d and 35c were selected. Alternatives 35b and 35 did not make the cut. Similarly, one low voltage breaker (9b, breaker ES06) makes the cut with testing and replacement as needed. As a result, these results are considerably different than those obtained in Figure 12.3, even though the same 35 equipment units are being considered, and even though the same reliability index (customer minutes) is being used as the goal. Important differences in the results are given below.

A 6.4% improvement in results has been obtained

For the same budget, an increase of 264,326 customer minutes of interruption expecting to be avoided has been obtained. Gaining this in the original optimization (Table 12.3), would have cost the utility an extra $187,500 (264,326 customer-minutes at 1.41 customer minutes per dollar).

Approval target is raised

Intra-project prioritization raises the bar for approval, because it gets more "bang" per dollar from some of the projects now being approved. Here, the approval limit increases by nearly 10%, from 1.41 to 1.55. Better projects are being implemented. This increase in the target value (10%) is greater than the increase in overall results (6.4%) because of the scree-shape of the top part of the project mix – the very best projects approved did not change.

More units of equipment receive some improvement

The number of units seeing some reliability improvement increases by over 30%. Maintenance effort is spread around the system more.

Smaller projects are being done

Since this method tends to "spread around" the money on more projects, that means the limited budget is splintered into more, but smaller, projects. On average, they are of greater average value compared to their cost.

The optimization selected "partial measures"

It is significant that in 40% of the equipment units selected for some maintenance, the RCM decided to maintain it by selecting only an interim measure. More significantly, in all but one case where it had the choice, it chose an interim measure.

Full Implementation would do even better

Table 12.10 applied prioritization using multiple alternatives for only thirteen of thirty-five units of equipment involved. If a multiple alternatives approach is applied to all projects on the list, the improvement in figure of merit obtained grows to about 10% or slightly more depending on details of the reliability index used and how thoroughly project alternatives are represented. Such an example is beyond the scope of what can be given here (it results in a list with over 110 entries). However, the concept of its execution is exactly as used above.

12.5 PRACTICAL ASPECTS FOR IMPLEMENTATION

The foregoing examples, while comprehensive, simplified several of the details needed when RCM is applied in actual practice in order to streamline the examples so that the major points could be made in a succinct manner. When applied to actual utility cases, the authors make the following recommendations for reliability-centered maintenance prioritization.

Cost and Benefits Should Be Treated as Expectations

As discussed in Chapters 6, 11, and elsewhere throughout this book, equipment failure is a random process that cannot be predicted with accuracy on a single-unit or small-lot basis except in a few very special cases. Similarly, both the cost of doing some types of maintenance and the benefits that maintenance provides, is probabilistic. For example, the category of service "Inspect and Make Repairs as Found Necessary" will vary in both cost and benefits from unit to unit depending on what is found when inspection is carried out. This cannot be predicted accurately in advance for any one unit of equipment (a specific pole, a particular power transformer) but it can be predicted en masse over a large group of units.

Reliability-Centered Maintenance should be applied in this context: the failure rates being dealt with, the benefits expected from maintenance, and the costs of that service are all expectations, accurate and dependably forecast-able on a large-scale basis but determinable on an individual basis only as probabilistic expectations.

Use Only Historically-Based Cost and Improvement Factors

The failure rates for equipment, as a function of condition as used for "before service" expectations of reliability, the improvement expected from various levels of service applied to equipment of various categories of age/condition, and the cost of various types of service should all be developed from historical data. For example, the entries in a table like Table 12.2 should be based, to the extent possible, on historical experience specific to the utility, or data from a large set of samples published in a reputable reviewed technical journal. If that is not possible, the data should be developed by consultation with experts who have experience in such work.[8]

Represent Improvements as Having Only Short-Term Impacts on Reliability

Even the most comprehensive service on a transformer, breaker, underground or overhead line will result in improvements that only last for a short time. Left unattended after the preventive maintenance, the device will return to its original "aged" failure rate condition in only a short period of time. The improvements in reliability used as the RCM benefit should reflect this in a realistic sense. Overestimating the length of time improvements make a positive impact will overemphasize the value of maintenance, leading to too much spending and a shortfall of results versus expectations.

Figure 12.5 illustrates the difference between the impact that service might make, and replacement. Here, a 30-year old transformer can either be replaced or re-built. Replacement results in "like new" condition and it will take 30 years for expected failure rate to return to the level of just prior to replacement.

Performing any type of inspection, testing, and preventive maintenance, however, has a different impact. First, it does not result in "like new" condition and failure rate expectation – the improvement is somewhat less. Secondly, the improvement does not last for 30 years, as replacement does, but for a much shorter time, as shown. Generally, noticeable improvements from service only older units last for only 3 – 7 years.

Use Category-Based Equipment Condition and Maintenance Types

Expected RCM costs, as well as the benefits (reliability improvement gained) from inspection, testing, and various grades of preventive service and the cost and improvement numbers assigned to each category represent expectations for equipment-service combinations in that category.

All equipment being considered for maintenance, and all maintenance efforts that are being considered to be performed, are classified by these category

---

[8] Many consultants can develop sets of good-looking numbers. Far fewer seem able to develop numbers that produce dependable results.

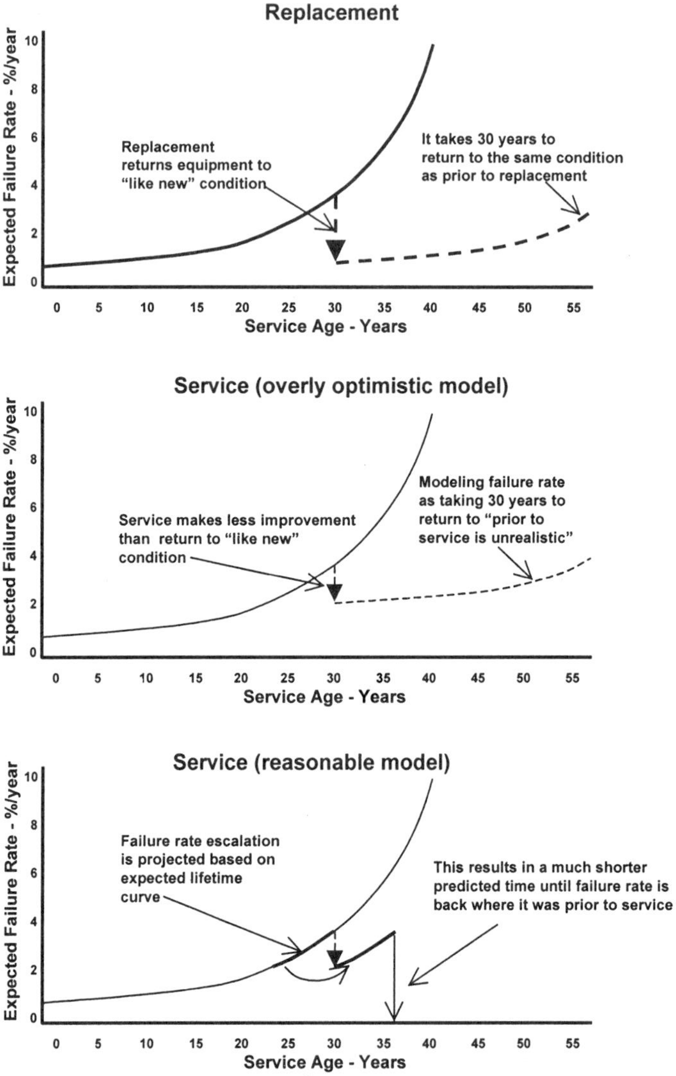

Figure 12.5 Top – replacement of equipment after 30 years returns a facility to "like new" condition. It will take 30 years for the equipment to age to the same condition as it was before replacement (assuming all other factors remain the same). Middle – anything less than replacement provides a somewhat lesser improvement in failure rate. It is a mistake to model that improvement as lasting for a long period of time (in this case 30 years). Bottom: a more realistic representation is to model the failure rate as escalating at the same rate as it historically does from the (lower0 rate after service, refurbishment, should be classified into categories for RCM planning and evaluation purposes, as shown.

Table 12.11 Example of Improvement Categorization by "Generic" Categories -- Improvement Factor I in Equation 12.3, for Power Transformers

Service Category		Condition Category – Equivalent Years in Service 0 – 5	10 – 20	20 – 30	30 – 35	35 – 40	40 – 45
	Age Category →	1	2	3	4	5	6
1	No service	0.00	0.00	0.00	0.00	0.00	0.00
2	Inspect/ test but do not follow up	0.02	0.03	0.05	0.10	0.11	.013
3	Inspect and replace as found needed	0.05	0.15	0.25	0.33	0.42	0.50
4	Inspect & test and replace as needed	0.15	0.24	0.33	0.40	0.48	0.55
5	Partial service - 1	0.22	0.30	0.38	0.45	0.52	0.59
6	Partial service - 2	0.29	0.37	0.44	0.50	0.57	0.63
7	Full service	0.52	0.57	0.62	0.66	0.71	0.75
8	"Field rebuild"	0.64	0.68	0.72	0.75	0.78	0.81
9	Remanufacture	0.70	0.75	0.80	0.85	0.85	0.85
10	Replace	1.00	1.00	1.00	1.00	1.00	1.00

definitions and the numbers assigned to each category are used in estimating the expected cost-improvement combination for that equipment, in *every* case. "Adjustments" or tailoring of data based on judgement is *not* permitted. As an example, one of several possible ways to implement this would be to:

1. Rate all equipment by condition based on "equivalent age" (see Chapters 6 and 11), and lump all equipment into categories by equivalent age group (see column headings in Table 12.11, above).
2. Model expected failure rate as a function of equivalent age or condition, rather than actual age (e.g., Chapter 6, Figure 6.3).
3. Group maintenance operations into several categories.
4. Any specific equivalent age category, by any specific category of maintenance, from a table of expected improvement factors (Table 11.11). In this example, the factor R (equation 8) was determined based on such categorization, using the formula

$$Fs = (Fp - Fc) \times M(a,s) \tag{12.3}$$

Where:

Fs = expected annual failure rate after service is performed

Fp = expected annual failure rate before service is performed

Fn = expected failure rate of new equipment

M (a,s) is the entry in Table 12.11, the maintenance improvement factor for age category a and service category s.

What to Do When Marginal Figures of Merit for Project Alternatives are not Strictly Increasing

In some cases, when the hierarchical alternatives chain for a particular project is developed (e.g. Figure 12.4) it will not have strictly decreasing marginal scores as one moves "down" the serial chain of decisions regarding how much maintenance to buy. Everything works well when costs are strictly increasing as one moves down the decision chain. Optimization may encounter problems when that is not the case. Adjustment of the list is required.

This problematic condition was the case with project 1 in the final version of the examples sued in Section 12.3. (Tables 12.9 and 12.10). The final option for transformer EC03 (moving to 1 from 1b) had a figure-of-merit of 5.28, whereas the alternative in the chain that was ahead of it in the "chain" (moving to 1b from 1c) had a lesser figure-of-merit, of only 4.28. As a result, in Table 12.9 option 1 was picked *before* option 1c.

In Table 12.10's optimization, this did not cause a problem, because alternative 1c was eventually selected (both 4.28 and 5.28 were nicely above the approval limit of 1.55). But, if the approval limit had fallen between 4.28 and 5.28, it would have meant picking 1d, 1c, and 1, without picking 1b –an impossibility- in the real world.

This is a complication that must be dealt with in the real world. The are two ways of adjusting the alternatives in such tables to avoid such computations in the tables, both requiring rather involved "programming" of the spreadsheet or database manipulation system being used. The first involves making the costs and improvements listed for the alternatives "active" functions of those which have already been selected (i.e. their values change depending on what other options have been selected). This necessitates an interactive spreadsheet application.

The second way is to use multiple entries representing both the entries in the chain shown in Figure 12.5, and combinations (e.g. moving from 1c directly to 1). Optimization requires a "post processor" to resolve ambiguities caused by two overlapping alternatives (i.e. both "move to 1b from 1c" and "move to 1 directly from 1c"). Either method works well and both give identical results when implemented correctly.

Real-World Application Requires a Computerized Approach

Implementation of rigorous RCM using intra-project alternatives evaluation as shown here does result in a large list of proposed alternatives to consider. The list of 35 projects in the example used in section 12.3 grows to over 110 entries if alternatives are used for all 35 units of equipment being considered. Addition entries or complications in database manipulation due to having to deal with non strictly-increasing marginal costs makes a further increase.

As a result, for a utility with a large metropolitan service area the list of possible alternatives for all significant distribution equipment in its system will easily run to hundreds if not thousands of possible entries. However, this is exactly what computers are made to do: manage and analyze involved lists. RCM as described here is implement-able with nothing more than an electronic spreadsheet and a set of well-designed and implemented organization procedures. The categorization of equipment by type, condition by equivalent age, maintenance by service category makes it possible to develop procedures that generate the entire alternatives database, as needed for RCM evaluation, from the raw equipment/conditions lists.

Multi-District or Multi-Department Application

One "trick" that helps a great deal in implementation of RCM at a large distribution utility is that the marginal cost of reliability – the "cutoff limit" in the optimization – can be used as both a general target to reduce planning and evaluation work. It can also be used as a tool to apply the RCM concept when O&M decisions are broken into areas or districts due to the utility's management structure. Figure 12.6 illustrates this concept. Three different regions of a utility each manage their maintenance budgets autonomously, using the same RCM procedure based on the same company-wide reliability index, economic factors and categorizations of equipment condition and maintenance service types.

All that senior management has to do to keep the overall use of maintenance resources "optimal" across all regions is to see that the marginal cost (approval limit) in each region is adjusted until all are the same. Adjusting the budget allocation for each region or department until the approval limit is the same for all regions and departments does this. Figure 12.7 illustrates this function. The utility has three regions that operate on a very autonomous basis. Each applies RCM, producing the prioritized lists that were shown in Figure 12.6. Based on preliminary estimated budgets, each has already picked an approval limit beyond which it cannot afford to spend (e.g. 3.95 for Region 1).

Note that in the top of Figure 12.6, region 3 has a limit that is 2.98, and region 1 is at 3.95. This means that the utility can pick up an improvement of .97, without spending anything more, by deleting a dollar from region 3's budget where it buys only 2.98 units of improvement, and transferring it to region 1, where it will buy 3.95 units of improvement. When it makes such a change, the approval limit in region 3 will rise slightly, and that in region 1 will gradually drop.[9] It can continue making such transfers, and the approval limits in the regions will rise and drop respectively, until they are equal. At that point spending is adjusted between the two regions: it is "optimal" in the sense that the utility is getting the most "bang for the buck" possible.

The company-wide budget limit determines where on the company-wide

---

[9] The approval limits are falling and rising as the budgets in regions 1 and 3 are respectively rising and falling, since they are respectively moving down and out and up and in on curves similar to that shown in Figure 12.3.

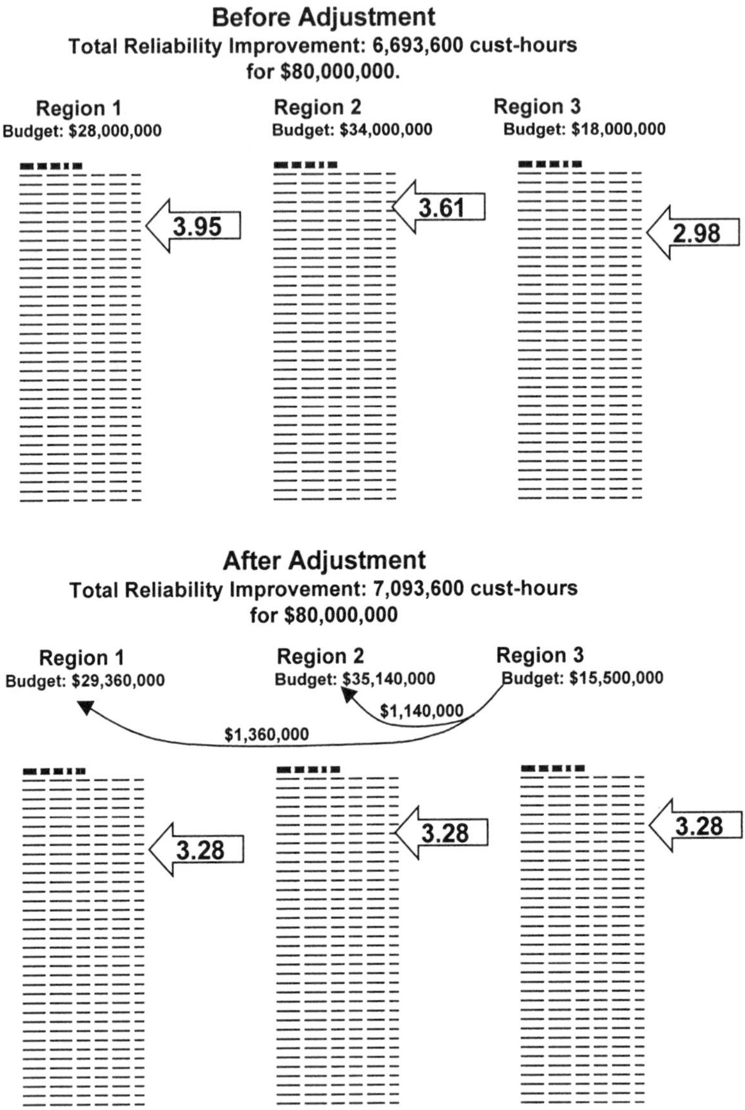

Figure 12.6 Three regions in a utility each autonomously manage their own maintenance budget using the same basically RCM procedure, locally administered by each regional office. Executive management achieves overall optimization by controlling budget allocation using the method shown in Figure 12.7, by using the individual regional projects lists (top) to determine a uniform company-wide approval score (bottom) which it applies to all three regions.

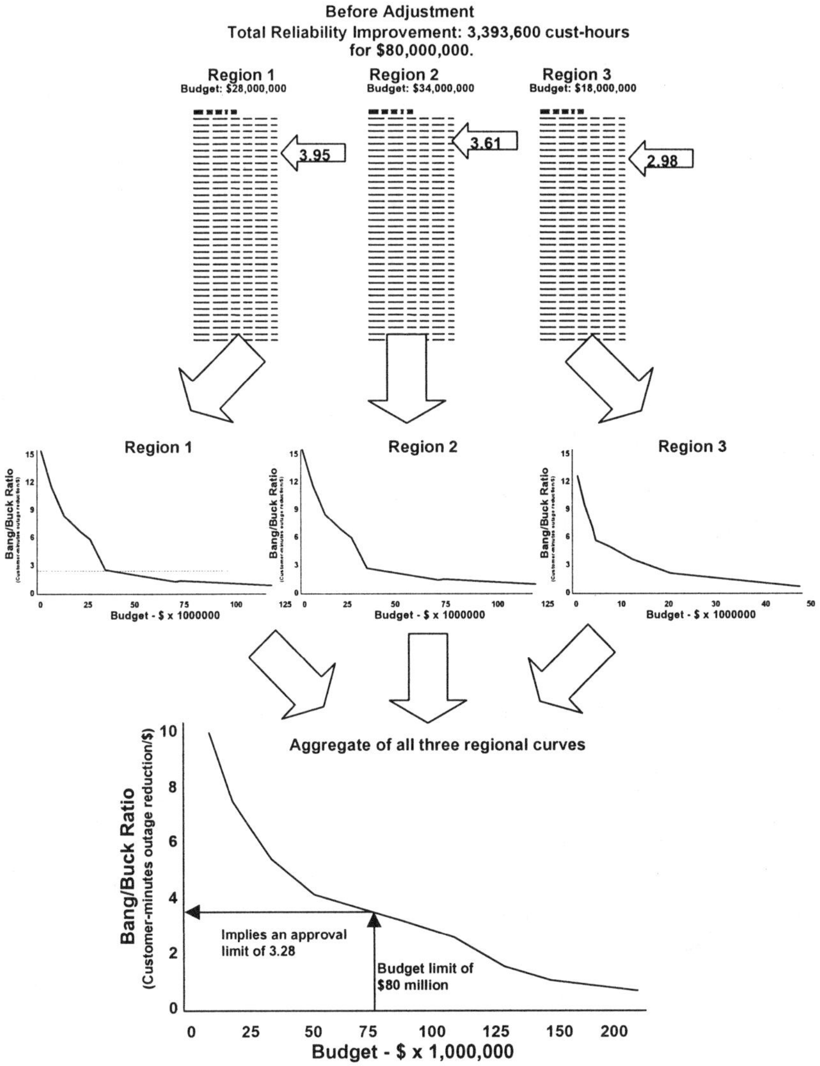

Figure 12.7 Executive management can easily control the overall maintenance budget to see both that maintenance budget is not exceeded, and that RCM is optimized over the entire utility, not just within regions, with the procedure shown above. Regions submit their complete project alternatives list (top), each of which is converted to the figure-of-merit vs. budget limit plot (similar to Figure 12.3) for each, and the curves summed (alternatively the lists can be added together, sorted, and plotted).

curve the utility must place the RCM approval limit. The individual curves then indicate what budget each region should get to fit that limit. Executive management does not even have to re-allocate budget – it simply approves only, and all projects above or equal to the limit – and the budget and its optimal allocation will work out correctly.

Rather than make a lengthy series of incremental adjustments to such budgets, the utility can determine the overall acceptance limit for all regions. As shown in Figure 12.7, the prioritized lists from all regions are used to produce an aggregate score-versus-budget limit curve. The utility's total budget ($80,000,000 in this case) is applied to this curve to determine the required approval limit (3.28 in this case). Executive management then uses the score vs. budget curve for each region to determine what regional budget corresponds to an approval limit of 3.28. The budgets determined in this manner for the three regions will sum to $80,000,000. Alternatively, executive management does not have to even re-adjust budgets – it can simply approve all projects with figures of merit above 3.28, but non below, regardless of regional budget limits. Company-wide, this will result in $80,000,000 in spending, but no more.

12.6 EXTENDING RELIABILITY-CENTERED PRIORITIZATION TO OTHER OPERATIONS PROJECTS

This section will briefly look at how the RCM concept can be extended to include activities and expenses that are not, strictly speaking, equipment maintenance. It will also demonstrate several nuances with respect to setting up comparisons of operations projects that differ in time or scope from one another

Example of a Much Longer Period of Benefit than Normal Maintenance: Computerized Trouble Call Analysis

Often a utility has options with respect to spending on reliability improvement that are not classified strictly as maintenance. For example, it can implement a computerized Trouble Call Management (TMS) system to facilitate the acceptance, processing, analysis of trouble reports from its customers, and optimize the dispatching of resources and equipment for restoration and repair. Such systems do little to reduce the occurrence of equipment outages and the interruptions they cause (i.e. SAIFI). However, in the authors' experience such systems directly reduce the duration of customer interruptions by about 15 minutes per year, and accommodate other changes (e.g., they permit more complicated and dynamic switching programs to be implemented for restoration) that make a further 5 minutes improvement. Computerized TMS systems are expensive – the advanced systems that are truly effective can cost a large utility as much as $16 million including hardware, software, training, system maintenance, and labor for operation. However, they provide long term benefits. If updated, maintained and used well, they do not lose their effectiveness at reducing reliability problems within only a few years, as do the preventive maintenance activities discussed earlier.

Table 12.12 Analysis of Long Term Expenses and Reliability Reduction for a Trouble Call Management System

Year	PWF	Customers	Expenses	Cust.-Min.	PW Cost	PW Impr.
0	1.00	1,137,000	$7,533,333		$7,533,333	-
1	0.90	1,150,644	$6,260,611	23,012,880	$5,634,550	20,711,592
2	0.81	1,164,452	$3,074,053	23,289,035	$2,489,983	18,864,118
3	0.73	1,178,425	$3,090,542	23,568,503	$2,253,005	17,181,439
4	0.66	1,192,566	$3,107,228	23,851,325	$2,038,652	15,648,854
5	0.59	1,206,877	$3,124,115	24,137,541	$1,844,759	14,252,977
6	0.53	1,221,360	$3,141,204	24,427,191	$1,669,365	12,981,611
7	0.48	1,236,016	$3,158,499	24,720,318	$1,510,700	11,823,651
8	0.43	1,250,848	$3,176,001	25,016,962	$1,367,164	10,768,982
9	0.39	1,265,858	$3,193,713	25,317,165	$1,237,310	9,808,388
10	0.35	1,281,049	$3,211,637	25,620,971	$1,119,829	8,933,480
11	0.31	1,293,859	$3,226,754	25,877,181	$1,012,589	8,120,534
12	0.28	1,306,798	$3,242,021	26,135,953	$915,643	7,381,565
13	0.25	1,319,866	$3,257,441	26,397,312	$827,998	6,709,843
14	0.23	1,333,064	$3,273,016	26,661,285	$748,761	6,099,247
15	0.21	1,346,395	$3,288,746	26,927,898	$677,124	5,544,215
16	0.19	1,359,859	$3,304,633	27,197,177	$612,355	5,039,692
17	0.17	1,373,457	$3,320,680	27,469,149	$553,796	4,581,080
18	0.15	1,387,192	$3,336,887	27,743,840	$500,849	4,164,202
19	0.14	1,401,064	$3,353,255	28,021,279	$452,975	3,785,259
20	0.12	1,415,075	$3,369,788	28,301,491	$409,688	3,440,801
21	0.11	1,427,810	$3,384,816	28,556,205	$370,363	3,124,591
22	0.10	1,440,661	$3,399,979	28,813,211	$334,820	2,837,441
23	0.09	1,453,626	$3,415,279	29,072,530	$302,694	2,576,680
24	0.08	1,466,709	$3,430,717	29,334,182	$273,656	2,339,883
25	0.07	1,479,910	$3,446,293	29,598,190	$247,409	2,124,848
26	0.06	1,493,229	$3,462,010	29,864,574	$223,683	1,929,575
27	0.06	1,506,668	$3,477,868	30,133,355	$202,237	1,752,247
28	0.05	1,520,228	$3,493,869	30,404,555	$182,851	1,591,215
29	0.05	1,533,910	$3,510,014	30,678,196	$165,326	1,444,983
30	0.04	1547715	$3,526,304	30,954,300	$149,484	1,312,189
					$37,862,950	216,875,180

Figure-of -merit score = 216,875,180/37,862,950 = 5.73

As an example, suppose Metropolitan Light and Power (see Chapter 8) is considering a system that is expected to provide an average 20 minutes reduction in outage time annually. Implementation will take 18 months, at a total cost of $11,300,000 for hardware, software, facilities, auxiliary equipment, labor, and training, which will be treated here as $627,778 per month average cost over the 18 month period. During that time the system provides no improvement in reliability. Subsequently, the system has a fixed operating cost of $1,700,000 per year (facilities rent and utilities, etc.) plus a variable cost estimated to average $1.38/customer/year.[10,11]

Table 12.12 lists these costs over a 31 year period, beginning with 12 months in year zero during which the project implementation begins. In year 1 the system is implemented at half year (thus there are half the variable costs and half the expected results in that year. There are no fixed costs in year 0. Customer count is expected to grow as shown (thus variable cost grow each year, too). Both future costs and future benefits are discounted using a present worth factor. As shown, the computed figure of merit is 5.73, enough to justify this TMS system against expenses in Table 12.8 or the example in Figures 12.6 and 12.7.

Figure 12.8 shows the "figure of merit so far" over time for the TMS system. When evaluating a long-term project with significant up-front expenses, it is sometimes useful to look at how good it looks in the short run, not just the long term. Since there are no improvements in year 0, and only one-half a year's improvements in year 1, the TMS system looks rather poor on a very short-term basis. Note however that Figure 12.8 shows that its figure of merit passes 3.28 (the acceptance limit in Figure 12.7) in only 4 years. Therefore, this looks good enough to gain approval within the example evaluations used in Figure 12.7, anytime it is evaluated over a four-year or longer period.

Example of evaluating very short-term impacts and comparing options for different impact periods: tree-trimming

Tree trimming is clearly a form of "maintenance" and yet not equipment maintenance in the strictest sense. At many utilities it is administered by a separate department or from a separate budget than equipment maintenance. Since it has an identifiable cost and a direct impact on reliability there is no reason it cannot be included in RCM evaluation, with evaluation used to determine if, how much, and where tree-trimming should be used and how much

---

[10] This analysis assumes that none of these costs are to be capitalized. If that were the case, adjustments to the cost basis for this project might have to be done following the guidelines shown in Chapter 15 with respect to comparing capital to operating expenses.

[11] There is no doubt that future improvement in hardware and software design will lower cost and improve the performance of such systems. However, the analysis is done over the long run using the costs and performance of the original system. This is what the utility is committing to, today. Future changes will justify themselves on their own merits.

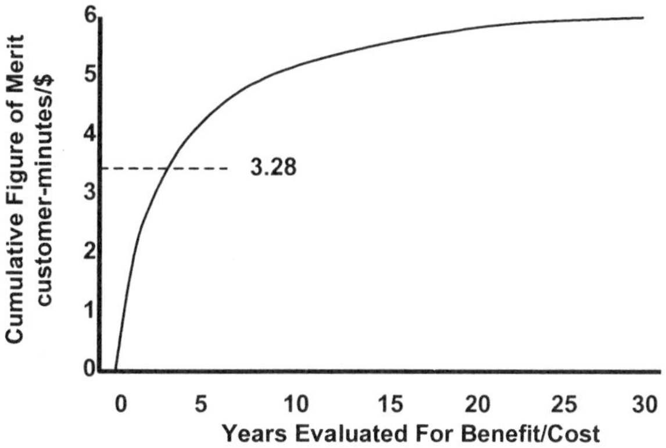

Figure 12.8 Figure of merit for the computerized trouble call system as a function of the time period over which the analysis of benefits and cost is carried out. The system exceeds the 3.28 acceptance limit in Figure 12.7's example in year four, meaning that if that utility is "looking ahead" even just four years, the TMS system's benefit justifies taking money from the three regional maintenance budgets to pay for the system.

budget should be allocated to it versus other maintenance issues.[12] Tree trimming generally makes only a relatively short-term improvement in distribution reliability. Some utilities manage vegetation control on a three-year cycle, but two years is considered the most effective by some experts.[13] Regardless, it often has a very good "bang for the buck."

Table 12.12 shows the evaluation of six tree-trimming alternatives for the eastern operating district of region 2 of the utility example used in Figures 12.6 and 12.7. The six alternatives involve two sets of three options, the sets being for programs that will provide respectively two- or three-years of benefit. Within each of these sets, there are three alternatives. These include: 1) trimming back vegetation from all primary and secondary overhead lines; 2) restricting trimming to just primary circuits or; 3) further restricting it to just along primary trunks and major branches (i.e. not along laterals).

As mentioned above, any of these three programs can be carried out with a three-year, or two-year duration program. Two- and three-year programs vary in how far back the trimming is done, and how much off-right-of-way and off-easement trimming is attempted (with customer permission). They differ in cost (the more comprehensive trimming in the three-year program costs more) and in

---

[12] Very often tree-trimming has a very high effectiveness per dollar and moves to near the top of the RCM list.

[13] See, for example, O. C. Seevers, *Management of Transmission and Distribution Systems,* Fairmont Press, Liburn GA, 1995, p. 170.

Table 12.12 Tree-Trimming Projects Evaluated for Comparison to RCM

Project Number And Area	Type of Work	Years Dur.	Reduced Cust-Min Out Yr. 0	Next	+2 yr.	PW Sum Cust-min	Cost (yr. 0)	Bang/ Buck
Tr 1a-2 E. distr. fdrs	Trim all primary/secondary	2	3500987	2345661		5612082	$1,002,158	5.6
Tr 1b-2 E. distr. fdrs	Trim only primary	2	2625740	1759246		4209062	$454,542	9.3
Tr 1c-2 E. distr. fdrs	Trim only primary trunks	2	2310651	1548136		3703974	$314,429	11.8
Tr 1a-3 E. distr. fdrs	Trim all primary/secondary	3	3579759	3161829	2455067	8554049	$1,517,500	5.6
Tr 1b-3 E. distr. fdrs	Trim only primary	3	3209639	2786258	2118362	7537975	$825,000	9.2
Tr1c-3 E. distr. fdrs	Trim only primary trunks	3	3000000	2550000	1980000	6898800	$600,000	11.5

a very subtle way, in their impact on public relations.[14] Table 12.12 compares these six alternatives in absolute terms (i.e., it is *not* set up as a set of marginal decisions). On its own merits, any of the six programs exceeds the acceptance limit in Figure 12.7. All look like "good projects."

Table 12.13 shows these programs ranked in order of increasing cost, with the impacts and benefits computed as marginal values. It is now a serial list of marginal decisions for use in RCM optimization, as described in section 12.3. The arrows at the right of the table show the serial decision chain, moving from one item to the next in the list, in this case as a series of single jumps down the list. Note that the second and fifth alternatives in Table 12.13 make little practical sense. The second on the list (Trlb-2) has a lower marginal figure of merit (3.6) than the step after it (18.5), indicating that perhaps it should be deleted from the serial list and the decision chain aimed to go around it to the next decision on the list. That would mean that the only decision to make is whether or not to take both it and the decision after it at the same time).

Similarly, item five on the list (Tr1a-2) has a negative marginal figure of merit and should not be selected. Although alternative Tr1a-2 looks good in absolute terms (Table 12.12 shows a figure of merit of 5.6), spending an additional $177,178 on it rather than the next less costly alternative (Tr1b-3) makes no sense because the additional money buys a reduction in improvement.

[14] The customer impacts differ in a way that makes it difficult to identify one program as better in this regard than the other, but on the other hand the difference is significant and should be considered. The three-year program garners a higher number of complaints about aggressive trimming, but it also involves a good deal more pro-active customer contact (requests to trim trees on private property which might fall during storms) and prior notification of trimming that, while legal, might concern homeowners. Surveys indicate this leads to a mostly positive public perception of the utility's efforts. (Table 12.12 is based on the experience of a utility in the southeast U.S. that decided on a three year program due to both its superior economics and because of a simple policy with regard to public relations: "all customer contact, if well-managed, is positive").

Table 12.13 Tree-Trimming Projects Ranked As Marginal Decision Chain

Project Number	Type of Work	Years Duration	PW Sum of Improvements	Cost (yr. 0)	Bang/ Buck
Tr 1c-2	Trim only primary trunks	2	3,703,974	$314,429	11.8
Tr 1b-2	Trim only primary	2	505,087	$140,113	3.6
Tr 1c-3	Trim only primary trunks	3	2,689,738	$145,458	18.5
Tr 1b-3	Trim only primary	3	639,175	$225,000	2.8
Tr 1a-2	Trim primary and secondary	2	-1925,893	$177,158	-10.9
Tr 1a-3	Trim primary and secondary	3	2,941,967	$515,342	5.7

Table 12. 14 Tree-Trimming Projects Ranked As Marginal Decision Chain with "Dumb" alternatives Deleted

Project Number	Type of Work	Years Duration	PW Sum of Improvements	Cost (yr. 0)	" Bang/ Buck"
Tr 1c -2	Trim only primary trunks	2	3,703,974	$314,429	11.8
~~Tr 1b-2~~	~~Trim only primary~~	~~2~~			
Tr 1c-3	Trim only primary trunks	3	3,194,826	$285,571	11.2
Tr 1b-3	Trim only primary	3	639,175.	$225,000	2.8
~~Tr 1a-2~~	~~Trim primary and secondary~~	~~2~~			
Tr 1a-3	Trim primary and secondary	3	1,016,074	$692,500	1.5

Table 12.14 shows the marginal decision table adjusted with items 2 and 5 in the original list deleted, and the marginal costs for items 3 and 6 changed to reflect the benefits and costs of a double jump in the table (the arrows indicate the decision sequence). There are now four items in the serial list with strictly decreasing marginal benefit/cost ratios as one moves down the list.

The optimal decision seems to be the second alternative on the list, Tr1c-3 – trimming only primary trunks, but for a three-year benefit period. This option has an absolute figure of merit of 12.5 (Table 12.12) and a marginal figure of 12.2 in this table. The next alternative in Table 12.14 has a marginal ratio of 2.8, below the utility's service territory-wide acceptance limit of 3.38.

Results are Valid even if Alternatives Cover Different Time Periods

The two foregoing examples bring up an important point. The evaluation method presented here is consistent with respect to different impact periods. For example, one aspect of the difference between the two- and three-year tree-trimming programs that was not addressed above was that in actual implementation, the two-year programs would be applied every two years, and the three-year programs every three years. That does not need to be considered.

In the evaluation method both cost and value are discounted over time, so the computed benefit/cost ratios are consistent to shifts in time. If one takes a two-year program and shifts it two years into the future (two years after doing it the first time we will repeat it) its figure of merit be the same. Similarly, shifting a three-year program three years ahead does not change its figure of merit.

As a result, it is not necessary to compare alternative programs of different equipment programs on the basis of the same time period. Readers interested in confirming this can use the data in Table 12.12–12.14 to build tables representing a six-year period (three two-year periods vs. two three-year periods) or a thirty year program (15 of the former, 10 of the latter). The figures of merit remain the same.

12.7 CONCLUSION AND RECOMMENDATIONS

It is difficult to argue against RCM, particularly in the case of a utility with significant aging power delivery infrastructure to manage. Depending on how one looks at it, RCM either maximizes the value a utility can obtain from a limited budget or minimizes the cost to obtain any targeted level of reliability improvement through maintenance.

Utilities with aging power infrastructures have need of more than the average amount of *results* from inspection, testing, preventive maintenance, and repair. Often they have fewer budget dollars than required for traditional maintenance approaches. As a result they need all the advantages possible with respect to getting the most "bang from the buck" from maintenance. RCM does precisely this. It is implementable with information that any distribution system owner (industrial or commercial or public utility) has, and uses only straightforward procedures (electronic spreadsheets, good record-keeping), although those procedures do have to handle large amounts of projects and alternatives to be effective. Therefore, it is strongly recommended.

Table 12.15 lists a number of recommendations for maintenance prioritization that generally hold true to all utilities but particularly those with aging infrastructures to manage. The rest of this conclusion discusses each item in the table in more detail.

Apply Prioritization to All Parts of the System, Not Just One Level or One Department's Venue

RCM can work at all levels and for all types of equipment at a utility. There is no reason to limit its scope to only one type of equipment (e.g. power transformers) or to within only one department (Substation Operations). In fact, as will be covered in Chapter 14–15, the basic concept should be extended to all activities, and all spending throughout the utility, not just restricted to maintenance.

Table 12.15 Guidelines for Maintenance Scheduling on Aging Infrastructure Systems

- ξ Apply the RCM concept using intra-project alternatives evaluation in the optimization
- ξ Use a reliability definition as close as possible to regulatory mandate or formula.
- ξ Zero-base all decisions.
- ξ Apply intra-project alternatives evaluation for RCM.
- ξ Use the same reliability definition in all regions and for all departments
- ξ Use generic categories of work and equipment along with pre-defined expectations and costs.
- ξ Do not confuse inspection period with improvement period
- ξ Rely on inspection and testing to carry a lot of the burden in any maintenance program

Zero-Base All Decisions

All good O&M planning begins with an assumption that nothing is sacred. Any action has to justify itself on the basis of the benefits it makes compared to the costs it requires. All possible O&M actions are in competition with one another for funding. The most effective will win.

Apply Prioritization Using Intra-Project Alternatives Evaluation

The improvement in "bang for the buck" wrought by applying the optimization to determine both what equipment should be maintained, and how maintenance on each unit should be done can be substantial. The added value is well worth the cost of the more comprehensive and detailed analysis required.

Minimize the Same Reliability Definition Everywhere

Similarly, unless regulatory mandates decree different definitions of reliability or "good customer service" be applied in different areas of the system, the same definition should be used in all RCM in all regions and all departments.

Use a Reliability Definition as Close as Possible to Regulatory Mandate

As discussed in section 12.3, while conceptually there may be indices (definitions of reliability measure) that make sense or seem fair from one perspective or another, ultimately a utility should let the regulatory process define "what is wanted." Once defined, this should be the performance index used in all decisions aimed at "improving reliability."

Use Generic Categories of Work and Equipment Along with Pre-Defined Expectations and Costs

As discussed in Section 12.3, use of categories for both equipment condition and service classes both simplifies application of RCM on a large scale, and keeps it within in a context of using cost and results factors in only a legitimate probabilistic, manner. Categorization within a rigorous format also discourages adjustment of data by individuals based on claimed "judgement" or special knowledge (i.e. fudging the data).

Do Not Confuse Inspection Period with Improvement Period

The periodicity needed for inspection, and the period during which improvements from service are expected to last are not the same. Section 12.3 (Figure 12.5) recommended that the period of improvement due to servicing a unit be modeled as relatively short-term. However, short as it may be, it is often longer than prudent periods for inspection. For example, "major service" might produce an improvement in expected reliability of a transformer that lasts five years. But the utility might decide it is prudent to inspect the unit annually, because problems with something other than what was serviced last time could develop.

Expect Inspection and Testing to Carry a Lot of the Burden

In any optimized program, inspection, testing, and diagnostics carry a good deal of the weight of improvement. In a very real sense, a good RCM plan is its own type of "just in time" system: frequent inspections and testing tend to uncover problems before they occur; a system based on these problems where repairs are only made to items that need to be replaced, and; periodic scheduled repair/replacement. On the other hand, something may be replaced that doesn't seem to really need it, just because it's time.

REFERENCES

J. J. Burke, *Power Distribution Engineering – Fundamentals and Applications,* Marcel Dekker, New York, 1994

EEI Transmission and Distribution Committee, "Guide for Reliability Measurement and Data Collection," October 1971, Edison Electric Institute, New York.

P. Gill, *Electrical Power Equipment Maintenance and Testing,* Marcel Dekker, New York, 1998

Institute of Electrical and Electronics Engineers, *Recommended Practice for Design of Reliable Industrial and Commercial Power Systems,* The Institute of Electrical and Electronics Engineers, Inc., New York, 1990.

H. H. Malik, A. A. AlArainy, and M. I. Qureshi, *Electrical Insulation in Power Systems,* Marcel Dekker, 1998

O. C. Seevers, *Management of Transmission and Distribution Systems,* Fairmont Press, Lilburn 1995

W. A. Thue, *Electrical Power Cable Engineering,* Marcel Dekker, New York, 1999.

13
Planning Methods for Aging T&D Infrastructures

13.1 INTRODUCTION

This chapter looks at planning, specifically electric power system planning, with some comments on planning of aging infrastructures. Readers who want more details, explanation, and example cases should consult the *Power Distribution Planning Reference Book* (see References).

Power system planning involves identification of the best equipment, along with its locations, manner of interconnection to the system, and schedule of deployment. Since cost is an important attribute in power planning, almost invariably one of the planner's chief goals is to minimize overall cost.

Aging T&D Infrastructures → Reliability Optimization

The primary negative effect created by aging T&D infrastructures is poor customer service reliability. All of the other impacts, which boil down to higher costs in one way or another, stem from the utility's attempts to control the situation or improve reliability. Thus, "Planning for Aging Infrastructures" is basically a process of planning reliability and optimizing the reliability gained per dollar and tailoring that reliability of service to consumer needs and utility targets.

This chapter begins with a look at the basic purposes of planning and the five steps of the planning process (Section 13.2) reviewing the most common pitfalls in each. The purpose of short-range and long-range planning and how they interrelate to aging infrastructures is then discussed in section 13.3. Section 13.4 then looks at the T&D planning process as implemented in most modern utilities, again emphasizing aspects that particularly interact with the aging infrastructure issues. Section 13.5 stresses the importance of the systems approach in all planning for aging T&D systems. Finally, section 13.6 summarizes key points with respect to aging infrastructure system planning.

13.2 PLANNING: FINDING THE BEST ALTERNATIVE

What is the Mission?

Planning is a decision-making process that seeks to identify the available options and determine which is the best for the situation at hand. Its purpose is to determine the best schedule of future resources and actions to achieve a company's, an individual's, or an organization's goals with respect to power. Usually, among the major concerns of planning are financial considerations – minimize cost, maximize profit, or some similar goal. In addition, service quality and reliability are almost always considerations. And often, other criteria are important, including environmental impact, public image, and a host of factors that a utility, energy services company, or energy consumer must weigh in making a selection regarding power supply. But regardless, planning is the process of identifying alternatives and selecting the best from among them.

Before beginning the planning step, it is best for a utility to review its mission statement – its overall definition of what guiding purpose and value-system defines its goals. This is particularly true with respect to meeting the challenge of aging infrastructures. Dealing with aging infrastructures, and getting what is desired in the outcome, means having focus and being realistic about constraints, goals, and capabilities. The mission statement is a good place to start. Given the realities of the situation, the aging system, a limited budget, regulatory pressures, is the current mission statement realistic? Making sure that the company's mission statement matches both what it wants to do, now that it has assessed its aging infrastructure problem, and making sure that those principles are realistic and feasible, are big steps toward success.

A company's "Mission Statement" succinctly defines the overall corporate objectives, while its economic goals explain its financial priorities. Usually, mission and financial goals have been incorporated in the economic guidelines and design procedures that the planner applies in his work. Regardless, while Missions Statements are often taken lightly, a review of the company's mission and financial goals provides the planner with strategic insight into its planning guidelines and economic evaluation formula. Table 13.1 lists the Mission Statements from four large metropolitan electric distribution utilities in the

United States. Planning goals for T&D, for DSM, for IRP, *will* be different among these utilities, because they have different priorities for what their executive management wants to accomplish. How the aging infrastructure problem will be handled in each utility and the priorities the utility puts of various approaches and solutions should be different as a result.

Mission statements are qualitative assertions of the overall philosophies and goals that lead to the quantitative engineering and economic criteria used in distribution planning. By understanding *what* their company is trying to do, the T&D planners have a better understanding of *how* to achieve those goals.

Table 13.1 Mission Statements of Four Electric Utilities

1. "Our company will be the premium regional provider of electric power."

   Recognizing that its current financial situation prohibits competing on the basis of price, this utility has decided to make quality and service its hallmark. Achieving lowest *possible* cost is not the goal; achieving low cost while meeting high service standards is.

2. "Provide economical electric power for the prosperity of the region."

   This municipal utility has a long-standing tradition of low rates, a way of attracting new industry (i.e. growth) to the region. Plans that invest a good deal to improve quality are simply "not with the program." Marginal quality improvements in a new plan are permissible, only if they lead to lower cost.

3. ". . . maximize return on equity."

   This criterion means the utility wants to earn the most on what it currently owns. Utilization of *existing* T&D facilities is the key. This utility is more reluctant to invest by borrowing than a utility that wants to maximize return on investment. Plans that call for massive capital spending may be unacceptable.

4. ". . . maximize return on investment."

   This mission makes no direct distinction between equity and debt – apparently the utility is willing to borrow, if it increases what it can earn on that investment. A plan calling for considerable capital outlay is permissible, if that investment yields a good return.

Planning: a Five-Step Process

Planning is a process of determining the best course of action or schedule of commitments that can be made to achieve one's goals. The planning process can be segmented into the five steps shown in Figure 13.1. Each one is an important part of the process for planning in any type of industry, but in particularly utility T&D planning.

Any of the five steps, poorly performed, can lead to poor decisions, a poor plan, and ultimately failure to attain those goals. Mistakes and lack of focus in any step will lead planners astray, although the types of mistakes that result will differ, depending which one of the five was the first misstep. Below, each of the five steps is described, along with a brief summary of the goals for this step and pitfalls to be avoided in each. A lengthier discussion, with examples, can be found in the *Power Distribution Planning Reference Book* (see References).

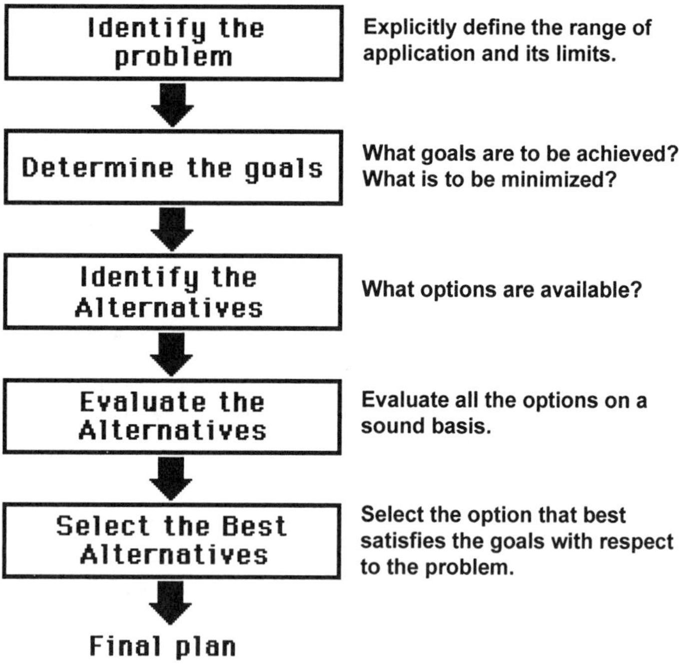

Figure 13.1 Planning involves the five steps shown. See text for details.

Step 1: Identify the Scope of the Problem

Before embarking on data collection, study, and analysis, every group of planners should identify their goal – exactly what they are trying to do. It is recommended that the planning problem and scope be formally written, as "this project involves planning of short- and long-term electric supply for all 821 homes planned for the new West Oaks subdivision." This written definition of the scope defines the *extent* of the problem to be solved. All subsequent parts of the planning process should focus on that, and not stray (as some planners are often wont to do) to solve other problems or work in other areas. Trying to solve some existing circuit overload problems in the nearby area at the same time this goal is achieved, is not part of this project.

Step 2: Identify the Goals

A common mistake made by many planners is to begin their activity without specifying the goals for the particular planning situation at hand, or identifying how these goals might have changed from prior planning situations. It may be one thing to identify the problem, but the goals must explain in detail to identify what the planners need to accomplish.

Goal definitions are quite different than mission statements. A mission statement lays out an organization's direction, principles and values. A goal defines explicitly what is desired to be accomplished in some situation. The former is strategic, the latter tactical. Time devoted to framing the planning situation at hand and explicitly defining the goal for it will be time well spent.

Again, as was the case with scope statement, it is best to produce a formal, written definition. This might read something like: "The goal is to develop a plan, able to be implemented within two years, which identifies all required equipment, labor, permits and other expenses for electric facilities. This will provide the lowest long-term cost when evaluated on a present worth basis, and provide West Oaks subdivision with expected SAIDI and SAIFI of 2.0 hours and 1.25 events respectively, with all system factors meeting company standards."

Aging Infrastructure Planning Often Encounters Severe Esthetic and "Fit-In" Constraints

Very often, particularly when dealing with aging infrastructure problems in already-developed metropolitan areas, there are additional goals unique to a particular plan or situation. For example, it might be important to accommodate the esthetic and land-use preferences of local community leaders, or to use equipment or rights-of-way and sites already in inventory. Usually, such special goals are unique to one particular case, created by some unusual situation or set

Table 13.2 Examples of Special Planning Goals or Criteria Often Encountered When Dealing with Aging T&D Systems

Fit within a specific budget.

> The utility may have a particular amount of money designated for aging infrastructure improvements. This money must be spent in an optimal manner, but further funds will not be available. Budget-constraining techniques (see Chapter 5) are often helpful in this situation.

Substation improvements must be accommodated within existing sites.

> The utility either cannot obtain, or management does not want to endure the political hassle, uncertainty, and cost of trying to gain more real estate at existing substation sites. As a result, innovative ways of cramming more capacity and more complex configurations into an existing, already-crowded site, such as use of GIS equipment, compact line designs, and small-footprint transformers, must be found.

No new substation sites and rights of way.

> This is occasionally a goal of management – to minimize the number of new substation sites, ideally to zero. New sites are expensive, subject to political and regulatory hurdles that make approval uncertain, and often create adversarial relationships within the community, which management judges it is best to avoid "at any cost." Lack of new sites does not cause a capacity limitation issue (lack of new room at existing sites does). It causes a problem with economic and capacity reach issues of the feeder system. So called long-reach planning techniques (*see Power Distribution Planning Reference Book,* Ch. 7-9) are often useful in these situations.

Expansion plan must include "targeted DG and DSM."

> Regulators may demand that any solutions include an objective use of any viable means to cut T&D capacity needs, including load reduction methods and on-site generation. Integrated resource T&D planning and DG planning methods are often needed in these situations.

of circumstances. Table 13.2 lists a number of special goals that are not uncommon to encounter when planning for an aging T&D system.

Step 3: Identify the Alternatives

This third step is often the most critical part of planning. This is identifying what *could* be done – all the alternatives open to the planner. What options are available? What variations on these options would be possible? This function encompasses determining the *range* of possibilities for solving the problem.

This is where the majority of great "planning disasters" occurs – the mistakes that lead, weeks later, to statements like, "Why didn't we think of that?" For, while apparently the least challenging part of planning, a good deal of skill, breadth of thinking, and time, are required to identify the range of possibilities. Time and resources are always limited, and the temptation is strong to assume that one can see all the options just by "a quick look at the problem." But this is seldom the case. A brilliant solution to an unusual problem most often comes from recognizing that there are also unusual options available to solve it.

The downfall of planners: failing to consider all the alternatives

This step is where a surprising number of computerized "automated planning procedures" fail. When looking at or reviewing their planning procedure, electric system planners should study if and how it assures them that it *does* consider all possible options. Just as planners need to remind themselves constantly to be open-minded, and not to have pre-conceived notions about where the solution lies, they need to assure that computer programs used are also not limiting their identification of the full range of options.

A common failing: too short a planning horizon

One way that electric utilities have traditionally failed to look at all alternatives, is by using too short of a planning horizon. Failing to look far enough ahead often limits the available options by default, contributing to poor planning. A common lament among T&D planners at some utilities is, "We could have done a much better job, but the good sites and land were already taken." What is usually left out of this complaint is that the planning period used for distribution expansion evaluation was only three to four years ahead. The utility is losing out to other people interested in acquiring sites for other purposes, people who are looking ahead more than three or four years.

By looking only a few years ahead, the utility reduces the number of options available to it. Yet despite the best sites and routes being committed to other purposes, the distribution planners *will* find a way to do the job, but the resulting plan will usually have *both* higher costs and lower reliability than would have been the case had better sites been obtained. Within reason, the recommended approach to any type of planning is to lengthen the planning period sufficiently to include all possible alternatives.

Having a sufficiently long planning period does not in and of itself assure good planning, but it is a necessary first move in that direction. For distribution planning, a *minimum* planning period of five to seven years is normally recommended. Within this, the utility can identify sites if and why and where it needs to obtain them early, before they are unavailable. Savings accrue both because the utility obtained the site early at a lower cost, but most important because the site at the correct location *is* obtained. Often, if the utility waits, a site at the best possible location is not available at any price. By using a lead time long enough to give it these options, the utility is performing the first function of the planning process correctly – assuring itself that it considers all possible alternatives.

"Do nothing" should always *be one of the alternatives considered*

One of the most serious mistakes made by planners is to assume that *something* must be done. Occasionally, it is both less expensive and better from a service standpoint to delay any commitment or change to the electric system – to do nothing. Even when this is not the case, "nothing" should always be included in a planning report as one of the reviewed options. This permits the planners to explicitly identify *why* something must be done, by showing that "doing nothing" would violate criteria or requirements, or lead to very poor economy.[1] Evaluation of "nothing" is a key element of justifying a plan's recommendations.

Step 4: Evaluating the Alternatives

All alternatives should be evaluated against a common and comprehensive evaluation standard, one that considers every aspect pertinent to any option, and one that addresses all the goals. For power delivery planning this means evaluating alternatives against criteria and attributes that represent the utility's requirements, standards and constraints. For DSM programs this means matching customer needs with marketing standards and guidelines.

Very often, the actual planning process will combine the evaluation and selection functions in a process designed to reject alternatives quickly and with minimum effort. Regardless, all alternatives should be evaluated completely with respect to everything that bears on the problem at hand. Planning methods should be examined carefully to make certain that they, too, like the various alternatives being considered, are complete.

Alternatives must be evaluated against both criteria and attributes. *Criteria* are requirements and constraints the plan *must* meet, including: voltage, flicker, other service standards, contingency rules, loading limits, safety and protection standards, and so forth. Such criteria only have to be satisfied: if standards

---

[1] A common argument against this approach is that evaluation of "do nothing" is not needed because the answer is so obvious that it does not need to be included. But if that is the case, it is simple to do, so why not include it and document its evaluation?

Table 13.3 Some Typical Criteria and Costs Evaluated in Electric System Planning

Criteria	Attributes
Voltage standards	Cost of equipment
Noise and emissions limitations	Cost of land and ROW
Safety and clearance standards	Cost of labor and construction
Electrical and fire protection standards	Cost of taxes, insurance, etc.
Pollutant/toxicity material restrictions	Cost of design and engineering
Aesthetic impact and standards	Cost for maintenance and operations
Maintenance access requirements	Cost of losses
Equipment loading standards	Cost of permits, licenses, etc.
Contingency margin requirements	Flexibility in future design
Equipment loading standards	Salvage cost (at end of life)
Construction method requirements	Fuel & fuel delivery cost

allow voltages down to .95 per unit, then a design alternative is acceptable if it has voltages of .95 per unit, but no lower.

On the other hand, attributes are values for which the planners wish to achieve the best performance possible within limits set by other aspects of the problem. An *attribute* is a quality that is to be *minimized* (or maximized) while still meeting all criteria. One attribute in nearly all electric supply planning is cost. And while voltage drop, protection coordination, and other requirements are quantities that the planners would *like* to see exceed standards by comfortable margins, they should be treated as criteria: good enough is good enough.

Traditional power delivery planning is *single attribute planning,* in that only one attribute (cost) is to be minimized. This evaluation step rejects the alternatives that do not meet all criteria. Among those left, the minimum cost alternative can be selected (in step 5). Cost is a multi-faceted attribute, *all* aspects of which must be included in the evaluation: equipment, site costs, taxes, operations and maintenance, and losses. Table 13.3 lists some of the more common criteria and cost factors that need to be assessed in electric supply evaluation at the distribution level.

Where mistakes occur in the evaluation step

Mistakes that take place in the evaluation function generally occur because the planners do not check to see if their methodology meets three requirements:

1. *It considers all criteria and factors that are important to the goals.* For example, if a goal is to achieve a certain level of expected

reliability, then the planning methods need to be able to assess the potential reliability of all plans on an equivalent basis.

2. *It must evaluate all criteria fully with respect to all resources.* For example, some planning methods evaluate reliability only with respect to outages and connectivity. When evaluating DG and DS, the reliability method must analyze and accommodate lack of fuel availability and potential for failure of the units themselves.

3. *It treats all options equitably.* An evaluation that considers all factors affecting some options but not all factors affecting others provides a biased comparison. For example, if planners include the cost of oil spill containment and fire prevention equipment when evaluating T&D alternatives that include oil-filled transformers, they should provide like detail in the assessment of the DG alternatives.

Step 5: Selecting the Best Alternative

In many planning procedures, this function is combined with the evaluation function, and it is difficult to identify where evaluation ends and selection begins. In fact, efficient computerized planning methods that help planning engineers minimize the time and effort required to complete the planning process tend to combine Steps 3 through 5 into one process. But regardless, of the most important point in the identification-evaluation/selection process is the *definition* of "best" must truly match the goals and value system being used. The evaluation/selection method being used must mirror the desires of the organization, and be capable of distinguishing between alternatives in a valid manner in these areas.

A surprising number of planning mistakes, or inefficiencies, occur in the final selection process, because the planners do not assess alternatives on the basis of what is truly important to their goals. A planning method's selection function should be examined to determine that it:

- *Uses the definition of "best" that matches the planning goals.* The fact that alternatives were evaluated on the basis of a particular attribute in Step 4 does not mean that it is weighted properly in the selection phase. For example, a common failure in T&D planning is to assess the value of electric losses in a distribution plan, but then fail to acknowledge their value in selecting the best design.

- *Can accurately distinguish between alternatives.* What are the limits of accuracy of the planning process? If it is only accurate to within 3%, then selection of a "best" alternative on the basis of its being 1% better than another is questionable. Error range in the evaluation and selection among close cases must be small enough to assure that the determination of "which is best" is accurate.

Many computerized planning procedures produce only approximate evaluations. Planners often fail to consider these limitations and interpret minor differences in computed evaluations as significant, when they are not.

13.3 SHORT- AND LONG-RANGE PLANNING

Lead-time – the time it takes to prepare, build, and install facilities – sets the minimum planning period for any endeavor. If it takes up to five years to order materials, obtain permits, survey, build, test, and put into service a new substation in the core of a major metropolitan area, then the power distribution utility has no choice but to plan at least five years ahead in order to make certain it will have all new substations by the time they are needed.

Thus, the minimum possible planning horizon is a function of the lead-time required to put solutions into place. Lead-time varies depending on the type of equipment, the level of the system, and the local planning approval process for electrical facilities, and other aspects of the situation. Table 13.4 gives some typical lead times by "level of the system."

Table 13.4 Typical Minimum Lead Times at Various Levels of the Electric Utility Power System

Level	Years Ahead
Central Station Generation	10
EHV Transmission (>500 kV)	9
Transmission	8
Sub-transmission (< 140 kV)	7
Primary Distribution Substations	5
Primary Voltage Feeders	3
Rebuilding existing substations	2
Rebuilding existing lines – UG	2
Rebuilding existing lines – OH	1
Lateral primary voltage taps, changes	1
Distributed generation (>10 MVA)	2
Distributed Generation (10 - 1 MVA)	1
Dispersed generation (< 1 MVA)	.4
Service level	.2

Short-Range Planning: Decisions and Commitments

The purpose of electric utility short-range planning is to make certain that the system can continue to serve customer load while meeting all standards and criteria. It is driven by the lead-time – by the fact that it takes time to get things done and that decisions must be made ahead of this lead-time. Short-range planning must identify the commitments and purchases that need to be made today in order to allow them to be implemented in time to maintain service-within-standards in the future. In the case of the substation given earlier, it takes five years to make all preparations. Therefore, the utility must plan at least five years ahead, so it knows when to "trigger" the process of creating that substation. Using a shorter lead-time means selecting "do nothing" by default.

The product of the short-range process is a series of *decisions* – the selection of various alternatives over other ones – for the expansion of the electric service facilities. Usually, these specify what facilities and equipment are to be bought, where they will be put, how they will be fit into the system, as well as what other commitments and plans need to be enacted. Short-range planning initiates a process which fills in the detail and produces the land acquisition requests, drawings, permit requests, materials lists and construction authorizations, hiring, re-organization, marketing, advertising, customer survey, and other efforts necessary to make it happen.

The "short-range plan"

The "short-range plan" is a schedule of additions, enhancements, reinforcements and changes that are to be made to the system and that have been authorized and committed. Although there is always some recourse in any plan, the short-range plan is composed of decisions which have, in effect, been "locked in." As such, it tends to be very "project oriented," as shown in Figure 13.2. The short-range planning process leads, ultimately, to a series of separately identifiable project authorizations, each committing to a particular addition(s) to the system, with its own scope of work, schedule, budget and justification. These project authorizations are the hand-off from the planning process to the engineering and construction process.

Long-range Planning: Focus on Reducing Cost

Long-range planning focuses on making certain that the equipment and facilities called for in the short-range plan provide lasting value and the lowest overall cost during their lifetime, unlike short-range planning, which seeks to identify

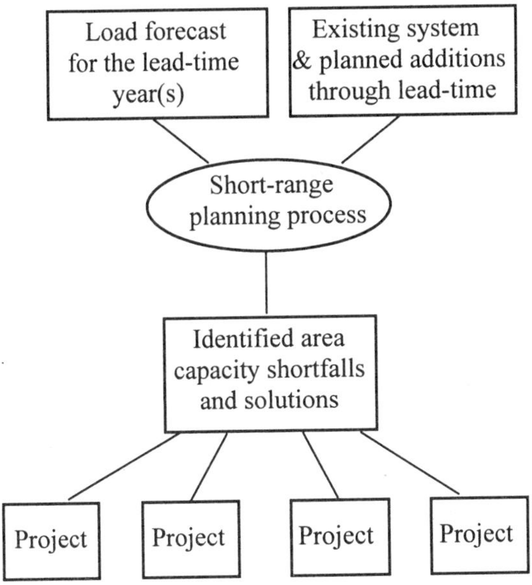

Figure 13.2 The short-range planning process consists of an initial planning phase in which the capability of existing facilities is compared to short-term (lead time) needs. Where capacity falls short of need, projects are initiated to determine how best to correct the situation. The load forecast is primarily an input to the short-range planning process, i.e., it is used to compare to existing capacity to determine if and where projects need to be initiated.

problems and solve them before they occur. The reason to look beyond the lead-time, and to plan for the long run, is to assure the decisions being made have the most lasting value.

How can a planner know if a new distribution line, DG unit, or other electrical addition is a good investment, with a low lifetime cost and a high utilization over that lifetime? That lifetime begins when the new line is completed and put in place, *which happens at the end of the lead-time.* Thus, long-range evaluation begins at the lead-time and assesses cost, value, and equipment contribution over a lengthy period thereafter. Its purpose is to determine if the short-term decisions will have long-term value. Given typical economic factors, and considering the uncertainties in predicting conditions over the very long term, a ten-year period is generally considered the *minimum* for such economic evaluations.[2]

A good long-range planning process is an end in itself

Long-range T&D planning is one of the few legitimate instances where a process' major goal is itself. The long-range plan, along with its accompanying budget estimates and indicators of economic factors for the future system, *is* the major goal of the long-range planning process. More importantly, it is crucial that this plan's development be ongoing, that it be maintained and continuously updated in order to provide long-range evaluation, guidelines, and direction to T&D planning and investment decisions.

To study a new transmission line, substation, distributed generator, or other planned expansion item, and determine its impact over a future period from five to 25 years ahead, the planners need to have a good idea of the future conditions throughout the time the new project will be in service. This includes a forecast of the future load levels it will be called upon to serve, as well as a description of the system and conditions within which it will operate. They need a *long-range plan,* not just for the particular facility they are planning, but as a backdrop to their evaluation of this project.

This long-range plan must lay out the economics (value of losses, etc.) and operating criteria (usually assumed to be the same as today's) that will be applied in future operations. It must specify the load that will be served (long-range load forecast), and it must specify *the other electric additions* that will be made during the period (e.g. other DG units planned for future years). This is so their interaction and influence on the performance and economy of any proposed short-range addition or change can be fully assessed over a good portion of its lifetime.

The long-range plan does not need great detail

A long-range plan needs only enough detail to permit evaluation of the economics and "fit" of short-range decisions regarding long-range system requirements. A long-range plan needs only a certain amount of detail, sufficient to its purpose. Effort beyond this is wasted.

---

2 When using typical discount rates (e.g. 11%) analysis of a ten-year period of service captures 66% of the present worth of a level investment with a 40-year lifetime, while a 15-year period captures 80% of its present worth. Thus 10- to 15-year periods that begin at the lead-time see a majority of the present worth aspects of an investment, but not all.

Other functions of the long-range plan

A good long-range plan provides several other useful functions, as listed below. The first is the long-term economic evaluation discussed above, the reason why a long-range plan *must* exist if good economy of investment is to be assured. But there are other valuable contributions that the long-range plan makes to the planning effort, and to the utility as a whole:

- *Forecast of long-range budget.* The long-range plan identifies capital costs, operations and maintenance costs, and losses costs to the point that they can be estimated for budget and financial planning purposes.
- *Identification of long-term direction and strategy.* A long-range plan provides a clear direction and long-term strategy for the electric system's future. It may not be the best direction or strategy, but good or bad, it shows engineer, manager, operator, marketer and executive alike, the current vision of how electric power will be supplied.
- *A basis for evaluation of new ideas or changes in procedure.* Proposals for changes in standards or system design are difficult to evaluate unless they can be compared on a fair basis against present standards and future needs. A long-range plan provides a "base case" to which new proposals can be compared, whether they suggest adopting a new primary voltage (23.9 kV versus the current 12.47 kV) or other changes in approach.
- *Coordination of planning among levels of the power system.* At some level, someone benefits if the plans for transmission, substation, distribution and distributed generation levels of the system fit together well. To assure this, their plans must be coordinated. This is best done by looking at the long-term implications of their interactions.

Table 13.5 Purposes of the Long-range Plan

Definition of "future" for evaluation of short-term planning decisions.
Identification of mismatches between levels – coordination of plans.
Forecast of long-range budget needs (for financial planning).
Identification of long-term direction and strategy.
A basis for evaluation of new ideas or changes in procedure.

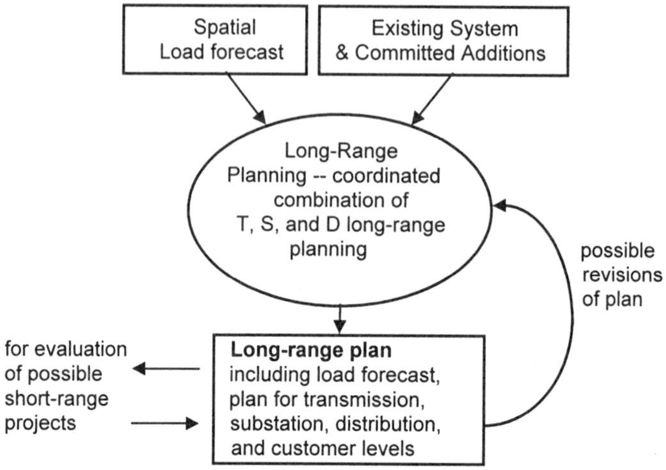

Figure 13.3 The coordinated multi-level long-range plan, where the major goal is the maintenance of a long-range plan that identifies needs and development of the system, so that short-range decisions can be judged against long-range goals. Here, the load forecast is a part of the plan, and there is no "project oriented" part of the activities.

This long-range plan must lay out the economics (value of losses, etc.) and operating criteria (usually assumed to be the same as today's) that will be applied by the utility in its future operations. It must specify the load (long-range load forecast) the system will be expected to meet, and it must specify how the T&D planners expect to handle that load growth – what new facilities will be installed in the long run to accommodate the future pattern of load growth. Most importantly, it should specify how the various long-range plans at the sub-transmission, substation, feeder, and customer levels will fit together. The economics of interaction of all four levels, leading to a least-cost system, is assured by evaluation of long-term PW value in a coordinated manner – Figure 13.3 shows the major flow in the long-range distributon planning process, while Table 13.5 summarizes its key purposes.

Handling Uncertainty: Multi-Scenario Planning

An important point to bear in mind is that long-range elements in their plan do not have to be built, committed to, or even decided upon, for years to come. This means that the utility can change its mind, or beyond that, be of two or more minds at the same time. *There can be more than one long-range plan.*

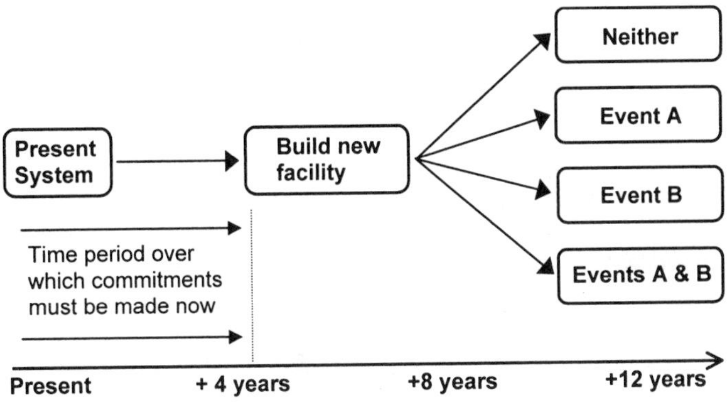

Figure 13.4 Multi-scenario planning uses several long-range plans to evaluate the short-range commitments, assuring that the single short-range decision, fits the various long-range situations that might develop. This shows the scenario variations for the case involving two *possible* future events.

Uncertainty in predicting major events is a major concern in long-range planning. Will a possible new factory (employment center) develop as rumored, causing a large and as yet un-forecasted increment of growth? Will the bond election approve the bonds for a port facility (which would boost growth and increase load growth)? Situations such as these confront nearly every planner. Those of most concern to distribution planners are those factors that could change the location(s) of growth.

In the presence of uncertainty about the future, utility planners face a dilemma. They want to make no commitment of resources and facilities for load growth that *may* develop, but neither can they ignore the fact that there are lead times required to put facilities in place, and that the events *could happen.* Given the reality of lead times, planners must sometimes commit without certainty that the events they are planning for will, in fact, come to pass.

Ideally, plans can be developed to confront any, or at least the most likely, eventualities, as illustrated in Figure 13.4. *Multiple long-range plans,* all stemming from the same short-range decisions (decisions that must be made because of lead times) cover the various possible events. This type of planning, called multiple-scenario planning involves explicit enumeration of plans to cover the various likely outcomes of future events. It is the only completely valid way to handle uncertainty in T&D forecasting and requirements planning.

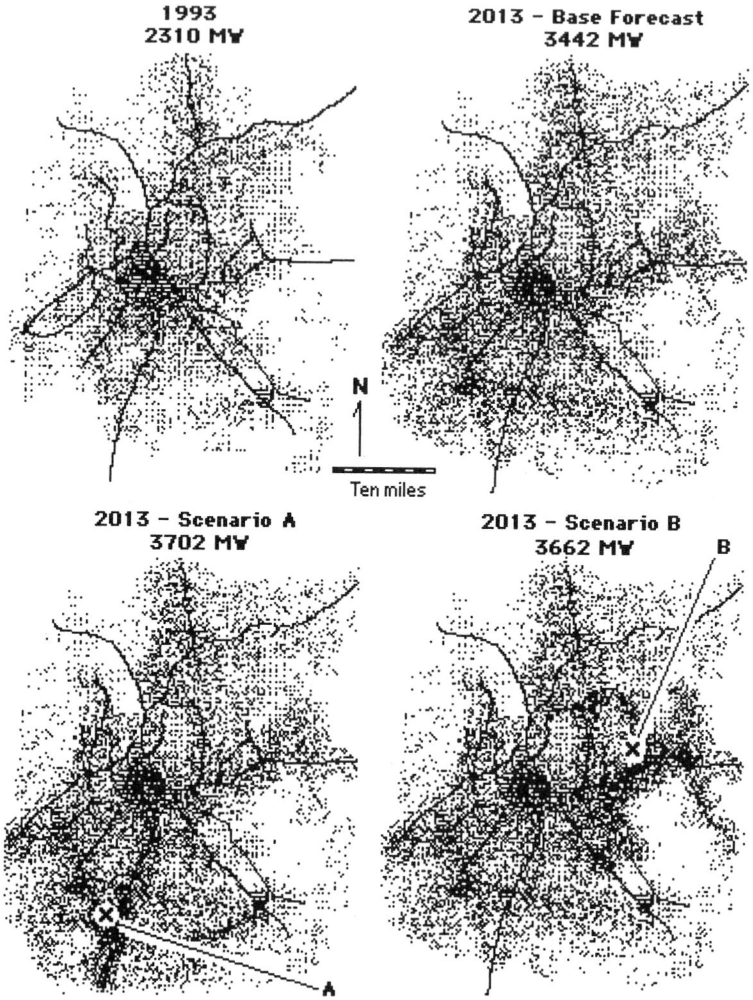

Figure 13.5 The base T&D load forecast (upper right) for an American city is based on recent historical trends. However, given voter approval of a $145 million bond issue, as well as approval by state and federal authorities, a "major theme park" will be built within five years at one of the two locations shown. The economic growth generated by the park could mean about 260 MW of additional load growth by 2013, in either spatial distribution shown, depending on the park's location. Utility planners should explicitly consider each of the forecasts, as described in the text, and develop plans accordingly (Forecasts A and B differ as to total amount of load because a good portion of the park's growth falls outside the utility's service area, to the east, in scenario B).

Uncertainty in T&D growth forecasts cannot be addressed by planning for the expectation of load growth

One of the worst mistakes that can be made in T&D planning, integrated or otherwise, is to try to circumvent the need for multi-scenario planning by using plan. This approach invariably leads to plans that combine poor performance (as evaluated with respect to both electrical and customer service) with high cost. "average" or "probabilistic forecasts" to develop a single "middle of the road"

As an example, the forecast in Figure 13.5 shows a detailed spatial forecast of electric load growth over a twenty year period for a large city (intermediate years were forecast, too, but are not shown). At the top are maps of load in 1994 and as projected 20 years later in the base case. The difference in these two forecast maps represents the task ahead of the T&D planners – they must design a system to deliver an additional 1,132 MW, delivered in the geographic pattern shown. This is a formidable task, even though they have a total of 20 years to implement whatever plan they decide upon.

At the bottom, in Figure 13.5 are two alternate forecasts that include a possible new "theme park and national historical center" to be built by an international entertainment mega-company. Given unpredictable municipal voter approval of certain tax credits, uncertain approvals and incentives from state officials, and hopefully continued commitment by the company that wants to build the park, the theme park/resort area would be built at either of two locations within the next few years. It would not only generate 12,000 new jobs but bring other new entertainment and tourism industries to the area, causing tremendous secondary and tertiary growth, and leading to an average annual growth rate of about 1% more than in the base forecast during the decade following its opening.

Thus, the theme park means an additional *260 MW* of growth on top of the already healthy expansion of this city. The maps in Figure 13.5 show where and how much growth is expected under each of three scenarios:) no park (base forecast, considered 50% likely); 2) site A for the theme park (25% likely); or 3) site B for the theme park (25% likely). As can be seen, the locations of the additional scenario growth will be very much a function of where the theme park is located. Hence, the where aspect, as well as the how much, of the utility's T&D plan will change, too. T&D equipment will have to be re-located and re-sized to fit the scenario that develops.

Some announcement about the decision on the theme park will be forthcoming within the next two to three years. Thus, the uncertainty will be resolved long before most of the facilities required are put into place. But the poor planning that would be done today, if the park were not taken into account, would make a significant impact on future costs and reliability. This impact

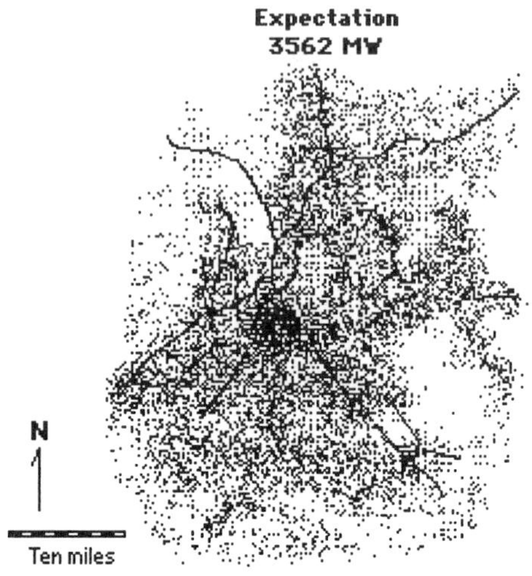

Figure 13.6 The probability-weighted sum of the three forecasts in Figure 13.6. This is perhaps the *worst* forecast the T&D planners could use to develop their plans.

could be significant and long-lasting.[3] Regardless of how they plan, whether they use the expectation of load growth or just the base forecast, the T&D planners in this case will have a chance to change direction once the uncertainty is resolved – they will patch up their plan and somehow "muddle through." But a much better way to handle uncertainty is to recognize that there *will* be a need to change direction and to plan for *that.* This is the essence of the multi-scenario approach and the concept that was illustrated in Figure 13.4.

Thus, the planners need to study the T&D needs of each of the three growth scenarios and develop a long-range plan to serve that amount and geographic pattern of load. Ideally, they will be able to develop a short-range plan that "branches" after the lead-time, to three different plans tailored to the three different patterns of load growth. This may not be possible, but the planners will never be able to do it unless they study the situation and try.

One thing the planners do *not* want to do is form a single forecast map based on a probability-weighted sum of the various possible forecasts, an "expectation of load growth" map. Such a map is easy to produce. Figure 13.6 is the

---

[3] Quantitative evaluation of the impact of incorrect forecasts or scenarios can be done using a method described in Willis, Powell and Tram, 1987. See References section at end of chapter.

probability-weighted sum of the three forecast maps shown in Figure 13.6.

While mathematically valid (from a certain perspective, at least) this forecast map will contribute nothing useful to the planning effort – it represents a future with 1/4 of a theme park at location A, 1/4 at location B and the other half somewhere else. That is one scenario that will *never* happen. Using it as the basis for T&D planning will lead to plans that spread resources too thinly over too much territory. In this case, it would lead the utility to plan additional T&D capacity to handle about half of the load difference represented by the theme park – capacity to serve about 140 MW of load – split between the south part of the system (site A) and the east (site B). No matter which scenario eventually develops, the system plan based on contain several major and potentially expensive flaws:

1) If the base case actually develops, then the planners have "wasted" facilities and resources capable of serving about 140 MW of capacity, installing them when and where they are not needed.

2) If Scenario A develops, the planners have wasted the capacity additions they made in the east part of the system (capable of serving about 70 MW). Additionally, they have put in place around site A facilities that fall short of the capacity needed there by 75%.

3) If Scenario B develops, the planners similarly have wasted the capacity additions made in the south (equal to about 70 MW). Additionally, they have put in place around site B facilities that fall short of capacity requirements by 75%.

Planning with the forecast shown in Figure 13.6 guarantees only one thing: the T&D plan will be inappropriate no matter what happens. In particular, if either of the alternate scenarios occurs, the plan is very non-optimal.

13.4 THE T&D PLANNING PROCESS

Figure 13.7 shows the T&D planning process as it is often represented, consisting of five stages, including a load forecast (customer needs), followed by coordinated steps of transmission, substation, distribution, and customer-level planning. The exact organization and emphasis of these individual-planning steps will vary from one utility to another. A particularly effective organization of these steps will be presented later in this chapter. What is important is that all are accomplished in a coordinated manner. Other chapters in this book will focus on the data, procedures, and analytical techniques needed in each step. Here, the emphasis is on what each of these steps is and how it fits into the overall planning process.

Of the five stages, only the load forecast involves a process over which the utility has no control. The other four concern resources over which the utility has considerable, although often not total, freedom of action and investment. Each of the five stages consists of both short- and long-range planning portions, of length shown in Table 13.4. For the load forecast, "short-range" planning consists of producing a single "base" forecast – a non multi-scenario "most accurate" forecast for use in all short-range T&D planning. This need cover no more than a decade ahead, that being the longest short-range planning period for any of the levels shown in Table 13.4. For the other four stages, short-range planning is aimed at achieving recognition of problems and definition of solutions within their lead time requirements.

Long-range planning is focused on evaluating the utilization and investment economics of all short-range projects, using a present-worth perspective or something similar. Short-range projects and decisions are judged against this sounding board of long-range performance. The load forecast can and often will

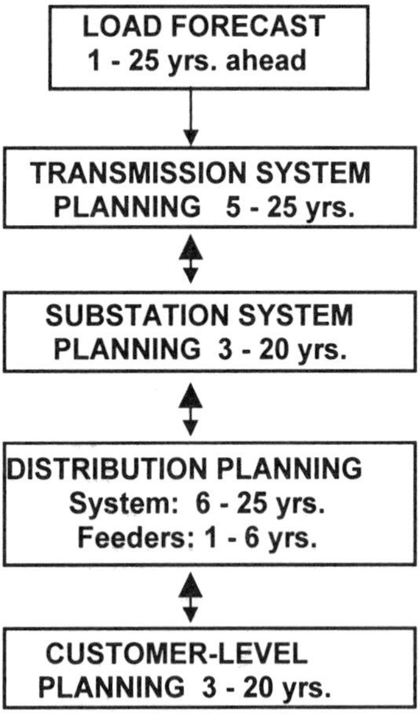

Figure 13.7 The basic T&D planning process involves five discrete stages.

Table 13.4 Typical Short and Long-range Planning Periods for T&D System Planning

System Level	Planning Period - Years Ahead Short-range	Long-range
Large generation* (> 150 MVA)	10	30
Small generation* (< 50 MVA)	7	20
EHV transmission (500 kV and up)	10	30
Transmission (1115 kV to 345 kV)	8	25
Sub-transmission (34 kV to 99 kV)	6	20
Distribution substations	6	20
"Feeder System" (service area) planning	6	20
Primary three-phase feeders	4	12
Laterals and small feeder segments	1	4
Service transformers and secondary	.5	2
Customer level (DSM, RTP)	5	20

* Generation time periods are for "traditional" planning in a vertically integrated utility planning within a least-cost monopoly-franchise regulatory structure. Generation in a competitive environment might be planned within a profit-payback structure with shorter time periods.

be multi-scenario. Working for that, the long-range planning steps of all four subsequent stages are merged into a single, *long-range T&D plan,* with appropriate and coordinated representation of all system levels.

The short-range planning periods shown in Table 13.4 are recommended for normal utility circumstances. The long-range planning at each level begins at that short-range lead-time and extends through a period at least twice and as far as four times farther into the future. *Coordination of focus, plan, and criteria among these five steps is the single most important quality-related issue in T&D planning.*

Each of the major levels of the power system – transmission, substation, distribution, customer, is connected to at least one other level. Its economy, performance, and flexibility of design interact with that other level, to the extent that no decision at one level is without noticeable, and often major, influence on the others. Coordination of method and criteria among the levels of the system is vital to successful, minimum-cost planning.

Spatial Load Forecast

The planning process is driven by a recognition that future customer demands for power, reliability, and services may differ from those at the present. The load forecast projects future customer demand and reliability requirements, which define the minimum capability targets that the future system must meet in order to satisfy the customer demands. Forecasting for T&D planning requires a spatial forecast, as was illustrated in Figure 13.5.

Spatial forecasts: critical to aging infrastructure planning, particularly when on a tight budget

Chapter 8's discussion of the problems traditional contingency-based planning encounters in dealing with modern utility systems demonstrated how one aspect of modern system interaction is a need for broader areas of contingency support. One result of this is that the interaction of capacity/load ratios in one area of a system, with that in other areas, is of greater importance in the overall performance of the power system. When a power system is loaded to modern (high) equipment utilization rates, and composed of aged (failure-prone) equipment, the reliability of service in one area of the system is sensitive to the quality of the planning done in nearby areas. All need each other for support, and a mistake in one area affects not just that area (as it did traditionally) but all its neighbors. The cost of making an error in matching capacity to peak load is much greater in contemporary power systems than it was in traditional power systems.

Spatial forecasts have always been an important element of good power delivery planning. But they take on an even more critical importance because of these aging infrastructure issues. The forecast of location of customer demand is as important in T&D planning as the forecasting of the amount of load growth: both are required in order to determine where facilities need to be located and what capacity they must have.

- *Timing.* Usually, spatial forecasts project annual figures (peak, etc.) over a period into the future or from three to five times the lead time of the equipment being planned – a spatial forecast for substation planning (five year lead time) would be carried out through 15 to 25 years into the future.

- *Spatial resolution.* Spatial forecasts are accomplished using the small area technique (Figure 13.8), dividing the utility service area into hundreds or perhaps tens of thousands of small areas and forecasting the load in each.

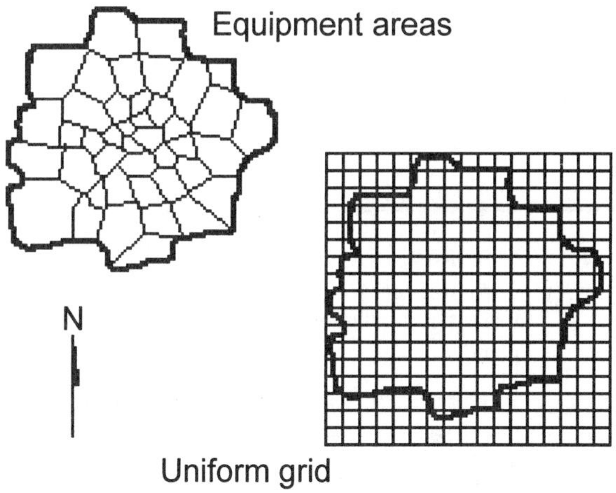

Figure 13.8 A load forecast for T&D planning (such as that shown in Figure 13.5 is accomplished by dividing the service territory into a number of small areas – either equipment service areas or the cells of a square grid – and forecasting customer demand in each.

As a general rule of thumb, the small areas must be no larger that about 1/10th the size of a typical equipment service area at that level, in order for the forecast to have sufficient "where" information for planning. Thus, spatial resolution needs are less at higher voltage levels – transmission can be planned with "small" areas that are perhaps 25 square miles in size, whereas substation planning must be done on a mile basis and distribution on a 40 acre (1/16 square mile) basis.

- *Quantities forecast.* A T&D forecast projects the peak demand (usually with respect to very specific weather called "design conditions") at every locality, since these local peaks are what the T&D system must satisfy. In addition, simulation methods forecast customer type (residential, commercial, and industrial) and annual energy requirements (kWh).

Table 13.5 The Spatial Load Forecast Step

Purpose:	To provide forecast of power delivery requirements for T&D planning
Timing:	Selected years to cover the period from 1 to 25 years ahead, usually 1, 2, 3, 5, 7, 10, 15, and 20 years ahead.
Products:	Small area forecast by year of peak demand, annual kWh, customer types, and other factors (e.g. end-use, reliability)
Coordinated with:	Corporate forecast, customer-level planning (DSM impacts forecast). Marketing plan. Critical input to distribution plan.
Tools used:	Spatial load forecast method (simulation or trending). End-use analysis of load/customer value. Corporate forecast database.

Methods available include:

- Regression-based extrapolation is good only for short range forecasting and proven less than ideally accurate if applicable, its chief advantage is low cost and low data requirements. Not recommended.
- Pattern-based extrapolation utilized non algebraic numerical methods to extrapolate future load based on repetition of past trends. Complicated math, but more accurate than regression.
- Customer-class based end-use location simulation (land-use based simulation). Very good representation and suitability to T&D planning. Good accuracy if well applied. Requires special skills, data, and is rather data intensive.
- Hybrid simulation/trending methods. These combine the accuracy and representation of simulation with lower data and skill requirements of trending.

Table 13.5 summarizes key points and methods for spatial forecasting. The book *Spatial Electric Load Forecasting* (H. L. Willis, Marcel Dekker, 1996) presents details and practical advice on spatial forecasting, and reviews a number of popular spatial forecasting techniques. Some spatial forecast methods include an end-use based small area analysis, in which usage of electric energy by end-use (heating, water heating, lighting, etc.) is used to provide estimates for marketing and demand-side planning. Advanced methods may use end-use analysis to estimate the customer value of reliability of service. Cumulatively, these forecast characteristics of load location, amount, customer type and number, reliability and service requirements set goals that the future system must meet. If the existing system cannot meet them, then additions or changes must be made to upgrade its capability.

Spatial forecast coordination with corporate planning

By far the most important aspect of the spatial load forecast, and good T&D planning, is that it be consistent with the corporate, or "Rate Department" forecast, which projects sales and revenues for the utility as a whole. The T&D forecast must be based upon the same assumptions and forecast base as that.

There are a number of different analytical approaches and computer algorithms that are applied to spatial forecasts for T&D planning. What is best for any given situation depends on planning needs, data, and other details specific to a utility.

Transmission Planning

In modern (de-regulated) utilities, "transmission" has two very distinct meanings applied to two very different parts of the power system:

> *Wholesale grid:* The regional grid, which interconnects generating plants and key bulk switching stations around the region, and connects the regional power system to other regions, is composed of high-capacity transmission lines, typically of voltages between 230 to 765 kV. This is the portion of the system that creates the wholesale market in electric power, and assures open access to merchant generators and buyers alike. System security, rather than reliability, is really the key factor in assuring "dependability" here.
>
> *Sub-transmission:* The term sub-transmission has been used for decades to designate those transmission voltage lines whose purpose is to deliver power to distribution substations. This is a mission quite distinction from that of the wholesale grid. Sub-transmission lines are generally of between 35 and138 kV, although some 230 kV lines exist primarily for this purpose.

The regional wholesale transmission grid is a key element in a healthy and economical regionally electric industry – it creates or enables the open access competitive market for electric power. The wholesale grid is usually planned and operated by an independent regional authority somewhat distinct, and often at arms-length from, the distribution utilities. Only part of the criteria and goals at this level relate to delivering power. Without doubt, one purpose of the wholesale grid is to deliver the required power to customers, but in addition, the transmission system is planned to provide:

1. *System Security and Stability* – the ability of the system to maintain interconnected integrity subsequent to any reasonably likely failure or event, and to react to transient events in a smooth manner.

2. *Generation access (de-regulation) or generation dispatch (traditional)* – provide sufficient electric tie strength between generation plants and the system to assure all plants can truly access the system – that transmission constraints do not limit choice of what generation can be run.

These two goals mean that a good deal of the transmission system is planned from the perspective of generation: permitting generation access the system, insuring system security no matter what happens, and providing stability against major disturbances. These aspects of transmission planning are not directly concerned with meeting power delivery needs, but with assuring that the generation-grid level of the system provides a stable bulk transmission capability and open access opportunity for all.

Planning of the "generation-grid" combination, which includes most of the high voltage transmission, is an activity quite apart from what might be termed "sub-transmission" planning – the planning of power transfer capability to the distribution substations, which is the only type of planning relevant to power delivery planning.

In general, sub-transmission planning reacts only slightly with grid planning – the grid is planned and presented as a "given" to the sub-transmission planning process. This is particularly true because of the "Chinese wall" that often has to exist between the wholesale (open access) and retail (local Distribution Company) under de-regulation.

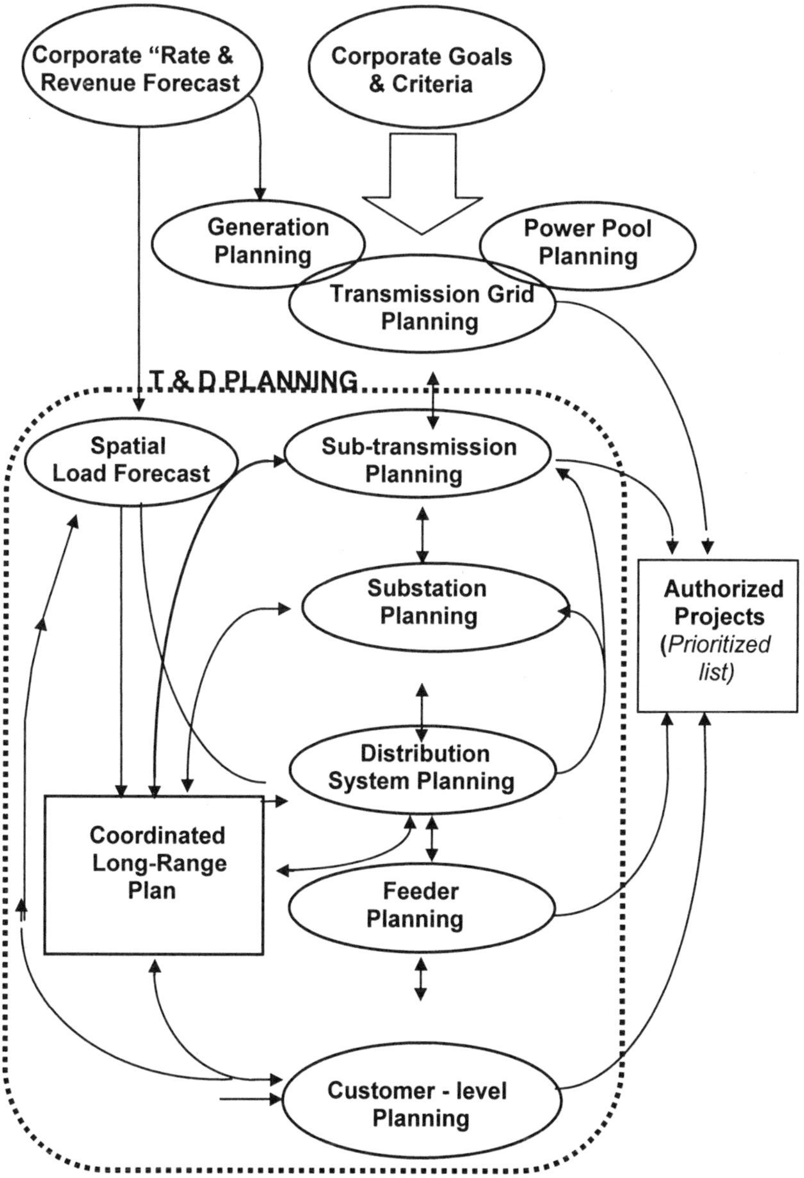

Figure 13.9 The overall T&D planning process includes load forecast, and planning functions for transmission, substations, distribution, and customer levels, as well as maintenance of a coordinated long-range plan encompassing all the levels. Its inputs are the corporate forecast & goals & criteria, its outputs are the short-range projects.

Thus, the interaction of the planning processes at the transmission-grid/sub-transmission-junction, has the most limited "coordination" of any of the junctions between adjacent levels of the power system. Sub-transmission is part of the power distribution planning process. It is best done as part of the distribution planning process. The sub-transmission – substation – feeder system ought to be planned as one system.

Figure 13.9 shows the recommended relation of sub-transmission planning to the other stages of utility system planning. Key aspects of these relationships include:

- *Transmission encompasses two planning functions:* as mentioned above, the "generation-grid" planning which focuses on establishing a strong grid with security and dispatch freedom and meeting all power pool requirements, and the "sub- transmission" planning, which includes all planning activities related to moving power to substations so that it can be distributed to customers. Only "sub-transmission planning" is overtly a power delivery planning function.

- *Forecast of future load for transmission planning* comes through the distribution planning process. Fully half of the statistical variation in future loads due to "load growth" is a function of distribution planning (Willis and Powell, 1985). To the transmission-sub-transmission planning process, the "loads" to be served by the system being planned are the high-side loads of the distribution substations. Exactly what and how much load is collected to a particular substation depends greatly on how the distribution planning is done – whether load growth ends up at one substation, or another, or is split between them, which depends on the distribution planning.

  This "high side substation load forecast" is best produced by starting with the spatial forecast, assessing the preferred patterns of distribution substation service areas in the distribution planning/substation planning stages, and then passing the target substation loads to the transmission planning process. If particular loads prove difficult to serve, transmission planning can provide feedback to distribution planning, asking if alternate possibilities can be considered.

- *Coordination with substation planning.* The transmission system must reach every substation. Therefore, clearly sub-transmission

Table 13.6 Sub-Transmission Planning

Purpose:	short and long-range planning of facilities to deliver power to the distribution substations.
Timing:	short-range – 3 to 7 years; long-range – 5 to 20 years.
Products:	short-range – sub-transmission project schedule coordinated with substation and transmission grid plans. Long-range – a base sub-transmission plan.
Coordination with:	"transmission system" planning (EHV, grid, generation), substation-level planning (must deliver power to substations).
Tools used:	transmission network load flow and short circuit, route optimization software, combined sub-transmission-substation optimization programs.

planning needs to be coordinated with the siting, capacity planning, and scheduling of substations (which of course has to be coordinated with distribution system planning, as shown in Figure 13.9). This coordination involves planning and scheduling of the sub-transmission and substation levels in a very coordinated manner in which cost impacts among the levels are "traded" back and forth. In many utilities, these functions are combined into one "sub-transmission level" planning process. Given that this is coordinated well with distribution planning, this is a good approach.

Table 13.6 summarizes the important elements of this part of the sub-transmission planning process.

Substation Planning

In many ways, the substation level is the most important level of the system, for two reasons. To a great extent this level fashions the character of the whole system. Substations are where the transmission and the primary feeder levels meet, the points to which the sub-transmission must deliver power, and from which the feeders must take it to neighborhoods throughout the system. The substation levels defines how and where cost, contingency, or capacity at one level can be traded or played against the other, and how expansion of both can be coordinated.

The substations sites set very definite constraints on the design and routing of the transmission and distribution systems – transmission lines must "end" at the substations and feeders must begin there. Substation location heavily impacts the cost and performance, and particularly the available design freedom, at both T and D levels. Substation capacity limits also heavily influence both transmission (how much incoming capacity is needed) and distribution (what amount of power can be distributed out of a substation).

As a result, regardless of whether substation planning is done as a separate planning function (about 20% of utilities), or as part of transmission planning (60%), or as part of distribution planning (20%), it should be interactively coordinated with both sub-transmission and feeder system planning. This means that communication should be explicit and frequent enough that glaring differences in marginal costs of expansion at the T, S, or D levels will come to the attention of the planning establishment in the utility and be dealt with by appropriate changes to the plan.

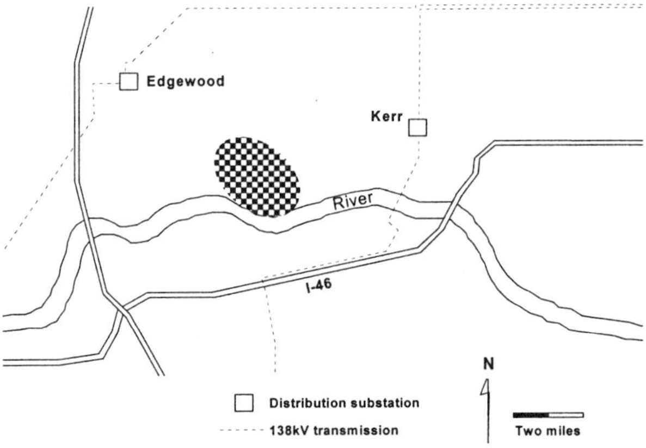

Figure 13.10 Development of new residential and commercial customers in the area shaded is expected to add 31 MVA in load over the next decade. Either Edgewood or Kerr substation could be expanded at about the same cost to serve this growth by adding a new 27 MVA transformer and associated buswork and control equipment. However, expanding the feeder system out of Kerr would be expensive – present feeder getaways are saturated, the system is underground and many duct banks are congested, and several geographic barriers make feeder runs quite lengthy. By contrast Edgewood's all-overhead feeder system would be straightforward to expand, costing $1,900,000 less. Despite this, the overall optimal alternative is to expand Kerr, because transmission limitations at Edgewood would cost more to overcome than their savings at the distribution level.

For example, in Figure 13.10, distribution planners have made plans to expand Edgewood substation from 55 to 90 MVA capacity in order to serve a peak load that is expected to grow from 49 to 80 MVA due to load growth in the area indicated. Both Kerr and Edgewood substations are close enough to the load growth to pick up this new load, and either could be expanded with roughly the same plan and cost – addition of another 27 MVA transformer and associated equipment, at a cost of $750,000 PW.

However, from the perspective of the distribution system, expanding the feeder system out of Kerr substation instead of Edgewood would cost an extra $1,900,000 PW. Thus, Distribution Planning selected Edgewood for addition of new feeders and equipment to serve the load growth. But suppose that expanding transmission delivery capability at Edgewood by the required 31 MVA will require extensive reinforcement of towers and conductor, at a cost of $2,600,000 PW, whereas the changes needed at Kerr to accommodate a similar increase in peak load are minor and would run only $325,000 PW. Then it is best for the utility to decide to serve the new load out of Kerr, since even though Edgewood is the least expensive distribution option, when overall T-S-D cost is considered, it comes to $600,000 PW less to serve the load growth from Kerr.

The planning process at many utilities can not accommodate this very simple, common sense approach to cost minimization. The procedural "hand-offs" between transmission, substation, and distribution planning groups does not convey enough information, nor allow for sufficient back-and-forth cooperative communication, so that cost differences like this are noted and the planning process can recognize if and how to adjust to them. At many utilities, the distribution planners will pass no details other than the projected total load for a substation – "We need 80 MVA delivered to Edgewood." At others the reverse happens, with transmission planners setting limits on substation capacity due to transmission constraints, without information of whether these could be eased if economically justifiable based on distribution costs.

Substation planning is important chiefly because of this requirement, that as the meeting place of transmission and distribution it defines their cost interaction. Planning procedures at this level ought to reflect this relationship and foster the required communication. There are at least three possible avenues to do so.

1. *Planning meetings* between transmission and distribution planning can be scheduled frequently and set up so that a free exchange of planning-details, not just plannning-results, occurs. The authors are well aware of the institutional barriers that exist within many large utilities to this type of communication. In the case of one large investor-owned utility, the managers of the transmission and distribution

groups, both of whom had been in their positions for over four years, had never met and had exchanged only a handful of formal "form reports" during their respective tenures. Under de-regulation, there could well be formal procedures defining the limits of and requiring such cooperation. Regardless, the results of such communication are significant savings, and worth considerable cultural adjustment. Situations like the Kerr-Edgewood example occur more often than is recognized.

2. *Use of methods and software* for the sub-transmission, substation, and distribution planning functions that accepts marginal or incremental cost data for the other level, or for the substation level that accepts such data for both. For example, some service-area based optimization algorithms applied to substation-distribution system planning can accept transmission system cost curves for incoming capacity, cost which is included in the optimization.

3. *A system of common planning procedures and nomenclature* that sets typical target values for every level of the system and reports the incremental cost of all projects at any level, to all planners at other levels in routine, periodic reports. This would distribute to distribution, substation, and sub-transmission planners information for their use on the costs at other levels of the system. This provides the information, but not the incentive, for cooperation.

Table 13.7 Substation Planning

Purpose:	short and long-range planning of facilities to control, route, and transform power from T to D.
Timing:	short-range – 3 to 7 years; long-range – 5 to 20 years.
Products:	short-range: substation project schedule coordinated with sub-transmission and distribution plans; long-range - long-range plan.
Coordination with:	sub-transmission planning, distribution planning.
Tools used:	AM/FM-GIS systems, substation selection optimization applications, combined sub-transmission-substation-feeder system optimization application.

Frankly, application of all three of the above measures is best so as to lead to the most coordinated planning, which will minimize the potential for planning errors. Table 13.7 summarizes substation planning.

Feeder Planning

There are two very different aspects of "feeder planning." The first is feeder system planning (Table 13.8), which involves the planning of the overall distribution feeder system in conjunction with the substations. The second, feeder planning, which involves the detailed layout of feeders and their specification is to a detailed engineering level of itemization and precision (Table 13.9).

Feeder system planning

What might be called "strategic" feeder planning involves determining the overall character of the feeder system associated with each substation, and with the role it plays in assisting in inter-substation contingency support. This includes determining the number of feeders from the substation, their voltage level, the overall service area served by the substation as a whole and the general assigned routes and service areas and peak loads for each. Additionally, operating problems (voltage, reliability, contingency switching) and overall costs may be estimated (budgetary).

Table 13.8 Feeder System Planning

Purpose:	Long-range planning of the feeder system, mainly so that feeder impact of substation-level decisions is assessed for substation planning.
Timing:	Five to 20 years.
Products:	Evaluation of feeder-level cost and performance impact of all substation-level decisions; long-range feeder level cost estimates for budgeting.
Coordination with:	Substation planning, customer-level planning.
Tools used:	Multi-feeder (multi-substation) optimization programs, combined substation-feeder system optimization application for capacity, reliability, economics

Generally, feeder system planning is done in the substation long-range planning period, usually from six to 20 years ahead, and with node resolution of perhaps only several dozen nodes per feeder – well beyond any lead time for feeders but with considerably less detail than needed for short-range feeder planning. Feeder system planning is *only* a long-range and strategic planning activity, but it is an essential one. Its purpose is to provide "distribution system" feedback to planning decisions or alternatives being considered at the substation and sub-transmission levels, and assess interactions of those levels.

As was mentioned earlier, the interplay of sub-transmission, substation, and feeder costs can be significant. In particular, the value of the primary distribution system associated with any particular substation probably outweighs the value of the substation itself. The 30-year PW of feeder losses on a heavily loaded system may exceed the value of both. Substation planning, particularly siting and sizing, can often be the economic equivalent of the tail wagging the dog. Small changes in location or planned capability can affect the cost of the feeder system dramatically and in some cases, a shift of one mile in the location of a suburban substation can have a PW impact of over $1,000,000.

Feeder system planning is often called multi-feeder planning to distinguish it from feeder planning which includes functions that can be done on a feeder by feeder basis. To be of maximum impact, multi-feeder planning should cover multiple substations, assessing how the feeder system interacts with substation capacities and costs – optimizing substation service areas while considering both substation and feeder costs (Figure 13.11). Proven techniques for simultaneous substation-feeder system planning exist and are described elsewhere in greater detail (Willis, 1996).

Feeder System Planning is a Key Element of Reliability Planning for Aging Infrastructures

Chapter 8 highlighted the need to use the feeder system to provide load transfers and contingency ties between substations. Feeder system planning, aimed at making certain that sets of feeders associated with each substation can accomplish the required target levels for transfers under normal and contingency situations, is essential to assure dependable backup capacity in an integrated T&D system plan. Overall policy decisions on what type of switching support to have both within each substation area, and for transfer of loads between substation areas, is a multi-feeder design aspect. While some single-feeder planning has to be done as a follow-up to the decisions made at this stage, assurance of reliability via good contingency withstand and support capability at the feeder level is done on a system basis.

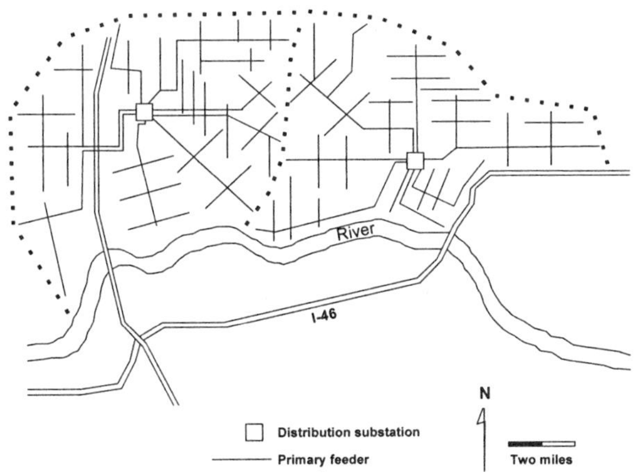

Figure 13.11 Multi-feeder system planning evaluates the performance of the entire feeder system associated with a set of substations. Its goal is to capture completely the interaction of substation level and feeder level constraints and costs to determine the overall substation-feeder optimum. A big part of this boils down to assigning "optimal service areas" to all substations while acknowledging capital and operating costs and electrical and geographic constraints at both the substation and feeder levels simultaneously, while determining the boundaries (dotted lines) between the substations.

Short-range feeder planning

Short-range feeder planning (Table 13.9) involves determining in detail the routing of each feeder and identifying its equipment to near engineering precision. In many utilities, feeder planning and feeder engineering are merged into one function – "Distribution Planning" produces completed feeder plans replete with detailed equipment specifications and route maps, pole locations, and all construction details. It also provides the final authorization and scheduling for construction.

Generally, feeder planning *can* be done on "a feeder at a time" basis, often using only load-flow and short-circuit analysis tools augmented by engineering judgment to determine alternative designs in a "trial and error" approach to find the best feeder layout. However, many modern CAD systems for feeder planning have load flow and short circuit applications that can be run on a number of feeders at once, and augment these with optimization utilities to help refine load balance and switching plan development among feeders. Feeder planning is done differently at every distribution utility, but there are standard recommended procedures that are well documented, (Engel, 1992, Burke, 1994, Willis, 1996). Table 13.9 summarizes its essential elements.

Table 13.9 Feeder Planning

Purpose:	Short-range planning of the feeder system, to produce project definitions and authorization for feeder additions and enhancements.
Timing:	One to 5 years.
Products:	Feeder system project specifications, schedule, and budget.
Coordination with:	Feeder system planning, substation planning, customer planning, and construction.
Tools used:	"Feeder design" CAD systems, feeder optimization programs (either single or multi-feeder), AM/FM and GIS systems. Reliability "load flows" for the design of optimized switching locations and branch and trunk tie capability.

Customer-Level Planning

Traditional supply-side-only planning included very little if any "customer-level" planning functions. Regulated utility integrated resource planning of the type widely practiced in the late 1980s and early 1990s includes a great deal – various DSM options optimized on a TRC, RIM, or other resource-value basis, (see Willis and Rackliffe, 1994). While the role of IRP and DSM, in particular, in a de-regulated power industry is unclear, the fact that some energy efficiency, advanced appliance, and load control/TOU technologies can prove cost-effective is not in doubt. For this reason, some sort of customer-level planning can be expected in the future. Under the right circumstances, DSM, DG, or DS may be of value to a regulated or unregulated electric supplier/distributor. Therefore this level of planning will be a part of many utilities, and will, at least obliquely, impact T&D planning. Table 13.10 summarizes key points in planning at the customer level.

The interaction may be much more than just oblique, because DSM and DG programs often have a very substantial impact on T&D, with more than half their potential savings being in avoided costs among the sub-transmission, substation, and distribution levels. Usually, optimization of the benefits of a DSM or DG program includes some assessment of T&D impacts and location.

Customer-level planning often has a very long planning period, much longer than distribution planning, and often as long as that of generation. There are four reasons why both its short- and long-range planning periods are longer than almost all other levels of the system:

1. *Lead-time.* While DSM programs can often be approved and implemented nearly immediately, it can take years of slow progress to obtain meaningful levels of participation – rates of 50% may take a decade or more to obtain. This "ramp up time" has the same effect as a long lead-time.

2. *Customer commitments.* In starting a DSM program, a utility is beginning a very visible customer-interaction. Such programs are not entered into lightly and are often studied over the very long-term to make certain the utility can feel confident it will commit to them for the long-term. Starting such a program and then canceling it several years later is not only costly, but leads to customer confusion and lack of confidence in the utility.

3. *Proxy for generation.* DSM is often traded against generation in integrated resource planning. Comparison of the two resources needs to be done over the time span used for planning of generation, which means a present-worth or similar economic comparison over a 20 to 30 year period.

Table 13.10 Customer Level Planning

Purpose:	Short- and long-range planning of customer-level resources including DSM, DG, DS, RTP and TOU rates.
Timing:	One to 25 years.
Products:	Schedule of customer-side resource projects/programs; long-range plan of DSM/DG/DS targets.
Coordination with:	All levels of planning, corporate forecast/rate plan, spatial load forecast. Note: this planning is often done by corporate or rate department planners.
Tools used:	End-use load models. Integrated resource (T&D or otherwise) optimization models. Customer response models. Value-based planning analysis. Econometric/demographic models. Spatial customer/load forecast models.

Customer-level planning generally begins by adding more detailed, end-use and customer value attributes to the load forecast – forecasts by end-use and appliance category and of reliability value functions. It then tries to optimize PW cost of providing end-uses to the customers by juggling various DSM and T&D options. Generally, optimization methods function by balancing marginal cost of the various customer-side options against marginal cost curves obtained for the supply side. DSM and DG evaluation may also be part of value-based planning in which the utility and customer value functions are jointly optimized.

13.5 THE SYSTEMS APPROACH

Throughout this chapter, discussion of T&D planning has stressed the need to coordinate the planning of the various levels of the power delivery (T&D) system. The various levels – transmission, substation, and distribution – are interconnected, with the distribution, in turn, connected to the customers (and hence a function of the local load density). The best equipment types, sizes, and designs at each level of the system are a function of the types of equipment, sizes, and designs selected for the other levels of the system. In general, the equipment is so interconnected that it is impossible to evaluate any one aspect of design without taking all others into account.

Consider the question of substation spacing – determining how far apart substations should be, on average, for best utilization and economy. Within any service territory, if substations are located farther apart, there will be fewer of them, saving the utility the cost of buying and preparing some substation sites, as well as reducing the cost of building a large number of substations. Thus, overall substation cost is a function of spacing, but how it varies depends on land and equipment costs, which must be balanced carefully. With fewer substations, however, each substation must serve a larger area of the system. Hence, each substation will have a larger load, and thus require a larger capacity, meaning it must have more or larger transformers.

The larger substations will also require a larger amount of power to be brought to each one, which generally calls for a higher sub-transmission voltage. Yet, there will be fewer sub-transmission lines required (because there are fewer substations to which power must be delivered). All three aspects of layout are related – greater substation spacing calls for larger substations with bigger transformers, and a higher transmission voltage, but fewer lines are needed – and all three create better economies of scale as spacing is increased. Thus, transmission costs generally drop as substation spacing is increased.

Nevertheless, there is a limit to this economy. The distribution feeder system is required to distribute each substation's power through its service area, moving power out to the boundary between each substation's service area and that of its neighbors. Moving substations farther apart means that the distribution system must move power, on average, a greater distance. Distributing power over these longer distances requires longer and more heavily loaded feeders. This situation increases voltage drop and can produce higher losses, all of which can increase cost considerably. Employing a higher distribution voltage (such as 23.9 kV instead of 13.8 kV) improves performance and economy, but regardless, it costs more to distribute power from a few larger substations farther apart, than to distribute it from many smaller substations close together. Feeder costs go up rapidly as substation spacing is increased.

The major point of this section is that all four of the above aspects of system design are interconnected: (1) substation spacing in the system; (2) size and number of substations; (3) transmission voltage and design and; (4) distribution feeder voltage and design. One of these factors cannot be optimized without close evaluation of its interrelationship with the other three. Therefore, determining the most cost-effective design guideline involves evaluating the transmission-substation-feeder system design as a whole against the load pattern, and selecting the best combination of transmission voltage, substation transformer sizes, substation spacing, and feeder system voltage and layout.

This economic equipment sizing and layout determination is based on achieving a balance between two conflicting cost relationships:

1. Higher voltage equipment is nearly always more economical on a per-MW basis.

2. Higher voltage equipment is available only in large sizes (lots of MW).

In cases where the local area demands are modest, higher voltage equipment may be more expensive simply because the minimum size is far above what is required – the utility has to buy more than it needs. How these two cost relationships play against one another depends on several factors: the load; the distances over which power must be delivered; and other factors unique to each power system such as the voltages at which power is delivered from the regional power pool, and whether the system is underground or overhead.

Figure 13.12 illustrates the difference that careful coordination of system design between levels of the power system can have in lowering overall cost. Shown are the overall costs from various combinations of T&D system layout for a large metropolitan utility in the eastern United States, all designed to achieve the same reliability targets. Each line connects a set of cost computations for a system built with the same transmission and distribution voltages (e.g. 161 kV transmission and 13.8 kV distribution) but varying in substation sizes (and hence, implicitly, their spacing).

In all cases, the utility had determined it would build each substation with two equally sized transformers (for reliability), with none over 75 MVA (larger transformers are too difficult to move along normal roads and streets, even on special trailers). Either 161 kV or 69 kV could be used as sub-transmission, either 23.9 kV or 13.8 kV could be used as distribution voltage. Any size transformer, from 15 MVA to 75 MVA, could be used, meaning the substation could vary from 30 MVA to 150 MVA in size. (Peak load of such substations can normally be up to 75% of capacity, for a peak load of from 23 to 100 MW). Substation spacing itself is implicit and not shown. Given the requirement to cover the system, determining transmission voltage, distribution, and substation size defines the system design guidelines entirely.

Overall, the ultimate lowest cost T&D system guidelines are to build 120 MVA substations (two 60 MVA transformers) fed by 161 kV sub-transmission and distributing power at 23.9 kV. This has a levelized cost (as computed for this utility) of about \$179/kW. In this particular case, a high distribution voltage is perhaps the most important key to good economy – if 13.8 kV is used instead of 23.9 kV as the primary voltage, minimum achievable cost rises to \$193/kW.

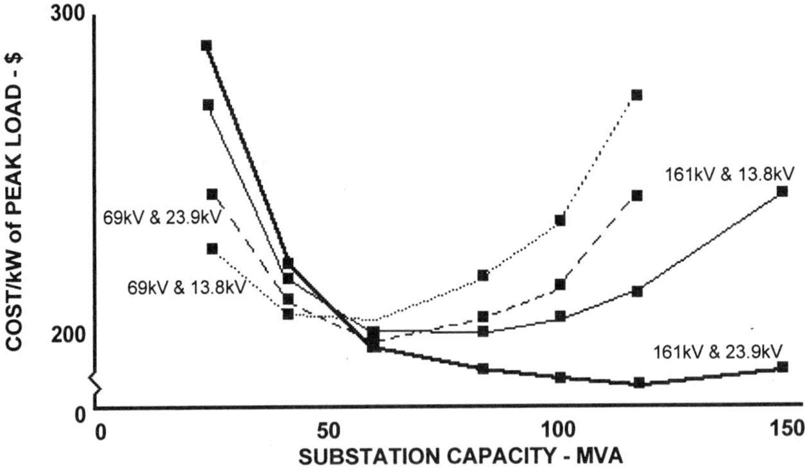

Figure 13.12 Overall cost of T&D system depends on the balanced design of the sub-transmission, substation, and feeder level, as described in the text. Cost can vary by significant margins depending on how well performance and economy at various levels of the system are coordinated.

The very worst design choices plotted in Figure 13.12, from an economic standpoint, would be to build 25 MVA substations fed by 161 kV sub-transmission and feeding power to 23.9 kV feeders ($292/kW). This would require many small substations, each below the effective size of both the transmission and distribution voltages, and lead to high costs. Overall, 161 kV and 23.9 kV are the correct choices for economy, but only if used in conjunction with a few, large substations. If substations are to be 25 MVA, then 69 kV and 13.8 kV do a much more economical job ($228/kW), but still don't achieve anything like the optimum value. The point is that achieving economy in power delivery involves coordinating the interactions, performance, and economies of the multiple system levels. Chapters 11 and 12 in the *Power Distribution Planning Reference Book* discuss techniques for such multi-level planning.

The Systems Approach Is Important for Aging Infrastructure Planning

The discussion and figure above showed the economic advantages that the systems approach can provide. These are certainly important to all utilities, including those facing aging infrastructure issues. However, there is another, and vital reason to apply the systems approach as a rule in all planning for aging T&D infrastructures:

> The systems approach helps work around constraints in siting, size, or topology by drawing on the resources of the other levels in the power system to solve problems present at another.

As mentioned in Chapter 8, very often the owner of the aging T&D system is "stuck" with too few, and too small substation sites. Sub-transmission rights of way may be narrow and non-optimal in routing. Expansion and addition may be costly, time-consuming, and politically unacceptable.

Regardless, in many cases the utility has little choice but to make the best of these assets, particularly in aging areas of its system where additional land and rights of way as very expensive. The best way to improve performance this is to seek out strengths in adjacent levels of the system in order to help support the weakness in another. Techniques discussed in Chapters 10, 11, and 14 can be applied to do just this, as was summarized in section 13.5. More detailed discussions of the "systems approach" necessary to optimize the synergism among the sub-transmission – substation – feeder system levels to truly maximize the reliability/cost ratio of the entire power system chain are contained in the *Power Distribution Planning Reference Book,* Chapters 7 through 12.

13.6 SUMMARY OF PLANNING FOR AGING T&D INFRASTRUCTURES

Planning involves the selection of the best alternative from among those at hand. In the power industry, that selection depends to a certain extent on the values of the group making the decision. *All* cost evaluation for power delivery planning are based in part upon economic evaluation that is consistent over the range of all alternatives. Usually, this means that present and future costs must be compared. Present worth analysis, levelized costing, or some similar method of putting present and future value on a comparable basis will be used. While methods and standards vary from one company to another, the important point is that planners apply their company's method consistently and correctly.

Traditionally, planning at many electric utilities has been something like a rhinoceros – never looking far ahead but instead focusing on only what is right in front of it, protective of its own turf, capable of surprising speed and powerful enough to get its way much of the time, but frequently content to move slowly and merely hold its position. Utilities with aging T&D infrastructures need to revise their planning focus, taking the following lessons to heart:

- *Good planning* is more important that ever. A system that has weaknesses or high equipment utilization ratios is especially intolerant of planning mistakes.

- *Optimization of reliability/cost* is a key element of planning for aging T&D infrastructures.

- *Constraints* in siting, size, and routing are usually the major challenge in planning.

- *Think of the long range, but act for the short range.* Money and resources should only be committed on the basis of good short-term benefits and efficacy, but every commitment should fit into a coherent long range plan.

- *Spatial forecasting* is more important that in normal T&D planning, important as it is in that venue.

- *Feeder system planning,* and coordination of feeder system strength to provide contingency support to the substation level, is a key factor in simultaneously achieving reliability and economy.

- *The systems approach* is a vital element of success in handling constraints in siting, sizing, and routing.

REFERENCES

D. Atanackovic, D. T. McGillis, and F. D. Galiana, " The Application of Multi-Criteria Analysis to a Substation Design," paper presented at the 1997 IEEE Power Engineering Society Summer Meeting, Berlin.

T. W. Berrie, *Electricity Economics and Planning,* Peter Peregrinus Ltd., London, 1992.

J. J. Burke, *Power Distribution Engineering – Fundamentals and Applications,* Marcel Dekker, New York, 1994.

M. V. Engel, et al, IEEE Tutorial on Distribution Planning, Institute of Electrical and Electronics Engineers, New York, 1992.

R. Orans et al., "Targeting DSM for Transmission and Distribution Benefits," Electric Power Research Institute, Palo Alto, CA, 1992.

F. Schweppe et al., *Spot Pricing of Electricity,* Kluwer Press, Cambridge, MA, 1988.

L. J. Vogt and H. L. Willis, "DSM: Transmission and Distribution Impacts," Volumes 1 and 2, Report CU-6924, Electric Power Research Institute, Palo Alto, CA, 1990.

H. L. Willis, *Power Distribution Planning Reference Book,* Marcel Dekker, New York, 1997.

H. L. Willis and R. W. Powell, "Load Forecasting for Transmission Planning," *IEEE Transactions on Power Apparatus and Systems,* August 1985, p. 2550.

H. L. Willis and G. B. Rackliffe, *Introduction to Integrated Resource T&D Planning,* ABB Guidebooks, Raleigh, NC, 1994.

World Bank, *Guidelines for Marginal Cost Analysis of Power Systems,* Energy Department paper number 18, 1984.

14
Reliability Can Be Planned and Engineered

14.1 INTRODUCTION

The problems presented by aging T&D systems boil down to one, minimizing the cost to provide reliability of service that will satisfy consumer and regulatory needs. The other issues that arise from aging infrastructures – higher inspection and restoration needs, headaches in obtaining replacements for old equipment, worries about flaws in traditional planning methods applied to modern system needs – all stem from this one primary issue: improve reliability and do so at the least cost possible.

Therefore, optimizing the performance/price in the customer service reliability arena is the key. In this respect, the first rule is that the customer deliverable is all that matters.

> The goal of good reliability-based planning, engineering, and operations is to maximize reliability of service as seen by the customer, not necessarily to improve reliability of equipment or service time on the system. Customer-level results are what count.

Generally, the best way to assure customers see good, reliable service is to make certain that the power system design itself is inherently reliable. But the point being made is that supply-side reliability is only a means to the important end. The planning and engineering process should focus on customer-level measures of service reliability.

Why is Reliability More Important Today than It Was Traditionally?

Distribution system reliability is driven by several factors including:

(1) The increasing sensitivity of customer loads to poor reliability, driven both by the increasing use of digital equipment, and changing lifestyles.

(2) The importance of distribution systems to customer reliability as the final link to the customer. They, more than anything else, shape service quality.

(3) The large costs associated with distribution systems. Distribution is gradually becoming an increasing share of overall power system cost.

(4) Regulatory implementation of performance based rates, and large-customer contracts that specify rebates for poor reliability, all give the utility a financial interest in improving service quality.

In the past, distribution system reliability was a by-product of standard design practices and reactive solutions to historical problems. In the future, distribution system reliability will be a competitive advantage for a distribution utility, in its competition with other forms of power and energy (DG, alternate fuels, and conservation). Reliability must be planned for, designed for, optimized and treated with analytical rigor.

Optimal Reliability isn't Necessarily Maximum Reliability

Planners must keep in mind that while reliability is important to all customers, so is cost, and that only a portion of the customer base is willing to pay a premium price for premium levels of reliability. The real challenge for a distribution utility is, within tight cost constraints to:

- Provide a good basic level of reliability.

- Provide roughly equal levels of reliability throughout its system, with no areas falling far below the norm.

- Provide the ability to implement means to improve reliability at designated localities or individual customer sites where greater service quality is needed and justified.

In section 14.2, this chapter will introduce an important concept: the engineering of reliability in a distribution system in the same way that other performance aspects such as voltage profile, loading, and power factor is engineered. Section 14.3 will summarize the analytical methods that can be used as the basis for such engineering, and discuss their application. Section 14.4 gives an example of reliability-based engineering as applied to a medium-sized power system, design for overall and site-specific targeted reliability performance. Financial risk, due to performance-based contracts and rates, is a big concern to a 21st century utility. Section 14.5 looks at the analytical methods needed to assess and optimize this, and shows an example analysis. Section 14.6 concludes with a summary of key points.

14.2 RELIABILITY CAN BE ENGINEERED

Figure 14.1 shows a screen from a computer program that can simultaneously assess voltage behavior, reliability, or cost of an arbitrary distribution system design input by the user. The reliability analysis is what is often termed a "reliability load flow" in the sense that it computes, in the same manner as a load flow, computes voltages and currents at every point in the system, frequency and duration of outages expected at every location in the system.

The program bases its computations on an arc-node structure model of the system, very similar in concept, construction, and data, to that used in a standard load flow program. By using failure rate data on system components, the program can compute the expected number of outages that occur, and the expected total minutes of outage, annually, at each and every point in the network. From these results, indices such as SAIDI, SAIFI, CAIDI and CAIFI, or other reliability indices, can be computed for the network.

The particular program whose screen display is illustrated in Figure 14.1 has several characteristics that are important to practical application:

- It represents the *impact of equipment failures* on the interconnected system around it.
- It can model the actions of *dynamic changes* associated with a contingency, such as a fuse blowing, a breaker tripping, a rollover switch operating, or a line crew repairing a damaged line section.

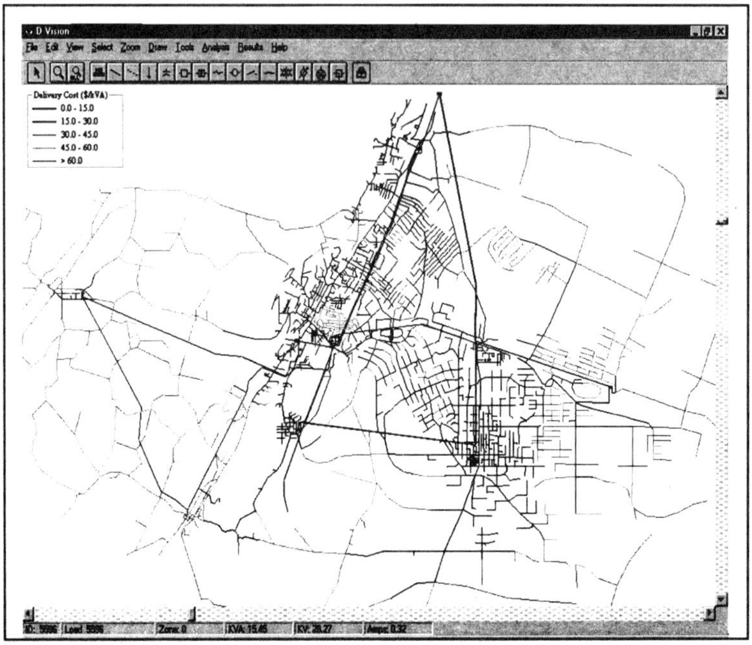

Figure 14.1 Screen display of results from a "reliability load flow" analysis of a distribution system in and around a small city. This type of analysis is the basic tool of good, dependable reliability-based planning for a power distribution system. Actual screen display is in color with color indicated level of SAIDI expected by location. See text for details.

- It is *self-calibrating.* Given historical data on outage rates by portion of the system, it computes the failure rates for equipment in each area until its depiction of the base system's reliability performance matches historical reality.

- It has a very *large system capability,* being able to analysis a distribution system composed of hundreds of thousands of nodes.

- Its model of the system itself (the circuit database) is in a *standard* load flow format compatible with most power system modeling.

- It is *easy to use* (graphic user interface).

Results are dependable

The results from this process, if done well, are roughly as dependable from an engineering standpoint as those obtained from a competent and well-applied load flow application. No one pretends that the results of either a load flow or a reliability assessment program are exactly accurate. In a load flow, actual loads are never known precisely, and line and equipment impedances are approximate within a small range. In a reliability analysis, failure rates are not known precisely. Generally, the reliability assessment has a *better* base of customer data than a load flow. It uses customer count data at the "load end" of the model (information that can be accurately determined) not kW load data (which are most typically estimates based on kWh sales).

But this whole issue is not a problem when using the best reliability assessment models, because they calibrate their data representations to actual historical results. Once the model for an existing system has been constructed, computed reliability performance may not match historical reliability performance. Any mismatch is corrected by adjusting default component parameters until model results match historical results. This process is often done by trial and error, but rigorous methods also exist.

Table 14.1 Reliability of Typical Types of Distribution System Components

Equipment	Failure Rate Per Year			Mean Time To Repair		
Type	Low	Typical	High	Low	Typical	High
Overhead Lines (per mile)						
Primary Trunk	0.020	0.100	0.300	2.0	4.0	8.0
Lateral Tap	0.020	0.160	0.300	2.0	4.0	8.0
Secondary & Service Drop	0.020	0.088	0.030	1.5	2.3	8.0
Pole Mounted Transformer	0.004	0.010	0.015	3.0	4.0	10.0
Disconnect Switch		0.014		1.0	1.5	4.0
Fuse Cutout				0.5	1.5	4.0
Line Recloser				3.0	4.0	10.0
Shunt Capacitor				0.5	1.5	4.0
Voltage Regulator				3.0	4.0	10.0
Underground Cable						
Primary Cable	0.003	0. 070	0. 100	2.5	10.0	30
Secondary Cable	0. 005	0. 100	0. 150	2.5	10.0	30
Elbows & Connectors	6e-5	0.0006	0.001	1.0	4.5	8.0
Cable Splices and Joints	6e-5	0.03	0.159	0.5	2.5	8.0
Padmount Transformers	0.001	0.010	0.049	4.0	6.5	7.8
Padmount Switches	0.001	0.003	0.005	0.8	2.5	5.0

Required component reliability data is not difficult to obtain

One of the primary concerns power system planners have with the use of a reliability-based engineering method is the data requirement. Such methods need data on the failure rates of various types of equipment, and repair and switching times for various situations. Generally, this data is available by culling through utility operating records. However, several sources of generic reliability data are available. These include default data obtained from industry standards (for example, Institute of Electrical and Electronics Engineers, International Conference on Large High Voltage Electric Systems), publications (such as books, journals, conference papers), and tests or surveys conducted by manufacturers. This information provides a good first guess and prepares the model to be calibrated against historical data. Table 14.1 lists typical values of failure rates for most common types of power system equipment.

As mentioned above, the best reliability-assessment programs have a self-calibrating feature: they "fit" their reliability analysis results to actual historical outage rates. Thus, lack of exact system data matters not at all. These methods start with "generic" failure rate data in internal tables and make various adjustments and re-computations of what failure rates in an area must be, based on the historical data. Self-calibration is the best way to assure that a reliability assessment model is accurate.

Results are useful even with only generic data

It is important to realize that a reliability-assessment method can help utility planning engineers improve the reliability performance of their system even if it is using approximate failure rate data. Most of the important decisions that are made with respect to reliability have to do with configuration, installation and location of protective and sectionalizing equipment, and switching. Even approximate data, used in accurate analysis of these issues, provides guidance on the wisdom of various decisions. Improvement can be made using only approximate (generic) failure rate and repair-time data.

Using A Reliability Load Flow

Planning engineers use a reliability assessment program in much the same manner that they use a load flow to design the feeder system. They begin with, performance targets - voltages and loadings with certain limits- for load flow planning, reliability within a certain range for reliability engineering. They then enter into the program a candidate system model, a data representation of the system as it is, or as they propose to build it.

The computer analysis then determines the expected performance of the system as represented to it. In a load flow, computing the expected voltages, flows, and power delivery at every point in the system does this. In a reliability assessment, computing the expected frequency and duration of outages at every point in the system does this.

The results are displayed for the user, usually graphically in a way that quickly communicates problems (results that are unacceptably out-of-range).

The planning engineers review the results. If they are less than desirable, they make changes to the system model, using the results as a guide to where improvements or changes should be made. The model is re-run in this manner until a satisfactory result is obtained.

Root Cause and Sensitivity Analysis

Planning engineers have several powerful tools they can use to augment their judgement and experience in determining what changes should be made to their candidate design in order to improve reliability results. These *are root-cause analysis* and *sensitivity analysis.*

A root-cause analysis determines the contribution of each component to poor reliability. For example, if reliability is measured using SAIDI, a root-cause analysis will identify the components that have the highest impact on SAIDI and the components that are having a low impact on SAIDI. The authors' particular program ranks all equipment in the system by its contribution to SAIDI. Results can be displayed graphically to quickly identify problem areas of the system.

In addition to knowing the contribution each component makes to the overall poor reliability of the system, it is desirable to know the impact that improving component reliability or design would have on the system's reliability. Will a change make a difference? For example, what would be the impact of reducing overhead line failures, reducing transformer failures, or reducing cable repair-time? A sensitivity analysis can be used to answer such questions.

This simple procedure adjusts a component parameter (such as its failure rate) by a small amount and records the corresponding change in system reliability.

Engineering the Design Improvements

In determining how to improve system performance, planning engineers focus on equipment or areas of the system that score high in terms of both root cause (they create a lot of the problems) and sensitivity (changes will make a difference on reliability). After a system has been modeled, calibrated, and examined with root cause and sensitivity analyses, potential design improvements can be identified and modeled. The focus should be on critical components – those with both high root cause scores and high sensitivity scores.

14.3 METHODS FOR DISTRIBUTION SYSTEM RELIABILITY ASSESSMENT

Power distribution reliability can be designed just like any other aspect of performance – voltage profile, loading. Doing so in a dependable and efficient manner requires an engineering planning method that can simulate how any particular distribution design will perform in terms of reliability of service. The system reliability will be as targeted by the planners and the effort exerted does not require special skills or undue levels of effort.

Several suitable analytical methods for this type of engineering exist. In the same manner that a power flow model can predict the electrical behavior of a distribution system (such as currents and voltages), a computer program called a *reliability assessment model*, based on one of these methods, can predict the expected reliability performance of any particular distribution system. As reliability becomes more important to electric utilities and electricity consumers, these reliability assessment models will equal or surpass power flow models in importance and usage. Reliability models allow distribution engineers to

- Design new systems to meet explicit reliability targets,
- Identify reliability problems on existing systems,
- Test the effectiveness of reliability improvement projects,
- Determine the reliability impact of system expansion,
- Design systems that can offer different levels of reliability, and
- Design systems that are best suited for performance based rates.

Distribution system reliability assessment is a rapidly developing field. Some of the first work on computer programs for this was done at Westinghouse Advanced Systems Technology, Pittsburgh, in the late 1970s. One of the first commercial distribution system reliability analysis programs, PREL, came from that group in 1981. This software was mainframe based and not widely adopted, largely because utilities in the 1980s did not put the heavy emphasis on reliability as they do 20 years later.

Increasing sensitivity of customer loads and impending deregulation led several utilities to develop their own distribution reliability assessment capabilities in the early and mid 1990. Demand for distribution reliability assessment software is continuing to grow and commercial packages are now available from several major vendors. Electric utilities that desire to apply such methods have no problem finding several suitable packages from which to choose.

Methodologies for Distribution System Reliability Analysis

Reliability assessment models work from an arc-node type database, very similar to that used in a distribution load flow program. Unlike a load flow, which uses data on each element to compute the interconnected flow of power through the system to the customer locations, the reliability assessment program uses data on each element to compute the interconnected reliability of service to the customer locations.

There are four common methodologies used for distribution reliability assessment: network modeling, Markov modeling, analytical simulation, and Monte Carlo simulation. These differ in their basis – in what they use as the foundation for their analysis of reliability. Each uses a difference approach to determining the reliability of the system. A brief description of each is provided below.

Network modeling

Network modeling is based on the topology of the system, which it translates from a physical network into a reliability network based on serial and parallel component connections. Serial components have additive failure rates and each one can disable the entire chain. Parallel components have redundancy (up to a point corresponding to capacity) and can cover outages of other parallel pathways.

Network modeling basically takes the system topology and translates it into a "formula" of serial and parallel reliability characteristics, which it then uses to compute the reliability of the system performance. In computer models this formula is a table-driven set of equation coefficients that implements a reliability computation. This computation computes the likelihood that a continuous interconnection between a source, and the demand point, remains in operation.

Network modeling is simple and straightforward to implement, and produces good results at a basic evaluation of reliability. A major disadvantage for distribution system analysis is that dynamic functions, such as switching, or sequential system responses to contingencies, are outside of its context. Its formula applies to a basic topology, anything such as switch operation that changes the topology, cannot be modeled. As such, it is not as widely used as other methods.

Markov modeling

Markov modeling is a powerful method for assessment of reliability, as well as for simulation of many other apparently random multi-state processes (such as wind and solar availability for renewable energy systems). Markov modeling is based on analyzing the states that a system could be in: states can be such things as "everything operating normally," or "component one has failed" or "component two has failed," etc. It focuses though, on analyzing the transitions between these states: it analyzes the likelihood that the system can move from the condition "everything is operating normally" to "component one has failed," as well as under what conditions, and how long, it takes to transition back.

Every condition, or state, that the system could be in is identified and enumerated. The analytical method focuses on computing the back and forth transitions between one state and another. It models the conditions and likelihood of these transitions, thereby mapping the amount of time that the system spends in each state. If one state is "everything operating normally" and others represent one failure model or another, then the model can be used to analyze reliability-of-service and its causes.

Markov modeling is excellent for representing systems where details of the transition between states are known or are important. For example, planners may want to study how different repair and replacement part policies would impact the availability of a DG unit. Having a repairman at the DG site, or the parts already there so they do not have to be spent for, would clearly reduce the time to transition from "broken" to "working." Markov models make it possible to focus in great detail on studying questions about how much such actions would impact reliability, because they focus on the mechanism of moving between states, not on the states themselves.

However, Markov models have some disadvantages for power distribution system applications. The first is that the states modeled are memory less (transition out of a state cannot depend on how the state was reached). This characteristic requires a type of duplication of states when system responses are a function of past events. Thus, if the system is in the state "working" because during the "broken" state a spare part was used, (and as a result there no longer is another spare part at the site), the system cannot model this unless a new state is introduced: "working but no spare parts available." Then, transitions between this state and "working" need to be established in the model (to represent delivery of the new part to the site). In practice, this complicates the application of a Markov model when used to analyze a distributed system such as a power distribution system.

The second limitation is computational. The matrix inversion required by Markov modeling requires that limits to the size of systems that can be represented and/or the complexity that can be represented to systems such as a few DG units. It also limits very simple distribution systems such as only the substation serving a small industrial site and its equipment.

Analytical simulation

Analytical simulation models each system contingency, computes the impact of each contingency, and weights this impact based on the expected frequency of the contingency. At first glance it appears to be very close to a form of contingency analysis, as was described in the Chapter 8 discussion on N-1 contingency analysis, except that likelihood's are assigned or computed for each contingency case. This is a big difference and not an easy one to add to N-1 analysis. Analytical simulation algorithms are built around the mechanism that computes the probability of each contingency case, rather than an explicit enumeration of contingency states, as in N-1 analysis.

Accurate reliability analysis using analytical simulation means computing the likelihood of each contingency based on failure rates and interconnection of equipment. Generally, dynamic functions of equipment such as fuses, breakers, rollover switches, etc., has to be modeled, too, hence the term "simulation." A great advantage of this method is that such types of dynamic activity can be modeled.

If the likelihood of each contingency is computed based of a good representation of the failure rates and modes that would lead to it, this type of method can accurately model complex system behavior and dynamically enumerate each possible system state. Aggregation of system states and their probabilities then permits detailed assessment of reliability (all states that lead to success vs. failure).

Monte Carlo simulation

Monte Carlo Simulation is similar to analytical simulation, but models' random contingencies based on probabilities of occurrence, rather than expected contingencies. This allows component parameters to be modeled with probability distribution functions rather than expected values.

Monte Carlo Simulation can model complex system behavior, non-exclusive events and produces a distribution of possible results rather than expected values (Brown et al, 1997). Disadvantages include computational intensity and imprecision (multiple analyses of the same system will produce slightly different answers). Additionally, Monte Carlo Simulation is not enumerative and may overlook rare but important system states.

For applications requiring determination of expected values of reliability, analytical simulation is the best method for distribution system assessment. It allows distribution engineers to quantify system reliability (SAIDI, SAIFI) over an entire system, and for individual customer locations, to calibrate models to historical data, to compare design alternatives, perform sensitivity analyses and to run optimization algorithms to maximize the expected results [See Brown, et al, 1998 - 2000]. Monte Carlo Simulation becomes necessary if statistical results other than expected values are required – analysis of such things as the distribution of expected SAIDI from year to year.

14.4 APPLICATION OF ANALYTICAL SIMULATION FOR DETAILED RELIABILITY ASSESSMENT

An analytical simulation method simulates a contingency, determines the impact of this contingency on system reliability, and weights the impact of the contingency by its probability of occurrence. This process is repeated for all possible contingencies, and results in the following information for each component, providing the results shown in Table 14.2.

Simultaneously with the analysis of each contingency and the results it produces, the analytical simulation method evaluates the probability of occurrence of this contingency – how likely is it to occur? Once all contingencies have been analyzed, their results are combined appropriately on a probability-weighted basis to produce aggregate results. In distribution system analysis this aggregation of results is based on system topology, keyed to adjacent equipment, branches, circuits, etc, as well as the entire system.

Modeling Each Contingency

A contingency occurring on a distribution system is followed by a complicated sequence of events. Each contingency may impact many different customers in many different ways. In general, the same fault will result in momentary interruptions for some customers and varying lengths of sustained interruptions for other customers, depending on how the system is switched and how long the

Table 14.2 Results Computed By an Analytical Simulation

- Expected number of momentary interruptions (per year)
- Expected number of sustained interruptions (per year)
- Expected number of interrupted hours (per year)
- Expected number of protection device operations (per year)
- Expected number of switching operations (per year)

fault takes to repair. The key to an analytical simulation is to accurately model the sequence of events after a contingency to capture the different consequences for different customers. A generalized sequence of events as modeled by analytical simulation is:

1. *Contingency*: A fault occurs on the system.
2. *Reclosing:* A reclosing device opens in an attempt to allow the fault to clear. If the fault clears, the reclosing device closes and the system is restored to normal.
3. *Automatic Sectionalizing*: Automatic sectionalizers that see fault current attempt to isolate the fault by opening when the system is de-energized by a reclosing device.
4. *Lockout:* If the fault persists, time overcurrent protection clears the fault. Lockout could be the same device that performed the reclosing function, or could be a different device that is closer to the fault.
5. *Automated Switching*: Automated switches are used to quickly isolate the fault and restore power to as many customers as possible. This includes both upstream restoration and downstream restoration. In upstream restoration, a sectionalizing point upstream from the fault is opened. This allows the protection device to reset and for restoration of all customers upstream of the sectionalizing point. In downstream restoration, other sections that remain de-energized are isolated from the fault by opening switches. Customers downstream from these points are restored through alternate paths by closing normally open-tie switches.
6. *Manual Switching*: Manual switching restores power to customers that could not be restored by automated switching (certain customers will not be able to be restored by either automated or manual switching). As in automated switching, manual switching has both an upstream restoration component and a downstream restoration component.
7. *Repair*: The fault is repaired and the system is returned to its pre-fault state.

The seven steps outlined above generate a set of system states for each contingency. Switches and protection devices being open or closed characterize these states. For each state occurring with frequency λ and duration δ, the accrued outage frequency of all de-energized components are incremented by λ

(if the component was energized in the preceding state) and the accrued outage duration of all de-energized components are incremented by $\lambda \times \delta$.

The analytical simulation sequence of events becomes more complicated if operational failures are considered. Operational failures occur when a device is supposed to operate, but fails to do so. The probability of such an event is termed *probability of operational failure, POF*. Operational failures cause the simulation sequence to split. One path assumes that the device fails to operate and has a weight of *POF*, the other path assumes that the device operates and has a weight of 1 – *POF*. This path splitting is illustrated in Figure 13.2, which shows the steps required when considering a fuse that is supposed to clear a fault.

The result of simulation path splitting is an enumerative consideration of all possible system responses to each contingency (in the context of operational failures). Enumerative consideration is important since some states may be rare, but have a major impact on the system when they do occur. During restoration, path splitting associated with the enumerative consideration of possible outcomes is important when intended switching fails and customers that would otherwise have been restored are not.

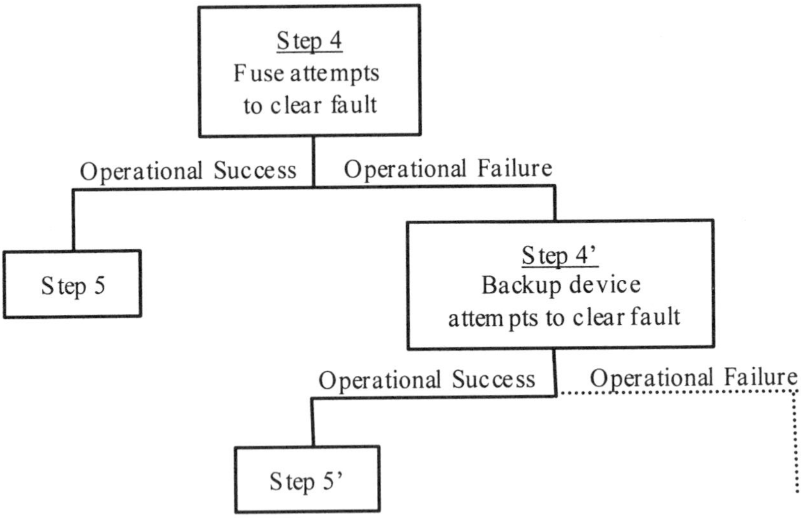

Figure 14.2 Simulation Path Splitting Due to Operational Failures

Incremental Studies of Improvement

An analytical simulation will produce identical results if an analysis is performed multiple times. In addition, small changes in input data will cause small changes in results (as opposed to some other methods). This allows the impact of small reliability improvements to be quantified for individual customers and reliability indices by running a series of studies of incremental changes in the system design. For example, the planning engineer can move the location of a sectionalizer or recloser up and down a feeder 1/8 mile at a time, seeing the results in total reliability, and the distribution of reliability results up and down the feeder, with each change. This capability also allows input parameters to be perturbed and result sensitivities to be computed. However, analytical simulation does not allow the uncertainty of reliability to be quantified. For example, while expected that SAIDI might be 2.0 hours per year, there is some probability that next year will just be an "unlucky year." To compute how often such unlucky years could occur, and how extreme SAIDI or other factors would be in those years, requires application of Monte Carlo techniques.

Example Application

An analytical simulation method was applied to the test system shown in Figure 14.1, which is based on an actual U.S. utility distribution system. The system model contains 3 voltage levels, 9 substations, more than 480 miles of feeder, and approximately 8000 system components. The model was first calibrated to historical outage data, then used to analyze candidate designs as described above. In addition the authors used an optimization module that maximized reliability results vs. cost. The diagram shown in Figure 14.1 was originally color coded (not shown here) based on computed outage hours, and shows the system plan that resulted from a combination of engineering study and use of optimization. Individual component reliability results can be easily used to generate a host of reliability indices. For this system, common indices include:

MAIFI (Momentary Average Interruption Frequency Index) = 2.03 /yr (1)

SAIFI (System Average Interruption Frequency Index) = 1.66 /yr (2)

SAIDI (System Average Interruption Duration Index) = 2.81 hr/yr (3)

Targets for the area were 2.00, 1.75, and 2.9 hours (175 minutes) respectively, thus this system essentially meets its target reliability criteria.

Differentiated Reliability Design

Realistically, the expected reliability of service over an entire distribution system cannot be the same at all points. Distribution systems use transshipment of power through serial sets of equipment. Thus there are areas of the system that are more "downstream", and being downstream of more devices and miles of line, have somewhat higher exposure to outage. Figure 14.1's example demonstrates this. Although SAIDI is 2.81, expected performance for the best 5% of customers is 1.0 hours and 4.02 for the worst 5%.

However, planning engineers can engineer a system to keep the range of variation in what customers in one area receive versus customers in another, with within a reasonable range. This involves the selected use of protective equipment such as fuses and breakers, sectionalizers, reclosers, and contingency switching plans, and in some cases, distribution resources such as on-site generation and energy storage, and demand-side management. Artful use of these measures, combined with application of sound distribution layout and design principles, should result in a system that provides roughly equitable service quality to all customers.

However, there are situations where a utility may wish to promise higher than standard reliability to a specific customer, or to all customers in an area such as a special industrial park. These situations, too, can be engineered. In fact, generally it is far easier to engineer the reliability of service to a specific location than it is to engineer an entire system to a specific target.

Performance-Based Industrial Contracts

Generally, such situations arise in the negotiation for service to medium and large industrial customers. Many of these customers have special requirements that make continuity of service a special consideration. All of them pay enough to the utility to demand special attention – at least to the point of having the utility sit down with them to discuss what it can do in terms of performance and price.

The actual engineering of the "solution" to an industrial customer's needs is similar to that discussed above for systems as a whole. Planning engineers evaluate the expected reliability performance of the existing service to the customer, and to identify weaknesses in that system. Candidate improvement plans are tried in the model and a plan evolves through this process until a satisfactory solution is obtained. The system design in Figure 14.1 includes three such customer locations, as shown in Figure 14.3. Special attention to configuration, fusing and protection coordination on other parts of the circuit, contingency (backup) paths and quick rollover switching, provided about 80% of the improvement in each case.

The other 20% came from extending the reliability assessment analysis to the

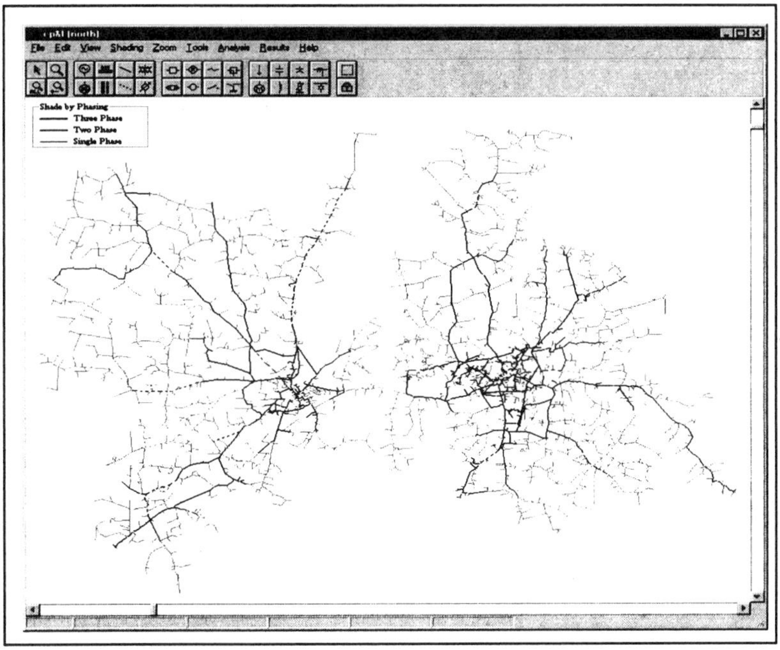

Figure 14.3 Service to three industrial customers in the example system were designed to higher-than-normal reliability performance targets.

Table 14.3 Reliability Targets for Three Special Customers

Customer	Industry Type	Annual Target		Three-Year Operating Avg.	
		Events	Total hours	Events	Total hours
System	Entire system	1.61	2.81	1.67	2.65
A	Tractor manufacturer	.33	.02	0	0
B	Synthetic fiber	1.	.5/event	.66	.085/event
B	Carpet manufacture	.25	-	.33	.42

customer's site. In each case, the customer's substation, all privately owned, was analyzed and suggested improvements in equipment or configuration were recommended to his plant engineer. In one case the analysis extended to primary (4 kV) circuits inside the plant. Table 14.3 gives the results for each. For site A, the target has been exceeded in practice to date because there have been no outages in three years of operation. However eventually performance will probably average slightly better than the target, no more.

In the case of site B, the customer's main concern was that the length of any interruption be less than 1/4 hour. Thus, the target is not aimed at total duration time, but on each event, something quite easy to engineer. Performance has greatly beaten this target – means to assure short outages such as automatic rollover and remote control restore power in much less time than required.

For customer C, any interruption of power resulted in loss of an entire shift's (eight hours') production. Thus, the major item of interest there was total count of events. Initial evaluation showed little likelihood of any interruption lasting that long. Reliability engineering focused on frequency only.

In all three cases, the performance-based contracts call for a price to be paid that can be considered a premium service price. It is typical for industrial customers in this class to negotiate price and conditions with the utility, but in all three cases the price included an identified increment for the improvement in service. Unlike some performance-based rate schedules, there is no "reward" or extra payment for good service – the premium payment is payment for the good service.

But performance penalties are assessed, not on an *annual* basis depending on annual performance, but on an *event* basis. If reliability of service is perfect, the utility has earned all the premium monies paid. Any event (any, not just those exceeding the limits) garners a rebate according to a certain formula. For example, customer B receives a $30,000 rebate for any interruption of power, but the rebate rises to $300,000 if the interruption lasts more than 30 minutes. (Customer B's average monthly bill is about $175,000).

14.5 USE OF A HYBRID ANALYTICAL SIMULATION – MONTE CARLO METHOD TO ANALYZE FINANCIAL RISK

If 100 identical distribution systems are built, the expected reliability of each will be identical. In a given year, some of these systems might be lucky and experience nearly perfect reliability. Others may experience near expected reliability, and others may be unlucky and experience far worse than expected reliability. This variation is natural, and is vital to understand when negotiating reliability-based contracts.

Variations in reliability can be examined using techniques referred to as risk assessment. A risk assessment model identifies all possible outcomes and the

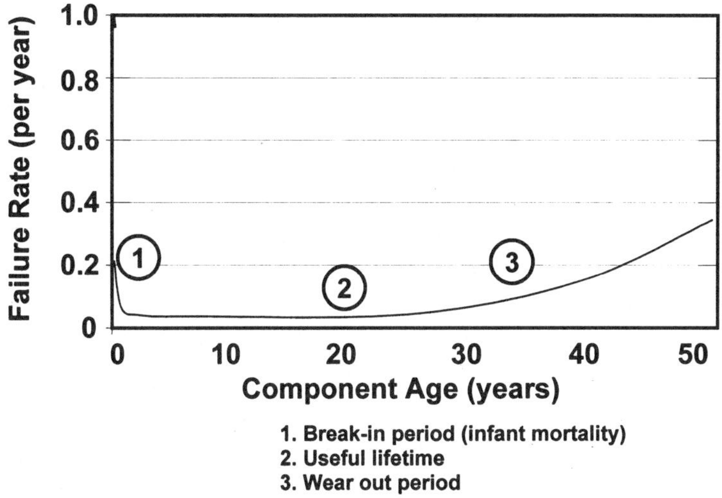

Figure 14.4 The traditional "bathtub" failure rate curve, as discussed in the text. Here, the device has a useful lifetime of about 30 – 35 years.

Monte Carlo techniques are required. This commonly takes the form of a probability of each outcome occurring. When possible, this is done through analytical methods such as function convolution. Usually this is not feasible and "Sequential Monte Carlo Simulation" where small time slices are modeled in succession and each component is tested for random failures during each time slice. This type of simulation is very flexible and can handle complex models, but is computationally slow and data intensive.

Many of the problems associated with a sequential Monte Carlo simulation can be overcome if components are assumed to have a constant failure rate over the course of one year. This is a reasonable assumption and can be visualized by "bathtub failure rate functions." These functions show that typical electrical components will have a high failure rate when they are first installed (due to manufacturing defects, damage during shipping, and improper installation). The failure rate will reduce after this "infant mortality" period and remain at a fairly constant level over the useful life of the component. The failure rate will gradually rise as the equipment wears out at the end of its useful life. A bathtub failure rate function is shown in Figure 14.4.

Components with constant failure rates follow a Poisson process. This allows the probability of the component failing a specific number of times in a year to be easily computed. If a component has a constant failure rate of λ times per year, the probability of it failing x times in a year is:

$$\text{Prob. of failing } x \text{ times} = \frac{\lambda^x e^{-\lambda}}{x!} \tag{14.1}$$

If constant failure rates are assumed, the analytical simulation described in the previous section can be modified to simulate a *random* year rather than an *expected* year. This is done by determining the number of times each component will fail *a priori*. For each component, a random number between zero and one is generated. If this number is less that $e^{-\lambda}$, no failure will occur in the year being simulated. If the random number is greater, the number of times that the component will fail is determined by Equation 1.

Once the number of times each component will fail is determined, an analytical simulation is performed that substitutes component failure rates with the number of times that they will fail in the random year being simulated. Using this process, many years can be simulated, a list of outcomes can be recorded, and distribution statistics can be computed. This methodology is referred to as an Analytical/Monte Carlo Hybrid Simulation. This particular hybrid simulation has two desirable features:

1. It requires no additional data beyond the requirements of an analytical model.
2. If there is confidence in the expected values generated by the analytical simulation, and there is confidence that component failure rates are constant over a single year; there is equal confidence in the results of the analytical/Monte Carlo hybrid simulation.

Analytical/Monte Carlo hybrid simulation is applied to the system being analyzed (i.e. the system in Figures 14.1 and 14.3) by performing 1,000 random simulations on the same system that was used to demonstrate the analytical simulation. The results are shown in Figure 14.5 as both a histogram (the bar chart) and an integral of the histogram (the continuous line). The statistical results of the simulation are (in hours per year): mean value = 2.81, standard deviation = .79, minimum value = .95, maximum value = 6.37.

The analytical/Monte Carlo hybrid simulation is computationally intensive, but provides statistical results not obtainable by using purely analytical methods. This statistical information is vital when assessing technical and financial risk associated with reliability based contracts. The remainder of this chapter demonstrates this by applying analytical/Monte Carlo hybrid simulations to distribution systems subject to performance based rates.

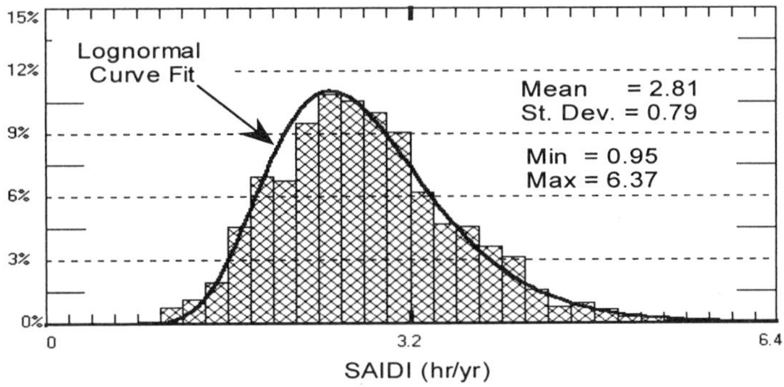

Figure 14.5 SAIDI Results for 1000 random years (hr/yr). Bars show a histogram of the expected annual results, which, while having a mean of less than 3 hours, have a good probability of being over 5 hours. Solid fits a log-normal curve to the data.

Analyzing the Risk from Performance-Based Rates

Deregulated utilities are reducing costs by deferring capital projects, reducing in-house expertise, and increasing maintenance intervals. As a direct consequence, the reliability on these systems is starting to deteriorate. Since these systems have been designed and maintained to high standards, this deterioration does not manifest itself immediately. System reliability will seem fine for several years, but will then begin to deteriorate rapidly. When reliability problems become evident, utilities often lack the necessary resources to address the problem.

Regulatory agencies are well aware that deregulation might have a negative impact on system reliability. In a perfect free market, this would not be a concern. Customers would simply select an electricity provider based on a balance between price and reliability. In reality, customers are connected to a unique distribution system that largely determines system reliability. These customers are captive, and cannot switch distribution systems if reliability becomes unacceptable. For this reason, more and more utilities are finding themselves subject to performance based rates (PBRs).

A PBR is a contract that rewards a utility for providing good reliability and/or penalizes a utility for providing poor reliability. This can either be at a system level based on reliability indices, or can be with individual customers. Performance is usually based on average customer interruption information. This typically takes the form of the reliability indices SAIFI and SAIDI.

A common method of implementing a PBR is to have a "dead zone" where

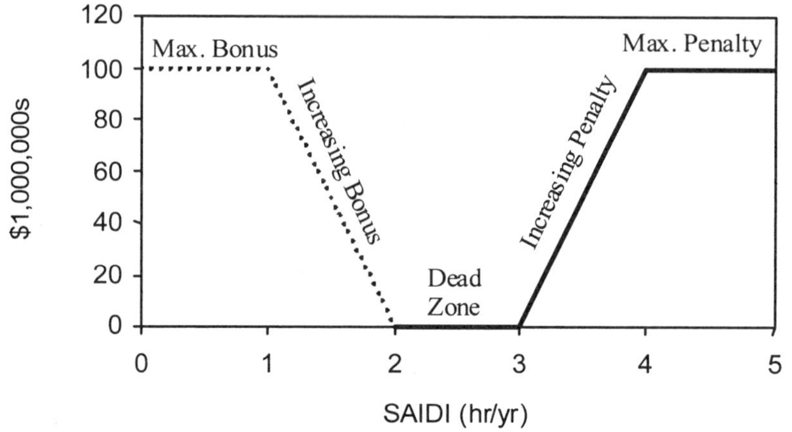

Figure 14.6 Performance based rate structure for the example in this section. There is no penalty or incentive payment for annual SAIDI performance that lies between 2 and 3 hours. Above that the utility pays a penalty of \$1,666,666 per minute, up to 60 minutes. Similarly, it receives a reward of the same magnitude for each minute below the dead zone.

neither a penalty nor a bonus will be assessed. If reliability is worse than the dead zone boundary, a penalty is assessed. Penalties increase as performance worsens and are capped when a maximum penalty is reached. Rewards for good reliability can be implemented in a similar manner. If reliability is better than the dead zone boundary, a bonus is given. The bonus grows as reliability improves and is capped at a maximum value. All PBRs will have a penalty structure, and some will have both a penalty structure and a bonus structure. A graph of a PBR based on SAIDI is shown in Figure 14.6.

Most initial PBRs will be negotiated by the utility to be "easy to meet." This means that business as usual will put the utility in the dead zone. This does not mean that a utility should do business as usual. It may want to spend less on reliability until marginal savings are equal to marginal penalties. Similarly, it may want to spend more money on reliability until marginal costs are equal to the marginal rewards. In either case, a utility needs a representative reliability assessment and risk model to determine the impact of reliability improvement and cost reduction strategies.

In order to make intelligent decisions about PBRs based on average system reliability, a probability distribution of financial outcomes is needed. This requires a PBR structure (such as the one shown in Figure 14.6), and a distribution of relevant reliability outcomes (like the histogram shown in Figure

4). It is also desirable to describe the reliability histogram with a continuous mathematical function. A good function to use for reliability index histograms is the log-normal distribution, represented by *f(x)*:

$$f(x)=\frac{1}{x\sigma\sqrt{2\pi}}\exp\left[-\frac{(\ln x-\mu)^2}{2\sigma^2}\right];\quad x\geq 0 \tag{14.2}$$

The parameters in this equation can be determined by a host of curve-fitting methods, but a reasonably good fit can be derived directly from the mean and variance. If a mean value, $\bar{x}$, and a variance are computed, log-normal parameters are:

$$\sigma=\sqrt{\ln\left(variance+e^{2\ln\bar{x}}\right)-2\ln\bar{x}} \tag{14.3}$$

$$\mu=\ln\bar{x}-\frac{1}{2}\sigma^2 \tag{14.4}$$

The log-normal curve corresponding to the test system is shown along with the histogram in Figure 14.5. The curve accurately captures the features of the histogram including the mean, median, mode, standard deviation, and shape.

Assume that a system analyzed earlier, characterized by the reliability shown in Figure 14.5, is subject to the PBR formula shown in Figure 14.6. Using these two functions, a financial risk analysis can be easily determined. For example, the expected penalty will be equal to:

$$Penalty=\int_0^\infty PBR(SAIDI)\cdot f(SAIDI)\,dSAIDI \tag{14.5}$$

The probability of certain financial outcomes can also be computed. For example, the probability of landing in the dead zone will be equal to the probability of have a SAIDI between 2 hours and 3 hours. This is mathematically represented by:

$$\% \; in \; Dead \; Zone = 100\% \cdot \int_{2}^{3} f(SAIDI) \; dSAIDI \qquad (14.6)$$

The expected outcome of the test system is $14 million in penalties per year. Thus, despite the fact that the expected annual SAIDI (2.81 hours) lies in the dead zone (no penalty – no reward) the utility will on average have to pay a fine amounting to about 8.4 minutes per year. Penalties will occur 32% of the time, bonuses will occur 12% of the time, and the utility will be in the Dead Zone 57% of the time. The distribution of all possible outcomes and their associated probabilities is shown in Figure 14.7.

For a cash-strapped utility, Figure 14.7 represents an unwelcome situation. And it is risky, because even though the average penalty of $14 million may be acceptable, that outcome will rarely happen. The $14 million average occurs because a penalty of $50 million or more will occur once every 7 years and the maximum penalty of $100 million will occur once every 12 years. Faced with this situation, a utility would be wise to negotiate a less risky PBR. Possibilities for adjusting the PBR to mitigate risk include:

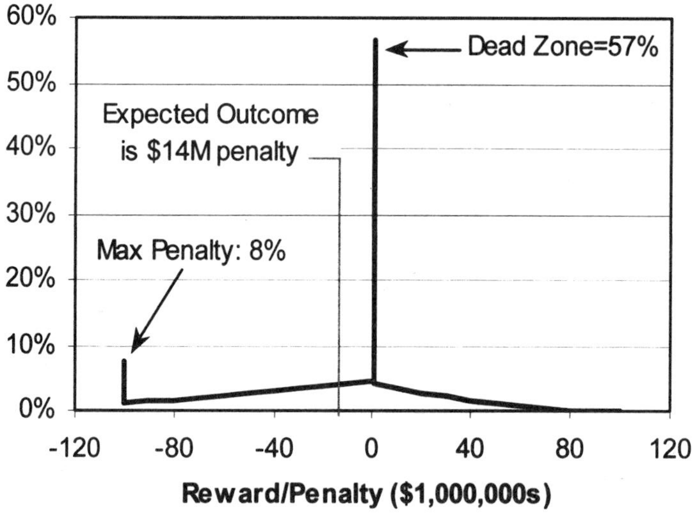

Figure 14.7 Distribution of financial outcomes for the example PBR problem. Although it is most likely the penalty/reward is zero (performance lies in the dead zone) the utility has a significant probability of getting hit with the maximum penalty (about once in every 12 years), while no chance of getting the maximum reward.

1. Make the reward and penalty slopes less steep,
2. Widen the dead zone boundaries,
3. Move the dead zone to the right, and
4. Reduce bonus and penalty caps. Each of these options can be used alone, or a combination can be negotiated.

A suggested change to the PBR rules to mitigate the risk for this utility system is shown in Figure 14.8. This change stretches out the bonus and penalty transition zones from one hour to two hours. The impact of this changed PBR on the utility's expected financial exposure is shown in Figure 14.9. On the surface, it may seem like not much has changed. Penalties will still occur 32% of the time, bonuses will still occur 12%, but total risk is reduced greatly, by $5 million. Utility companies can use predictive reliability assessment tools when planning *and* designing their systems and their responses to regulatory rules. This will allow the reliability of existing systems to be assessed, the reliability impact of projects to be quantified, and the most cost-effective strategies to be implemented. In most cases, an analytical model (Chapter 14) will be sufficient. This will allow utilities to compare the expected performance of various systems and various design options. Analytical/Monte Carlo hybrid techniques become necessary when technical and financial risk needs to be examined. This chapter has shown methodologies to serve both of these need categories.

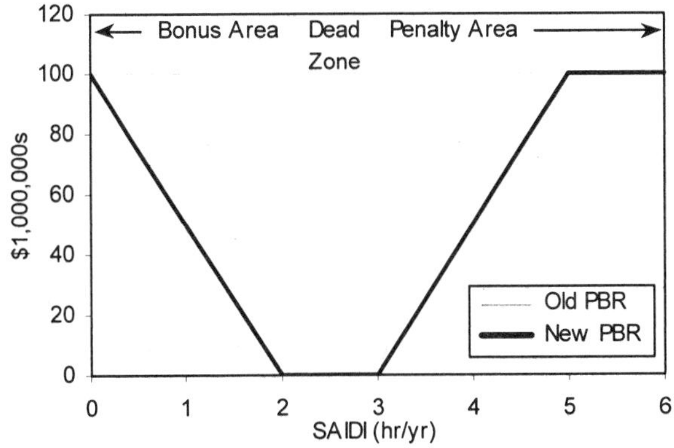

Figure 14.8 Changed PBR schedule that reduces risk. The schedule has a milder rise in both penalty and reward sides, and is a "fair" proposal for the utility to make to regulators because it treats both penalty and reward sides equally.

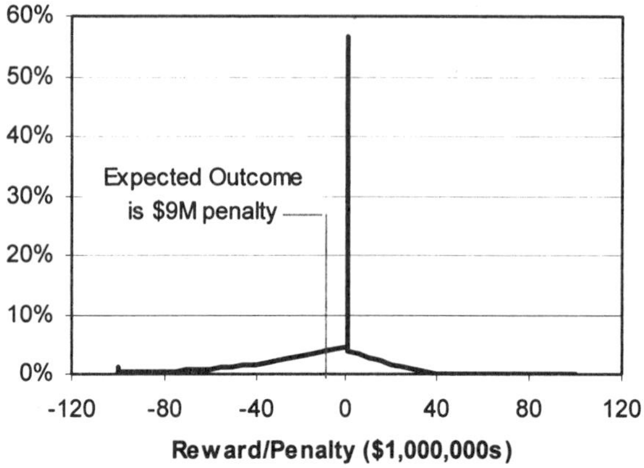

Figure 14.9 Financial outcome with the revised PBR formula. The utility's expected average is still a penalty, but only $9 million, an improvement of $5 million. More important, the possibility of it being hit with a large (over $50 million) penalty has decreased greatly.

14.6 CONCLUSION AND KEY POINTS

The reliability of service of a power distribution system can be designed to meet specific goals with respect to reliability criteria such as SAIDI and SAIFI. It can also be designed to minimize financial risk whether that is measured in terms of utility and/or customer costs, or includes performance-based rates and contracts as well. Of four possible methods, all capable of computing distribution reliability, the authors believe analytical simulation method is most applicable to reliability-based planning.

Analytical simulation computes the expected reliability of each component based on system topology and component reliability data. It does so by simulating the system's response to each contingency, determining the impact of this contingency to each component, and weighting the impact of each contingency by its probability of occurrence. The result for each component is the expected number of momentary interruptions per year, the expected number of sustained interruptions per year and the expected number of outage hours per year. Results can be displayed in tables, visualized spatially or aggregated into reliability indices.

To assess technical and financial risk, the analytical simulation method is expanded to an analytical/Monte Carlo hybrid simulation. Such a method assumes that component failure rates are constant over a specific year and models many random years rather than a single expected year. The result is a probability distribution of possible results that can be used to obtain information such as variance, median, modality, shape, and extremes. It also allows rigorous risk analyses to be performed on various types of performance-based rates and customer contracts.

After negotiating a performance-based rate, a distribution utility will find itself in a radically new paradigm. An explicit dollar value has been placed on reliability, and systems can be designed to balance the cost of reliability with the cost of infrastructure. If reliability measures are based on indices such as SAIFI and SAIDI, it may be beneficial to the utility to allow the reliability of a few customers in areas where it would be expensive to "fix" problems to decline to levels far below average. At the same time, improving reliability in "easy to work" areas to far above average. This "gaming" the regulatory performance requirement will be viewed as unacceptable by most commissions, sparking more complicated PBR formulas and regulations designed to prevent such situations. Ironically, "deregulation" may therefore result in regulation of a previously unregulated area.

Regardless, the key point of this chapter is that *reliability can be engineered.* Key aspects of reliability-based engineering for distribution systems are:

- *Focus on a customer-oriented metric.* What really counts is the reliability of service the customer sees not reliability as measured on the power system itself. Reliability-based engineering should target customer-level reliability performance.

- *Use a good reliability-assessment method.* Such methods apply a rigorous analytical method, compute expected reliability on a detailed (nodal) basis, and use a method that provides consistent evaluation of results so that the effect of incremental design changes can be seen.

- *Use differentiated reliability where necessary* to best meet customer reliability needs and price sensitivity. Premium areas (higher reliability and price) can be designed to specific targets for those areas.

- *Calibrate the base model* to historical outage data as seen in actual operation of the system.

- *Engineer reliability by improving design* in response to weaknesses identified by the reliability assessment method, in existing and proposed system designs.

- *Minimize financial risk,* using a method appropriate to the evaluation of the probability distribution of expected results

REFERENCES AND FURTHER READING

P. F. Albrecht and H. E Campbell, "Reliability Analysis of Distribution Equipment Failure Data," EEI T&D Committee Meeting, January 20, 1972.

R. N. Allan, et al., "A Reliability Test System for Educational Purposes – Basic Distribution System Data and Results," *IEEE Transactions on Power Systems,* Vol. 6, No. 2, May 1991, pp. 813-821.

D. Atanackovic, D. T. McGillis, and F. D. Galiana, "The Application of Multi-Criteria Analysis to a Substation Design," paper presented at the 1997 IEEE Power Engineering Society Summer Meeting, Berlin.

R. Billinton, and J. E. Billinton, "Distribution System Reliability Indices," *IEEE Transactions on Power Delivery,* Vol. 4, No. 1, January 1989, pp. 561-568.

R Billinton R., and R. Goel, "An Analytical Approach to Evaluate Probability Distributions Associated with the Reliability Indices of Electric Distribution Systems," *IEEE Transactions on Power Delivery,* PWRD-1, No. 3, March 1986, pp. 245-251.

R.E. Brown, J.R. Ochoa, 'Distribution System Reliability: Default Data and Model Validation,' *IEEE Transactions on Power Systems,* Vol. 13, No. 2, May 1998, pp. 704-709.

R. E. Brown, S. S. Venkata, and R. D. Christie, "Hybrid Reliability Optimization Methods for Electric Power Distribution Systems," *International Conference on Intelligent Systems Applications to Power Systems*, Seoul, Korea, IEEE, July 1997.

R. E. Brown, S. S. Gupta, R. D. Christie, and S. S. Venkata, "A Genetic Algorithm for Reliable Distribution System Design," *International Conference on Intelligent Systems Applications to Power Systems*, Orlando, FL, January 1996, pp. 29-33.

R.E. Brown, S. Gupta, S.S. Venkata, R.D. Christie, and R. Fletcher, 'Automated Primary Distribution System Design: Reliability and Cost Optimization,' *IEEE Transactions on Power Delivery*, Vol. 12, No. 2, April 1997, pp. 1017-1022.

R.E. Brown, S. Gupta, S.S. Venkata, R.D. Christie, and R. Fletcher, 'Distribution System Reliability Assessment Using Hierarchical Markov Modeling, *IEEE Transactions on Power Delivery*, Vol. 11, No. 4, Oct., 1996, pp. 1929-1934.

R.E. Brown, S. Gupta, S.S. Venkata, R.D. Christie, and R. Fletcher, 'Distribution System Reliability Assessment: Momentary Interruptions and Storms,' *IEEE Transactions on Power Delivery,* Vol. 12, No. 4, Oct 1997, pp. 1569-1575.

J. B. Bunch, H.I Stalder, and J.T. Tengdin, "Reliability Considerations for Distribution Automation Equipment," *IEEE Transactions on Power Apparatus and Systems,*

PAS-102, November 1983, pp. 2656 - 2664.

"Guide for Reliability Measurement and Data Collection," EEI Transmission and Distribution Committee, October 1971, Edison Electric Institute, New York.

S. R. Gilligan, 'A Method for Estimating the Reliability of Distribution Circuits,' *IEEE Transactions on Power Delivery,* Vol. 7, No. 2, April 1992, pp. 694-698

W. F. Horton, et al., "A Cost-Benefit Analysis in Feeder Reliability Studies," *IEEE Transactions on Power Delivery,* Vol. 4, No. 1, January 1989, pp. 446 - 451.

Institute of Electrical and Electronics Engineers, *Recommended Practice for Design of Reliable Industrial and Commercial Power Systems,* The Institute of Electrical and Electronics Engineers, Inc., New York, 1990.

G. Kjølle and Kjell Sand, 'RELRAD - An Analytical Approach for Distribution System Reliability Assessment,' *IEEE Transactions on Power Delivery,* Vol. 7, No. 2, April 1992, pp. 809-814.

A. D. Patton, "Determination and Analysis of Data for Reliability Studies," *IEEE Transactions on Power Apparatus and Systems,* PAS-87, January 1968.

N. S. Rau, "Probabilistic Methods Applied to Value-Based Planning," *IEEE Transactions on Power Systems,* November 1994, pp. 4082 - 4088.

A. J. Walker, "The Degradation of the Reliability of Transmission and Distribution Systems During Construction Outages," Int. Conf. on Power Supply Systems. IEEE Conf. Publ. 225, January 1983, pp. 112 - 118.

H. B. White, "A Practical Approach to Reliability Design," *IEEE Transactions on Power Apparatus and Systems,* PAS-104, November 1985, pp. 2739 - 2747.

W. Zhang and R. Billinton, "Application of Adequacy Equivalent Method in Bulk Power System Reliability Evaluation," paper presented at the 1997 IEEE Power Engineering Society Summer Meeting, Berlin.

15
Strategy, Management and Decision-Making

15.1 INTRODUCTION

First and Foremost, a Power Delivery Company is a Business

Overall, the recommended perspective for all segments of a power delivery company is the business-case perspective. This is not to say that the company's role as a public servant, its obligation to serve, or its importance and role in the community should be ignored. However, a business-case perspective assures focus on objectives and the reasons for them, and assures efficient management and use of resources. Doing things in a "businesslike manner" means doing them efficiently, and using labor, equipment, talent, and commercial or technical advantages as effectively as possible. It means weighing alternatives carefully, based on the truly important criteria, when prioritizing scarce resources. Ultimately, it is the best perspective when one wants results, whatever those results may be.

Figure 15.1 on page 494 shows the difference between the business perspective presented in this chapter (bottom) and the traditional paradigm within which utilities operated (top). Traditionally, utilities took a rules-based approach to planning, engineering, and operations, mostly through the use of fixed standards and guidelines which were rigidly applied to their system design and operation. An engineering case approach drove the budget.[1] Obligations

---

[1] The term "engineering" here means all system-related concerns including planning, engineering, design, maintenance, and operations.

and desired results were interpreted, through the application of corporate and engineering processes, into a plan that provided those results, met all the rules, and had as low a cost as possible within those constraints. That paradigm worked well within the traditional power industry.

But that paradigm has failed to deliver the results that are needed in today's business environment. This is due to a mismatch between the traditional procedures and rule bases that planners, engineers, and operators used, and the *way* that power delivery businesses need to optimize their financial and customer service needs today. In both the traditional and the new order, obligations the company must meet, and desired results of executive management, drive spending. The exact perception of what is desired as results is different under the present de-regulation and competition than it was in the traditional utility industry. But that is rather unimportant to this discussion. What is important is the different *process* that must be used in the planning, engineering, and operation of the system (bottom of Figure 15.1), whenever the business needs of a power delivery utility are driving the overall corporate focus. That is the reality in the 21st century.

In the next section, 15.2, this chapter presents a discussion of the typical historical pattern of an "aging infrastructure utility," using a simple corporate-system model the authors developed based on fourteen large domestic investor owned utilities. This is used to identify certain relationships between investment, customer service, and procedures. Basically, a utility cannot simultaneously meet financial and customer service expectations unless it changes the efficiency of its investment towards improving customer service: it must change.

Section 15.3 outlines how and what new approaches can help accomplish the utility's mission. The Results-Driven Management (RDM) approach that is outlined there is recommended as an overall approach. Something like this approach must be used if a power delivery company is to succeed in achieving maximum possible reliability within its given budget, or conversely, if it wishes to minimize the revenue required to provide a targeted level of service. The authors recommend this approach because they have seen it work well many times. Section 15.4 then discusses changes and paradigm shifts in the details of delivery system planning, engineering, and operations that then flow out of this new overall approach. Section 15.5 concludes with an overall summary.

15.2 THE BUSINESS PERSPECTIVE ON THE PROBLEM

This section presents the results of a rather simple model of an electric utility as both a business and an operating system. These values in the model (not given in the interests of brevity) and the graphs shown here, are from the hypothetical Metropolitan Power and Light, MP&L (See Chapter 8). MP&L is a power delivery company (a wires company) whose data base was created by averaging business and system data from fourteen large investor owned utilities in the U.S.

The customer population served by the sample companies ranged from 1.2 million to 4.3 million. Linear regression models relating growth of the system to growth of the customer population are used to drive the model. Additional relationships were developed from the data to model the operational and financial behavior of this hypothetical company. Data from the Bureau of Labor Statistics and the Census Bureau supplement the company data and tie the model to the economic environment of the United States over the last fifty years. Relationships developed by the authors or published sources provide the tie between aging equipment and other operational variables. Although the model can be used to investigate the effects of changing many of the parameters affecting financial and operational performance, the only parameter varied for this comparison is the target utilization factor.

Looking for Profits in All the Right Places

Metropolitan Power and Light is a "modern" utility in at least one sense for this example. It has had to operate with a very tight business focus – on the bottom line – for some time. Conditions forced it to tighten spending and improve financial performance. In this model, this change occurred in 1980 (a little ahead of most utilities in the industry, but giving the example here more time to develop the consequences). MP&L felt it had no choice if it wanted to be competitive. Improved performance had to be found. As MP&L's president put it, "I will not put the company at risk to avoid putting the system in jeopardy." Costs were cut.

Spending was cut in all categories beginning in 1980. Capital spending was cut drastically. Increasing target utilization ratios for major equipment did this. Beginning in 1980, MP&L increased target substation utilization ratio from its traditional .68 to .85.

Figure 15.1 illustrates the MW of substation capacity installed by year for a base case representing no change in utilization and the change case representing the 85% ratio. The only difference between the two cases is this target utilization ratio for substation capacity. An underlying assumption of the model is that all of the load growth can be accommodated without building new substations if there is adequate capacity. This is a simplification made to keep the model simple, the causes and effects directly attributable and "clean" so it can be contained to half a chapter. The only new transformers added in this high-utilization ratio case are replacements for failed units.

When the target utilization rate changed, MP&L found 'released capacity" in its existing system, enough to absorb 13 years of load growth without installing additional substation capacity. This outcome closely parallels the actual experience of companies that have pursued this strategy. Several large IOU's through the U.S. experienced roughly a one-to-one correspondence between percent increases in utilization ratios and number of years that capacity additions were cut to near zero. Thus, for a time this company was able to avoid the capital requirements for installing substation capacity.

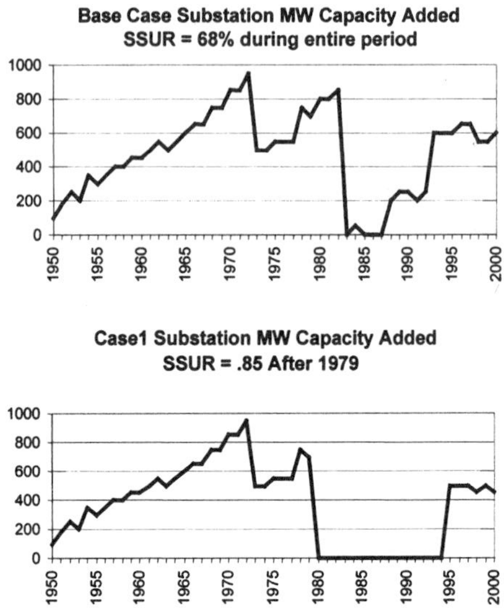

Figure 15.1 Top – substation capacity additions under the base case, where the utility clings to its traditional substation utilization ratio (SSUR). Bottom: capacity additions when the target SSUR is raised from 68% to 855 in 1980.

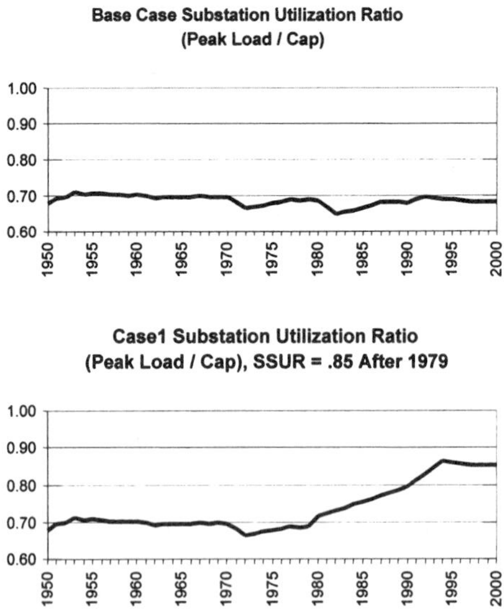

Figure 15.2 Substation utilization ratios in the two scenarios.

Eventually, however, new capacity is required. In the hypothetical situation of this model: 1) no new capacity need be added for 13 years, 2) at the end of that period, the capacity that is required annually is about 100 MW less than in the base case. Figure 15.2 shows the result of substation utilization. System average utilization begins to climb beginning in 1980, taking 13 years to reach its target on a system basis.

The effect of this one change in system engineering standards – utilization ratio – on the utility's financial status is profound. Figures 15.3 through 15.5 compare several financial measures of the effects of changing the target utilization rate. To begin, as the substation utilization ratio increases, the ratio of net to gross plant falls in the change case since less new equipment is installed in the period from 1980 onward (Figure 15.3). The return on assets (Figure 15.4) exhibits a noticeable improvement when substation capacity additions are suspended in the change case. Profitability is up substantially. The gap begins to narrow after new substation capacity installation is resumed in the high-utilization case beginning in 1994, but is never closed, because that case always gets "more" from each MVA of capacity purchased.

Impact on Customer Service

While financial performance of the utility's investment went up, customer service quality went down. An obvious side effect of the increase in utilization, and something that is by now a quite familiar topic to the reader, is the average age of substation transformers. Figure 15.5 shows the difference in the base and high-utilization case. In the base case a good deal of new capacity is added, with many old units being retired before failure because they are under-capacity where they are, but too old to be put back into service elsewhere. While average age increases, it increases only slightly.

By contrast, no new units are added to the system in the high utilization case, except for replacements due to failures. The general nature of failure replacement on average equipment age is easy to determine (a model as used here is really not needed). Failure rate for older units is about 2%, that for new less than .6%, with an average of about 1.0 %. Thus, 1.0% of the units are replaced each year because they fail. Assuming the average failed unit is the average age (about 33 years), then 1.0% of the system is replaced by new units, subtracting 33 years x 1.0%, or .33 year overall. The system "ages" at 2/3 year per year in this case. This is quite clear in Figure 15.5, where age increases from about 35 years to 45 years in the 15-year period 1980 – 1994.

System outage frequency (SAIFI) is affected almost immediately, because average age of equipment is increasing and consequently failure rate is increasing. Units are both left in service longer (instead of being upgraded) and they are more heavily loaded. Failure rates are greater due to both reasons. SAIFI begins to rise almost immediately, although the trend may not be noticeable for what it is for three or four years.

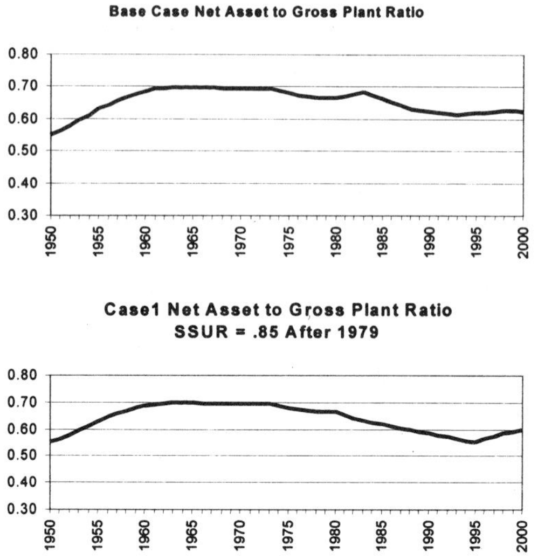

Figure 15.3 Ratio of Net Assets to Gross Plant for base (top) and high-utilization (bottom) cases.

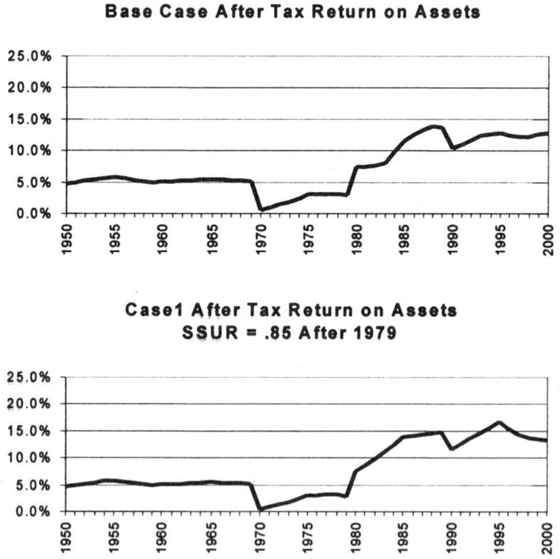

Figure 15.4 After tax term of assets for the base (top) and high-utilization (bottom) cases. The difference may not look like much when plotted in this manner, but it represents as much as a 30% improvement in financial return.

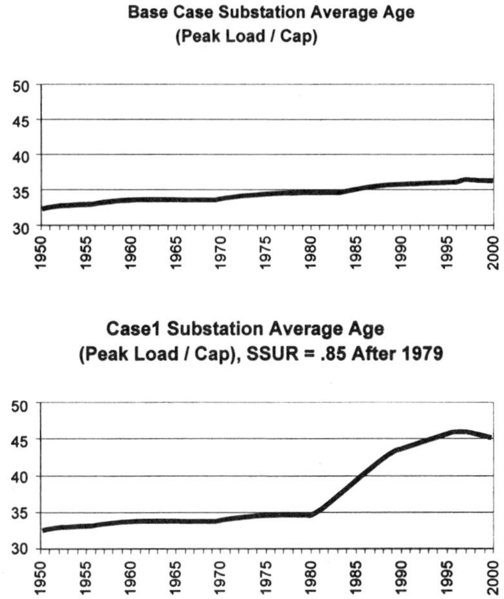

Figure 15.5 Average age of substation transformers in the base and high utilization case.

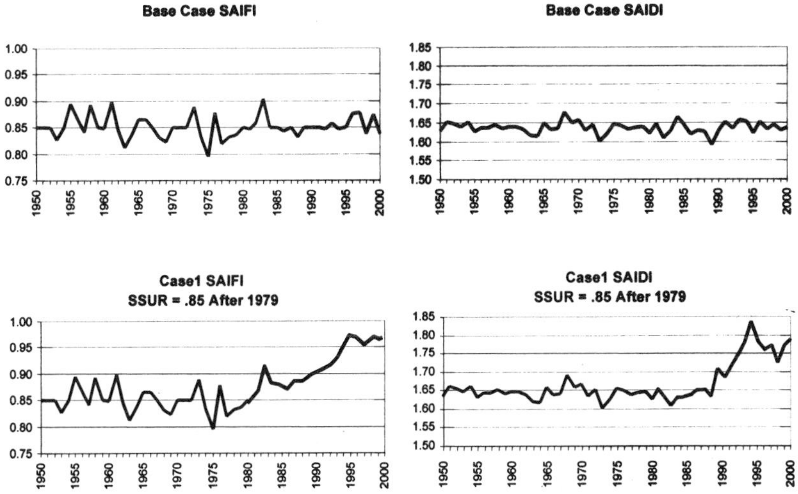

Figure 15.6 Reliability of the two scenarios. Top: base case SAIFI and SAIFI. Bottom: comparable measures for the high-utilization case.

The effect on system outage duration (SAIDI) is barely noticeable in the short term but becomes quite dramatic after a few years. This is a general result: in system where utilization rate is pushed upward and the various alternatives to mitigate reliability problems are not taken, or in aging infrastructure areas, SAIDI degradation generally lags SAIFI degradation by five to seven years, as it does here.

The reason this pattern usually appears is that when SAIFI begins to increase (generally within five years after the utilization rate is increased) most of the additional outages are in places where there are still traditional levels of contingency margin. Utilization ratio has yet to climb by a noticeable amount. Analytical work performed by one of the authors' colleagues (Brown) has confirmed a generally held belief, that there is a strong inverse relationship between the utilization ratio of the system and the average duration of outages (but it is more complicated that one might assume).

Thus, initially the number of events (SAIFI) increases as failure rates creep upward, but the duration of the outage is not noticeably different. The impact on SAIDI is small. After a while, as the utilization ratio throughout the system gradually moves upward and erodes the contingency margin, the average duration of outages increases. Suddenly SAIDI increases rapidly, first because it is proportional to the duration of these ever-longer outages, and second because the number of outages is increasing every year, too.

Figure 15.6 shows the pronounced effect that the increase in asset utilization can have on outage duration as measured by SAIDI at the level of utilization modeled in the change case.

Summary of Overall Effects

Almost immediately after the increase in utilization rates, MP&L's financial performance improves. System performance, as measured by customer service quality, begins to erode. Equipment age increases. If recent events at actual utilities are any guide, customer satisfaction also declines.

A Rock and a Hard Place

The example above represents well, if in slightly simplified form, the situation many utilities face. Their system matches the overall statistics given here but the reliability impacts are not spread evenly over the system. Rather they are concentrated on one or more aging areas. In addition to the problems detailed above, the utility faces another inequality of service. Some areas of the system, those with new equipment, have adequate service quality. Other parts, in aging areas, have really poor service.

At this point, the utility is faced with two unsavory alternatives. It can maintain its recent suitable financial performance, but it cannot improve customer service if it does so. Or, it can improve customer service, but to do so it must spend money to the extent that financial performance will suffer. Worse, much the same effects seen in the reliability degradation (Figure 14.6) will occur in reverse as the utility seeks improvement. SAIFI will improve almost immediately upon initiation of a program to cut failure rates, replace old equipment, and improve reliability. SAIDI – what counts most to many people – will improve noticeably only after several years as utilization ratios decline. Financial performance drops and reliability does not immediately improve. This is not a desirable situation.

The third alternative

We believe that there is another alternative. The utility can "cheat" – it can change the rules, or in this case the relationship, between cost and reliability. This means it must innovate. Incremental change is not enough. New methods are necessary and not just any new methods will do. It must make changes to planning, engineering, operations and management methods that work together and that are compatible with the financial goals of executive management. Change can, and must take place. The utility planners and engineers have already proven they can change. In the example above, as has happened at many utilities, the planners and engineers made a tremendous change – they increased their utilization rate, among the most sacred of standards, by one quarter, from 68% to 85%.

But the changes needed this time are more fundamental. Instead of changing a rule (a standard), planners, engineers and operators need to move to system or results-based standards, rather than rule-based. This means that the utility maintains strict standards, but that they are defined and applied as definitions of the results that must be obtained. This is not necessarily a relaxation of the traditional standards levels. Sometimes the new paradigm may call for increases above traditional limits.

15.3 RECOMMENDED MANAGERIAL APPROACH: BUSINESS-CASE BASED, RESULTS-DRIVEN MANAGEMENT

Standards for the Result, Not the Approach

The modern business-case approach (bottom of Figure 15.7) differs from the traditional paradigm in several respects; the most important of which is that it uses *results-based standards.* This business approach does not mean that a utility turns its back on the use of standards or relaxes them. In fact it may mean that certain design criteria or standards are tightened.

But it does mean that a utility's standards and guidelines are defined based on the result, rather than rules about how it will be obtained. A traditional utility basically set rules for how it did things: rules that dictated precisely how and when something was done. A modern power delivery company must set standards that define what the result is to be, and vary the way it gets the job done to best fit its business needs. An example of this will be given later in this section, with regard to transformer loading.

However, this is a great departure from the past, one that makes some engineers uncomfortable. The business case, rather than any engineering perspective, is the basis for all managerial decisions and spending justifications. It drives the entire budget and priority process. Often, this process is budget constrained. That means that the financial case, which specifies the budget, sets it at an amount considerably less than needed to do things in the traditional way. Planning, engineering, operations and maintenance have to fit within that budget.

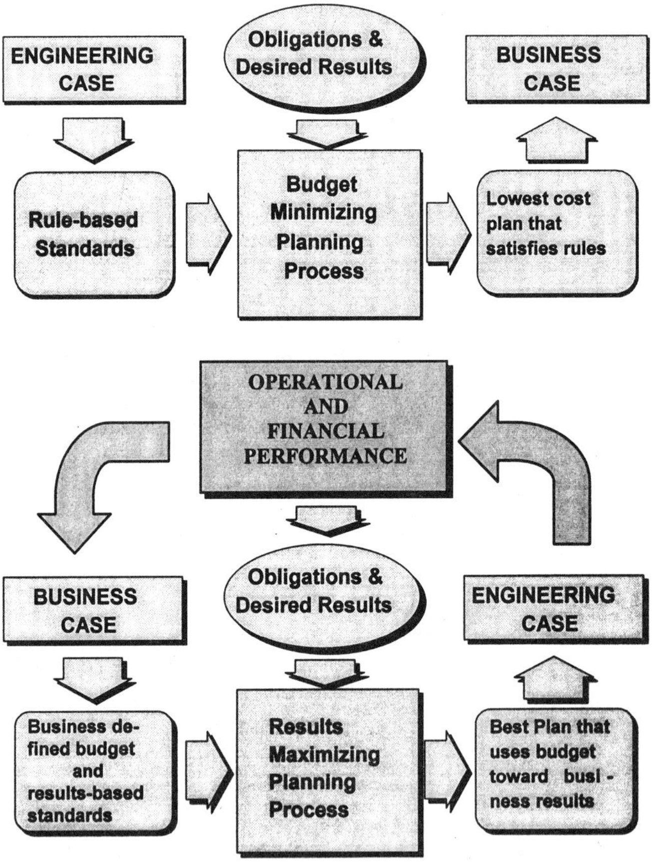

Figure 15.7 A utility that wants to break the bonds of aging infrastructure performance must adapt, and one way is to evolve to a uniform business-case approach to all decisions, using an RDM (Results-Driven Management) approach.

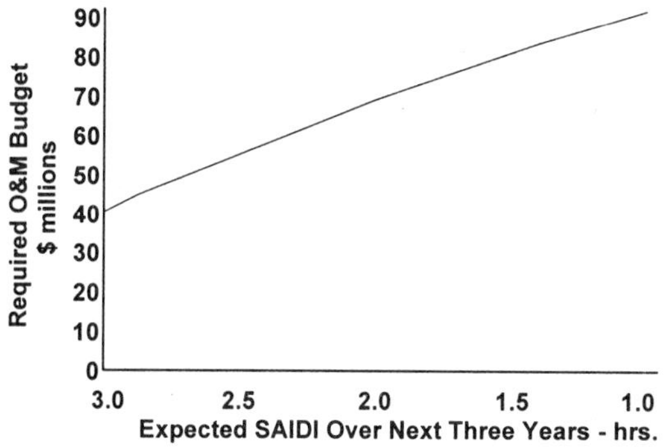

Figure 15.8 One critical element for success in a business-driven RDM environment is that engineering and operations planning provide detailed information to management on the cost of results. Only with such information can a result-focused process work, and only with this information can executive management determine what results and what budget to target. Here, operations planning has determined the relationship between SAIDI (a desired result) and annual O&M spending. The curve represents the optimal results obtainable from any budget, or conversely, the minimum budget needed to obtain any particular level of results) and relies on optimization methods discussed in the text.

A crucial element for success of this approach is a feedback mechanism that provides executive management with information on the consequences of depending decisions. In order to obtain the results it wants with the budget it decides to spend, it must know what the relationship is between results and budget. A planning method must have a way of giving upper management a curve something like shown in Figure 15.8. "These are the results you will get depending on what you decide to spend." "These are the risks that the company will face."

Results-Driven Management

Here, the acronym RDM is used to mean "Results-Driven Management." RDM means uniformly focusing on reliability and prioritizing projects on the basis of reliability/dollar in all departments and across all functions in a power delivery organization. To a certain extent, this is like reliability-centered maintenance, except that it extends that concept of maximizing "bang for the buck" across all departments and all functions of the power delivery company. In addition, it encompasses planning and engineering tools and prioritization methods for planning, engineering, operations and maintenance, appropriate for that purpose and compatible with one another when RDM is integrated across all departments. The basic concept is simple:

Spend money where it will do the most good from the

standpoint of improving customer service. Be willing to move funds from one budget to another, from one function to another, and to spend it in non-traditional ways if those changes will buy more reliability for the same money.

Chapter 12 presented a lengthy example of reliability-centered maintenance applied to decisions about what equipment to maintain, and how to maintain it – a comprehensive but traditional view of "RCM." There, alternative actions that could be taken, from inspection to rebuilding of equipment, were all evaluated on the basis of the ratio of expected improvement in reliability of service to expected cost. The most cost-effective actions were approved. Others were not. The key factors in such analysis are:

- Results are measured against what the utility wants to accomplish specifically the type of customer service it wants to provide.
- Costs are evaluated in terms of what the utility wants to accomplish financially, specifically reduction in revenue requirements or increases in profitability.
- The ratio of the two is the measure of efficiency of any proposed project or action: those with the highest ratio of results to costs are the "winners."

Similarly, reliability-based planning and prioritization (Chapters 13 and 14) assess capital spending on the same basic basis – evaluate "bang for the buck" and approve the most effective projects. Replacement of equipment, either in kind or with different types or sizes of units, can be evaluated from a similar standpoint, as will be shown below.

Why this perspective? It is simply the best way to get the most results from limited resources (i.e. money) and a utility with an aging infrastructure will never have enough money to afford the obvious solution: re-build the entire system with new equipment, in a new, modern configuration. Spending, whether for new capital projects, equipment replacement, refurbishment and upgrades of older areas, preventive maintenance, repair and restoration, new software systems, or for performance-based customer service penalties, should be evaluated on the same basis. It absorbs revenues, lowers financial efficiency and potentially reduces the bottom-line (profits). Therefore, spending on these measures must be coordinated. Money should be spent on one, at the expense of spending on others, only if it generates more improvement in what is important to the utility.

Quantitative target for results

Results-Driven Management focuses on a *measurable objective* for customer service quality, such as SAIDI, SAIFI, some combination of the two, or another similar type of customer service index. The "what" doesn't matter to the process, which can work with any measurable index.

The actual "tool" used is inter-project alternatives optimization. This was covered, with comprehensive examples, in Chapter 5 (section 5.5) for capital decision-making, and in Chapter 12 (sections 12.3 and 12.4) for preventive maintenance, tree trimming, and spending on other operational resources. The approaches described in those two chapters are identical, simply applied to decisions among projects that are respective by all capital, and all operating, costs. In each, the marginal "results per dollar" from various alternative ways to do the project is evaluated against the marginal results per dollar of alternative ways to do other projects. Project alternatives are ranked on the basis of cost effectiveness and the budget is spent on the most effective project alternatives.

This method can work for evaluation and prioritization within each function in the company (i.e. preventive maintenance, operations, replacement and repair, capital projects, reliability-augmentation projects, transmission, substation, distribution departments). What is necessary in order to assure that the overall utility budget (sum of all departments and functions) is spent optimally, is

- All must use the same definition of results – in the cases discussed here, reliability of customer service.
- All must compute cost on a consistent basis.
- All must be evaluated together.
- Managers must be accountable for spending the budget according to the plan – not on pet projects.

Executive management enforces consistency in evaluation rules: all departments use the same definitions of "good results" and consistent definitions of costs, all compatible with upper management's goals. Results are measured in terms of the type of customer service quality desired, and costs are evaluated in the manner that fits the corporate business model. This is not how many utilities currently evaluate results and costs: different departments use different priorities or rules, and cost bases are different and often incomplete from one department to another.

Loop Process of Information and Approval

This method can enforce what is essentially an optimal use of the entire utility budget while still permitting departments a certain amount of autonomy in the details. Figure 15.9 illustrates how. On a preliminary basis, each of the largely autonomous departments submits both total and marginal result vs. budget curves. They do not have to submit their actual project list, only the cumulative curves of budget vs. result and marginal result as a function of budget amount. Figures 15.9 and 15.10 look quite similar to Figures 12.7 and 12.6, and in fact describe a process similar in concept and method. In Chapter 12, the discussion

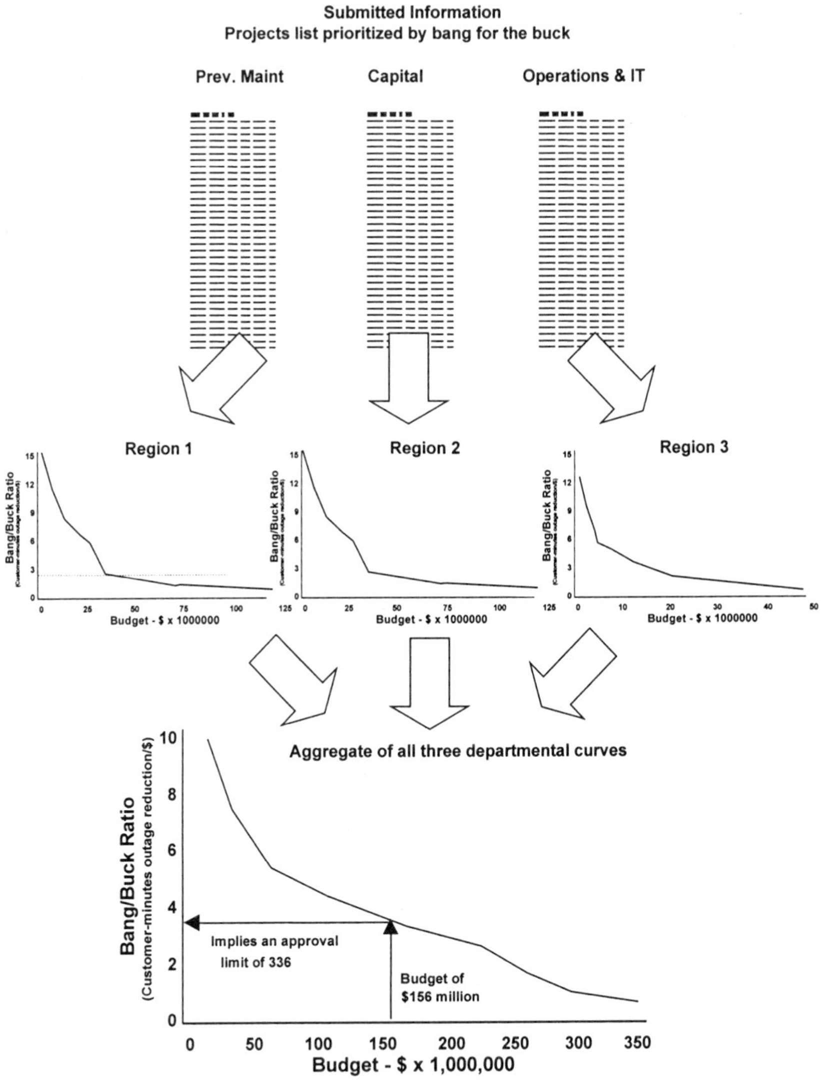

Figure 15.9 The annual budget optimization process begins with each department or function submitting the results vs. budget curve for its project or spending list (top). This leads to an aggregate curve of results vs. spending for the utility, and a decision on setting the approval limit (marginal reliability per dollar limit).

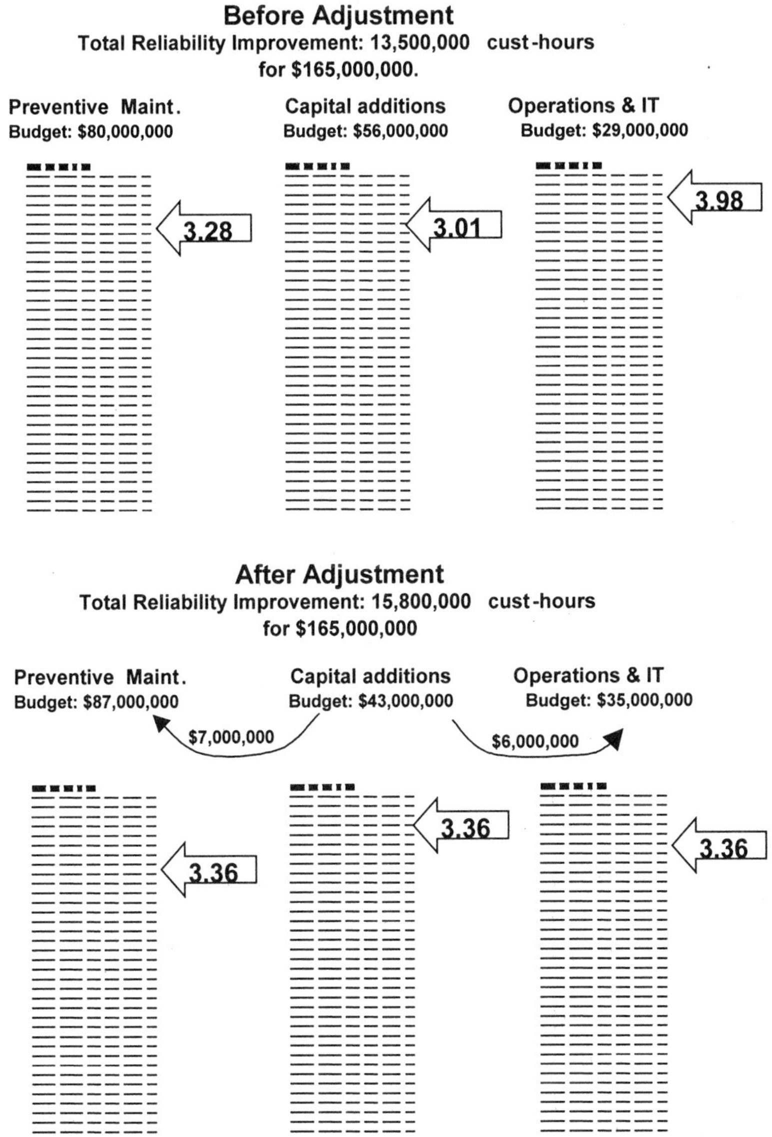

Figure 15.10 Budgets are effectively changed by implementation of a common marginal result/dollar figure across all functions.

was restricted to implementing maintenance prioritization across different regions; here the same basic idea is extended across different functions. Note that the preventive maintenance budget that is "optimized" in Figures 12.6 and 7 is the same one submitted here for the entire preventive maintenance function. This process can be applied in a hierarchical manner, if desired. Departments submitting to divisions, which optimize their numbers, division data then submitted to the overall corporate-wide prioritization-coordination process.

Using the submitted information on overall "bang-for-the-buck" expected from each department, the central aggregation process assembles an overall curve for the entire utility (Figure 15.1). Executive management can then determine how much "bang" it wants to buy for how many "bucks" – setting the overall-spending pattern. At this point the spending pattern is not a budget because capital and operating monies are different, but the constraint on the total has been set.

Management then uses the marginal result vs. spending curves, in aggregate (Figure 15.2) to determine the approval limit for the entire company. This sets the results/dollar at the same for all departments and functions taking into account legitimate differences in capital and operating monies as viewed by the financial basis for decision-making (none are shown here, in order to simplify the example). Essentially, management uses the absolute data (Figure 15.1) to determine what it wants to do: it uses the marginal results/budget curve to determine how to implement that goal.

At this point, executive management can either assign a budget to each department or function (e.g. $80 million for preventive maintenance and tree trimming, $43 million for capital projects), or it can simply distribute the approval limit for projects. No project with a marginal result/dollar figure less than the limit (3.36 in this case), will be approved. Either way it is implementing the same plan. Key elements that permit this system to work well in the real world are:

1. A common definition and formula for "bang" and "buck,"
2. A set of generic categories of capital projects maintenance work and equipment classes along with pre-defined expectations and costs, and
3. A way of tying the absolute result vs. budget (Figure 15.1) and the marginal result vs. budget (Figure 15.2) functions together.

Item 1 assures uniformity of focus of results and accounting for resources in all departments. Item 2 makes certain that uniform standards of the expected results are used throughout (no one can "cheat" by claiming their projects will produce better results, just because that is their opinion – see Chapter 12, page 366). Item 3 makes certain management has the information consistency so they can do their job. Once the marginal result/cost approval target and the budgets are determined, the largely autonomous departments implement their budgets, prioritizing their projects using this same basis (no other method will give them better performance per dollar).

Of course, no plan ever goes perfectly. There will always be the occasional unexpected project that is needed – something not in the original budget. One advantage of this approach is that the corporate-wide use of an approval limit (3.36 in this example) provides a useful measuring stick for new projects or needs. Any new project that is proposed is really consistent with corporate policy only if it has a marginal result/dollar cost greater than the approval limit. Make no mistake; implementation requires total commitment by all levels of management. There is no room for pet projects when the financial survival of the enterprise is at stake.

Practical and Effective, but not Trivial to Implement

This method works well. The authors have seen it reduce capital and O&M by between 10% – 30% while improving or keeping reliability performance at target levels. The method is not particularly difficult to implement from a technical standpoint, but it does take considerable organization commitment, and some changes.

Without doubt, the greatest challenge is culture change. Professionals in many different departments must: 1) learn a new set of priorities; 2) cooperate in a way that is quite foreign; and 3) work within a system that has much more obvious accountability. A utility cannot take a trivial or "band aid approach" here. This is not an easy evolution.

There are also detail changes in functions within departments and shared services within the utility. Engineering and planning need to use some form of reliability-based budget-constrained planning methods (see chapters 5, 13, and 14).

Considerable change in cost-computation and tracking – generally a shift to ABC (Activity-Based-Cost evaluation), an accounting method that tracks all costs on a function basis – is a pre-requisite to effective RDM. The authors strongly recommend this – it is worth the effort because total costs of projects and services must be comparable across departmental and functional boundaries. This is often not the case.

For example, certain costs that can be (but shouldn't be) viewed as fixed are often not included in intra-department cost evaluations. The costs of vehicles and line trucks, tools, fittings, test equipment, and training are often not included in the costs of preventive maintenance projects as calculated at some utilities. These costs are a "given" and evaluation of cost effectiveness of different preventive maintenance actions is evaluated on the basis of only the labor, outside services, and parts required for one versus the other.

While that may be a sound approach to deciding among projects in that department in the short run, it is not valid for long term planning and comparison with outside service providers. For example, how can these expenses truly be compared to contract tree trimming, in which the utility has no fixed costs of comparable type? Maybe the utility should buy fewer line trucks and tools, etc., and transfer funds to more tree-trimming. It can't tell unless all costs are included. Also quite possible is that analysis against other departments might show more investment in those resources is needed. The point is that costs-comparisons across departments and functions need to include all costs. In the long run all costs are variable.

Incentives for Performance

What prevents department managers or proponents of one particular function from "cooking their numbers" and submitting data that will get them an "unfair" amount of the overall budget? The answer is *incentives* and *accountability* to deliver what has been promised. No one will cheat by inflating his or her numbers if held accountable for the submitted improvements: "You were given so much money because you promised a certain amount of results in return. Where are they?" Fairness also dictates that results be normalized to the environmental conditions for which they were established. Managers should be neither punished no rewarded for events outside of their reasonable control.

A visible program of accountability and incentives for performance must be put in place and enforced. Incentives are based on achieving the results at the projected cost. Good incentives can be:

- *Personal*: Recognition awards, perks, and bonuses that are a significant portion of salary.
- *Team:* Recognition awards, special site and team bonuses.
- *Public:* Very visible "trophies". As an example, the top department or district manager can be given a premium company car to drive for the next year, the lowest performer - a "turkey."

Zero-Base All Budgets

This approach effectively zero-bases all functions. No assumed budget of spending pattern is built into the system. All expenses, for all departments, should be justified on the basis of the same cost effectiveness and nothing should be sacred or excluded from this comparison.[2]

Some departments will be very unhappy

Efforts to implement an effective and optimized RDM program *will* result in movement of money from one budget to another and from one function to another, often resulting in an allocation of funds to departments that is considerably different than the traditional pattern. The authors have worked

---

[2] The authors fully realize that the utility will be obligated to a good deal of spending that does not fit this evaluation directly. For example, equipment and construction is often required purely for safety reasons and to meet code, zoning, and other requirements. The authors do not imply that such expenses should be denied, just that they should not be excluded from consideration.

with utilities within which the marginal effectiveness of spending differed by a factor of 8 to 1 from one department to another. The utility could improve the effectiveness of dollars already in its budget by a factor of eight, by cutting one department's budget and giving the money to another department to fund expenses that could not fit within its original budget. This periodic flow of funds between functional areas will be most effective and most easily accomplished when the fixed costs of the functions are minimized.

In the authors' experience similar differences in marginal effectiveness are common, although perhaps not to the degree of the above example. Institutionalized spending patterns still in effect at many utilities were for the most part established under regulation and prior to many of the changes in technology, and need, that characterize today's power industry. It is almost certain that one or more departments will feel very cheated in this process, and that conversely a few will feel vindicated – they always knew their function was more worthwhile. This is a necessary aspect of cultural change which executive management must accept and manage.

15.4 AGING EQUIPMENT AND SYSTEMS – A BUSINESS PERSPECTIVE

It is not necessary to revert to the capacity utilization levels of the 1960s and 1970s. Nor is it necessary to replace all equipment at the end of its original design life if it can be refurbished and kept in good operating condition. Today there are tools that engineers can use to assess the risk level of the power delivery system and determine the most efficient way to manage that risk. In today's environment managing the risk level of the delivery system is just as necessary as managing the risk level of the company's energy and financial investment portfolios or assuring adequate voltage for consumers.

The airline industry and the electric power delivery industry have some interesting characteristics in common. They are both capital intensive. They both have "loads" that vary over time and with the economic health of the nation. The loads in both industries contain diurnal, weekly and seasonal variability as well. They both need to monitor and manage the performance of their systems in real time and as well as over the days, months and years. Both will suffer if they do not maintain adequate levels of operational and financial performance.

Equipment purchased by an electric power distribution company has *value*, so that if used well it provide *service*, its contribution to the overall performance of the system. It has a cost, both initial capital and continuing operating. An important financial goal of any business is to get the highest value/cost ratio possible from equipment. All equipment purchased by a utility will eventually fail or be replaced due to concern that it will fail, or because of obsolescence, or due to changes that mean it is no longer needed. All equipment. The vast majority will be replaced when it fails. Therefore, a somewhat simple, but effective business perspective for equipment can be established as: equipment requires an investment and continuing costs, in return it provides a certain value, dependent on its utilization, up to the end of its lifetime.

Value can be defined in several fundamentally different ways, each ultimately measured in dollars. For example, the recovered price of power delivery from the wholesale grid feed point to the customers can be determined, and allocated to all equipment along the chain in proportion to the loading of each, or levelized cost, or some other basis of allocation. Or, the value of the service provided by the equipment can be determined based on avoided costs of other means: if the cost of the least costly other alternative to get the same performance is X, then the value of the unit is X, regardless of its cost.

Exactly what basis should be used by a utility depends on the financial basis picked by its executive management and financial officers, and is not relevant in this discussion. The important point is that value depends on what the unit does or how it is used, and has nothing to do with cost.

Optimizing Equipment Value

The cost of most power equipment (lines, transformers, switchgear) varies only slightly depending on use. Losses and required maintenance go up as loading goes up, but only slightly. However, the value obtained from a unit of equipment is roughly proportional to its utilization. Thus, if a utility decides to double the power transmitted through a particular transformer, its annual levelized cost increases by a few percentage points (losses costs rise). The utility may feel obligated to inspect and service the unit more often since it serves more load, but the value obtained annually from the unit doubles.

However, it would be shortsighted to assume that doubling the loading on a transformer will double the total value obtained from it over its lifetime, because the higher loading might reduce its lifetime significantly. In doubling the loading, the utility obtains more value annually, but can expect that annual value fewer years into the future. At some point, increasing the loading results in a lower lifetime present worth value, because expected lifetime begins to fall more than loading rises.

Figures 15.11 and 15.12 illustrate this concept. Here, a 32 MVA transformer is considered for various levels of loading.[3] That loading will affect the expected lifetime of the unit. As shown in Figure 15.12, the annual levelized

[3] This rating is what the authors called design-lifetime rating, and is not necessarily "nameplate," nor does it correspond closely with the rating many utilities would give it based on their own internal rating methods. Since capacity rating methods differ so much throughout the industry, the authors have decided to use their own. Design-lifetime rating is the peak load level the unit can serve in this situation with an expected service lifetime of 40 years (without rebuilding or major refurbishment). The actual rating of the unit is irrelevant in this example. It could (and would) be called anything from a 24 to a 38 MVA unit depending on the utility or methods being used to rate it. The numbers would still work out exactly the same.

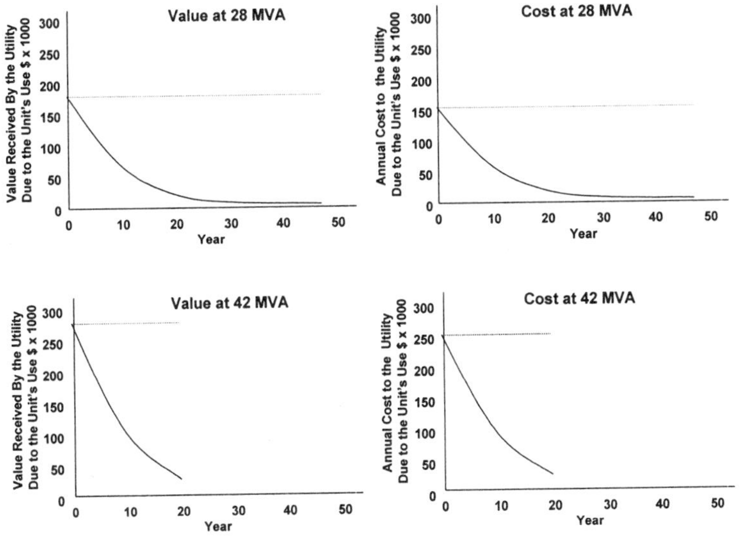

Figure 15.11 Value versus cost. Upper left: annual value (dotted line) and PW value (solid line) provided by a new "32 MVA" transformer used to serve a 28 MVA (peak) load, expected to last 47 years. Lower left: same values if the unit serves a peak load of 42 MVA, in which case expected lifetime is only 19 years. Plots at the right show annual and PW costs. Areas under the curves are the total PW value and cost for each scenario.

value received and the annual cost (which includes losses) are higher if the unit is loaded to a higher peak load, but in that case both value and cost cover a shorter period of time due to the decreased lifetime.

Value in this particular case is determined as simply its allocated share of total revenues for power delivery using an average cost/kWhr factor developed as an average for the entire system, using a standard regulated-utility cost-recovery formula. Figure 15.12 shows the ratio of value/cost determined, not just for the two load levels shown in Figure 15.11, but for all values between 0 (the unit provides no annual value at all, but provides this over an "infinite" lifetime) and 50 MVA (it lasts only about a year in service).

The evaluation and computations shown here are very straightforward and can be carried out with only simple spreadsheet analysis. The alert reader will note that the values and costs do not need to be evaluated as present-worth figures in order to compute the value/cost ratio. Both denominator and numerator are sums over the same PW reduction series by year, using a constant value and constant cost levelized over the expected lifetime.

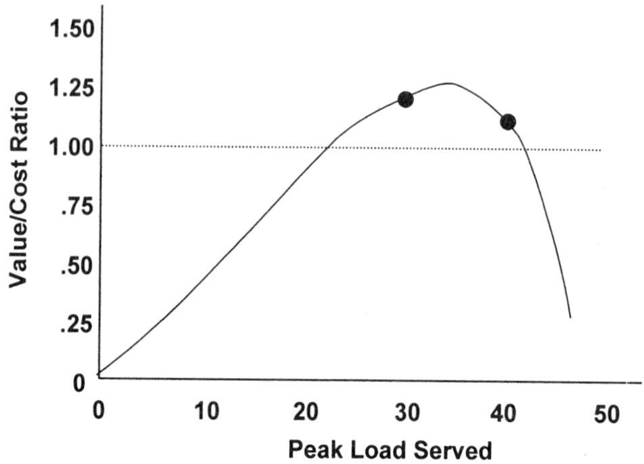

Figure 15.12 Value to cost ratio for the transformer in Figure 15.11 as a function of the peak load level it will serve. Dots indicate the two scenarios in Figure 15.11. The unit provides a positive contribution if loaded anywhere from 22 MVA (peak) and 43 MVA (peak) at which loads it is expected to last 71 and 14 years respectively. Optimum load is a peak load of 36 MVA, or a 13% overload, at which expected lifetime is 34 years.

However, this method using PW of future one-time costs is quite useful. Figure 15.13 shows the evaluation carried out with the addition of a failure penalty cost. In this case the total failure penalty cost is assumed to be \$700,000, a cost that results from the sum of supply-side failure costs (the utility's overtime and higher-than-best prices paid for emergency change-out[4]) and customer rebates or PBR penalties for the interruptions this outage causes. [5] This is high, but not unreasonable.

Advanced applications of value/cost optimization

The example given here illustrates the major concept, that the value of an investment is judged against the cost of buying it and maintaining it in service.

---

[4] This cost does *not* include the minimum cost of a replacement transformer or the normal cost of its scheduled installation. This is a cost judged against the replacement unit's value, and estimated here at \$150,000.

[5] It is assumed here that the transformer causes a "four-hour or more outage" to 8,000 residential and small business customers, at a rebate of \$75/customer, for a total of \$600,000.

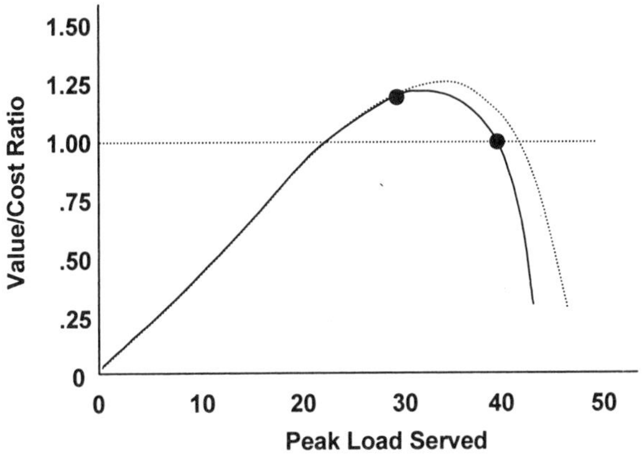

Figure 15.13 Ratio of the value/cost for the transformer in Figure 15.12 as a function of the peak load level it will be used to serve, computed assuming its failure causes an \$800,000 cost. Dotted line shows original (Figure 15. 12) curve. Optimum loading level drops to 32 MVA at which point the expected lifetime is 40 years. Dots show the points for the two loading levels in Figure 15.12 – same loads shown in Figure 15.12.

Optimization then means finding a loading (and consequently a target service lifetime) that maximize value/cost or value-cost, as the financial goal may be. The evaluation is carried out over the expected lifetime of the unit, using levelized or present-valued costs.

A number of improvements or detailed additions can be made to improve this basic concept. They will not be illustrated here in the interest of space and because each is fairly straightforward. These are:

- *Costs for future maintenance,* such as major servicing, rebuilds, etc., can be added to the analysis along with increases in the lifetime as expected of the unit. For example, perhaps a \$200,000 rebuild in year 30 (of 35 expected years of service), will extend lifetime to 45 years. That cost can be added in year 30, and the value and operating costs extended out another ten years.

- *Probabilistic analysis* can be done. Expected lifetime is not a single deterministic value but a probability distribution. Since PW costs are skewed toward the present, the net result of taking a probabilistic approach (the unit might last longer than expected, but it could fail sooner) is that the high-point of the curve moves to the left slightly.

- *Load growth can be included.* Loading is seldom constant over time. The analysis can trend the loading upward over the unit's life to reflect an increasing load. This tends to make the eventual peak load target higher (overloads are in the future, where the increased failure likelihood has a lower PW).

- *Avoided costs can be added to the value.* For example, suppose that this transformer is one of two at a substation. The utility is deciding how to best handle transformer contingencies, whether to: 1) implement feeder-level switching for substation transformer contingencies; 2) increase the transformer capacities (i.e. buy bigger units); 3) drastically overload the remaining unit when one fails or; 4) simply pay the outage rebates and not try to delivery good customer service.[6]

The expected number of times, amounts, and hours of overload over each of the two unit's lifetimes can be calculated using probabilistic methods. The lifetime reduction resulting from this expected added stress can be applied to the unit. Since option 4 is normally not a viable alternative, the only alternatives are to buy larger units or implement feeder switching options. Either involves capital expenses *now*. Thus the unit provides considerable additional value. The loss of value from the decision to "abuse" the unit for contingency purposes lies in the future, and is modeled by removing years at the end of its lifetime, which have very little present worth.

Interestingly, adding an expected contingency support role to this unit increases the optimum target peak load (under non-contingency situations). To understand why, the reader should note that not only does the value the transformer contributes to the utility go up when it is planned for a contingency support role, but the value per normal peak kW served does too. Costs vary little. Thus, there is more incentive to use the unit to serve more kW of peak load. Table 15.1 summarizes several scenarios of target loads and lifetimes for this transformer as evaluated for this example.

Life Extension

Continuing or periodic investments in the right type of inspection and preventive maintenance can sometimes extend the lifetime of equipment. First and foremost, a system's deterioration due to aging may be partially mitigated through the application of sound *life extension technologies.* For example, treated wood distribution poles have an expected life of approximately 25 - 35 years. If the poles are inspected for insect damage and rot periodically, say once every 10 years after the first 15 years in service, their expected life can be extended to 50 years. Some will undoubtedly need to be replaced sooner, but others will also be in sound condition after 60 years under favorable conditions.

---

[6] From a pure business-case perspective this is a case that should be examined. If it wins the evaluation, then the regulatory PBR cost for interruptions is too low, or the utility believes it has heard from its customers (whether they said it or not) that indicates they care more about costs than service.

Table 15.1 Target Peak Loads and Resulting Expected Lifetimes for Various Applications of a "32 MVA" transformer

Scenario Number	Comments on Eval.	Optimum Target Peak Load – MVA	Planned Cont Overload - %	Expected Lifetime - Yrs	Value/Cost Ratio
1	Failure cost not incl.	35	0	37	1.25
1g	#1 assuming 1% growth	41	0	35	1.25
1p	#1 using prob. analysis	33	0	39	1.23
2	Failure cost included	32	0	40	1.22
3	#2 plus rebuild year 28	34	0	50	1.23
4	#1 with cont. support	37	100%	34	1.27
5	#2 with cont. support	33	100%	36	1.24

Other examples of life extension are the periodic rebuild and refurbishment of substation circuit breakers, cleaning and refurbishment of substation bus structures, rewinding and/or re-tanking of substation transformers, rehabilitation of distribution transformers and periodic rehabilitation of pole-mounted and substation class reclosers. The list is not endless, but it is long. With appropriate maintenance and refurbishment, the life of power distribution equipment can be extended beyond its initial design life. By employing these techniques, the operator of a power distribution system can postpone the replacement of equipment for a decade or more by spending money on maintenance activities rather than capital investment. This approach has been employed for many years. It does not alter the basic principals upon which the system was designed, and properly done, should not affect system performance negatively.

The design parameters for the distribution systems of most utilities were established in the 1950s and early 1960s. The engineering and financial arms of the companies developed an integrated approach to system planning and design that optimized financial and technical performance subject to the political and regulatory realities of the time. Simplistically, utilities were allowed to earn a market return on their investments as long as the system performed adequately, expenditures were judged prudent by the regulatory agency and the company did not request increases too frequently.

Since economies of scale were applicable to the generation and bulk power transmission systems of the time, the investment in the distribution system was rarely an issue. The design parameters established in the 1950s and early 1960s, created a system that met the needs of the utilities and their customers symbiotically. Some of the parameters established for planning and design are listed in Table 15.2. Unfortunately, while utilities have reviewed and updated the technical elements of their planning and design criteria, few have performed a top to bottom integrated financial and technical optimization of their planning and design criteria since the early 1960s.

In the meantime, the technical, social and political environment in which utilities operate has changed. As discussed in Chapter 1, today's consumers take for granted a wide variety of home appliances, office equipment and industrial processes that are far more sensitive than what was in use prior to 1970. While power delivery companies are unlikely to build a "10 9s" system as envisioned by some Silicon Valley companies because of cross-subsidy issues, they must be sensitive the needs of the majority of their customers

Other changes that affected utility design and operating practices appeared in the 1970s and early 1980s. Intervenors had for some years been claiming that electric power systems were overbuilt. "Gold-plated" was a term often used. As rates for electricity began rising after years of steady or falling rates reasons were sought to explain the "sudden" change of direction. Falling rates were driven largely by the economies of scale in generation and the growth in the per-capita use of electricity in the decades following World War II, which spread the fixed cost of the system over a growing number of kilowatt hours. The decades following the war were also a period of relatively low inflation. It was in this environment that the search for the guilty began.

If the system was "gold-plated" by building more facilities than necessary, then it would be possible to harvest the gold by imposing a moratorium on delivery system capacity additions in existing service areas until the existing "unused" capacity was fully employed. Actually, the majority of the excess investment that was in the system was probably due more to "custom" designs for what should have been standard products than it was to the construction of unnecessary delivery system capacity. As a simple example, consider medium voltage breakers used in distribution substations. One can buy the standard type of breaker or one can specify a custom design based on requirements unique to the individual utility. Some of the more frequently seen custom requirements are for paint, hinges, access panels, safety barriers, signage, etc. The operating mechanism is not affected. One can question if the additional investment really

Table 15.2 Items That Need to Be Re-Examined in a Major Revision of "Standards"

- Spacing between Substations
- Substation Transformer Size
- Maximum Substation Capacity
- Distribution Voltage Level
- Distribution Wire Sizes
- Distribution Transformer Sizes
- Substation Buss Topology
- Application of Automation

Table 15.3 Equipment Subject to "Custom Design" At Most Utilities

• Substation Transformers
• Substation Layout
• Substation relaying and control schemes
• Breakers
• Switchgear
• Fused disconnects
• Cable
• Meter raceways
• Service transformers
• Software

is necessitated by these customized designs, and if it is worth the money. Nevertheless, over the years many utilities developed their own standards for major equipment to meet what they perceived as their unique requirements. Equipment often customized by each utility is listed in Table 15.3. What is obvious from this tabulation is that only wood products, wire, switches and line hardware are bought "off-the-shelf." The potential for savings to the industry that could be obtained by buying vendor standard equipment was, and is, significant. The way to capture these savings is by changing the specification for the product, however, not increasing the age of the system or increasing system loading

Substation transformer utilization at time of peak in the 1970s was in the range of 68%. At the same time the ratio of installed distribution transformer capacity to the coincident system peak load was approximately 4:1. Practicing engineers realize that these utilization ratios were the result of the (N-1) design criteria applied to the substations to ensure reliability and the effective loss of diversity as one moves closer to the load. Effectively, each component must be able to serve the maximum demand placed on it by the load it serves even though that individual load may not have its peak occur at the same time as the system peak.

These utilization ratios were the direct outcome of the design criteria that electric power system planners had been following to ensure the reliability of service to the customers served by the system. However, as seen by various interest groups, these measures were considered clear and convincing evidence that the system was overbuilt. "Gold-plating" now referred to building more delivery system capacity than was necessary. Proponents of the "use what you have before building more" philosophy assured senior managers that they could easily go at least ten years without building additional delivery system capacity in addition to that necessary to connect new customers. In one respect they were correct. The capacity was there to serve the load under the normal conditions for which the system was designed. What they failed to predict was the effect of the higher utilization rates on the performance of the system during the outage of equipment that resulted in an interruption of service to customers.

In the early to mid-1980s regulators and advocates of the free market were beginning to discuss deregulation of the electric power industry. After the initial round of protests and attempts to slow the inevitable move to a more competitive market structure, utility managers began to look at how they could prepare their companies to survive in the new environment in which they envisioned competing. One of the actions that many companies took was to look for a way to improve their financial performance vis-à-vis companies in competitive industries. A measure often used to benchmark financial performance was Return on Assets (ROA). Efforts to improve ROA provided added emphasis to the efforts already underway to increase the utilization ratio.

In and of itself, increasing the utilization of the system is neither good nor bad. As with many other choices in life the outcome depends on what else one does or does not do. In the case of increasing the utilization of the power delivery system, the choice made by the majority (perhaps all) of the companies that followed this path was to keep the other variables in the performance equation constant. Thus the engineers continued to apply (N-1) planning criteria (suitably modified to create higher utilization factors) at the substation and sub-transmission levels. They did not change the design of substations nor the equipment used in them. Similarly, the basic layout and design of distribution feeder systems connected to these substations remained unaltered.

Since "sympathetic changes" to accommodate the higher utilitization had not been made, most systems operated with reduced flexibility and a higher incidence of operating problems. Forced by limited budgets to cut back on what could be done, planners responded the only way they knew how: They prioritized the work identified under the formerly unified operational and financial planning criteria and built what the budget allowed.

This approach further increased the age and loading of the power delivery system. Adding to the risk implied by an aged system was the issue that system operators often did not fully appreciate the implications of operating the system at these higher utilization levels. Over time system utilization at the time of peak increased. The authors believe that many of the recent problems plaguing electric power delivery systems are a direct result of these decisions. Power delivery systems can be operated at higher than historic utilization levels, but only if all of the design, maintenance and operating practices are modified to accommodate the characteristics of an aged system operating at these higher loading levels.

15.5 CONCLUSION AND SUMMARY

The real problem many utilities face is not that their systems are aging, nor that they have increased utilization too much, nor that they have cut spending too much. The real problem, as outlined in the foregoing sections, is that traditional engineering, planning, and project prioritization methods are *incompatible* with

Table 15.4 Some of the Detailed Changes that Accompany the Shift to a Results-Driven Management Perspective.

• Old	• New
– Delivery paid for through energy charge	– Delivery paid for based on demand placed on system
– Keep everyone in service all of the time	– Identify customers willing to tolerate outages and pay them when they don't have service
– Everyone gets the same service (fiction)	– Basic "plain vanilla" level for everyone - those that want more, pay more
– Build for maximum peak demand	– Utilize customers that are willing to be interrupted to optimize capacity and return on assets
– Point forecasts (load, generation) – a deterministic world	– Probability distributions - a stochastic world
– Budgets set without explicit knowledge of risk	– Iterative process - senior management must know the level of risk
– Sense of equipment stewardship	– Use it up - plan for equipment failure
– Interruption = failure	– Optimize the use of the system - interrupt willing customers
– Regulator and utility set reliability level	– Customers buy level of reliability for which they are willing to pay
– Consumer is the only customer	– Consumer is a customer, but not all customers are consumers
– Reliability is a by-product of capacity	– Plan and engineer explicitly for both capacity and reliability
– Define reliability in terms of equipment	– Define reliability as experienced by the consumer

the needs of the new, financially driven executive management paradigm. Utilities (and regulators) must recognize that the new economic/social/technical-environment in which power delivery systems must operate require new approaches to the design of the system. Reduction of capital and operating budgets without an understanding of the risk being created is foolhardy. Similarly for the engineers, failure to adopt new tools and techniques that allow them to design to a level of risk acceptable to senior management will only serve to exacerbate the problems they are already experiencing. They can not operate in a vacuum and assume that senior management understands the level of risk associated with the funding levels that have been provided unless they have provided that information to senior managers.

Those who believe that the level of risk will not increase as the system ages and loading levels increase have been proven wrong over time. Above all, the environment in which today's power delivery companies must operate mandates that the issues of operational and financial performance are addressed with a *systems* perspective. The two are interrelated and they must be addressed with that realization firmly in mind.

This perspective demands nothing less than a re-evaluation of the basic planning and designed criteria established in the 50s and 60s. A re-optimization of the combined financial and technical performance of the system is necessary to re-synchronize the financial and technical performance of the system to the realities of the current technical, social and regulatory environments in which power delivery systems will be operating in the future. This re-optimization will result in a new paradigm for the design and operation of the power delivery system. Key paradigm shifts that will be required are listed in Table 15.4. The tools and techniques discussed in this book will play a critical role in the development of a new planning and design paradigm for the power delivery system of the 21st century.

REFERENCES

J. J. Burke, *Power Distribution Engineering – Fundamentals and Applications,* Marcel Dekker, New York, 1994

P. Gill, *Electrical Power Equipment Maintenance and Testing,* Marcel Dekker, New York, 1998

H. H. Malik, A. A. AlArainy, and M. I. Qureshi, *Electrical Insulation in Power Systems,* Marcel Dekker, 1998

O. C. Seevers, *Management of Transmission and Distribution Systems,* Fairmont Press, Lilburn 1995

W. A. Thue, *Electrical Power Cable Engineering,* Marcel Dekker, New York, 1999.

16
Guidelines and Recommendations

16.1 INTRODUCTION

This final chapter summarizes and integrates the key points made about aging power delivery infrastructures and the technical and managerial methods covered in the previous fifteen chapters, into a coherent overview of the problem and its solution. It provides guidelines and recommendations for utilities that must improve performance despite the challenges these systems present.

Section 16.2 begins by summarizing the four main "causes" of aging infrastructure problems and highlighting how they interact among themselves and with other aspects of a utility's operations. It then summarizes the four key points of solutions – methods that can mitigate the effects of aging infrastructures and improve the performance of the distribution utility. Section 16.3 then presents a series of recommended approaches, not all of which will be relevant to every utility, but all work well at some utilities. These are organized based on two criteria: how quickly they can be implemented and how much impact they can make. Section 16.4 gives guidelines on setting up and managing an aging infrastructure program at a utility.

16.2 FIVE INTERRELATED FACTORS

An aging power delivery infrastructure is an area of an electric utility system that is mostly old equipment near the end of its lifetime, configured in a layout that is itself quite old and not completely compatible with modern needs. For a variety of reasons, the aged equipment and layout create reliability,

Table 16.1 Characteristics of the Aging Infrastructure Areas (same as Table 1.4)

The system layout and design was first put in place more than forty years ago.
The majority of equipment in the area is more than forty years old.
The system is well engineered and fully meets minimum engineering criteria.
The area is seeing steady, if perhaps low, growth in load.
The area is plagued by above average equipment failure rates. Overtime is high due to large amounts of unscheduled repair and restoration.
SAIFI began rising some years ago; SAIDI is beginning to rise now.
Major interruption events always occur due to a bizarre series of outages.
Things go bad very often.

maintenance, and budgeting challenges that, if not anticipated and brought under control, will eventually overwhelm even the most effective and efficient utility company.

The most easily identifiable characteristic of aging delivery infrastructures is that the equipment in the area is old. It will have an *average* service age greater than its design lifetime. Situations vary from one location to another, but most aging infrastructure areas share similar characteristics in other areas, exhibiting a type of "signature" as listed in Table 16.1. This "aging infrastructure signature" was discussed in section 1.1 and chapters 7 – 10 and 15.

Aging infrastructures are not just about old equipment. There are actually five interrelated, contributing factors at work in these areas, all of which work against efficient, reliable operation. These are:

1. *Old equipment,* which requires high O&M efforts, and whose higher than normal failure rates degrade customer service quality. Old equipment is discussed in Chapter 7. Methods to measure, manage, and mitigate its effects are discussed chapters 6, 10, and 14.

2. *Obsolete system layouts.* Many metropolitan utilities have been unable to obtain new substations or ROW into the central core of their metropolitan areas for decades. As a result they must serve the high levels of both power and reliability that modern society demands, from systems that are far less than ideal with respect to topology and interconnection/configuration strength for contingencies. This exacerbates problems caused by aging by limiting the system's strength to tolerate the loss of equipment when it fails. This issue is discussed in Chapter 8. Methods to measure,

manage and mitigate problems caused are covered in Chapters 13 – 15.

3. *Old engineering methods.* Many T&D utilities are using engineering methods that worked well in the 1970s, but cannot fully guarantee reliable operation in a world where substation and line loadings are pushing beyond traditional levels and equipment is old. Traditional tools need to be augmented by new ways of searching for problem areas in a system's design and in identifying and prioritizing how and where money is spent to bolster reliability of the system. This problem is discussed in Chapter 9. Methods to fix the situation are covered in Chapters 12 -14.

4. *Uncoordinated and non-optimal use of distribution.* Few utilities coordinate the reliability economics and planning of the sub-transmission – substation level and the primary distribution feeder level optimally or even near optimally. As a result, the distribution system is often an underutilized reliability resource. Although it involves more difficult planning methods,

Table 16.2 Contributing Factors to Aging Infrastructure Problems and Their Major Impact

Cause	SAIFI	SAIDI	Costs
Aging Equipment			
1a. Equipment failures	X		
1b. Higher repair and restoration efforts		x	X
1c. More inspection and testing			X
Obsolete System Layouts			
2a.Accelerated aging	x		x
2b.Limited contingency capability		X	
2c. Higher delivery operating costs			x
Outdated Engineering Methods			
3a. Inadequate criteria application	X	X	
3b. Weather normalization		X	
3c. Load forecasting		X	
Outdated Engineering Methods			
4a. Traditional, inflexible design rules	X	x	X
4b Lack of rigorous reliability design	x	X	
4c. Poor plan coordination with other levels		X	
Old Cultural Identities			
5. Inappropriate paradigms			X

coordinated multi-level, reliability – based planning of a power system results in much better reliability and much better use of both capital investment and operating budget.

5. *Old cultures and ideas.* Institutionalized procedures and values – what might be termed corporate culture – at many utilities still tends to operate in terms of the culture and goals appropriate for purely regulated, vertically integrated power industry in the last quarter of the 20th century. Goals are different today and are continuing to evolve. A fresh perspective must be taken by executive management, and professionals in all areas of a utility, to re-examine the most basic concepts throughout the organization, and to apply innovation and creativity to the methods used to manage the utility, in both aged and new areas of the system. Discussion and methods falling into this category are discussed in Chapter 1, throughout the book, particularly in Chapter 14, and in this chapter.

Table 16.2 shows the impacts these five contributing causes and their sub-areas make on the utility's reliability and cost. The size of the "x: under the three categories of impact shows the degree to which the cause affects that category.

16.3 RESULTS-DRIVEN MANAGEMENT

Number One Recommendation: Implement Cross-Functional and Cross-Departmental Results-Driven Management

The authors use RDM here to designate the term "results-driven management," as discussed in Chapter 15, the uniform application of reliability-focused prioritization of spending in a coordinated manner throughout the utility. This is very highly recommended. "Results-driven Management" means taking that same basic concept, which lies at the heart of reliability-center maintenance, budget-constrained capital prioritization, and reliability-based optimization, and integrating its use in all aspects of the utility's operation. The basic concept is simple:

> Spend money where it will do the most good from the standpoint of improving customer service. Be willing to move funds from one budget to another, from one function to another, and to spend it in non-traditional ways if those changes will buy more reliability for the same money.

Chapter 12 presented a lengthy example of RCM applied to decisions about what equipment to maintain, and how to maintain it – a comprehensive but traditional view of RCM. There, alternative actions that could be taken, from inspection to rebuilding of equipment, were all evaluated on the basis of the ratio of expected improvement in reliability of service to expected cost. The most cost-effective actions were approved. Others were not. The key factors in such analysis are:

- Results are measured against what the utility wants to accomplish specifically the type of customer service it wants to provide.
- Costs are evaluated in terms of what the utility wants to accomplish. Financial, specifically reduction in revenue requirements or increases in profitability.
- The ratio of the two is the measure of efficiency of any proposed project or action: those with the highest ratio of results to costs are the "winners."

Similarly, reliability-based planning and prioritization (Chapters 13 and 14) assess capital spending on the same basic basis – evaluate "bang for the buck" and approve the most effective projects. Replacement of equipment, either in kind or with different types or sizes of units, can be evaluated from a similar standpoint (Chapter 15).

The standpoint of overall financial efficiency of the utility, spending, whether for new capital projects, equipment replacement, refurbishment and upgrades of older areas, preventive maintenance, repair and restoration, or for performance-based customer service penalties, is all the same and should be evaluated on the same basis: it absorbs revenues, lowering financial efficiency and potentially reducing the bottom-line (profits). Therefore, spending on any of these measures should be coordinated. Money should be spent on one, at the expense of spending on others, only if it generates more improvement in what is important to the utility.

Implementing RDM

Reliability-centered approaches have been so effective in improving results that they are standard practice in other industries where reliability is an important consideration (e.g. airlines, nuclear materials processing). Their effectiveness in the power industry is well proven. Therefore RDM, as described here, is highly recommended. It gets results and it minimizes costs.

Implementation can follow the approach discussed in Chapters 11 and 14. Evaluation and prioritization within each function in the company (i.e. preventive maintenance, operations, replacement and repair, capital projects, reliability-augmentation projects) is evaluated on the basis of "bang for the buck" as described above and detailed in Chapters 11–14. Executive management enforces consistency in evaluation rules: all departments use the same definitions of "good results" and consistent definitions of costs, all compatible with upper management's goals. Results are measured in terms of the type of customer service quality desired. Costs are evaluated in the manner that fits the corporate business model. This is not how many utilities currently evaluate results and costs: different departments use different priorities or rules, and cost bases are different and often incomplete from one department to another.

Utilities can not take a trivial or "band aid approach" here. This is not an easy evolution. This is because total cost must be comparable across departmental and functional boundaries. For example, certain fixed costs that can be (but shouldn't be) viewed as continuing are often not included in intra-department cost evaluations. The costs of vehicles and line trucks, tools, fittings, test equipment, and training are often not included in the costs of preventive maintenance projects as calculated at some utilities. These costs are a "given" and evaluation of cost effectiveness of different preventive maintenance actions is evaluated on the basis of only the labor, outside services, and parts required for one versus the other.

While that is a sound approach to deciding among projects in that department, it is not valid for extension and comparison outside the department. For example, how can these expenses truly be compared to contract tree trimming, in which the utility has no fixed costs of comparable type? Maybe the utility should buy fewer line trucks and tools, etc., and transfer funds to more tree-trimming. It can't tell. Also it's quite possible that analysis against other departments might show more investment in those resources is needed. The point is that cost-comparisons across departments and functions need to include all costs.

Considerable change in cost-computation and tracking – generally a shift to ABC (Activity-Based Cost evaluation), an accounting method that tracks all costs on a function basis – is a pre-requisite to effective RDM. The authors strongly recommend this since it is worth the effort.

Zero-base all expenses

Thus, the way to implement this includes zero-base budgeting, at least for a period of time until the overall pattern of spending for the "new paradigm" of RDM is identified and appears to be stable. All expenses, for all departments, should be justified on the basis of the same cost effectiveness and nothing should be sacred or excluded from this comparison.[1]

Some departments will be very unhappy

Efforts to implement an effective and optimized RDM program *will* result in movement of money from one budget to another and from one function to another, often resulting in an allocation of funds to departments that is considerably different than the traditional pattern. The authors have worked

[1] The authors fully realize that the utility will be obligated to a good deal of spending that does not fit this evaluation directly. For example, equipment and construction is often required purely for safety reasons and to meet code, zoning, and other requirements. The authors do not imply that such expenses should be denied, just that they should not be excluded from consideration.

with utilities within which the marginal effectiveness of spending differed by a factor of 8 to 1 from one department to another. The utility could improve the effectiveness of dollars already in its budget by a factor of eight, by cutting one department's budget and giving the money to another department to fund expenses that could not fit within its original budget.

In the authors' experience something similar is common, although perhaps not to that degree in most utilities. Institutionalized spending patterns still in effect at some utilities were for the most part established under regulation and prior to many of the changes in technology, and need, that characterize today's power industry. It is almost certain that one or more departments will feel very cheated in this process, and that conversely a few will feel vindicated – they always knew their function was more worthwhile. This is a necessary aspect of cultural change which executive management must accept and manage.

16.4 RECOMMENDED SOLUTION ACTIONS

This section presents ideas and concepts – action programs, each of which will individually mitigate some of the aging infrastructure-related service reliability problems. But they will be most effective if applied within a well-managed program or coordinated RDM program as described in section 15.3. That will achieve synergies among both their improvements and optimized spending.

Voltage Reduction on the Primary Distribution System

Chapters 7 and 11 pointed out that the application of a voltage, aside of any current that flows, creates a type of "voltage stress" on insulating materials which eventually leads to deterioration, and over enough time, will alone cause failure. Voltage stress contributes directly to aging – other factors of age are multiplied by its effects. Unlike thermal stress caused by current flow, which occurs only during peak conditions and then only when peak loads last for a considerable period of time, voltage stress of roughly the same magnitude occurs all 8760 hours of the year. In fact, voltage stress is less during peak conditions (because of an increase in voltage sags, voltage is lower).

The relationship between voltage level and deterioration rate (and hence lifetime) is highly exponential. In the case of some types of UG cable, a 5% reduction in applied voltage will decrease voltage-induced degradation by 35%. This means expected failure rate from voltage stress is reduced by 35%. This will not render a 35% decrease in overall failure rate, because voltage alone is not the only factor causing deterioration. However, it will make a substantial difference – perhaps as much as a 15%-20% improvement.

Therefore, one means to reduce the stress, and increase the expected lifetime of UG cables and the insulation in transformers and voltage regulators, etc., is to reduce voltage to the lowest level consistent with operation within required voltage standards. The utility can implement a type of Conservation Voltage Reduction (CVR) as is recommended (and required) by some state regulatory commissions for energy conservation purposes, but aimed at lifetime extension. This is implemented at the primary voltage distribution level, by adjusting the boost on transformers, tap changers, and voltage regulators at the substation.

Methods of analysis and implementation for CVR are applicable for this purpose [Willis, 1997, page 143]. The only modification of traditional CVR methods needed for this application is to analyze the system on a feeder by feeder basis to determine where voltage reduction would in fact have a worthwhile impact on equipment lifetimes. This will tend to be on feeders with substantial UG cable and/or very old, rather highly loaded service transformers (a good TLM program can be used to determine that).

Another factor that needs to be taken into account in assessing when, where, and how to apply voltage reduction for lifetime extension is the impact on revenues. Generally, demand levels are proportional to voltage. A 3% reduction in voltage will render roughly a 3% reduction in demand, although feeders vary greatly in this respect.

Due to the type of loads they serve, some feeders (usually those in industrial areas or residential areas with high saturation of AC and electric heat pump loads) would see no reduction in demand and revenues from a 5% reduction in voltage. A few others (mostly residential areas with high saturation of gas appliances) will see a 10% reduction in demand and revenues. Utility management needs to assess this impact, as it will occur on some feeders. It should be balanced against the expected savings due to the reduction in equipment failure rates and customer interruptions.

Planners and managers should not also use the demand reduction achieved in the pursuit of extended equipment life to justify deferral of required upgrades due to load growth. For example, one can calculate that if the voltage reduction rendered a 5% reduction in load, and load growth is 1.66% per year, then the reduction "bought three years" of deferral on any expansion needs, delaying the time until load exceeds capacity by three years. In truth this does not happen. The current that it carries thermally limits most equipment. Since a reduction in voltage can effect at most a proportional change in current, reducing needed investment in capacity will defeat the desired life extension effects of voltage reduction.

While the above may seem to involve a complicated set of evaluations, it is in fact straightforward. In addition, voltage-reduction can extend the lifetime and reduce the failure rates of equipment that is often both very expensive to replace and virtually certain to cause customer interruptions when it fails. It can be implemented at low or no capital cost, relatively quickly. A very comprehensive assessment can be done using traditional CVR planning methods [Willis, Tram, Powell, 1985], coupled with a detailed spreadsheet analysis of present and future costs and savings on a PW basis. The optimal operating voltage for each feeder is that which produces the most effective impact on the utility's bottom line: expected costs due to voltage-stress related deterioration

Adopt Reliability-Based Planning and Engineering Methods

Chapters 8 and 9 discussed why modern power systems, particularly those in aging areas, are particularly prone to customer interruptions due to equipment failures, and how traditional engineering methods cannot assure that all necessary and sufficient steps have been taken in their design to assure reliable operability.

The authors do not recommend that traditional planning tools (i.e., N-X contingency evaluation loads flows, etc.) be abandoned. Such tools are useful and cost little to use. They provide a good perspective on how the system will deal with each contingency and are often useful for examining what are called post-contingency conditions and determining if recommendations for special operating restrictions should apply after a failure has occurred.

However, utilities must augment their use of those tools with modern planning and engineering tools that work directly with expected reliability of service, tools that can assure measures taken to provide reliability of service are sufficient to reach the levels of customer service quality targeted by the utility. Such tools include a "reliability load flow" analysis of any of several types, and possibly a reliability-based system optimization (see Chapters 12 –13).

This use of such tools will have three positive effects on the results that the utility obtains from both its system and all new capital investments:

1. *Improved customer service quality.* Application of these methods does result in better reliability of service. Planning and Engineering produce plans that are more robust with respect to contingency strength and tolerance to equipment failures, and void of sensitivities to relatively likely combinations of multiple failures. Very simply, the power delivery system is better designed from a reliability standpoint.
2. *Lower cost.* Reliability-based planning and engineering methods can be used to optimize cost effectiveness of capital spending for reliability. In fact, a considerable improvement in this regard is an almost inevitable result of their use. In a very real and practical sense, money spent on improvements can be optimized to buy the most "bang for the buck" in terms of improving reliability where that is most important to the utility. Money is both used more effectively and targeted more accurately at problem areas.
3. *Improved predictability of results.* The dependability of planning and engineering projections of performance from a reliability standpoint will improve dramatically. Traditional tools deal with reliability implicitly (as capacity margins for contingency backup). Modern reliability-based tools deal directly with reliability of service as a calculated design parameter and thus are much more accurate in determining what to expect in terms of results. Management can be more confident that it will obtain results and can better anticipate where and why problems might develop.

Use Budget-Constrained Planning Methods (Marginal B/C Ratio Prioritization)

Generally, prioritization of capital projects, the selection and approval phase of the planning and engineering process, needs to be done with so-called Budget-Constrained Planning (BCP) methods (see Chapters 5 and 12, and Willis, 1997, Chapter 18). BCP methods should be used in order to fit within both the business model approach recommended here, and to permit valid cross-departmental comparisons of cost that balance capital and operating costs in an effective and completely valid manner.

Budget-Constrained Planning is an approach to prioritization, not engineering or planning method per se, and can be implemented along with planning and engineering tools at any level of analytical rigor and detail. It works best, however, when applied in conjunction with reliability-based methods.

The basic concept behind BCP was described with examples in section 5.5. Alternatives for each project will be evaluated against one another and against alternatives for other projects on the basis of the same marginal bang/buck target ratio. This is exactly the same approach described for reliability-center maintenance in section 11.4.

Budget-constrained planning has proved effective for both large and medium sized utilities. It has successfully worked within Engineering departments that use only traditional tools and have very limited engineering resources, as well as those that used reliability-based evaluation methods in conjunction with fairly strong labor resources and skill levels in their Planning and Engineering departments. However, it works best in the latter environment.

The authors have seen BCP render capital-spending reductions of up to 30% while still permitting the utility to obtain necessary targeted levels of reliability performance in its system.

In many cases, executive management should expect that when BCP is implemented within the results-driven management structure recommended in section 15.3, it will result in a noticeable (10% or more) increase in staffing and spending for the engineering and planning functions. BCP dramatically improves the results obtained from engineering effort (the marginal payback at one large IOU in the central US was over 20:1!). Generally, money spent on more effort and better tools devoted to improving reliability and optimizing (cutting costs while improving performance) is very effective at obtaining results.

Optimize Against PBR Rules

Performance-based rates and penalties or incentives set down by regulatory authorities and/or customer contracts define precisely the costs and value to the utility of customer service performance. If interruptions of service cost $1/customer-minute, then the distribution utility would be less than maximally prudent from a financial standpoint if it were to spend more than $1 to reduce customer interruptions by one customer-minute.

Optimizing spending, engineering, and operation policies to minimize expected pay-out/maximize expected gain under PBR is sometimes referred to as "gaming" the PBR rules, but it is not a game and there is nothing fundamentally unethical or wrong in doing so. Through PBR, the regulatory process identifies the value that customer interruptions have to the utility. If regulators are displeased with the results of a particular PBR formula, they can respond by changing the PBR formula to provide more or less incentive, or to target a different definition of "good."

Utility Efforts under PBR should be directed at Two Goals

1. *Optimization of results.* First, the utility should implement the type of optimization – minimization of risk – described in section 13.5. That example was based on an actual investor-owned utility case. There a hybrid analytical simulation method, admittedly a fairly advanced, highly skilled method for engineering, was used to determine policies and prioritize spending in order to minimize the utility's total pay out under a PBR penalty formula. However, the basic concept will work to good extent with less sophisticated engineering methods.

   Proper implementation requires that this be coordinated with the overall RDM approach: the definitions of reliability used throughout all departmental evaluations under RDM must be identical to the definition of reliability or service problems used in the PBR. Ultimately, the PBR sets the utility wide marginal cost of reliability.

2. *Consistency of regulatory rule..* The utility should make one effort it can to influence regulatory policy, so that rates and PBR rules are consistent with one another. If, through PBR, regulators put a certain cost on reliability of service (or lack of it) they really should approve a base rate structure that gives the utility the required amount of money to implement these changes. Marginal cost of reliability sets the optimum decision-making point for RDM: it is the value used as the target in prioritization and its use does assure cross-functional optimality in decision-making value. That is essentially defined by the PBR rules.

   But the required budget required to implement reliability improvements up to that level is not necessarily a function of that marginal value. The marginal cost of reliability is the derivative (rate of change) of reliability versus spending. Required budget is the total spending up to that point. Two utilities can reach the same marginal cost with very different budgets, as shown in Figure 16.1.

   The point is that total allocated spending, and PBR defined penalties, can be set up so they are consistent. Ideally they should be. The authors have seen cases where that is not the case, and typically quite controversial issues develop as a result. The reason is that in such cases the regulatory

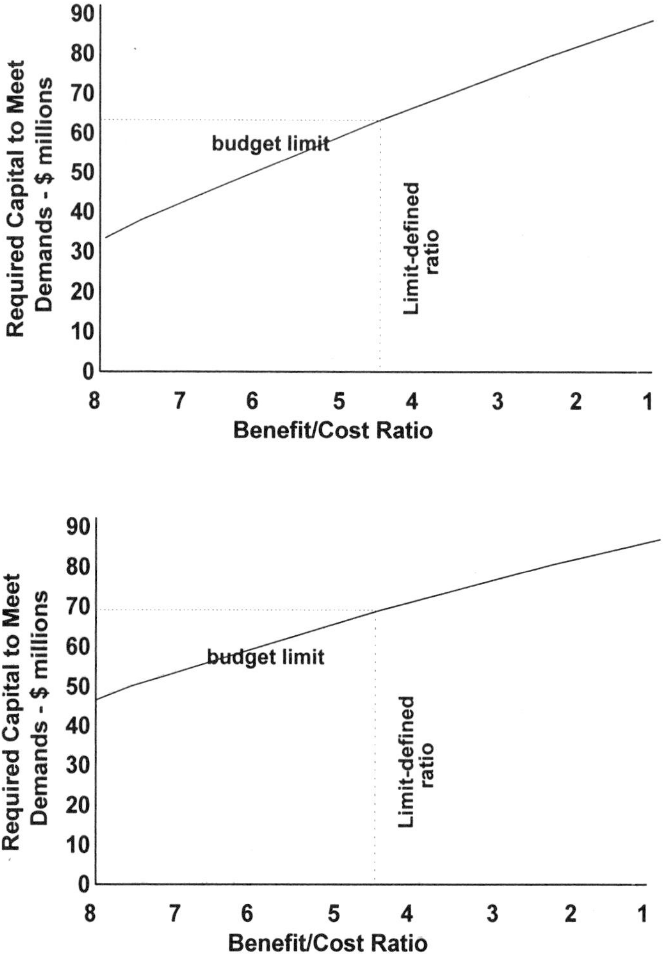

Figure 16.1 Marginal cost of reliability. Basically defined (very much a function of) the PBR formula, does not necessarily correspond to any specific budget level. Here, optimized budgets for two utilities are shown here . A theoretical, but realistic, example shown that one needs $62 million, the other - $69 million, to work up to the same level of effectiveness in spending for reliability. The reasons for this are numerous. One system may been traditionally allowed (by utility and regulator) to be less reliable than the other and thus need more reinforcement effort. One system could simply be bigger than the other (more customers) and thus need more money. Or, work in one system could be more expensive, for any wide range of reasons, than corresponding work in the other system. The important point is that one utility here needs about 10% more allowable expenses to be covered by rates than the other in order to adhere to the intent of the PBR. That happens only through effort to assure rates and PBR formula are consistent.

rules as a whole can be "gamed" in the worst sense of the word. The utility can make decisions which are fully within the rules, yet that benefit some interested parties at the expense of others, in a way not intended by the regulatory process. This might mean "unfairly" favoring industrial and commercial customers over residential, or vice versa, generating extra profits at the expense of customer service levels the regulators expected to be attained, or favoring capital spending (investment that generates future profits) over operating expenses (which are only recovered). In such cases, the financially sound procedure for the utility to follow is to act in its own business interests (that usually means in favor of its stockholders) rather than its customers.

Re-Examine and Revise All Standards and Guidelines

De-regulation, competition, changes in customer expectations, new planning and engineering tools, the operational IT tools discussed above, and PBR, all mean that the basis for priorities and decisions has changed dramatically since its standards, guidelines, and policies were last revised. The changes occurring in the period 1995 to 2005 are more dramatic than those that occurred during the five decades from 1945 to 1995. Despite this, many utilities are following basically the same approach they traditionally used. A few have not made anything but incremental changes since the early 1970s.

Make real changes

A rigorous and zero-based approach needs to be taken here. Innovation coupled with a good deal of skepticism both for the need to maintain traditional standards, and for the claims made for new ideas, needs to be applied throughout.

Outside help makes a difference

The authors' strongly recommend that this examination be made in part by people who are quite apart from those people within the utility who use the standards and work with the policies. Utilities who perform this re-examination using internal resources only – professional personnel who "grew up" in the utility – tend to make only small changes. It is to be expected that people who have long used the existing standards both have faith in them and know little about competing possibilities. Those utilities who seek the input of experts from outside the company tend to examine more objectively the methods used elsewhere (including outside the US) and ultimately make much more sweeping, and effective, changes.

IT Should Focus on Customer Service

During the 1980s and 1990s, electric distribution utilities began implementing large information systems in the management of their distribution facilities, engineering, and operations. This included AM/FM (Automated Mapping and

Facilities Management), ALS (Automated Line Switching), AMR (Automated Meter Reading), CAD (Computerized Automated Design), OMS (Outage Management Systems), WOM (Work Order Management, RDM systems (which means, in this instance, Results-Driven Maintenance scheduling) and integrated systems called DMS (Distribution Management System) s that linked some or all functions together. Some of these systems (OMS) directly affect customer service quality. Others indirectly affect it and are mostly justifiable on the basis of cost reductions (AM-FM), or efficiency improvements (WOM).

> Decisions on what systems and how much IT to implement, and how to use it most effectively, should be based upon a uniform customer-service focused basis, using exactly the same bases for "bang for the buck" as applied in all other departments and to all other functions.

This is fairly obvious in the context of previous recommendations in this chapter. What may not be obvious is the priorities assigned to these various software systems will change greatly depending on the importance given to the customer service aspects of their use. How their priority in the eyes of executive management will rank depends greatly on whether the utility's focus is mostly on reducing SAIFI or SAIDI. Where aging infrastructures are a major concern, this depends on how advanced the problem has become. As mentioned in Chapter 1 and demonstrated with example in Chapter 15, usually degradation in SAIFI leads that of SAIDI by several years. This is due to a number of interactions of failure rates, utilization ratios, and contingency flexibility (or lack thereof) in a traditionally laid out power system.

Therefore, the priority associated with each of the IT systems shown in Table 16.3 will depend on whether the utility's immediate concern focuses mostly on reducing SAIFI, SAIDI, or some combination of both. For example, OMS systems help reduce duration of outages much more than they can influence frequency of outages (they only come into play once an outage has occurred. By contrast, RCM (reliability-centered maintenance) software tends to work effectively against failure rate (hence SAIFI), as does improved engineering and protection analysis software (CAD), whereas ALS reduces SAIDI, and if quick enough, SAIFI too.

The reason that SAIFI can be reduced by ALS is that reliability definitions often distinguish between momentary interruptions, which are those less than some arbitrary limit of time, and those that are longer. Events less than this limit (the IEEE standard being considered at the moment is five minutes), are classified as momentary interruptions (MAIFI) and do not "count" as events in SAIFI or SAIDI calculations. In such a situation, SAIFI can be reduced noticeably if automated switching within five minutes restores outages that would take more than five minutes to restore. In such cases they do not "count" with respect to SAIFI statistics. Actual SAIDI as experienced by customers will also fall slightly due to the faster restoration capability of the system as a whole.

Table 16.3 Power Delivery System Related IT Systems and Their Interaction with Major Aspects of Customer Service and Utility Cost

IT Systems	SAIFI	SAIDI	Costs
Automated Line Switching (ALS)	x	X	
Automated Mapping/Facilities Management (AM/FM)*	x	x	X
Automated Meter Reading/Power Monitoring (AMR)		X	X
Computerized Planning and Engineering (CAD)	X		X
Dynamic Equipment Loading (DEL)	x	x	x
Feeder automation (FA)		X	x
Online Equipment Monitoring (OEM)	X	x	
Outage Management & Trouble Call (OMS		X	
Reliability Centered Maintenance (RCM)	X		
Transformer Load Management (TLM)	X	X	X
Work Order Management (WOM)			X

Table 16.3 lists some common distribution-related IT systems and their interaction with SAIDI, SAIFI, and cost control. Size of the X indicates degree of involvement.

Transformer-Load Management System

Transformer Load Management (TLM) systems are one of many IT systems which can be applied to power delivery systems. However, they deserve special attention because they can be used quite effectively to improve customer service, they can directly contribute to lowered failure rates among service transformers, and they have the among the highest payback ratios of power delivery IT applications.

TLM programs correlate monthly-metered sales to customers with the transformers through which the sales to those customers were made. High levels of sales relative to a transformer's capacity indicate a possible overloaded transformer, and put it on a "jeopardy list" that is sent to preventive maintenance, so the units can be checked and replaced if found to be overloaded. Good TLM programs use formulae based on type and number of customers, kWh sales, seasonal changes in load, and weather to estimate stress on each transformer and whether it is likely in jeopardy. They produce very accurate results (about 90%) in the sense that units predicted to be overloaded are in fact overloaded, and few false positives (units predicted to be overloaded that are not). They also miss only a few overloaded units (estimated to be fewer than 5%). As a result, change out of units that will fail and have to be changed anyway can be scheduled at the utility and customers' convenience. This reduces cost and improves reliability.

Effective TLM "programs" are much more than just software. They must include:

- *Good procedures for use*, including monthly inspection of the jeopardy list by field crews and replacement of units found overloaded.

- *Proven feedback from the field.* Information on units that were predicted to be overloaded and prove not to be, and on units that failed due to suspected overloads but were not on the jeopardy list, should be used to adjust the TLM analysis formulae.

- *Adjustments* of the jeopardy list functions to accommodate the corporate-wide marginal cost of reliability.

Extending TLM to include performance management of service

The concept of using monthly operating records in an analysis of possible trouble spots can be extended in two ways into what is called a Distribution Performance Management (DPM) system.

First, the analysis is extended to include not just transformers, but the entire lower delivery system (laterals, service transformers, and service level circuits and drops). Using facilities data, as well as metered data, trouble records, weather and other "monthly" records, a DPM can infer where trouble (failure) is likely to occur, as well as where poor quality problems (excessive voltage drop) is likely to occur in any part of that system. Really "smart" programs even use the results of primary distribution voltage analysis to determine the voltage stress on service transformers. This adds only a slight amount of improvement in the resulting analysis but in an integrated software system, costs next to nothing to add.

Secondly, the reporting of a DPM system is extended beyond what are traditional TLM reports, which include only the jeopardy list and perhaps also a list of severely under-utilized transformers (those where no-load losses are high enough to justify switching out to a much smaller size). DPM typically adds a series of management reports that include:

- *Predictive jeopardy lists* – a "smart" program can look ahead several months. The utility can schedule during the spring change-outs that need to be made before summer.

- *Asset management lists* for the service transformer set, including utilization of units, average monthly loss of life (during peak months the average unit may lose more than one month's life, during off peak, less, etc.), predicted

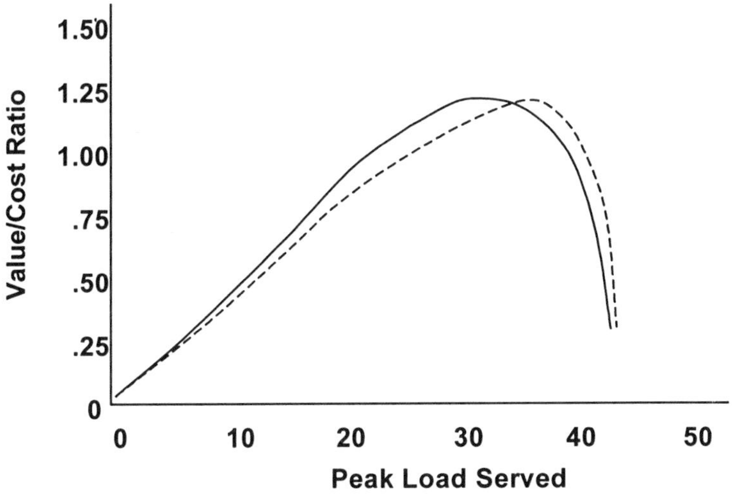

Figure 16.2 The value obtained per unit of investment in a transformer depends tremendously on the way it is used. Here, the ratio of the value/cost for two 32 MVA transformers is shown as a function of the peak load level it will be used to serve. Solid line shows the transformer example from section 15.4. Value varies by 22% over what could be called the useful range for this transformer (24 – 41 MVA – those loads over which value > cost). Optimum loading is 32 MVA. The dotted line shows the curve for the same type of 32 MVA transformer but at a different site with a different annual load curve shape. Value curve obtained is different as is the optimum peak loading, 36 MW.

failures and replacements over the entire set, etc.

- *Predictive trends* about future failures, replacements, loss of life, and low voltage events.

TLM and DPM programs often have very high economic paybacks – on the order of less than a year. The major expense is in developing a reliability data link between the customer records and the transformer records (tying each customer to the correct transformer).

Optimize Equipment Loading to Provide Maximum Value

The loading of equipment such as transformers, cables, and lines should be determined based on evaluation of the *value* obtained, rather than by arbitrary standards for utilization ration. Value is measured in terms of results obtained versus cost. Figure 16. 2 shows the variation in value obtained from investment that can occur as a function of loading. This is from Chapter 15, section 15.4, which gave a detailed example of how outlined how the concept is applied.

Chapter 8 emphasized how crowded facilities, particularly substations in the central part of the system, often limit the utility's options with respect to expanding its system in downtown and high-density load areas. The result is very high equipment utilization ratios, marginal configurations and contingency capability, and ultimately, a potential reduction in reliability in these areas.

At the heart of this issue is the fact that the substations in the central part of many utility systems cannot be expanded because there is no room for more equipment, an important topic covered in Chapter 8. However, this is almost always the case *only* if the utility stays with its traditional way of laying out substations. There are a number of approaches to building "compact" or *high-density substations* that have proven very effective in the US and Europe, and in fact are standard practice at some European utilities. In terms of increasing likelihood of relieving aging infrastructure reliability problems, these are:

- *Minimum clearance engineering.* Many utilities build substations with considerably more clearance in places than required this being what their own standards dictate. Others add a margin in case they wish to upgrade voltage in the future (e.g., a 69 kV station built to 138 kV clearances. All proposed reductions in clearance should be considered carefully from safety, maintenance access, insulation coordination, reliability (increased exposure of fire taking out multiple units) and convenience standpoints. Nevertheless, there are times when a relaxation to minimum industry standards may be the best way to achieve an increase in capacity at a constrained site. This approach can often increase capacity at a site by as much as 15%. It is costly, but in cases where considerable rework has to be done anyway (old equipment is going to be replaced anyway) this option should be examined.
- *Compact-footprint equipment and design.* It is possible to build transformers that are "tall instead of wide." For example, a 56 MVA transformer can be built that requires only the ground space of 40 MVA unit. Such units are considerably taller than standard units and require revision of buswork, etc. In addition, they cost more than standard units of equivalent capacity and they have slightly higher losses than equivalent normal units.

  Beyond this, substation designs based on "vertical buswork" and designs that push equipment higher, not wider as capacity is added, can be developed for special situations. Typically the capacity at an air-insulated substation can be increased by up to 35% using this approach. Frankly, the authors' experience with this approach has not been entirely satisfactory, but there are occasionally situations when this approach is feasible and the additional cost is justifiable.
- *Vertical construction.* The concept above, of "going high" can be extended to include "two-story" substations, which can permit capacity extensions of up to 60% at a constrained site. Such a design was used

by the authors in one very challenging aging infrastructure area, and provided a big relief from very severe capacity constraints. Again, this costs a great deal more than traditional design and is often not workable, but it deserves consideration.

- *Gas Insulated Switchgear (GIS).* In the authors' experience this is by far the most effective means of solving capacity constraint problems in aging infrastructure areas. Use of Gas-Insulated Switchgear and substation layouts built around it permit up to a *times four* (400% of normal) improvement in the capacity that can be squeezing into a congested substation site. It also permits location of high capacity substations in places where "building upward" is not feasible due to height restrictions.

A big advantage of GIS over other compact approaches is that it does not require custom design. While GIS is not commonly used by electric utilities in the U.S., it is a standard feature in utility systems throughout Europe, where it is used in a *majority* of new urban/suburban substations. As a result, the initial (capital) cost of GIS substation equipment is only about 15% - 20% more than that for standard types of air-insulated substations (which would require much, much more space) because it is standard "off the shelf" equipment.

Lifetime cost of GIS substations is actually less than equivalent AIS substations in some cases – a "substation in a bottle" requires less maintenance and sees less weather damage. Although GIS is not commonly used by electric utilities in the U.S., a good many industrial plant substations throughout North America use GIS. Large industries, who are always bottom-line oriented, are attracted to it, due to both the improved reliability it gives and the lower lifetime cost it provides, even when evaluated over the typically short (8-10 year maximum) periods commonly used in competitive business analysis. A majority of GIS in the U.S. is in privately owned industrial facilities.

In fact, in downtown areas, the initial cost of GIS substations is often noticeably lower than that of traditional air-insulated alternatives of equal capacity, when the cost of land is taken into account. As mentioned above, a majority of distribution substations used in downtown and urban areas throughout Europe employ gas-insulated instead of air-insulated switchgear. The reasons are a combination of cost, compatibility with urban constraints, and reliability.

Figure 16.3 Where is the substation? GIS equipment has permitted an entire substation to be located in this old building in a historic part of downtown Heidelberg, Germany.

As an example, Figure 16.3 shows a substation recently added in a historic part of downtown Heidelberg, Germany. The local utility bought a 400-year old building, gutted the inside, and put the substation inside the existing shell. Esthetic covenants that required any buyer to preserve the traditional appearance of the site (e.g., leave the existing building, with its exterior unaltered) worked in the utility's favor. The site, so restricted, was less expensive than might first be expected for a key location in the heart of a thriving city.

Figure 16.4 gives another example, in which the utility serving Orensa, Spain worked with city planners and local neighborhood leaders to convert a portion of downtown to both a new substation and a community park. The compact characteristics of GIS permitted it to be included quite invisibly in the overall plan. Most passers by are unaware that they are within feet of a major electrical facility.

Both of these examples demonstrate willingness on the part of the utility to explore non-traditional ways of dealing with constraints that the community and urban congestion place on expansion of electric facilities. It is particularly

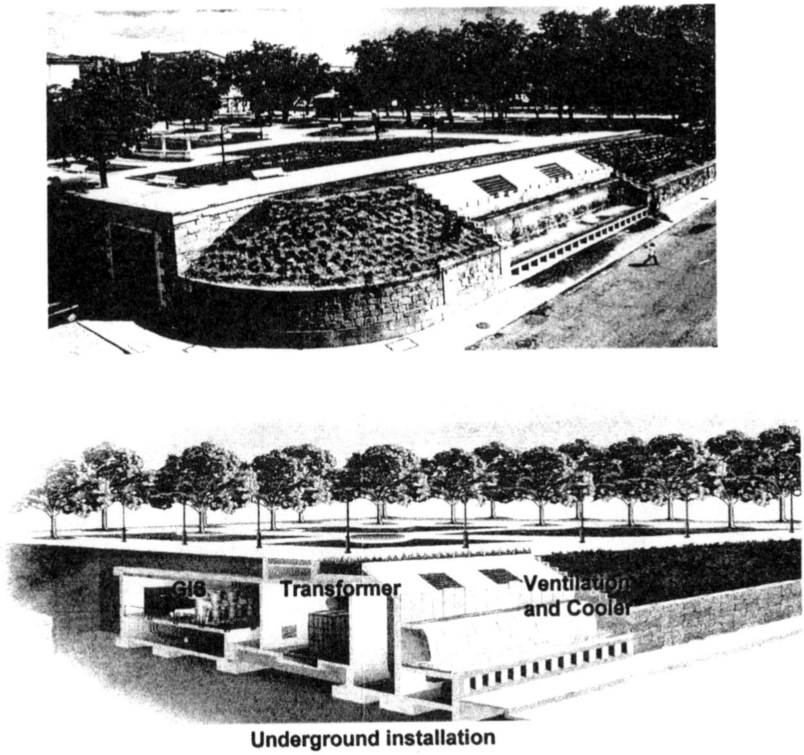

Figure 16.4 A GIS substation built underneath a park donated to the community of Orensa, Spain. Working with city planners the utility both improved the electric service capability in a growing portion of the metropolis, and also provided increased recreational facilities to the community. The waterfall provides cooling for the substations transformers.

noteworthy that the application of advanced technology, in this case GIS switchgear, not only permitted these two utilities to work within the very tough barriers that urban congestion and esthetics place in front of electrical expansion, but in fact use them to promote quick and enthusiastic municipal approval. Planners in Heidelberg wanted to find a responsible party to act as steward of that old building – its use as an electric substation assured its exterior would be well maintained. City planners and municipal leaders in Orensa wanted more recreational space in the central part of the city. In both situations the utility offered the community a "win-win" solution.

Cost of good solutions is not high

Equipment and labor costs for some of the high-density solutions discussed above are greater than for traditional, standardized AIS substations. However in many cases, lifetime costs may be less, particular when the cost of land is considered, as mentioned earlier. But this is neglecting a key point, that in most cases there may be not choice – options involving more land are not available. And what has become the "traditional" response in this situation, to do without and rationalize a solution through involved contingency plans at high utilization, is a solution that can, and has in some cases, been pushed too far. Finally, in many cases total cost evaluated over ten or more year period will be least for the GIS options due to its lower maintenance and higher reliability.

Table 16.4 compares costs for five configurations for a congested substation site in an aging infrastructure region within a major metropolitan area in the central US – the original (estimated cost to re-build all new) and four proposed compact designs. The area has a particularly flat load curve in which summer peak conditions can last over eight hours. The table shows these various combinations evaluated for reliability (using the explicit reliability analysis methods covered in s 14 and 15) against the actual 1999 peak demand level (127 MVA) and projected five-years ahead of peak (140 MW). Performance of scenarios where the substation would pick up load from neighboring stations, serving up to 300 MVA, is also shown. All alternatives use the same size site. As can be seen, the costs per kW for any of the compact designs are greater than for AIS, but when O&M costs are considered the costs for GIS may be less, depending on how O&M is taken into account in determining lifetime cost.

Table 16.4 Five Configurations the Same Size Site a Distribution Substation the Downtown Area of a Major US City

Substation Type	Peak Load MVA	Number & Size of Units	High-Side Bus/Breakers	Reliability SAIFI	Reliability SAIDI	Relative Cost/kW Capital	Relative Cost/kW O&M
Existing (1999)	127	4 x 30 MVA	None	.43	2.3	1.00	1.00
	140	4 x 30 MVA	None	.90	6.7	1.00	1.12
"Small footprint"	127	4 x 42 MVA	None	.34	.80	1.24	1.22
	140	4 x 42 MVA	None	.52	1.7	1.24	1.28
"Two-story"	127	4 x 30 MVA	Ring	.12	.62	1.41	1.31
	140	4 x 30 MVA	Ring	.16	.96	1.41	1.39
GIS low-cost	127	4 x 30 MVA	Ring	.07	.50	1.14	.67
	140	4 x 42 MVA	Ring	.11	.57	1.38	.75
GIS -- big capacity	250	4 x 65 MVA	Dble bus	.08	.22	2.31	.88
	330	4 x 83 MVA	Dble bus	.08	.25	3.00	.94

16.5 CONCLUSION AND SUMMARY

Aging power delivery infrastructure areas present a challenge, often a very difficult challenge, to an electric utility trying to keep costs low and customer service level high. In effect, there are two issues that must be coordinated: the areas will be composed of equipment with below average component and configuration reliabilities, and reliability will usually be an issue because it is currently too low. Thus a big part of aging infrastructure management boils down to artful management of reliability improvement in the area. This is quite easy if money is not an object of concern, but for any real situation it is often a major concern, as important as good customer service is in the minds of upper management.

There are no miracle cures, but there are effective ways of dealing with the situation created by aging power delivery infrastructures and of maximizing the performance that can be obtained from both the existing system and new investment in it. Overall, the single most important guiding principles are:

1. Apply Reliability Driven Management and prioritization throughout the organization, using the same coordinated focus on results in all departments and for all functions.

2. All spending, whether for prevent maintenance refurbishment, replacement, IT, professional staff and training, or capital additions, should be prioritized based on its marginal benefit/cost ratio. Intra-project alternative evaluation and optimization should be used in all cases.

3. Re-examination of the basic tenets of current standards, guidelines, and corporate values needs to be made based on the business case for good service and the needs of modern customers and power system technology.

4. Innovate, particularly with regard to employing proven methods developed elsewhere. "Not invented here" is a big barrier to success in improving performance. Keeping to traditional methods will assure traditional levels of performance/dollar.

REFERENCES

R E Brown, et al, "Spatial Load Forecasting Using Non-Uniform Areas" in *Proceedings of the IEEE T&D Conference,* April, 1999,

R E Brown, et al, "Reliability and Capacity: A Spatial Load Forecasting Method for a

H. L. Willis, *Spatial Electric Load Forecasting,* Marcel Dekker, New York, 1996.

H. L. Willis, *Power Distribution Planning Reference Book,* Marcel Dekker, New York, 1997.

H. L. Willis, H. N. Tram, and R. W. Powell, "A Computerized, Cluster-Based Method of Building Representative Models of Distribution Systems," IEEE paper 85 WM 090-6, 1985 Power Engineering Winter Meeting, and also appearing in IEEE Transactions on PAS, 1986.

Index